Lecture Notes in Civil Engineering 627

Lecture Notes in Civil Engineering (LNCE) publishes the latest developments in Civil Engineering—quickly, informally and in top quality. Though original research reported in proceedings and post-proceedings represents the core of LNCE, edited volumes of exceptionally high quality and interest may also be considered for publication. Volumes published in LNCE embrace all aspects and subfields of, as well as new challenges in, Civil Engineering. Topics in the series include:

- Construction and Structural Mechanics
- Building Materials
- Concrete, Steel and Timber Structures
- Geotechnical Engineering
- Earthquake Engineering
- Coastal Engineering
- Ocean and Offshore Engineering; Ships and Floating Structures
- Hydraulics, Hydrology and Water Resources Engineering
- Environmental Engineering and Sustainability
- Structural Health and Monitoring
- Surveying and Geographical Information Systems
- Indoor Environments
- Transportation and Traffic
- Risk Analysis
- Safety and Security

To submit a proposal or request further information, please contact the appropriate Springer Editor:

- Pierpaolo Riva at pierpaolo.riva@springer.com (Europe and Americas);
- Swati Meherishi at swati.meherishi@springer.com (Asia—except China, Australia, and New Zealand);
- Wayne Hu at wayne.hu@springer.com (China).

All books in the series now indexed by Scopus and EI Compendex database!

Sergey Vasil'yevich Klyuev ·
Nikolay Ivanovich Vatin ·
Karina Rashidovna Nabiullina ·
Venera Maratovna Yumagulova
Editors

II International Scientific Conference "Recent Advances in Architecture and Construction" 2024

Springer

Editors
Sergey Vasil'yevich Klyuev
Belgorod State Technological University named after V.G.Shukhov
Belgorod, Russia

Karina Rashidovna Nabiullina
Kazan Federal University
Kazan, Russia

Nikolay Ivanovich Vatin
St. Petersburg Polytechnic University
St. Petersburg, Russia

Venera Maratovna Yumagulova
Kazan Federal University
Kazan, Russia

ISSN 2366-2557 ISSN 2366-2565 (electronic)
Lecture Notes in Civil Engineering
ISBN 978-3-031-82937-6 ISBN 978-3-031-82938-3 (eBook)
https://doi.org/10.1007/978-3-031-82938-3

This Springer imprint is published by the registered company Springer Nature Switzerland AG
The registered company address is: Gewerbestrasse 11, 6330 Cham, Switzerland

Contents

Enhancing the Load-Bearing Capacity and Stability of Weak Soils: An Evaluation of Soil Improvement Techniques

M. Y. Yerzatova[1(✉)], K. K. Mukhambetkaliev[1,2], A. Jumabayev[1], and A. Tlegenov[1]

[1] L.N. Gumilyov Eurasian National University, Astana 010000, Republic of Kazakhstan
yerzatova@gmail.com, K.mukhambetkaliyev@qazjolgzi.kz, atagali@list.ru

[2] JSC "Kazakhstan Road Research Institute", Astana 010000, Republic of Kazakhstan

Abstract. Weak soil conditions present significant challenges in various civil engineering projects, including construction, slope stability, and infrastructure development. Weak soils are found in several regions, many loess and silty soils are most widespread in the central and southern Kazakhstan regions. This study investigates and evaluates various methods for enhancing soil load-bearing capacity and stability. Through a detailed analysis of the literature and experimental studies of the geotechnical properties of weak soils, this research explores conventional and innovative techniques employed for soil improvement. Laboratory tests, such as liquid, plastic limit tests, and chemical analysis, are performed to examine soil composition and characteristics. Experimental investigations are conducted to assess the soil properties and identify most effective methods. The outcomes provide valuable insights into the sustainability, and applicability of soil strengthening methods. This research advances the understanding of soil strengthening techniques and provides criteria for selecting methods to address weak soil challenges in civil engineering projects. The findings aim to assist engineers and researchers in optimizing construction on weak soils, enhancing durability, and sustainability.

Keywords: Weak Soils · Methods · Geotechnical Properties · Soil Improvement

1 Introduction

The challenges posed by weak soils in civil engineering due to their low bearing capacity and high susceptibility to deformation can significantly compromise the stability and longevity of structures. Regions like central and southern Kazakhstan with prevalent loess and silty soils are particularly vulnerable to foundation failures and slope instabilities. Addressing the issues associated with weak soils requires understanding of geotechnical properties and behavior under various loads. Effective methods for soil improvement are critical to enhancing the load-bearing capacity, thereby ensuring the safety and durability of engineering projects [1]. Globally, weak soil conditions pose

S. V. Klyuev et al. (Eds.): ICRAAC 2024, LNCE 627, pp. 1–8, 2025.
https://doi.org/10.1007/978-3-031-82938-3_1

major challenges in civil engineering. In Mexico City, buildings tilt and sink due to the highly compressible lacustrine clays of its ancient lakebed, while Shanghai faces similar settlement issues on soft alluvial soils. Rapid urbanization and construction in cities have exacerbated these issues, impacting infrastructure [2]. Astana, Kazakhstan's capital, is built on weak, clayey soils, leading to settlement issues. The high water table and winter frost worsen these problems, requiring pile foundations and soil stabilization. In contrast, Shymkent in southern Kazakhstan has loess, fine-grained, porous soils that can collapse when saturated [3].

Despite significant advancements, the lack of region-specific data can lead to the inefficient application of methods that may be optimal for local conditions [4]. Many studies focus on the immediate improvements in post-treatment but do not adequately address the long-term sustainability [5]. This study investigates and evaluates various methods for strengthening weak soils. Through a detailed analysis of existing literature and experimental studies, it explores the properties of weak soils to identify the most suitable strengthening techniques. Laboratory tests, including liquid, plastic limit, and chemical analysis, are conducted to assess soil properties. Soil samples were collected from a region in central Kazakhstan dominated by loess and weak soils. By examining the liquidity and plasticity characteristics, this research seeks to identify the most effective existing methods for enhancing soil strength and stability.

2 Methods and Materials

The investigation evaluates existing literature and conducts experimental research. It combines measurement techniques and laboratory investigations to assess the properties of weak soils and identify suitable strengthening methods. A system-structural approach based on soil improvement techniques and property analysis guides the experimental research. By utilizing modern research methods and advanced testing equipment, the study ensures high reliability in its findings.

To identify the most suitable soil improvement method, it is essential to establish rigorous criteria for evaluation. The first criterion is the effectiveness in enhancing critical soil properties, such as load-bearing capacity, shear strength, compressibility, and settlement reduction. The selected method should significantly improve the soil's ability to support loads without excessive settlement and increase its resistance to shear forces, thereby reducing the risk of failure. The soil improvement technique should be well-suited to the soil types being addressed and should effectively manage the soil's response to moisture changes. The method must maintain its effectiveness over an extended period, ensuring sustainability and minimizing the need for repeated interventions. Additionally, it should be resilient to local environmental conditions, including frost action, seismic activity, or salinity, to ensure lasting performance.

Proper laboratory testing of soils to determine their physical properties is an integral part of the design of structural foundations, the improvement of soil properties, and the quality control of soil compaction works. The natural soil deposits often exhibit a high degree of nonhomogenity. The fundamental theoretical equations that are developed in soil mechanics can be properly used in practice if the physical parameters used in those equations are properly evaluated in the laboratory [6].

According to SP RK 1.02-105-2014, weak soils are defined as those exhibiting fluid-plastic or fluid-powdery characteristics, including clay soils with high porosity and moisture levels [7]. These soils are marked by low strength, high compressibility, prolonged consolidation times during compaction, and reduced creep strength. To identify the type of soil and its state, liquid plastic limit tests were conducted [8].

3 Results and Discussions

According to SP RK 1.02-105-2014, saturated foundational soils exhibit high compressibility, with a total deformation modulus not exceeding 5 MPa. The angle of internal friction for clay soils in a saturated state ranges from 5 to 12 degrees, while adhesion values are 0,01–0,03 MPa, indicating low strength [4]. Additionally, high porosity coefficients suggest a loose soil and high hydraulic capacity [9].

3.1 The Liquid and Plastic Limit Analyses

Table 1. Liquid and plastic limit test results.

The cup №	The cup mass (g)	The moisture soil mass (g)	The dry soil mass (g)	Number of blows	Moisture content (%)
Liquid limit test					
17	7,6	18,9	16,2	33	31,4
47	7,7	27,2	22,5	16	31,8
2	7,5	26,1	21,6	28	31,9
324	7,9	19,3	16,6	34	31,0
8	12,4	29,0	25,0	14	31,7
			Liquid Limit:	25	31,56
Plastic limit test					
97	7,7	14,6	13,5	-	18,9
3	7,7	15,1	14	-	17,5
				Plastic Limit:	18,2

Table 1 shows liquid and plastic limit test data [10]. The liquid limit test records the moisture content of the soil at various numbers of blows, providing key insights into soil behavior under moisture conditions. In the corresponding diagram, the trend line indicates a decreasing moisture content with an increasing number of blows. The liquid limit is equal to 31,56% (Fig. 1).

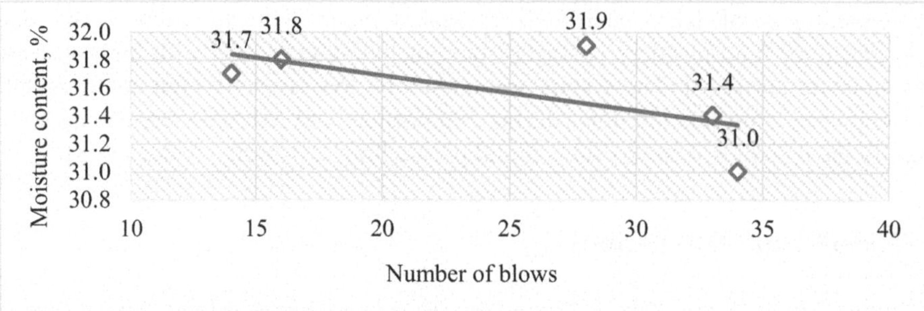

Fig. 1. Liquid limit diagram.

3.2 Statistical Analysis

The determination of normative and calculated values of soil characteristics for IGE and CSE is conducted in accordance with GOST 20522–96. The normative value X_n of soil is equal to the arithmetic mean X and calculated by the formula [11]:

$$X_n = \overline{X} = \frac{1}{n}\Sigma X_i = \frac{1}{5}(31,4 + 31,8 + 31,9 + 31,0 + 31,7) = 31,56\% \quad (1)$$

Where n is the number of definitions, X_i are the values of the property. A statistical check is done to eliminate any errors. The X_i meeting the condition is excluded [11]:

$$|X_n - X_i| > \nu \cdot S; Ex : |31,56 - 31,4| < 1,92 \cdot 0,36, 0,16 < 0,69 \quad (2)$$

There is no value of X_i for which the condition is met, so no errors have occurred. Here v is a statistical criterion, which is 1,92 for $n = 5$, and with a two-way confidence probability of $\alpha = 0,95$, S is the mean square deviation calculated by the formula [11]:

$$S = \sqrt{\frac{1}{n-1}\sum_{i=1}^{n}(X_n - X_i)^2} = \sqrt{\frac{1}{5-1} * 0,532} = 0,36 \quad (3)$$

The coefficient of variation V and the accuracy index ρ_α are calculated [11]:

$$V = S/X_n = 0,36/31,56 = 0,0114 \quad (4)$$

$$\rho_\alpha = \frac{t_\alpha V}{\sqrt{n}} = \frac{2,13 * 0,0114}{\sqrt{5}} = 0,011 \quad (5)$$

Where t_α is the coefficient taken according to GOST 20522–96, depending on one-sided confidence probability and the number of degrees of freedom $K = 4$. The reliability coefficient γ_g and calculated value are calculated using the formulas:

$$\gamma_g = 1/(1 \pm \rho_\alpha) = 1/(1 - 0,011) = 1,01 \quad (6)$$

$$X = X_n/\gamma_g = 31,56/1,01 = 31,25\% \quad (7)$$

3.3 Determination of the Soil Denomination

To identify the type of soil, it was necessary to determine the characteristics of the soil, such as plasticity I_p and liquidity index I_L, by results taken from limit tests [12].

$$I_p = (\omega_L - \omega_p)100\% = (0,316 - 0,189)100\% = 12,7\% \quad (8)$$

$$I_L = (\omega - \omega_p)/(\omega_L - \omega_p) = (0,1789 - 0,189)/(0,316 - 0,189) = -0,08 \quad (9)$$

Where ω is natural soil moisture, ω_L is at the yield point, and ω_p is at the rolling edge. Since $7 < I_P \leq 17$ and $I_L < 0$, the denomination is "Solid Loam" [13]. The denomination according to ASTM D 2487 is lean clay with an index CL [14].

3.4 Chemical Analysis

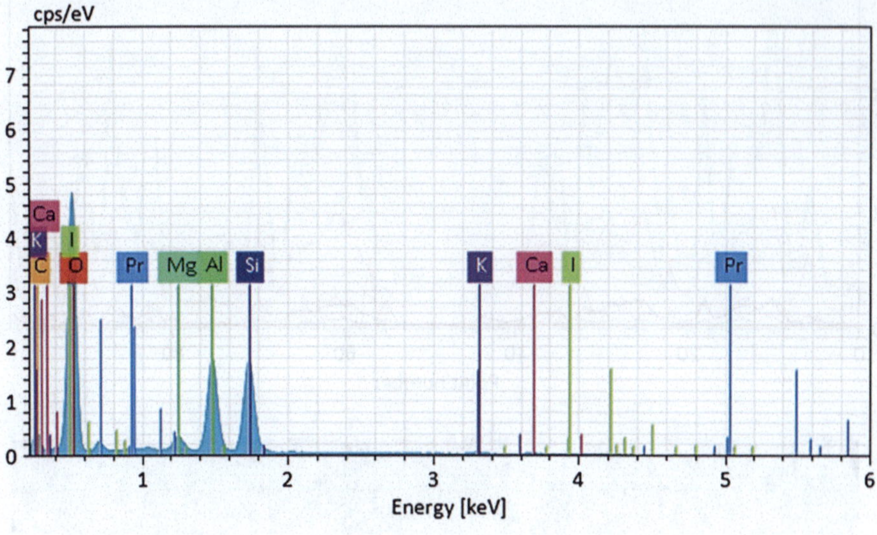

Fig. 2. X-ray fluorescence (XRF) analysis graph of elements in a soil sample.

Figure 1 illustrates the results of XRF analysis, which identifies elements in a soil sample by measuring the X-rays they emit. Low energy levels of oxygen and carbon suggest the presence of carbonate minerals or organic material. Calcium shows two peaks, likely from gypsum or carbonates. Potassium has peaks at 0,3 and 3,3 keV, while aluminum peaks at 1,5 keV, likely from feldspars or clays. Iodine and praseodymium double peaks at 4,5 and 2,0 indicate contamination (Figs. 2, 3, and 4).

The table and graph summarize the elemental composition of the soil from a chemical study. Oxygen, a key component of oxides and silicates, is the most abundant element, comprising 57,43% of atoms and 29,14% of the sample's mass. Magnesium, present in minerals like MgO, accounts for 1,36% of atoms. Silicon, common in SiO2, has

Element	At. No.	Netto	Mass [%]	Mass Norm. [%]	Atom [%]	Comp.	Sto. [%]	Sto. Norm. [%]	abs. error [%] (1 sigma)	rel. error [%] (1 sigma)
C	6	4373	0.00	0.00	0.00		0.00	0.00	0.00	0.00
O	8	72138	93.90	29.14	57.43		0.00	0.00	2.97	3.16
Mg	12	5843	3.37	1.05	1.36	MgO	5.59	1.73	0.18	5.37
Al	13	37322	23.94	7.43	8.68	Al2O3	45.24	14.04	1.07	4.47
Si	14	40174	55.08	17.09	19.19	SiO2	117.83	36.56	2.44	4.43
K	19	1321	4.97	1.54	1.24	K2O	5.99	1.86	0.49	9.81
Ca	20	950	8.25	2.56	2.01	CaO	11.54	3.58	1.07	13.01
I	53	19113	113.22	35.13	8.73		113.22	35.13	4.12	3.64
Pr	59	3791	19.54	6.06	1.36	Pr2O3	22.87	7.10	0.82	4.17
		Sum	322.26	100.00	100.00		322.26	100.00		

Fig. 3. X-ray fluorescence (XRF) analysis table of a soil sample.

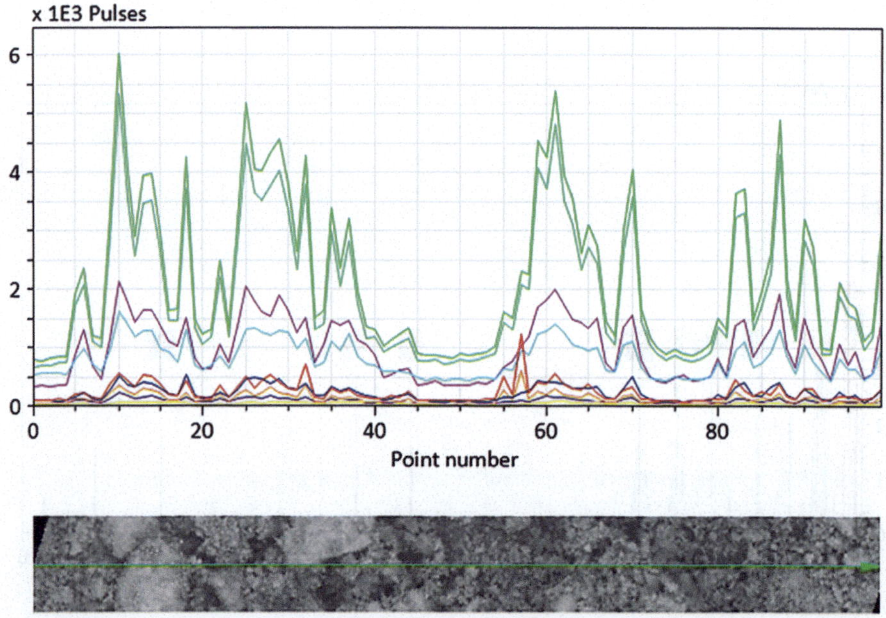

Fig. 4. Fluctuation graph of the X-ray fluorescence (XRF) analysis of a soil sample.

19,19% atomic presence. Potassium, from feldspars, contributes 1,54% mass. Iodine with a high mass percentage (8,73%), suggests contamination. Praseodymium, a rare element, shows 6,06% mass, indicating mineral inclusions.

3.5 The Recommended Soil Improvement Methods

Early studies show that conventional and innovative soil improvement techniques significantly enhance weak soils' geotechnical properties. Field compaction, a widely used method, is carried out with various rollers. Smooth-wheel rollers are suitable for proof-rolling subgrades and finishing fills with sandy or clayey soils, offering 100% coverage but may not compact thick layers uniformly. Pneumatic rubber-tired rollers, weighing 2000 kN, provide better compaction with 600–700 kN/m^2 contact pressure and 70–80% coverage. Sheepsfoot rollers, with drums featuring projections, are best for cohesive soils, exerting 1500–7500 kN/m^2 contact pressure. In contrast, the vibroflotation process uses a vibroflot with an eccentric weight that enables horizontal vibration. The vibroflot is raised in 0,3-m and held vibrating for 30 s per lift to compact the soil to the desired density [15]. Admixtures like lime, cement, and lime-fly ash are frequently used to stabilize fine-grained soils. Lime types used include hydrated high-calcium lime, dolomitic quicklime, applied in amounts ranging from 5–10%. Lime stabilization in the field can be performed in three ways. In situ material is mixed with the appropriate amount of lime, followed by compaction after moisture is added. Soil is combined with lime and water at a processing plant and transported for compaction. Lime slurry is pressure-injected into the soil to a depth of 4 to 5 m. Cement is increasingly used as a soil-stabilizing material in highway and earth dam construction. Similar to lime, it reduces the liquid limit and enhances the plasticity index and workability of clay soils. Cement stabilization is most effective for clayey soils with a liquid limit below 45 to 50 and I_p under 25. Granular and clayey soils with low plasticity are particularly well-suited for such stabilization. While calcium-rich clays are more readily stabilized with cement, expansive clays containing sodium or hydrogen respond better to lime stabilization [15].

4 Conclusions

In order to improve the stability and load-bearing ability of soils, this study investigated and assessed several strengthening methods, particularly in central and southern Kazakhstan. The research confirms that both conventional and innovative soil improvement techniques significantly enhance the geotechnical properties of weak soils by reducing plasticity, increasing strength and stability. By evaluating various techniques, this research offers guidance for optimizing design and construction on weak soil sites and improves the understanding of soil behavior, aiding in predicting settlement and stability issues, which contributes to more durable infrastructure. The findings provide a comparative analysis of soil improvement methods and practical guidance for engineers, emphasizing the relevance of selecting techniques based on specific soil characteristics. Future studies could incorporate a practical analysis of methods like compaction, pre-compression, and stabilization using admixtures, focusing on a broader range of soil conditions and evaluating long-term performance. Ultimately, these findings are expected to enhance project safety, durability, and sustainability in infrastructure development on weak soil sites.

References

1. Sánchez-Garrido, A.J., Navarro, I.J., Yepes, V.: Evaluating the sustainability of soil improvement techniques in foundation substructures. J. Cleaner Prod. **351** (2022)
2. Zhussupbekov, A., Shakhmov, Zh., Shin, E.C., Krasnikov, S.: Challenges for transportation geotechnics in extreme climates of Kazakhstan and Korea. 2nd International Conference on Transportation Geotechnics (ICTG), 655–660 (2012)
3. Utepov, Y., et al.: Investigating the influence of initial water pH on concrete strength gain using a sensors and sclerometric test combination **7**(12) (2022)
4. Baitova, A.A., Zhambakina, Z.M.: Features of engineering and geological properties of Kazakhstan soils. Innov. Sci. Res. Netw. Publication **2–1**(4), 188–196 (2021)
5. Zhussupbekov, A.Zh., Utepov, Ye.B., Shakhmov, Zh.A., Ling, H.I.: Model testing of piles in a centrifuge for prediction of their in-situ performance. Soil Mechanics and Foundation Engineering 50, 92–96 (2013)
6. Awwad, T., Shakhmov, Z., Lukpanov, R.E., Yenkebayev, S.B.: Experimental Study on the Behavior of Pile and Soil at the Frost Condition. In: El-Naggar, H., Abdel-Rahman, K., Fellenius, B., Shehata, H. (eds.) Sustainability Issues for the Deep Foundations, Sustainable Civil Infrastructures, pp. 69–76. Springer, Cham (2019)
7. SP RK 1.02-105-2014. Engineering survey in construction. Basic positions. JSC «KazNIISA», LLP «ZTs ATSE», Astana (2015)
8. GOST 12248-2010. Soils. Laboratory methods for determining the strength and strain characteristics. NIIOSP im. N.M. Gersevanova, Moscow (2011)
9. GOST 23161-78. Soils. Laboratory method for determination of subsiding characteristics. NIIOSP im. N.M. Gersevanova, Moscow (1979)
10. Gunaratne, M. (ed.): The foundation engineering handbook, 1st edn. CRC/Taylor & Francis, Boca Raton (2006)
11. GOST 20522-96. Soils. Statistical treatment of the test results. NIIOSP im. N.M. Gersevanova, Moscow (1997)
12. Krawczyk, D., Flieger-Szymańska, M.: The value of plasticity index (IP) and liquidity index (IL) of North Polish ablation boulder clays and varved clays depending of the method of its determination. Scientific Review Engineering and Environmental Sciences **27**(2), 167–174 (2018)
13. GOST 25100-2020. Soils. Classification. NIIOSP im. N.M. Gersevanova, Moscow (2020)
14. ASTM D2487-17. Practice for Classification of Soils for Engineering Purposes (Unified Soil Classification System). D18.07, (2020)
15. Das, B.M.: Principles of foundation engineering, 7th edn. Cengage Learning, Stamford (2010)

Modification of Concrete with Chemical Additives

S.-A. Yu. Murtazaev[1,2], M. Sh. Salamanova[1,2(✉)], M. S. Saidumov[1], and M. G. Pletnev[3]

[1] Grozny State Oil Technical University Named After Acad. M.D. Millionshchikov, Grozniy, Russia
madina_salamanova@mail.ru
[2] Kh. Ibragimov Complex Institute of the Russian Academy of Sciences, Grozniy, Russia
[3] State University of Management, Moscow, Russia

Abstract. More deep chemical interactions mechanism research with the subsequent formation of new formations in the structure of concrete stone; undiscovered possibilities for modifying the matrix system with organic and mineral additives; development of filled highly active binding materials; use of enriched aggregates of improved quality; the combination of concrete, reinforcement and dispersed reinforcement and other technological techniques contributed to the evolution of the development of modern concrete science. The studies were carried out in accordance with regulatory documents GOST 12730.3–2020 Concrete. Methods for determining water absorption; GOST 10180–2012 Concrete. Methods for determining strength using control samples; GOST 12730.1–78 Concrete. Methods for determining density. In support of the above, the work investigated the integrated use of organic and mineral additives, which contributed to the accelerated formation of the structure and strength of the stone without the use of thermal effects. An improvement in rheotechnological characteristics has been recorded, a water-reducing effect has been established, and an increase in the shelf life of the concrete mixture has been established, which is no less important when concreting large-sized structures or transporting semi-finished products over long distances. The proposed research results will make it possible to successfully implement projects of global importance, creating stable, strong and durable concrete structures.

Keywords: Mineral Powders · Superplasticizer · Concrete Scrap · Flask · Microsilica · Volcanic Ash · Filler · Structure · Strength Concrete

1 Introduction

Currently, modern concrete technologies have undergone radical changes and have made this material one of the most popular and indispensable in construction. The prerequisites for the creation of new composites were more in-depth studies of the mechanism of

The study was supported by the Russian Science Foundation grant No. 24–69-00043, https://rscf.ru/project/24-69-00043/.

S. V. Klyuev et al. (Eds.): ICRAAC 2024, LNCE 627, pp. 9–17, 2025.
https://doi.org/10.1007/978-3-031-82938-3_2

chemical interactions with the subsequent formation of new formations in the structure of concrete stone; undiscovered possibilities for modifying the matrix system with organic and mineral additives; development of filled highly active binding materials; use of enriched aggregates of improved quality; combination of concrete, reinforcement and dispersed reinforcement and other technological methods [1–4].

The correct choice of concrete compositions plays a key role in the successful implementation of projects of global importance, being a guarantor of strength, durability and stability of structures. Depending on the specific task, it is necessary to model the compositions and technology of concrete of a certain class, which has certain properties and features of molding in a structure.

For the construction of rigid frames of high-rise buildings and structures, bridges and other critical areas, High-Strength Concrete (high-strength concrete) with a strength of up to 150 MPa is designed; Ultra High-Strength Concrete (heavy-duty concrete) with a strength above 150 MPa is used in the construction of critical structural elements of objects; impermeable and dense concrete W12 and higher have found their recognition in the construction of underground parts of buildings and structures, tunnels; monolithic massive structures are made of Shrinkage Compensated Concrete to avoid deformation disturbances; concreting of densely reinforced structures with complex configurations is successfully carried out using self-compacting concrete - Self-Compacting Concrete, which makes it possible to avoid compaction; enclosing structures and small-piece elements can be produced - Reactive Powder Concrete (powder concrete) and Macro Defect-Free Concrete (fine-grained concrete without defects), which are characterized by minimal porosity of up to 1% and fairly high strength in the range of 150–250 MPa. The listed new generation concretes demonstrate the strengths of the innovative material, and are the subject of discussion at almost many scientific sites around the world.

By modeling the formulations of multicomponent concretes, introducing modifiers of various natures into the system, it is possible to control structure formation, starting from the processes of dispersion of the solid phase in water to the formation of a durable crystalline body in the end. Environmental problems associated with the decarbonization of the economy, and in particular the construction sector, prompt an expansion of the range of binding materials through the partial use of technogenic waste instead of the expensive and resource-intensive clinker component. By using a recycled product obtained by processing fragments of destroyed concrete buildings and structures as a filler, it is possible to solve the growing shortage of natural exhaustible raw materials [5–8].

Thus, all of the listed technological approaches for obtaining high-quality multicomponent concrete will make it possible to implement new architectural solutions that require the implementation of all untapped reserves of this unique material. In order to obtain concrete of the B25-B40 class line, compositions of binders and composites were developed using mineral additives from concrete scrap, microsilica, volcanic ash, and flask. Attempts to use recycled coarse aggregate from the dismantling of destroyed buildings to obtain class B25 concrete were considered. And of course, in combination with mineral additives and mixing water, water-soluble organic substances were introduced into the mixture, which made it possible to control rheotechnological properties, colloidal, physical and chemical phenomena, modifying the structure at the micro- and macrolevels [9–11].

2 Methods and Materials

According to the terminology, GOST 31108 – 2020 General construction cements. Technical conditions, mineral additives used in the research can be classified as pozzolanic, since in the presence of CaO these powders will exhibit hydraulic activity, and also do not have a negative effect on the water requirement of the cement paste and strength characteristics. Pieces of rock of sedimentary origin - flasks - were first passed through a jaw crusher for subsequent fine grinding in a grinding unit. Mineral additives (mortar part of concrete scrap, volcanic ash and flask) were prepared by grinding in a laboratory roller mill MLR-15 for 10 min. The specific surface area of the resulting powders was determined on a PSH-12 device and amounted to the following numerical indicators: S_{sp} concrete scrap 571 m^2/kg; S_{sp} of volcanic ash 542 m^2/kg; S_{sp} of the flask 592 m^2/kg. The microsilica additive in its natural form has a high specific surface area and was 21 000 m^2/kg.

The main carrier of astringent properties is Portland cement grade CEM I 42.5N, manufactured by JSC «Chechencement», at the time of testing, the activity of cement, determined using the activity indicator IAC-04M, was 48.4 MPa. Inert material from the Alagirsky quarry from granite-diabase rocks was used as fine and coarse aggregates. Quartz sand has a stable granulometric composition and belongs to the coarse class, the particle size modulus was 2.9, the PGI content was 0.82%, and the bulk density was 1470 kg/m^3. Crushed stone fr. 5–20 mm meets the requirements of GOST 8267–93 Crushed stone and gravel from dense rocks for construction work. Technical specifications, crushability grade M1000, PGI content - 0.61%, frost resistance F 300, bulk density 1480 kg/m^3. To obtain class B25 concrete, recycled coarse aggregate fr. 5–20 mm, having approximately 10–15% of the mortar part, which will only enhance the contact strength in the "binder – recycled crushed stone" zone and will have a positive effect on the strength of concrete in general.

Particular attention in the work was paid to the organic component; in the selection of concrete formulations, various modern modifiers were studied, both domestic and imported, available and at approximately the same cost. The chemical additives under study have a super water-reducing effect, a complex composition based on polycarboxylates and lignosulfonates, and by reducing the water-cement ratio, they increase strength by 20–25%. The dosage of the superplasticizer was chosen taking into account the manufacturer's recommendations and varied from 0.1 to 1.8% by weight of cement.

The following additives were studied as an organic modifier: RHEOPLAST LS 01 (Reoplast LLC, Vsevolozhsk), Stark EW (Starkhaus, Germany), MegaKhimTrade sealing additive (MegaKhimTrade LLC, Tula), Cryoplast (Poliplast LLC, Novomoskovsk), Stark RM (Starkhaus, Germany), TECHNONICOL ICE 611 and TECHNONICOL READY-MIX 111 (TECHNONICOL Corporation).

3 Results and Discussions

Using the above raw materials, a large number of experimental batches of concrete mixtures were carried out and samples were manufactured and tested, as a result of which optimal dosages were established for all additives under study. The compositions of the designed concretes and test results are given in Tables 1 and 2. In compositions 1 – 7, recycled coarse aggregate fr. 5–20 mm, mineral filler from scrap concrete and microsilica additive.

Compositions 8–21 used high-quality crushed stone from granite-diabase igneous rocks with a fraction of 5–20 mm. Modification of cement stone with mineral and organic additives had a beneficial effect on the rheotechnological properties of the concrete mixture and the strength properties of concrete samples. Mineral powders from concrete scrap, flask, volcanic ash and microsilica have made it possible to reduce the consumption of Portland cement.

According to the obtained results, the most advantageous combination of binary powders is "volcanic ash - microsilica"; the additives are a product of heat treatment, both natural and artificial. The mineralogy and particle dispersion of these additives contributed to achieving the design strength with a margin, reducing porosity and improving the rheology of the mixture. The additives Cryoplast and TechnoNIKOL READY-MIX 111 at a dosage of 1.1% of the mass of cement in terms of dry matter create a deflocculation effect in the medium, increasing the contact surface of cement grains with the mixing water, which has a positive effect on the increase in the number of neoplasms. The numerical values of the compressive strength of concrete samples at 7 days of age confirm this; an increase in strength indicators of 104–106% of the brand value is observed.

The use of organomineral additives contributed to the accelerated formation of the structure and strength of the stone without the use of thermal effects. All superplasticizers under study exhibit a complex electrostatic and steric mechanism of action; polycarboxylate ester molecules are adsorbed on the active centers of crystallization of cement and mineral powder, reduce internal friction, surface tension between the solid and liquid phases, and intensify the dispersion of particles. It should be noted that in addition to the water-reducing effect, the preservation of the concrete mixture was present, which is no less important when concreting large-sized structures, or transporting the semi-finished product over long distances.

The developed compositions of concrete classes B25, B35 and B40 almost all correspond to the declared strength. For a more in-depth analysis of the effect of complex modification of cement stone, concrete samples of composition No. 11 using the binder "Portland cement - volcanic ash - opoka" and the additive Cryoplast (1.1% by weight of cement) on the 7th day of hardening were examined on a Quanta 200 scanning electron microscope 3D (Fig. 1).

Comparing microphotographs of samples of 7 days of hardening without and with organomineral additives, one can observe a noticeable compaction of the structure of the cement stone, and this occurs due to an additional source of transformation, the interaction of hexagonal crystals of portlandite with the amorphous component of the fine powder, leading to the formation of tubular and needle-shaped hydrate formations. Complex modification of the cement stone matrix leads to an increase in the strength

Table 1. Compositions of modified concretes.

compositions	SP and dasage	W/C	Consumption, kg per 1 m^3 of concrete								
			C	W	CS	S	SC	MS	VA	Fl	A
Concrete class B25 CEM I 42.5N											
1	RHEOPLAST LS 01 (0.9%)	0.52	340	176	1070	830	30	11	-	-	3.0
2	Stark EW (0.9%)	0.52	340	176	1070	830	30	11	-	-	3.0
3	MegaChimTrade (1.0%)	0.52	340	176	1070	830	30	11	-	-	3.4
4	Cryoplast (0.9%)	0.52	340	176	1070	830	30	11	-	-	3.0
5	Stark PM (0.9%)	0.52	340	176	1070	830	30	11	-	-	3.0
6	TechnoNIKOL ICE 611 (1.0%)	0.52	340	176	1070	830	30	11	-	-	3.4
7	TechnoNIKOL READY-MIX 111 (1.0%)	0.52	340	176	1070	830	30	11	-	-	3.4
Concrete class B35 CEM I 42.5N											
8	RHEOPLAST LS 01 (1.1%)	0.34	410	140	1100	745	-	-	25	25	4.5
9	Stark EW (1.1%)	0.34	410	140	1100	745	-	-	25	25	4.5
10	MegaChimTrade (1.1%)	0.34	410	140	1100	745	-	-	25	25	4.5
11	Cryoplast (1.1%)	0.34	410	140	1100	745	-	-	25	25	4.5
12	Stark PM (1.1%)	0.34	410	140	1100	745	-	-	25	25	4.5
13	TechnoNIKOL ICE 611 (1.1%)	0.34	410	140	1100	745	-	-	25	25	4.5
14	TechnoNIKOL READY-MIX 111 (1.1%)	0.34	410	140	1100	745	-	-	25	25	4.5

(*continued*)

Table 1. (*continued*)

compositions	SP and dasage	W/C	Consumption, kg per 1 m^3 of concrete								
			C	W	CS	S	SC	MS	VA	Fl	A
Concrete class B40 CEM I 42.5H											
15	RHEOPLAST LS 01 (1.1%)	0.32	460	148	980	750	-	10	30	-	5.0
16	Stark EW (1.1%)	0.32	460	148	980	750	-	10	30	-	5.0
17	MegaChimTrade (1.1%)	0.32	460	148	980	750	-	10	30	-	5.0
18	Cryoplast (1.1%)	0.32	460	148	980	750	-	10	30	-	5.0
19	Stark PM (1.1%)	0.32	460	148	980	750	-	10	30	-	5.0
20	TechnoNIKOL ICE 611 (1.1%)	0.32	460	148	980	750	-	10	30	-	5.0
21	TechnoNIKOL READY-MIX 111 (1.1%)	0.32	460	148	980	750	-	10	30	-	5.0

Note: *C – cement; W – water; CS – crushed stone; S – sand; A – additive; SC – filler made from scrap concrete; MS– microsilica; Fl – flask; VA – volcanic ash; CS – compressive strength;* C_S *– cone spread;* ρ_C *– concrete density*

properties of concrete, and this is due to the enhanced synthesis of hydrate compounds of silicates, aluminates, sulfoaluminates and calcium aluminoferrites in the body of the concrete composite [13–16].

4 Conclusions

Modern transformations, providing for the quantitative and qualitative expansion of the components of concrete composition: cement, several types of mineral powders, organic additives, high-quality fillers and effective technological methods for preparing mixtures, open up new opportunities in concrete science, and this without increasing the consumption of energy-intensive Portland cement; on the contrary, the use of pozzolanic additives will allow develop environmental technologies and manage the properties of composites in the right direction [17–19].

The use of multicomponent molding mixtures with a high content of amorphous components and chemical modifiers reduces the consumption of mixing water, promotes complex chemical transformations with the formation of a dense and impermeable structure, an intensive increase in the strength of concrete and many important indicators. The research results proposed in this work will make it possible to successfully implement projects of global importance, creating stable, strong and durable concrete structures.

Table 2. Properties of modified concrete.

compositions	SP and dosage	$\rho_б$, kg/м^3	R_{cs} at the age, day, MPa		Water absorption, %	CS, sm
			7	28		
Concrete class B25 CEM I 42.5N						
1	RHEOPLAST LS 01 (0.9%)	2326	31.8	42.8	6.4	65
2	Stark EW (0.9%)	2340	34.5	45.2	6.1	67
3	MegaChimTrade (1.0%)	2371	34.3	43.8	6.3	66
4	Cryoplast (0.9%)	2372	36.2	46.8	5.9	68
5	TechnoNIKOL Stark PM (0.9%)	2337	29.5	41.1	6.6	64
6	TechnoNIKOL ICE 611 (1.0%)	2308	29.1	41.4	6.6	64
7	TechnoNIKOL READY-MIX 111 (1.0%)	2331	35.9	46.9	6.0	67
Concrete class B35 CEM I 42.5N						
8	RHEOPLAST LS 01 (1.1%)	2330	42.3	46.9	6.3	72
9	Stark EW (1.1%)	2342	44.6	49.2	6.0	73
10	MegaChimTrade (1.1%)	2373	44.1	47.8	6.2	74
11	Cryoplast (1.1%)	2378	48.5	52.9	5.7	75
12	Stark PM (1.1%)	2338	39.6	46.0	6.5	70
13	TechnoNIKOL ICE 611 (1.1%)	2310	38.4	45.9	6.5	69
14	TechnoNIKOL READY-MIX 111 (1.1%)	2339	47.5	51.2	5.8	75
Concrete class B40 CEM I 42.5H						
15	RHEOPLAST LS 01 (1.1%)	2332	52.9	54.0	6.2	73
16	Stark EW (1.1%)	2344	53,8	55,0	5,8	74
17	MegaChimTrade (1.1%)	2376	53.2	54.9	6.1	74

(continued)

Table 2. (*continued*)

compositions	SP and dosage	$\rho_б$, kg/м^3	R_{cs} at the age, day, MPa		Water absorption, %	CS, sm
			7	28		
18	Cryoplast (1.1%)	2380	55.3	61.2	5.2	75
19	Stark PM (1.1%)	2340	48.0	51.2	6.4	71
20	TechnoNIKOL ICE 611 (1.1%)	2320	46.5	51.0	6.4	70
21	TechnoNIKOL READY-MIX 111 (1.1%)	2341	54.2	58.6	56	75

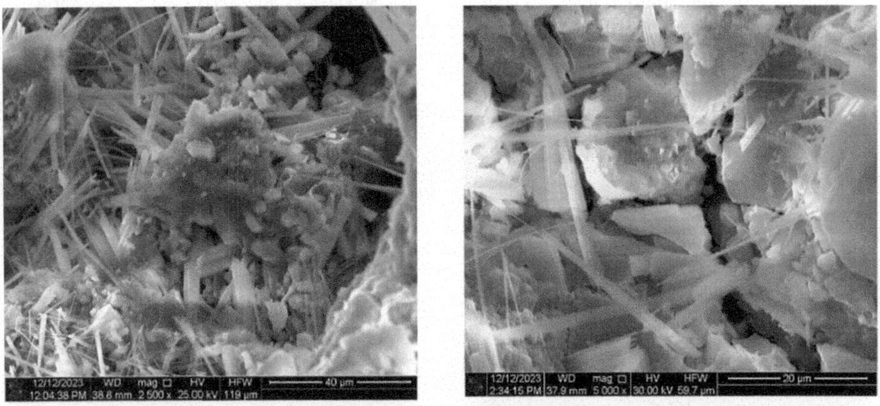

samples without additional ones

samples modified with organomineral additives

Fig. 1. Microphotographs of the studied samples, magnified. 2500, 5000 times.

References

1. Lesovik, R.V., Klyuyev, S.V., Klyuyev, A.V., Netrebenko, A.V., Kalashnikov, N.V.: Fiber concrete on composite knitting and industrialsand KMA for bent designs. World Appl. Sci. J. **30**(8), 964–969 (2014)
2. Lesovik, R.V., Klyuev, S.V., Klyuev, A.V., Tolbatov, A.A., Durachenko, A.V.: The development of textile fine-grained fiber concrete using technogenic raw materials. Res. J. Appl. Sci. **10**(10), 696–701 (2015)
3. Lesovik, R.V., Klyuyev, S.V., Klyuyev, A.V., Netrebenko, A., Yerofeyev, V.T., Durachenko, A.V.: Fine-grain concrete reinforced by polypropylene fiber. Res. J. Appl. Sci. **10**(10), 624–628 (2015)
4. Lesovik, R.V., Klyuyev, S.V., Klyuyev, A.V., Netrebenko, A.V., Durachenko, A.V.: High-strength fiber-reinforced concrete containing technogenic raw materials and composite binders with use of nanodispersed powder. Res. J. Appl. Sci. **9**(12), 1153–1157 (2014)
5. Mailyan, D.R., Nesvetaev, G.V., Kolleganov, N.A.: To determine the crack resistance of reinforced concrete beams from various types of concrete. Eng. Bull. Don **1**(97), 533–548 (2023)
6. Chepurnenko, A.S., Nesvetaev, G.V., Koryanova, Y., Yazyev, B.M.: Simplified model for determining the stress-strain state in massive monolithic foundation slabs during construction. Int. J. Comput. Civ. Struct. Eng. **18**(3), 126–136 (2022)
7. Lesovik, V.S., Popov, D., Fediuk, R.S., Usanova, K.: Composite binders with superabsorbent polymers. Constr. Unique Build. Struct. **3**(108), 10803 (2023)
8. Alaskhanov, A.Kh., Lesovik, V.S., Tolstoy, A.: "Green" composites based on technogenic raw materials. AIP Conference Proceedings 153–158 (2023)
9. Bazhenov, Yu.M., Demyanova,V.S., Kalashnikov, V.I.: Modified high-quality concrete. Moscow: Association of Construction Universities 289 (2006)
10. Kalashnikov, V.I.: Industry of non-metallic building materials and the future of concrete. Constr. Mater. **3**, 20–24 (2008)
11. Murtazaev, S., Salamanova, M.: Prospects for the use of thermally activated raw materials of aluminosilicate nature. Volga Sci. J. B. **46**(2), 65–70 (2018)
12. Kaprielov, S.S.: Modified high-strength concrete of classes B80 and B90 in monolithic structures. Constr. Mater. **3**, 9–13 (2008)
13. Salamanova, M.Sh., Saidumov, M.S., Murtazaeva, T.S-A., Khubaev, M. S-M.: High-quality modified concrete based on mineral additives and superplasticizers of various natures. Innovations and investments **8**, 159–163 (2015)
14. Murtazaev, S., Salamanova, M., Aliev, S.A., Bisultanov, R.G.: Rocks of volcanic origin as fillers for producing lightweight concrete. Scientific Review **7**, 105–113 (2015)
15. Salamanova, M.Sh., Bisultanov, R.G., Movsulov, M.M.: Promising possibilities for producing high-quality fine-grained composites. Scientific and technical journal Vestnik GGNTU. Technical Science **XIX 2**(32), 94–101 (2023)
16. Murtazaev, S-A.Yu., Salamanova, M.Sh., Koryanova, Yu.I.: Development of a polymorphic binder system based on a carbonate additive. Scientific and technical journal Vestnik GGNTU. Technical Science **XIX 3**(33), 96–103 (2023)
17. Yang, J., Qu, D., Hu, J., Song, L., Cheng, B.: Research on singular value detection method of concrete dam deformation monitoring. Measurement **179**, 109457 (2021)
18. Chen, J., Jia, Q., Xu, S., Fan, P.: The PDEM-based time-varying dynamic reliability analysis method for a concrete dam subjected to earthquake. Structures **33**, 2964–2973 (2021)
19. Gowripalan, N., Shakor, P., Rocker, P.: Pressure exerted on formwork by self-compacting concrete at early ages: A review. Case Stud. Constr. Mater. **15**, 00642 (2021)

Improvement of the System of Incoming Control During Construction of Facilities with Application of AI in the Arctic Conditions

Azariy Lapidus, Tembot Bidov(✉), Alan Khubaev, and Dmitry Nikolenko

National Research Moscow State University of Civil Engineering, 129337 Moscow, Russia
BidovTH@mgsu.ru

Abstract. The article discusses in detail the specific aspects of the organisation of incoming inspection during the construction of facilities in Arctic conditions. Special attention is paid to the unique challenges associated with extreme climatic conditions, remoteness of construction sites and limited time windows for work. These factors significantly complicate traditional methods of quality control of incoming materials, structures, equipment and machinery, which requires new approaches and solutions. The paper identifies information gathering and analysis techniques that form artificial intelligence (AI) models aimed at automating and optimising input inspection processes. These technologies can not only improve the accuracy of inspections and reduce data processing time, but also minimise the impact of the human factor, which can be particularly critical when access to skilled professionals is limited. The authors also look at the potential of AI to predictively analyse and forecast potential defects, enabling early detection of problems and prompt corrective action. The application of AI in the incoming inspection system also helps to reduce overall construction costs by optimising logistics, reducing downtime and increasing resource efficiency. The article provides examples of successful AI implementation at real construction projects in the Arctic region and analyses the effectiveness of the proposed solutions in comparison with traditional quality control methods.

Keywords: Input Control · Organisational and Technological Solutions · Arctic · Extreme Conditions · Artificial Intelligence · Automation

1 Introduction

In recent years, the active use of artificial intelligence (AI) is affecting more and more spheres of human activity. Research related to the use of AI in the exploration and development of mineral resources in the extreme conditions of the Arctic is widespread and is already being actively implemented, which allows to significantly improve the efficiency and safety of work. Nevertheless, the construction industry is still not fully utilising the potential of innovative technologies, in particular AI, which remains an understudied and underdeveloped area [1].

S. V. Klyuev et al. (Eds.): ICRAAC 2024, LNCE 627, pp. 18–28, 2025.
https://doi.org/10.1007/978-3-031-82938-3_3

Traditional methods of incoming inspection based on manual inspection, spot checks and human error are becoming insufficient in the harsh Arctic climate. Mistakes made at the incoming inspection stage can lead to catastrophic consequences: from construction delays and budget overruns to accidents and structural failure.

In turn, the complex climatic conditions of the Arctic impose special requirements on the construction of facilities: from the selection of materials to quality control of work. This is where the introduction of AI can play a key role, helping to adapt processes to the unique Arctic conditions. In particular, improving the system of incoming inspection at construction projects using AI can be a significant step in improving the reliability and safety of construction.

This article discusses the relevance and prospects of using artificial intelligence to optimise the entry control system for construction projects in the Arctic. It analyses the potential of AI to ensure a high level of quality and compliance with the strict norms and standards required in Arctic construction projects [2–4].

2 Methods and Materials

In order to study and summarise the experience of incoming inspection in sub-zero temperatures, this paper used the tools of qualitative research method. This approach was chosen because it allows for deeper analysis of complex processes, identifying key problems and finding possible solutions. During the research process, special attention was paid to the specifics of construction conditions in the Arctic: extreme climate, limited construction season and complex logistics. These factors have a direct impact on the effectiveness of the incoming inspection system, making them critical to the development of effective quality control methods.

The preparatory phase for the organisation of the study included a detailed conceptualisation and strategy that defined the objectives of the study and the tasks required to achieve them. This phase highlighted the main challenges faced by the entry control system in the Arctic environment.

1. Extreme climatic conditions such as low temperatures, high winds and frequent snowstorms have a significant impact on the condition of materials and equipment. This requires careful checking of their quality and compliance with established standards, as any deviations may lead to their premature failure [5, 6].
2. a limited construction season creates the need to strictly adhere to delivery deadlines and respond quickly to any logistical disruptions. During the short construction period, it is important to minimise any delays that will affect the timely completion of construction [7].
3. The difficult accessibility of sites and the distance from production bases make on-site inspection difficult. Continuous inspection is often impossible, requiring the introduction of remote monitoring and control methods, including the use of AI [8, 9].
4. Increased demands on the quality of materials force more thorough inspections upon receipt, as well as monitoring the condition of materials during all stages of transport and storage. Materials and equipment must be resistant to extreme conditions to ensure the reliability and durability of facilities [10, 11].

5. Difficulties in providing continuous monitoring and control due to the remoteness of facilities make traditional control methods less effective. This requires innovation, such as automated monitoring systems that can operate at remote sites without constant human presence [12].

Methods for collecting and analysing information were also identified, including documentation analysis, observation of incoming inspection processes at construction sites in the Arctic. Criteria for evaluating the information and results obtained by observing and analysing experience were set, taking into account the requirements for the quality of construction materials, equipment and machinery, as well as the efficiency of logistics processes in the Arctic conditions [13–16].

The main documents allowing to assess the qualitative and quantitative characteristics of certain stages of incoming inspection were the following documents:

Journal of incoming inspection—designed to record data on incoming materials, identify defects and mark to which warehouse or construction site the products were sent.

Quality certificates—confirm the compliance of materials and products with the established standards and requirements.

Technical data sheets—contain information on the characteristics of materials and equipment.

Test protocols—results of laboratory tests that demonstrate that the materials meet the necessary requirements.

Transportation documents—confirm that the materials were delivered in accordance with the terms of the contract.

General work log—this is the main primary document for recording the performance of construction and installation works.

The introduction of artificial intelligence (AI) technologies opens up new opportunities to improve the efficiency and automation of the entry control system for construction in the Arctic. AI allows solving key tasks of incoming inspection, such as:

1. Developing methods and procedures to verify the quality of materials from production to delivery to the site. This includes using AI to automate inspection processes and analyse data [17, 18].
2. Optimise the process of controlling the arrival of equipment and machinery to ensure their reliability and suitability for operation in the extreme climatic conditions of the Arctic. AI systems can help in predicting breakdowns and preventing possible equipment failures [19].
3. Improve logistics efficiency by implementing AI systems to track and predict possible problems in the supply chain. This will minimise delays and ensure that materials and equipment arrive on time [20–22].
4. Monitoring the condition of materials during transport and storage, which is especially important in low temperature and high humidity environments. AI systems and sensors can monitor these parameters in real time, enabling rapid response to deviations from the norm [23].

3 Results and Discussion

The results obtained in the research are systematised and on their basis the scheme of input control with the established interrelation between the elements with the artificial intelligence system integrated into it, which helps in decision making and control of all parts of the process Fig. 1 is developed.

Also for each element of the decision-making system the control factors are developed, on the basis of which an information model is built to help control each stage and make a decision on the acceptance of materials and equipment to the construction site or their rejection Table 1 [24].

Table 1. Control factors.

N	Decision-making system element	Control factor
1	Obtaining data on materials and equipment	Input of data on material characteristics (certificates, laboratory tests, etc.)
		Data on the status of equipment and materials at the delivery stage
2	Analysing data using AI	Verification of compliance with project requirements
		Risk forecasting (quality, durability, climate resistance)
3	Logistics risk assessment	Analysing weather conditions and delivery routes
		Forecasting delays and optimising logistics
4	Monitoring the condition of materials during storage and transport	Collecting data from sensors (temperature, humidity, etc.)
		Notification of critical deviations
5	Automated visual inspection	Use of drones and AI cameras to inspect materials and equipment
		Detection of defects or violations of storage conditions
6	Decision on approval of materials and equipment for construction	Decision making based on AI analysis

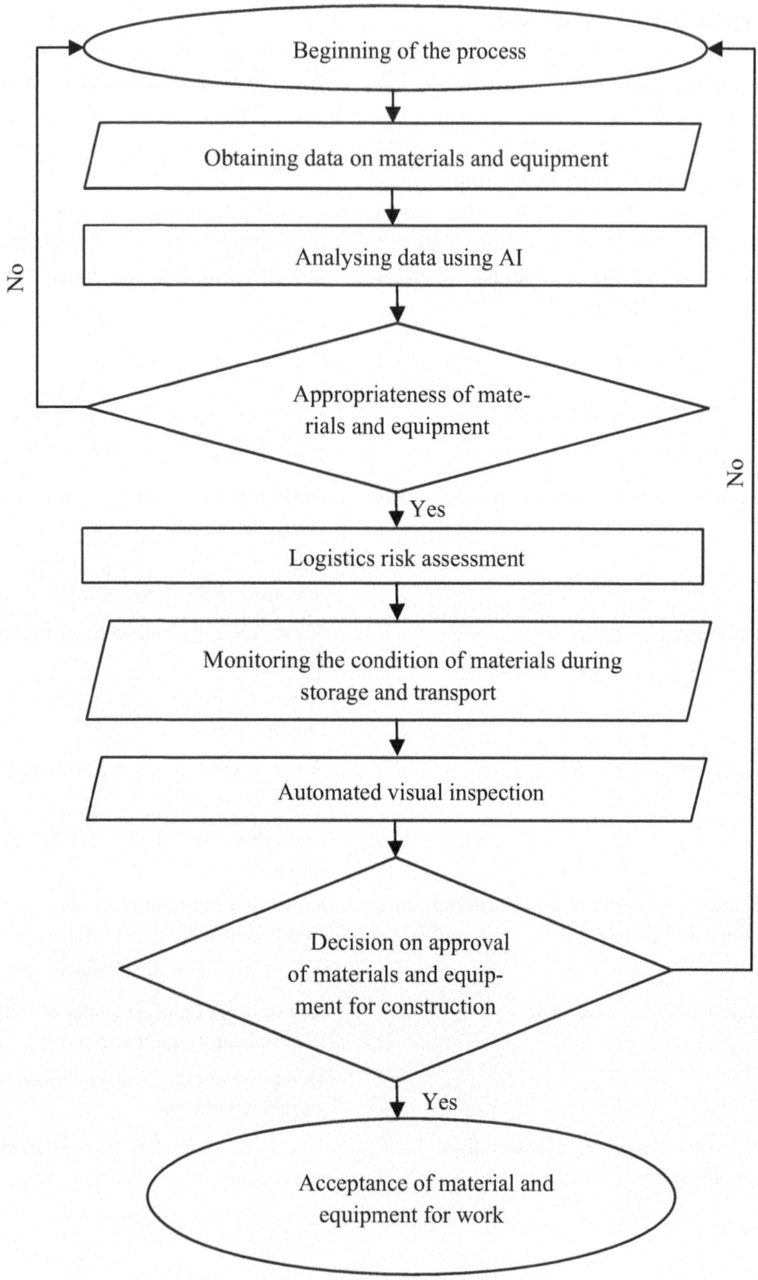

Fig. 1. Block diagram of input control.

An artificial intelligence architecture has been developed to form an input control system for Arctic conditions Fig. 2. A centralised data management platform integrates information on materials, equipment, logistics and climatic conditions, stores data on previous projects for AI training, ensures data security and encryption, and provides APIs for integration with external systems and sensors [25, 26]. The materials analysis module compares performance against design requirements, analyses quality certificates and assesses suitability for Arctic conditions. The condition prediction module monitors the condition of materials, predicts problems (corrosion, brittleness) and provides recommendations for storage and handling. The logistics module optimises routes and delivery times, analyses weather conditions and predicts delays and risks. The climate analysis module forecasts weather conditions at the construction site, assesses the impact of climate on materials and equipment, and provides recommendations for adapting work processes. The quality control module automatically checks material conformity to specifications, identifies deviations and defects, and generates quality reports with recommendations for corrective action.

Implementation of the system of input control of construction materials in the Arctic conditions using the developed system based on artificial intelligence (AI) is aimed at improving the efficiency, accuracy and reliability of processes related to the acceptance control of materials. The expected results of such implementation can be divided into several key aspects and compared with the results of the analysis of the current state obtained by generalizing the current situation of incoming inspection in the construction industry and the experience of applying similar monitoring systems in other areas of science and production Fig. 3.

The application of AI in entry control systems in Arctic conditions has demonstrated high efficiency, but there are also certain challenges. First, significant adaptation of AI technologies to the extreme temperatures and weather conditions of the Arctic is required. Data collection equipment must operate in low temperatures, high winds and limited visibility, which requires specialised solutions.

Secondly, an important aspect is the training and adaptation of personnel to new technologies. Despite the automation of many processes, humans continue to play a key role in managing and interpreting results, so special training programmes are needed for specialists working with AI.

It is also worth noting the need to create a unified database that will accumulate information about the behaviour of materials in Arctic conditions, which will allow AI systems to learn and improve, making predictions more accurate.

In addition, implementing AI requires a significant initial investment, which can be an obstacle for small and medium-sized companies. Nevertheless, given the long-term benefits, such as improved quality and lower costs, this investment can be justified.

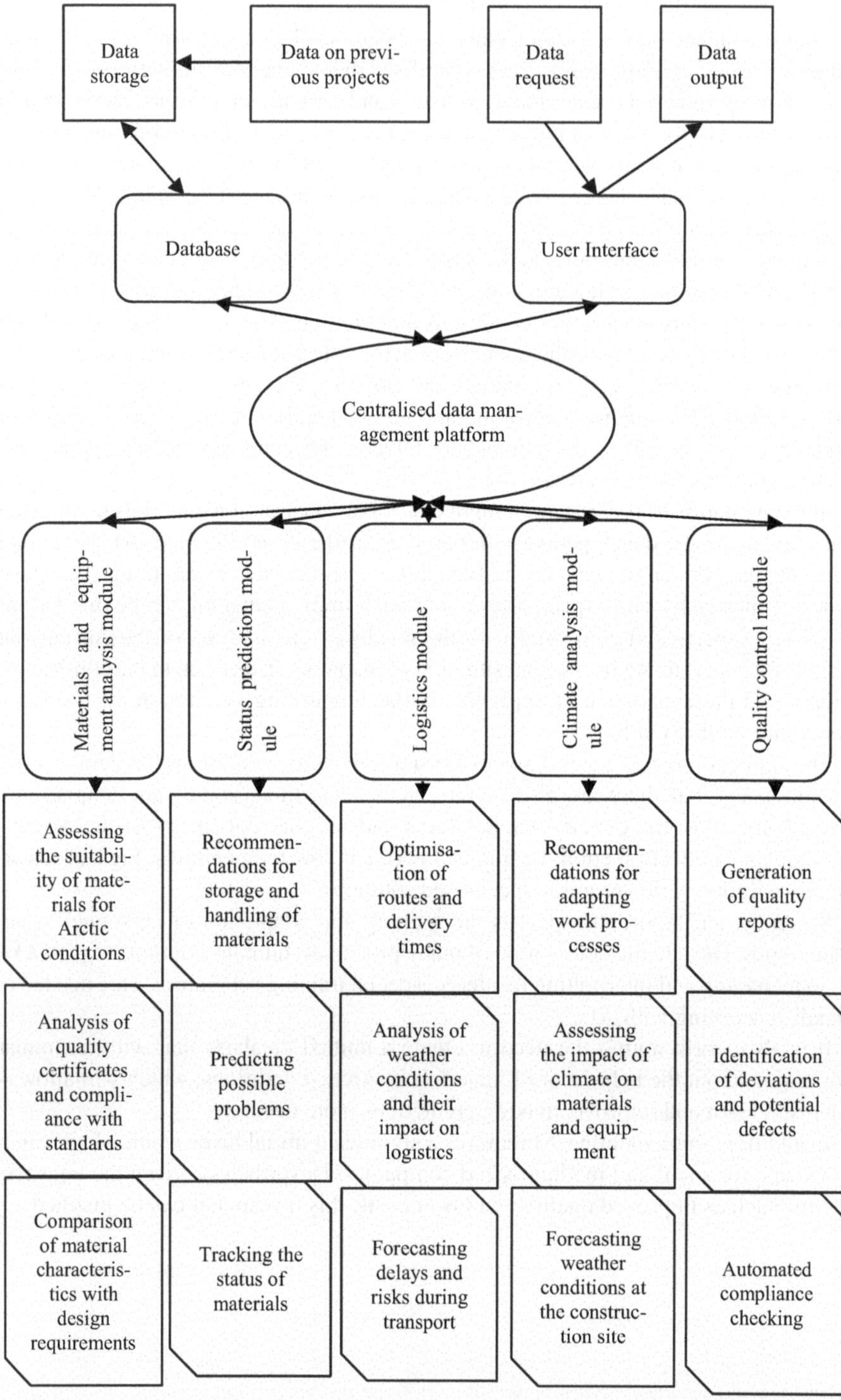

Fig. 2. AI architecture for the system of input control during construction of facilities.

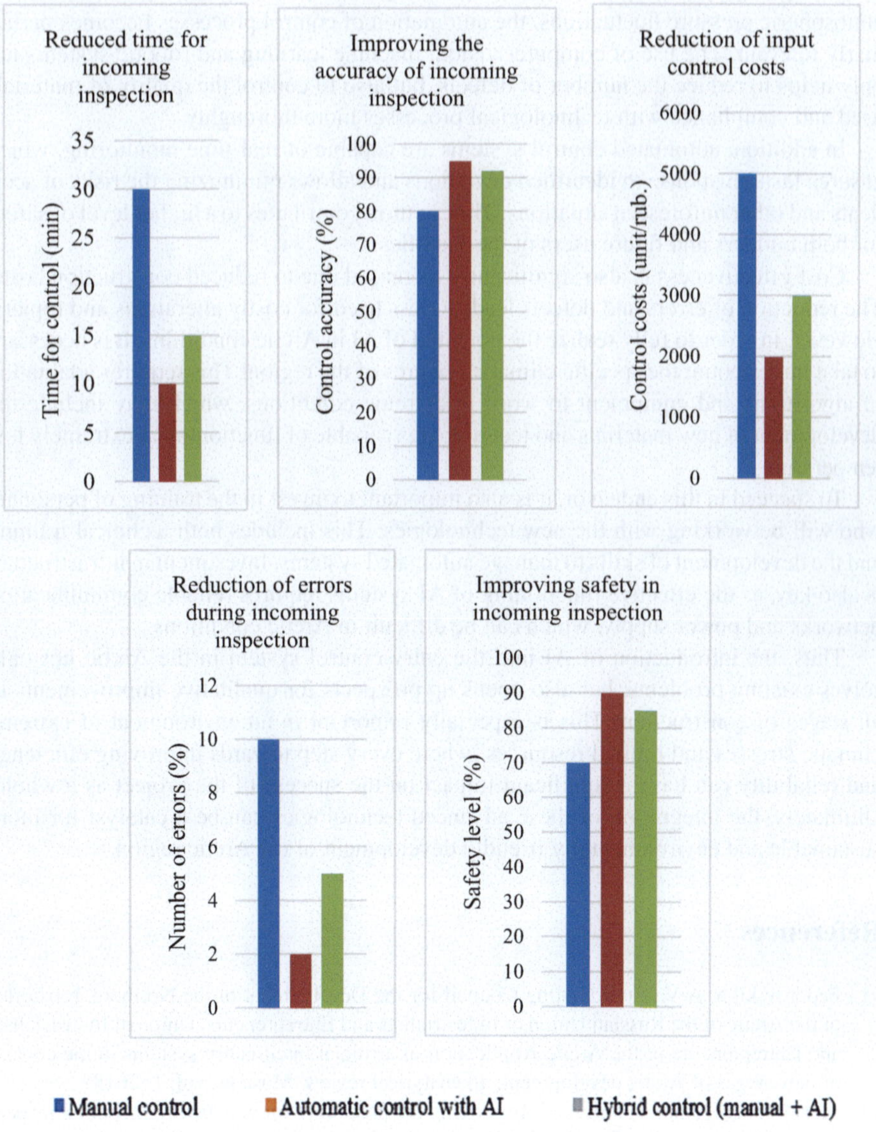

Fig. 3. Planned result of AI implementation for the input control system.

4 Conclusions

Improving the system of incoming inspection during construction of facilities in the Arctic with the use of AI really opens up new horizons for increasing the efficiency and reliability of construction. In the Arctic, where climatic conditions pose special challenges to construction workers, such as low temperatures, strong winds and significant atmospheric pressure fluctuations, the automation of control processes becomes particularly relevant. The use of computer vision, machine learning and robotic systems not only helps to reduce the number of defects, but also to control the quality of materials used and compliance with technological processes more thoroughly.

In addition, automated control systems are capable of real-time monitoring, which ensures faster response to identified deviations and allows minimizing the risks of accidents and other unforeseen situations. This, in turn, contributes to a higher level of safety for both builders and future users of the facilities.

Cost-effectiveness is also significantly increased due to reduced construction costs. The reduction of errors and defects leads to less need for costly alterations and repairs. However, in order to fully realize the potential of AI in Arctic conditions, it is necessary to take into account the specific climatic features of the region. This requires adaptation of algorithms and equipment to work in extreme conditions, which may include the development of new materials and technologies capable of functioning at extremely low temperatures.

To succeed in this endeavor, it is also important to invest in the training of personnel who will be working with the new technologies. This includes both technical training and the development of skills to manage automated systems. Investment in infrastructure is also key, as the effective functioning of AI systems requires reliable communication networks and power supply, which can be difficult in Arctic conditions.

Thus, the introduction of AI into the entry control system in the Arctic not only solves existing problems, but also opens up prospects for qualitative improvements in all stages of construction. This is especially important in an environment of extreme climatic stresses and limited resources, where every step towards improving efficiency and reliability can have a significant impact on the success of the project as a whole. Ultimately, the integration of these advanced technologies can be a catalyst for more sustainable and environmentally friendly development of the Arctic region.

References

1. Fedotovskikh, A.V.: Coordinating Council for the Development of the Northern Territories of the Arctic of the Russian Union of Industrialists and Entrepreneurs, Union of Industrialists and Entrepreneurs of the Arctic Application of artificial intelligence systems in the context of new stages of Arctic development: an analytical review, Moscow, vol. 1 (2018)
2. Motylev, R.V., Karpushkin, A.S.: Improving the procedure for conducting incoming inspection of incoming materials in construction and recording its results (part 1). Constr. Prod. **1**, 19–27 (2022). https://doi.org/10.54950/26585340_2022_1_19
3. Motylev, R.V., Karpushkin, A.S.: Improving the procedure for conducting incoming inspection of incoming materials in construction and recording its results (part 2). Constr. Prod. **1**, 27–30 (2022). https://doi.org/10.54950/26585340_2022_1_27
4. Lapidus, A.A., Yan, Ts.: Improving construction supervision in the construction of high-rise buildings using a neural network model. Compon. Sci. Technol. Prog. **1**(91), 51–58 (2024)
5. Lapidus, A.A., Yan, Ts.: Analysis of international experience in the quality of BIM technologies for construction control of high-rise construction. Constr. Prod. **2**, 46–51 (2022). https://doi.org/10.54950/26585340_2022_2_46

6. Lapidus, A.A., Kuzmina, T.K., Bolshakova, P.V.: Improving the procedure for interaction between participants in the construction control system in order to increase labor productivity. Constr. Mater. Prod. **7**(4), 1 (2024). https://doi.org/10.58224/2618-7183-2024-7-4-1
7. Ginzburg, A.V., Ryzhkova, A.I.: Possibilities of artificial intelligence to improve the organizational and technological reliability of construction production. Bull. of MGSU **13**(1), (112), 7–13 (2018). https://doi.org/10.22227/1997-0935.2018.1.7-13. - EDN XCIOMJ
8. Shishkina, D.N.: Assessment of the economic efficiency of using artificial intelligence in construction: choosing the optimal method. Bull. of MGSU **19**(9), 1550–1561 (2024). https://doi.org/10.22227/1997-0935.2024.9.1550-1561
9. Oluleye, B.I., Chan, D.W.M., Antwi-Afari, P.: Adopting Artificial Intelligence for enhancing the implementation of systemic circularity in the construction industry: a critical revie. Sustain. Prod. Consum. **35**, 509–524 (2023). https://doi.org/10.1016/j.spc.2022.12.002
10. Baduge, S.K., Thilakarathna, S., Perera, J.S., Arashpour, M., Sharafi, P., Teodosio, B. et al.: Artificial intelligence and smart vision for building and construction 4.0: machine and deep learning methods and applications. Autom. Constr. **141**, 104440 (2022). https://doi.org/10.1016/j.autcon.2022.104440
11. Bidov, T.Kh., Khubaev, A.O., Shabanova, A.A.: Organizational and technical modeling of an integrated system for producing concrete works in winter during the construction of residential buildings. Constr. Geotech. **12**(2), 15–25 (2021). https://doi.org/10.15593/2224-9826/2021.2.02
12. Bidov, T.Kh., Khubaev, A.O., Pomytko, E.A., Kotelnikova, A.D.: Methodology for assessing factors influencing the efficiency of implementing a construction project by a general contractor in the Fit-out direction. Bull. of MGSU **19**(9), 1562–1569 (2024). https://doi.org/10.22227/1997-0935.2024.9.1562-1569
13. Khubaev, A., Bidov, T., Bzhienikov, A., Nesterova, V.: Modern Russian high-tech construction materials and their application in domestic construction industry (on example of metal-ceramic panels Hardwall). IOP Conf. Ser.: Mater. Sci. Eng. **365**, 032005 (2018). https://doi.org/10.1088/1757-899X/365/3/032005
14. Fatullaev, R.S., Bidov, T.Kh., Shesterikova, Ya.V., Nikolenko, D.M.: Analytical study of the actual cost of the completed overhaul of multi-apartment residential buildings in Russia. E3S Web of Conf. **389**, 06029 (2023). https://doi.org/10.1051/e3sconf/202338906029
15. Lapidus, A., Topchiy, D., Kuzmina, T., et al.: An integrated quality index of high-rise residential buildings for all lifecycle stages of a construction facility. Appl. Sci. (Switz.) **13**(3), 2014 (2023). https://doi.org/10.3390/app13032014
16. Fedosov, S.V., Lapidus, A.A., Narmaniya, B.E., Ayzatullin, M.M.: Cauchy problem for modeling of unsteady mass transfer processes in an unbounded plate by the integral Laplace transform method. Constr. Mater. Prod. **7**(5), 4 (2024). https://doi.org/10.58224/2618-7183-2024-7-5-4
17. Fatullaev, R., Bidov, T., Kagazezhev, A.: Analysis of the labour of roof repair in different countries. AIP Conf. Proc. 050019, (2023). https://doi.org/10.1063/5.0178176
18. Lapidus, A., Khubaev, A., Bidov, T.: A. Regression analysis of the calculation of organizational and technological potential for the production of cold-weather concreting. IOP Conf. Ser.: Mater. Sci. Eng. 072033, (2020). https://doi.org/10.1088/1757-899X/869/7/072033
19. Khubaev, A.O.: Determination of the effectiveness of the production process of winter concreting based on field studies. E3S Web of Conf. **389**, 06012 (2023). https://doi.org/10.1051/e3sconf/202338906012
20. Oleynik, P., Efimov, V.: Modeling volumetric block types in residential building construction. Appl. Sci. (Switz.) **14**(9), 3565 (2024). https://doi.org/10.3390/app14093565
21. Fokina, O., Ivanovskaya, Zh., Mottaeva, A., Khubaev, A.: The conceptual role of leadership in project management. E3S Web of Conf. **458**, 04020 (2023). https://doi.org/10.1051/e3sconf/202345804020

22. Lapidus, A., Khubaev, A., Bidov, T.: Development of a three-tier system of parameters in the formation of the organizational and technological potential of using non-destructive testing methods. E3S Web of Conf. **97**, 06037 (2019). https://doi.org/10.1051/e3sconf/20199706037
23. Lapidus, A.A., Kuzmina, T.K., Bolshakova, P.V.: Developing administrative and process-related solutions for the benefit of residential construction projects. Constr. Mater. Prod. **7**(6), 2 (2024). https://doi.org/10.58224/2618-7183-2024-7-6-2
24. Lapidus, A., Khubaev, A., Bidov, T.: Organizational and technological solutions justifying use of non-destructive methods of control when building monolithic constructions of civil buildings and structures. MATEC Web of Conf. **251**, 05014 (2018). https://doi.org/10.1051/matecconf/201825105014
25. Lapidus, A., Khubaev, A., Bidov, T.: Development software for the non-destructive control of monolithic structures in housing construction. E3S Web of Conf. (2021). https://doi.org/10.1051/e3sconf/202125809003
26. Bidov, T.Kh., Khubaev, A.O., Pomytko, E.A., Kotelnikova, A.D.: Methodology of evaluating of factors affecting the efficiency of construction project implementation by general contracting organization in Fit-out direction. Vestnik MGSU **19**(9), 1562–1569 (2024). https://doi.org/10.22227/1997-0935.2024.9.1562-156

Carbonation Studies in Alkali-Activated Concrete

P A Fedorov[1], S O Yamaeva[1], V M Latypov[1], S. V. Klyuev[2](✉), and E V Lutsyk[1]

[1] Ufa State Petroleum Technological University, Ufa, Russia
[2] Belgorod State Technological University Named After V.G. Shukhov, Belgorod, Russia
Klyuyev@yandex.ru

Abstract. The development of carbon footprint reduction technologies in construction materials science will allow successful utilization of large-tonnage industrial waste. A promising direction is the use of geopolymers and alkali-activated concrete in construction. However, due to the poor study of the problem, questions remain about ensuring the durability of these materials in various aggressive environments, as well as ensuring a stable composition of binders based on industrial waste. This article describes the results of a study of the features of the mechanism of accelerated carbonization of alkali-activated concrete based on dust-carryover from a cupola of mineral wool production. The binder is a low-base dust-carryover from the gas cleaning system of a cupola of mineral wool production. An aqueous solution of NaOH with a concentration of 6M is used as a liquid activator. Fine-grained concrete samples with water-cement ratio of 0.45, 0.55, 0.60 were used for testing. It was found that the kinetics of carbonation of the samples has a fading nature and is expressed as a power function of the depth of carbonation over time. The exponent fluctuates from 0.4 to 0.57 depending on the density of the concrete. The carbonation front develops uniformly with small breakthroughs. At the initial stage, the front is controlled by kinetic limitation, subsequently it passes to internal diffusion limitation.

Keywords: Alkali-activated concrete · Accelerated carbonation · Strength · Nakholite · Density · Water-binding deviation

1 Introduction

In connection with the global transition from traditional technologies to low-carbon ones, one of the promising areas in construction is the use of concretes based on alkaline and alkaline earth binders (for example, geopolymers and alkali-activated concretes) [1–4]. They help reduce greenhouse gas emissions due to the low carbon footprint formed during their manufacture.

Alkali-activated concretes have a higher durability compared to Portland cement concretes, but there are a number of problems, such as efflorescence and accelerated carbonation, which require comprehensive research. These issues are especially relevant for cement systems obtained from large-tonnage industrial waste with an unstable composition.

S. V. Klyuev et al. (Eds.): ICRAAC 2024, LNCE 627, pp. 29–35, 2025.
https://doi.org/10.1007/978-3-031-82938-3_4

One of the direct indicators of the durability of concrete and reinforced concrete is resistance to carbonation [5]. Carbonation is usually understood as a chemical reaction between CO_2 and alkaline phases of the cement matrix with the formation of carbonates, bicarbonates and carbonic acid. The mechanism of carbonation in concretes based on Portland cement has been studied in detail, which is confirmed by the review [6]. During carbonation of cement concretes, the hydrogen index (pH) of the pore solution decreases from 12.5 to 9 or less. Currently, the ideas about the mechanism of carbonation of alkali-activated concretes are the subject of discussions. This is primarily due to the difference in the raw materials used for the binders. The most studied from the standpoint of the influence of carbonation are concretes based on blast-furnace slag and fly ash activated by sodium metasilicate or caustic alkali. In [7], it is noted that slag-alkali concrete activated by a two-component activator of liquid glass and hydroxide solutions is less resistant to carbonation compared to cement concrete. In this case, tests were carried out for 70 days at a CO_2 concentration of 10 - 20% by volume and a humidity of 70%. In [8], when testing fly ash and blast furnace slag-based geopolymers for carbonation under natural conditions after 8 years of observation, it was found that compositions without liquid glass showed similar results to Portland cement concrete. In turn, concrete with the addition of liquid glass showed worse resistance to carbonation compared to Portland cement concrete. In [9], it is indicated that high-strength concretes with reduced and especially low permeability for acid gases can be produced based on slag-alkali binders activated only by liquid glass. The main objective of this study is to study the carbonation rate of concretes based on alkali-activated binders from mineral wool production waste under conditions of exposure to high concentrations of carbon dioxide. The paper presents the results of accelerated carbonation tests with CO_2 concentration recording, and the kinetics of the process at an early stage of exposure during 14 days of aging.

2 Methods and Materials

The compositions of fine-grained concrete based on dust-carryover from the gas cleaning system of a mineral wool production cupola and a single-component liquid activator from an aqueous solution of caustic soda with a molar concentration of 6 mol/l were adopted for the study. The water-binding ratio was adopted as 0.45, 0.55, 0.60.

The consumption of raw materials is given in Table 1.

The binder is a fly ash from the gas cleaning system of a mineral wool production cupola (OOO Agidel, Republic of Bashkortostan). The binder is grey. The average specific surface area of the fly ash is 730 kg/m2. The true density of the fly ash is 2.952 g/cm3. The content of CaO oxide is 14%. The mineralogical composition of the fly ash includes siliceous and aluminosilicate formations such as quartz, ackermanite, gehlenite and other minerals.

For the preparation of the alkaline activator, technical flaked sodium hydroxide produced by JSC Bashkir Soda Company (Sterlitamak, Russia) and drinking water according to GOST 23732-2011 were used [10]. The molar concentration of the alkaline solution was 6 mol/l, which corresponds to a solution density of 1.2162 g/cm3.

The fine filler was polyfractional quartz sand, composed in accordance with the requirements of GOST 30744-2001 [11] from quarry sand from the Kulushevskoye

Table 1. The quantitative content of the components per 1 m^3 of the mixture.

Type of composition	Alkaline activator			Binder	W/B	Sand, kg/m^3
	Molar concentration, mol/l	Caustic soda (dry) kg/m^3	Water (W), kg/m^3	Dust removal (B), kg/m^3		
I	6	50	203	400	0.45	1350
II	6	61	248	389	0.55	1350
III	6	67	270	383	0.60	1350

deposit (Karmaskaly settlement, Republic of Bashkortostan). Bulk density 1300 kg/m3, true density 2.691 g/cm3, content of dust and clay particles - no more than 0.5% by weight.

Hardening of samples was carried out using isothermal heating in accordance with the recommendations of the authors of the work [12]. Isothermal heating was adopted as a single-stage with a total duration of 12 h. The maximum heating temperature was 90°C. Accelerated carbonization tests were carried out according to GOST R 52804-2007 [12] under the following conditions of the air environment in the chamber: CO2 9.15 ± 0.25% by volume; relative humidity 75 ± 5%; temperature 20 ± 5 °C. Concentration measurements were carried out using a PKU-4/1-Щ-1P-1A-USB gas analyzer with a sensor up to 10% by volume.

For testing, prism samples measuring 40x40x160 mm were used. The number of samples for accelerated carbonization testing was 3 pcs. of each composition.

To simulate one-sided exposure to a gas environment, before testing, the samples were covered with rosin-paraffin grease, leaving one working (lower) side. Then the samples were kept in an environment with a given temperature and humidity regime until equilibrium humidity was established.

The exposure in the accelerated carbonation chamber was carried out for 14 days. To determine the depth of carbonation of the samples, after 1, 7 and 14 days the samples of each type of composition were split and the fresh chip was treated with an indicator of 1% alcohol solution of phenolphthalein [14]. The measurement of the depth of carbonation of concrete was carried out taking into account the recommendations for accounting for changes in the carbonation front [15, 16].

3 Results and Discussions

Chippings of samples made 1 and 14 days after exposure in an accelerated carbonization chamber with subsequent application of a 1% alcohol solution of phenolphthalein are shown in Fig. 1.

The studied samples showed intense gas absorption. The greatest amount of absorbed gas corresponded to composition III, and the least to composition I. The average rate of decrease in gas concentration in the air was 0.03% v/min for composition I, 0.02% v/min for II, and 0.06% v/min for III. In the initial period, saturation of concrete with carbon

Fig. 1. Chips of the studied samples with an indicator applied 0.1% phenolphthalein solution.

dioxide occurs intensively and is characterized by a rapid increase in the threshold value of CO_2 concentration.

Figure 2 shows a graph of changes in CO_2 absorption by samples during the first 24 h. All samples are characterized by an increase in the amount of gas absorption within 1–2 h. The greatest absorption is characteristic of composition II.

After 3 h and up to 11 h of exposure, the gas-air environment stabilizes. It is obvious that during this period, the final saturation of the pores with gas occurs due to CO_2 diffusion, as well as chemical processes, including dissolution in the pore liquid and carbonization of hydrated phases. The third stage is characterized by a gradual decrease in gas absorption depending on the capillary-porous structure of the samples.

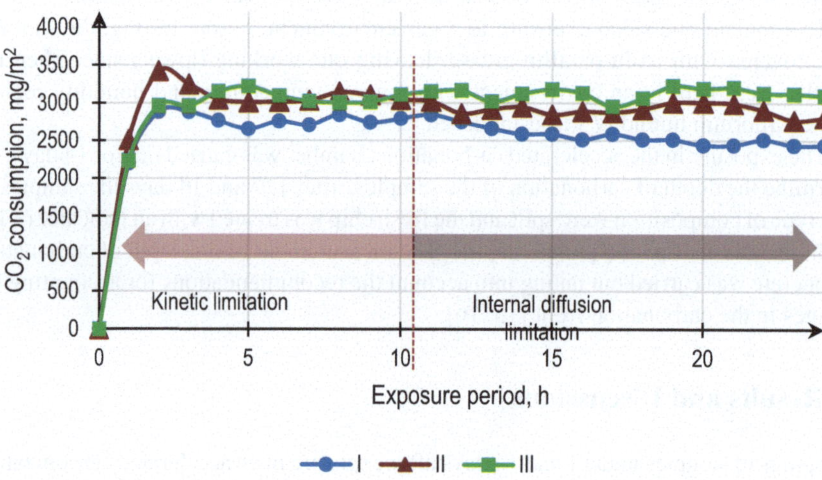

Fig. 2. Dynamics of CO_2 absorption by the cement matrix in the first day after exposure of samples in the chamber.

The results of changing the depth of concrete carbonation are shown in Fig. 3. All samples showed a fading nature of carbonation. The maximum depth of carbonation for composition I was 14 mm, II - 23 mm and III - 32 mm. Regression analysis showed that the obtained values correspond to the power function.

The obtained results on the kinetics of carbonation of alkali-activated concretes based on dust from mineral wool production showed that one of the factors controlling the rate of carbonation is the water-binder ratio, which forms the capillary-porous structure of concrete. This conclusion is confirmed by the previously obtained results of studies on testing slag-alkali concretes by Nguyen T.N. et al. [17]. In addition, the carbonation process itself is consistent with concretes based on Portland cement. However, due to the low content of CaO in the binder, the accelerated carbonation of the studied compositions occurred much faster compared to concretes using Portland cement, described in work [16]. This explanation is confirmed by the results of the studies of the authors of works [18–20].

Taking into account the results of the studies, the mechanism of carbonation in the initial period can be represented as follows. After the start of the supply of carbon dioxide, the walls of the surface pores are saturated with it. In this case, the rate of the chemical reaction of CO_2 interaction with the surface of the cement matrix is less than the rate of gas supply, since the surface pores are not filled or only partially filled with pore fluid. This period is characteristic of the kinetic limitation described by Alekseev S.N., Rosenthal N.K. [21]. The total duration of this period for the considered compositions and a CO_2 concentration of 9.15% v/v is about 10–11 h.

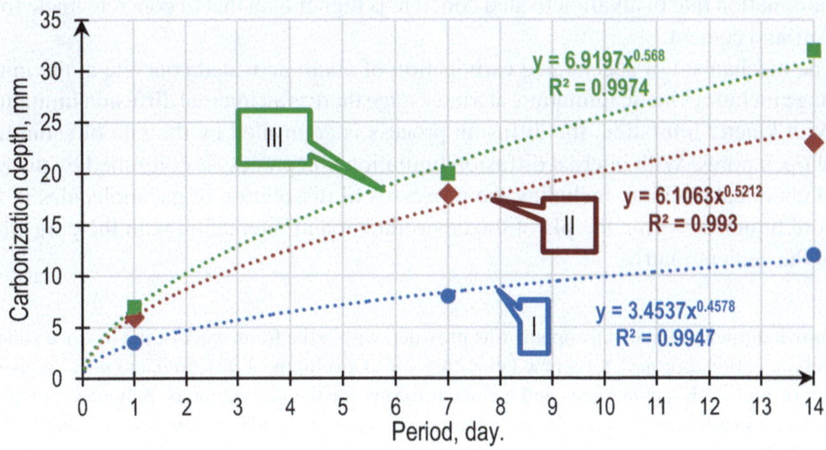

Fig. 3. Dynamics of changes in the depth of carbonation of the studied samples.

After kinetic limitation, internal diffusion limitation occurs, in which carbon dioxide diffuses through the capillary system of the cement matrix filled with pore fluid. Dissolution of gas in the pore solution leads to a decrease in the pH value and precipitation of carbonates in the pores.

In denser samples, this process is inhibited due to a larger number of pores completely filled with pore fluid, as well as due to the compaction of pores by carbonation products. Obviously, this fact, as well as in concretes based on Portland cement, leads to a deviation from the law of "square root of time", describing the kinetics of carbonation of cement concrete. The results of the studies show that the carbonation rate is described by a power function with a degree in general form "1/n". In this study, this ratio was 0.42… 0.57. These results are consistent with the results obtained by Li Z., Li S. [19], where for alkali-activated concrete based on fly ash and blast furnace slag the values of "1/n" were obtained equal to 0.31 and 0.22.

It should be noted that there is apparently a dependence of the "1/n" indicator on the porous structure of concrete. Research has established that compositions with a water-binder ratio of 0.5…0.55 correspond to the law of "square root of time". In the case of a decrease in the porosity of concrete, the value of the "1/n" ratio becomes less than 0.5. This phenomenon was also established for concretes based on Portland cement [22].

4 Conclusions

Based on the research conducted, the following conclusions can be drawn:

1. The carbonation rate of slag-activated concrete based on low-basic waste from mineral wool production depends on the porous structure and type of alkali activator. The dependence of the carbonation depth on time is of a fading nature and is described by a power function of the form $y = kx^{1/n}$. For the compositions under consideration, the ratio 1/n fluctuates from 0.46 to 0.57 depending on the density. The accelerated carbonation rate of alkali-activated concrete is higher than that of concrete made from Portland cement.
2. The mechanism of accelerated carbonation of alkali-activated concrete at the initial stage includes kinetic limitation, at a later stage there is an internal diffusion limitation. With kinetic limitation, the diffusion process is controlled by the rate of saturation of open pores. With internal diffusion limitation, the process is controlled by the rate of chemical reaction, including the processes of dissolution of gas molecules in the pore liquid, lowering the pH of the environment and interaction with the pore walls of the cement matrix.

Acknowledgment. Financial support was provided within the framework of the Science and the University national project for a new laboratory for «Development, research and pilot-industrial testing of high-tech technologies and technical means for the production of polymer-containing composite mixtures and products from technogenic organomineral components» (project FZWN 2024-0002).

References

1. Fediuk, R., Amran, M., Klyuev, S., Klyuev, A.: Increasing the performance of a fiber-reinforced concrete for protective facilities. Fibers **9**(11), 64 (2021)
2. Klyuev, S., et al.: Phase formation of mortar using technogenic fibrous materials. Case Studies in Construction Materials **16**, e01099 (2022)
3. Klyuev, S., et al.: Technogenic fiber wastes for optimizing concrete. Materials **15**(14), 5058 (2022)

4. Amran, M., Fediuk, R., Klyuev, S., Qader, D.N.: Sustainable development of basalt fiber-reinforced high-strength eco-friendly concrete with a modified composite binder. Case Studies in Construction Materials **17**, e01550 (2022)
5. Von Greve-Dierfeld, S., et al.: Understanding the carbonation of concrete with supplementary cementitious materials: a critical review by RILEM TC 281-CCC. Mater. Struct. **53**, 136 (2020). https://doi.org/10.1617/s11527-020-01558-w
6. Bakharev, T., Sanjayan, J.G., Cheng, Y.-B.: Resistance of alkali-activated slag concrete to carbonation. Cem. Concr. Res. **31**, 1277–1283 (2001). https://doi.org/10.1016/S0008-8846(01)00574-9
7. Pasupathy, K., Berndt, M., Castel, A., Sanjayan, J., Pathmanathan, R.: Carbonation of a blended slag-fly ash geopolymer concrete in field conditions after 8 years. Constr. Build. Mater. **125**, 661–669 (2016). https://doi.org/10.1016/j.conbuildmat.2016.08.078
8. Novgorodsky, V.I.: Fundamentals of durability of reinforced concrete structures. Sputnik+ Publishing House, Moscow, Russia (2015)
9. GOST 23732-2011: Water for concrete and mortars. Specifications. Standartinform, Moscow, Russia (2012)
10. GOST 30744-2001: Cements. Methods of testing with using polyfraction standard sand. Gosstroy of Russia, Moscow, Russia (2001)
11. Gluhovskij, V.D., Pahomov, V.A.: Slag-alkaline cements and concretes. Publishing house "Budivelnik," Kyiv, USSR (1978)
12. GOST R 52804-2007.: Protection against corrosion of concrete and reinforced concrete constructions. Test methods. Standartinform, Moscow, Russia (2008)
13. CPC-18 Measurement of hardened concrete carbonation depth. Materials and Structures. 21, 453–455 (1988). https://doi.org/10.1007/BF02472327
14. EN13295.: Products and systems for the protection and repair of concrete structures-Test methods-Determination of resistance to carbonation=Продукты и системы защиты и ремонта бетонных конструкций. Методы испытания. Определение стойкости к карбониза ции (2004)
15. Fedorov, P.A., Anvarov, B.R., Latypova, T.V., Anvarov, A.R., Latypov, V.M.: On the mathematical dependence describing the process of neutralization of concrete. Bulletin of the South Ural State University. Series: Construction and Architecture. 15, 13–15 (2010)
16. Nguyen, T.N., et al.: Microstructural evolution and its impact on the mechanical strength of typical alkali-activated slag subjected to accelerated carbonation. Developments in the Built Environment. **19**, 100519 (2024). https://doi.org/10.1016/j.dibe.2024.100519
17. Hossain, M.M., Karim, M.R., Elahi, M.M.A., Islam, M.N., Zain, M.F.M.: Long-term durability properties of alkali-activated binders containing slag, fly ash, palm oil fuel ash and rice husk ash. Constr. Build. Mater. **251**, 119094 (2020). https://doi.org/10.1016/j.conbuildmat.2020.119094
18. Li, Z., Li, S.: Carbonation resistance of fly ash and blast furnace slag based geopolymer concrete. Constr. Build. Mater. **163**, 668–680 (2018). https://doi.org/10.1016/j.conbuildmat.2017.12.127
19. Pasupathy, K., Sanjayan, J., Rajeev, P.: Evaluation of alkalinity changes and carbonation of geopolymer concrete exposed to wetting and drying. Journal of Building Engineering. **35**, 102029 (2021). https://doi.org/10.1016/j.jobe.2020.102029
20. Alekseev, S.N., Rosenthal, N.K.: Corrosion Resistance of Reinforced Concrete Structures in Aggressive Industrial Environment. Stroyizdat, Moscow, USSR (1976)
21. Fedorov, P.A., Anvarov, A.R., Lutsyk, E.V., Latypov, V.M., Latypova, T.V.: Kinetics of fine concrete carbonation in humid operational environment. Int. J. Appl. Eng. Res. **11**, 7439–7445 (2016)

Determination of Temperature Stresses During Construction of a Dock Wall

Vasilina Turina[1], Anton Chepurnenko[1](✉), Konstantin Novikov[1], Vladimir Kuznetsov[2], and Mikhail Petrov[1]

[1] Don State Technical University, Rostov-On-Don 344000, Russia
anton@chepurnenk.mail.ru

[2] Kh. Ibragimov Complex Institute of the Russian Academy of Sciences (CI RAS), Grozny City 364051, Russia

Abstract. The article considers the problem of modeling temperature fields and stresses during the construction of a massive monolithic reinforced concrete structure in the form of a dock wall. The structure under consideration has a pier inside. The calculation is performed by the finite element method in a three-dimensional formulation using the ANSYS software package. The calculation takes into account the dependence of the elastic modulus of concrete, as well as its strength characteristics on the time and temperature of hardening. The relationships determining the kinetics of heat release of concrete, as well as the dependence of its properties on the degree of maturity are presented. Graphs of temperature changes in the thickness and on the surface of the structure, as well as graphs of stress changes at dangerous points are given. It is established that at the age of up to 4 days, the maximum tensile stresses exceed the tensile strength of concrete, which results in the formation of vertical cracks. The calculation results are confirmed by field survey data.

Keywords: Temperature · Stresses · Finite element method · Numerical modeling · Early Age cracking · Mass concrete

1 Introduction

For massive reinforced concrete structures, the problem of early crack formation associated with their uneven heating is relevant [1–3]. Heat release of concrete during hardening and heat exchange with the environment leads to a temperature gradient between the center of the structure and the surfaces, which results in the occurrence of thermal stresses [4–6].

One of the tools for analyzing the causes of crack formation and choosing measures to prevent this negative phenomenon is computer modeling [7]. The finite element method is the main method for modeling temperature fields and stresses during the construction of massive monolithic structures [8–10]. At the same time, most publications on this topic relate to foundation slabs [11, 12], walls [13, 14] and dams [15, 16], which have a relatively simple geometry.

S. V. Klyuev et al. (Eds.): ICRAAC 2024, LNCE 627, pp. 36–42, 2025.
https://doi.org/10.1007/978-3-031-82938-3_5

This article presents the results of modeling the thermal stress state of a fragment of a massive dock wall with a pier inside. The concrete block has a length of 21980 mm and a height of 4710 mm. During the inspection of the block, cracks were found, the location of which is shown in Fig. 1.

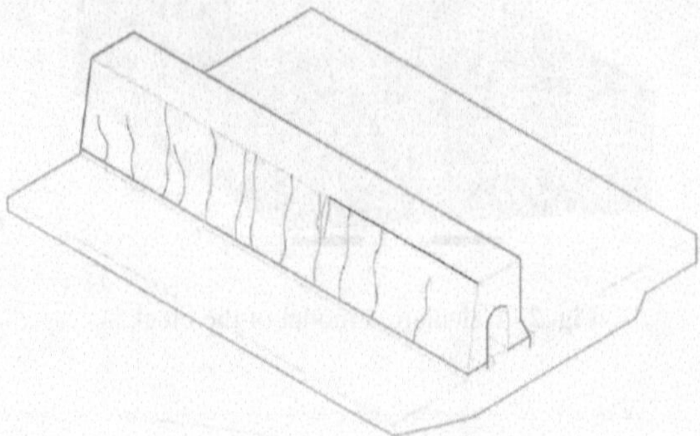

Fig. 1. Location of cracks.

The purpose of this work is to analyze the causes and develop recommendations for preventing crack formation during construction.

2 Methods and Materials

The calculation of the thermal stress state is performed by the finite element method in a volumetric formulation using the ANSYS software package. Due to symmetry, half of the block is considered, the second half is taken into account by specifying the plane of symmetry. The calculation is performed together with a fragment of old concrete. The size of the finite element mesh is taken to be 0.2 m. The calculation model is shown in Fig. 2.

The calculation is performed using the following initial data:

1. Concrete B45W18F300 according to Russian standards.
2. Initial temperature of concrete mix: 15 °C.
3. Average daily ambient temperature: 10 °C.
4. Heat transfer coefficient on surfaces covered with formwork: 10 W/(m2•°C).
5. Heat transfer coefficient on surfaces not covered with formwork: 25 W/(m2•°C).
6. Poisson's ratio of concrete: ν = 0.2.
7. Modulus of elasticity of old concrete: E = 37·103 MPa.

The heat release function of concrete in accordance with [17] is given by the formula:

$$Q(t) = Q_{28} \cdot \exp\left[k \cdot \left(1 - \left(\frac{28}{t}\right)^x\right)\right] \quad (1)$$

where t is the time in days, $Q_{28} = 158.8$ MJ/m^3 is the total heat release by the time of 28 days, the coefficients $k = 0.002042$ and $x = 1.564$ determine the kinetics of heat release. The graph $Q(t)$ is shown in Fig. 3.

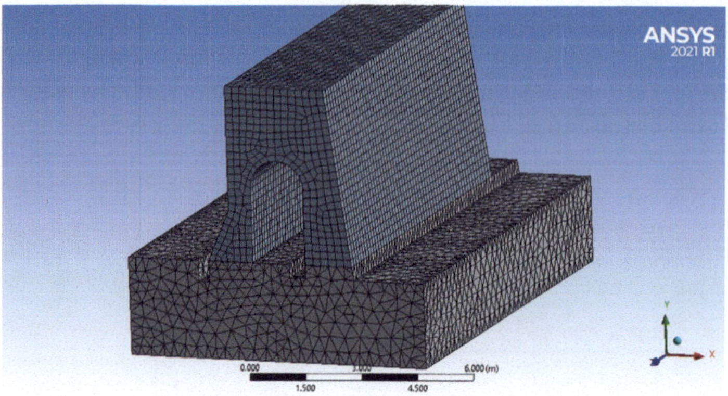

Fig. 2. Calculation model of the block.

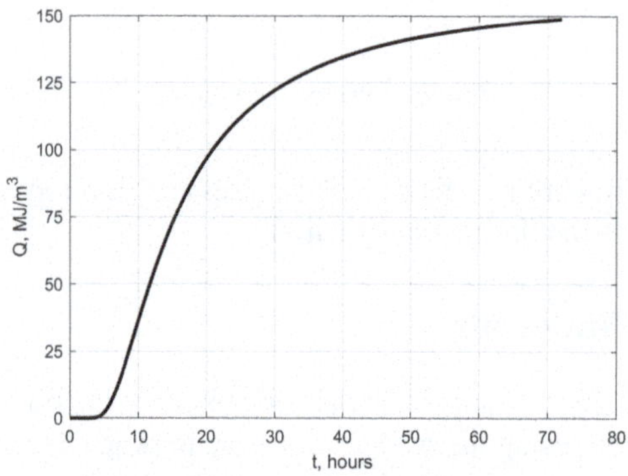

Fig. 3. Heat release curve of concrete.

The calculation takes into account the dependence of the strength and deformation characteristics of hardening concrete on the degree of its maturity, determined by the integral [18]:

$$DM(t) = \int_0^t T(\tau)d\tau \tag{2}$$

where $T(\tau)$ is the temperature of concrete at time τ.

The compressive strength of concrete R at time t is determined by the formula [18]:

$$\frac{R}{R_{28}} = \exp\left(0.35\left[1 - \left(\frac{15800 - 122.5\overline{T}}{\overline{T}t}\right)^{0.55}\right]\right) \tag{3}$$

where $R_{28} = 58$ MPa is the strength of concrete at the age of 28 days, $\overline{T} = DM/t$, t is the age of concrete in hours.

The modulus of elasticity of concrete aged more than 1 day is represented by a function of its compressive strength [19]:

$$E(R) = 1000\frac{0.04R + 57}{1 + \frac{29}{3.8+0.8R}}, \text{ MPa} \tag{4}$$

At $t \leq 1$ day, the dependence of the elastic modulus on time is represented by the equation:

$$E(t) = E_{24} \cdot \exp\left(1.348\left[1 - \left(\frac{24}{t}\right)^{1.438}\right]\right) \tag{5}$$

where t is the time in hours, E_{24} is the modulus of elasticity of concrete after 24 h of hardening, calculated on the basis of formulas (3)-(4) taking into account the time and temperature of hardening.

Taking into account the change in the modulus of elasticity of concrete over time during calculations in the ANSYS is performed according to the method given in [17]. The tensile strength of concrete R_t is taken as a function of the compressive strength based on [18].

$$R_t = 0.29 \cdot R^{0.6} \tag{6}$$

Thermophysical characteristics of concrete used in calculating the temperature field: density $\rho = 2500$ kg/m^3, specific heat capacity $c = 1000$ J/(kg·K), thermal conductivity coefficient $\lambda = 2.67$ W/(m·K), coefficient of linear thermal expansion $\alpha = 10^{-5}$ 1/°C.

3 Results and Discussion

Figure 4 shows the graphs of the change in time of the maximum temperature obtained as a result of the calculation, as well as the temperatures at points 1–3 on the surfaces of the structure, the location of which is shown in Fig. 5. The calculated section, where the maximum stresses occur, is located at a distance of approximately 1/3 of the block length from its end. The maximum temperature according to the calculation results was 67.8 °C, which is in good agreement with the data of field measurements ($T_{max} = 67.3$ °C).

Figure 6 shows a graph of the change in normal stresses σ_z over time (the z axis is perpendicular to the plane of the design section) for point 1. The dashed line on this graph shows the current tensile strength of concrete R_t. Figures 7–8 are the same for points 2 and 3.

From Figs. 6–8 it is evident that for point 1 the tensile stresses in concrete exceed its current tensile strength at 72 h, for point 2 – at 90 h, for point 3 – at 25 h.

4 Conclusions

It follows from the calculation results that with the initial data considered, the formation of vertical cracks going from the bottom to the top of the block is inevitable. To prevent crack formation during construction, it is recommended to reduce the length of the

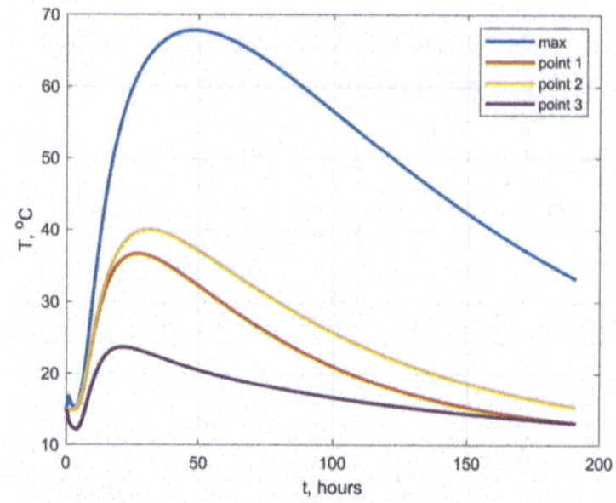

Fig. 4. Change in temperature of the structure over time.

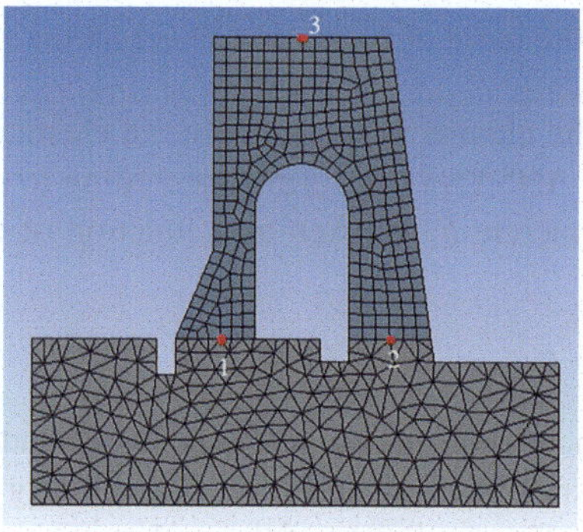

Fig. 5. Location of characteristic calculation points.

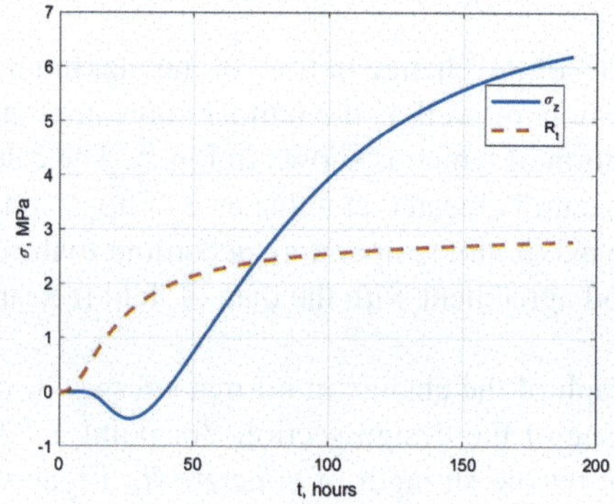

Fig. 6. Change in time of normal stresses σ_Z for point 1.

concreted block by 3 times. Also, to reduce the temperature gradient between the internal points and the surface of the block, it is recommended to use insulated formwork and insulate the upper surface upon completion of the block concreting.

Note that the calculation does not take into account concrete shrinkage. Taking it into account will lead to an even greater increase in tensile stresses.

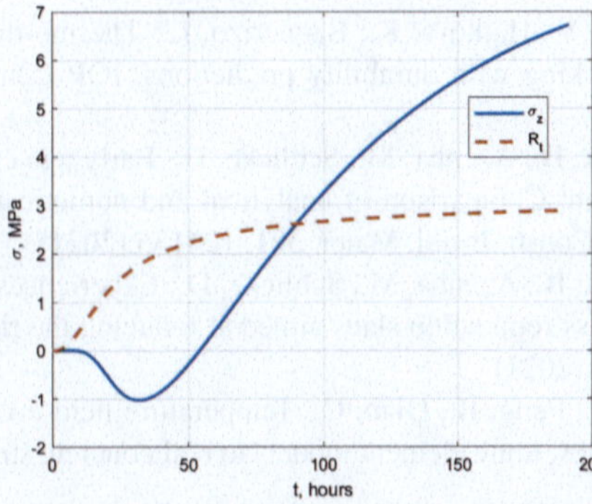

Fig. 7. Change in time of normal stresses σ_Z for point 2.

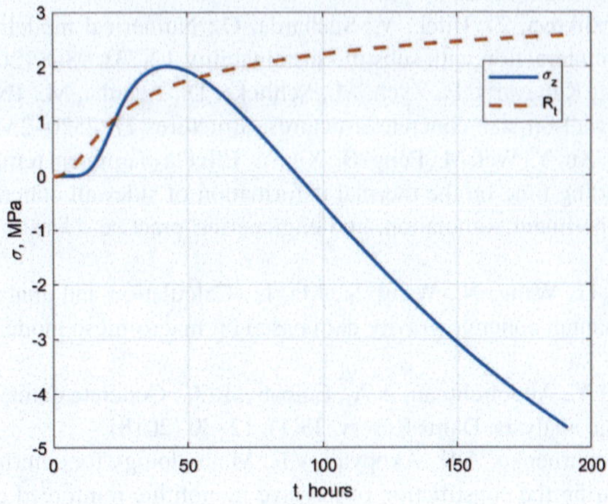

Fig. 8. Change in time of normal stresses σ_Z for point 3.

References

1. Litvinov, S.V., Yazyev, B.M., Kuznetsov, V.V., Belyugin, N.V., Avakov, A.A.: Study of the concordance between various concrete deformation models and experimental data for uniaxial compression cases. Constr. Mater. Prod. **7**(5), 6 (2024). https://doi.org/10.58224/2618-7183-2024-7-5-6
2. Klemczak, B., Knoppik-Wróbel, A.: Early age thermal and shrinkage cracks in concrete structures–description of the problem. Arch.-Civ. Eng.-Environ. **4**(2), 35–48 (2011)
3. Safiuddin, M., Kaish, A.A., Woon, C.O., Raman, S.N.: Early-age cracking in concrete: Causes, consequences, remedial measures, and recommendations. Appl. Sci. **8**(10), 1730 (2018)
4. Nama, P., Jain, A., Srivastava, R., Bhatia, Y.: Study on causes of cracks & its preventive measures in concrete structures. Int. J. Eng. Res. Appl. **5**(5), 119–123 (2015)
5. Maruyama, I., Lura, P.: Properties of early-age concrete relevant to cracking in massive concrete. Cem. Concr. Res. **123**, 105770 (2019)
6. Pulyaev, I., Pulyaev, S., Bazhenov, Y., Fetisova, A., Shcherbeneva, O.: Effect of thermal induced stress of concrete on performance characteristics of constructions. E3S Web of Conferences **97**, 06026 (2019)

7. Havlásek, P., Šmilauer, V., Hájková, K., Baquerizo, L.: Thermo-mechanical simulations of early-age concrete cracking with durability predictions. IOP Conf. Ser.: Mater. Sci. Eng. **236**(1), 012052 (2017)
8. Smolana, A., Klemczak, B., Azenha, M., Schlicke, D.: Early age cracking risk in a massive concrete foundation slab: Comparison of analytical and numerical prediction models with on-site measurements. Constr. Build. Mater. **301**, 124135 (2021)
9. Smolana, A., Klemczak, B., Azenha, M., Schlicke, D.: Experiences and analysis of the construction process of mass foundation slabs aimed at reducing the risk of early age cracks. J. Build. Eng. **44**, 102947 (2021)
10. Xie, Y., Du, W., Xu, Y., Peng, B., Qian, C.: Temperature field evolution of mass concrete: From hydration dynamics, finite element models to real concrete structure. J. Build. Eng. **65**, 105699 (2023)
11. Chepurnenko, A.S., Turina, V.S., Akopyan, V.F.: Processing of nonlinear concrete creep curves using nonlinear optimization methods.Constr. Mater. Prod. **7**(1), 2 (2024). https://doi.org/10.58224/2618-7183-2024-7-1-2
12. Cajka, R., Marcalikova, Z., Bilek, V., Sucharda, O.: Numerical modeling and analysis of concrete slabs in interaction with subsoil. Sustainability **12**(23), 9868 (2020)
13. Jędrzejewska, A., Kanavaris, F., Zych, M., Schlicke, D., Azenha, M.: Experiences on early age cracking of wall-on-slab concrete structures. Structures **27**, 2520–2549 (2020)
14. Du, W., Qian, C., Xu, Y., Wei, M., Peng, B., Xie, Y.: Effects of ambient temperature, formwork type, and demolding time on the thermal deformation of sidewall concrete in underground engineering: Experiment, simulation, and engineering practice. Constr. Build. Mater. **353**, 129083 (2022)
15. Jiao, Y., Cheng, L., Wang, N., Wang, S., Ma, L.: Calculation and analysis of temperature damage of Shimantan concrete gravity dam based on macro-meso model. Materials **15**(20), 7138 (2022)
16. Sayed-Ahmed, E.Y., Abdelrahman, A.A., Embaby, R.A.: Concrete dams: thermal-stress and construction stage analysis. Dams Reserv. **28**(1), 12–30 (2018)
17. Turina, V.S., Chepurnenko, A.S., Akopyan, V.F.: Methodology for determining true temperature stresses during the construction of massive monolithic reinforced concrete structures. Constr. Mater. Prod. **7**(3), 5 (2024). https://doi.org/10.58224/2618-7183-2024-7-3-5
18. Chepurnenko, A., Nesvetaev, G., Koryanova, Y., Yazyev, B.: Simplified model for determining the stress-strain state in massive monolithic foundation slabs during construction. Int. J. Comput. Civ. Struct. Eng. **18**(3), 126–136 (2022)
19. Chepurnenko, A.S., Turina, V.S., Akopyan, V.F.: Artificial intelligence model for predicting the load-bearing capacity of eccentrically compressed short concrete filled steel tubular columns. Constr. Mater. Prod. **7**(2), 2 (2024). https://doi.org/10.58224/2618-7183-2024-7-2-2

Finite Difference Method for Problem of Forced Vibrations of Rod

L. A. Baragunova, M. M. Shogenova(✉), L. V. Kanukoeva, and A. G. Ezaova

Department of Building Structures and Mechanics, Kabardino-Balkarian State University Named After H.M. Berbekov, Nalchik, Russia
baragunoval@mail.ru, shogenova_mar@mail.ru

Abstract. The paper deals with forced vibrations of rods. Transverse distributed vertical load is applied. In the longitudinal direction, the rod is stretched by a concentrated force. The technical theory for bending vibrations of beams is involved. The flat cross-section hypothesis is applied. Bends and angles of rotation of sections are small values. The longitudinal axis deformation is not taken into account. Longitudinal rod fibers do not press on each other. Longitudinal force does not change during time fluctuations. Oscillating rods are now widely used in robotic and mechatronic systems. Rod oscillations can be used to create walking robots. Oscillating rods can be realized with electromagnetic, piezoelectric or other types of batteries. One example of the use of oscillating rods is the robot snake. Vibrating rods allow the robot to overcome obstacles and climb to different heights Oscillating rods can be used in various mechatronic systems. The solution to the problem is sought by the method of partitioning variables. The pivotally operated rod system is considered.

Keywords: Rods forced oscillations · Distributed load · Study of oscillations · Growth of machine capacity · Concentrated force

1 Introduction

The rods vibration theory is one of the most important areas covering many fields of research for specialists in natural sciences and technology. Currently, the study of vibrations is of particular importance due to the capacity of machines, the speed of movement of their units, a decrease in relative mass, an increase in their durability and reliability, ensuring the stability and controllability of systems [1, 2]. Problems of forced vibrations of elastic systems triggered by periodic forces belong to quite developed sections of the fluctuation theory [3, 4].

Oscillating rods are now widely used in robotic and mechatronic systems [5–9]. Oscillating rods can be used in robotic and mechatronic systems to generate motion and control. For example, rod vibrations can be used to create walking motion of a robotic structure, as well as to control robotic manipulators and other mechanisms. Oscillating rods can be realized with electromagnetic, piezoelectric or other types of

S. V. Klyuev et al. (Eds.): ICRAAC 2024, LNCE 627, pp. 43–49, 2025.
https://doi.org/10.1007/978-3-031-82938-3_6

batteries. They may be manageable or unmanageable, depending on the application. One example of the use of oscillating rods is the snake robot's motion across uneven surfaces. The vibrations of the rods allow the robot to overcome obstacles and climb to different heights. In addition, oscillating rods can be used in various mechatronic systems such as flight control, stabilization and automatic control systems. They can be useful for creating more effective and accurate mechanisms that can perform complex tasks.

2 Methods and Materials

The oscillating rod (Fig. 1) can be stretched or compressed. Let the tensile force P act in the longitudinal direction and the distributed load in the vertical direction. Here we apply the technical theory of bending vibrations.

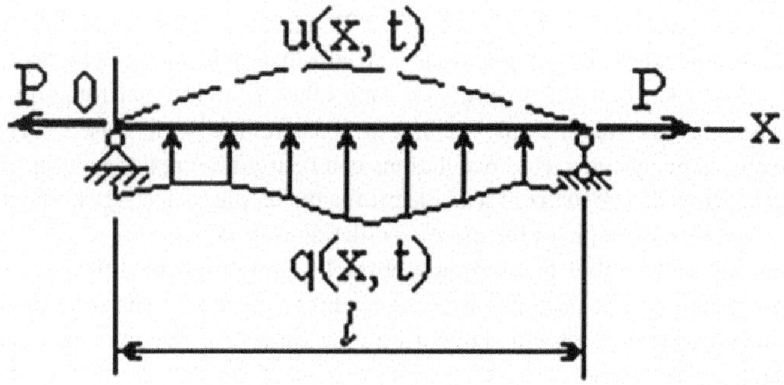

Fig. 1 Calculation scheme

The procedure for deriving the vibration equation in this problem produces

$$(b(x)u'')'' - Pu''m(x)\ddot{u} + \eta m(x)\dot{u} = q(x,t), \quad x \in (0,l), \quad t > -\infty \tag{1}$$

Example 1. Given: I-beam No. 14, length l = 6 m, material density $\rho = 7800\ kg/m^3$, elastic modulus $E = 200\ GPa$, cross-sectional area $S = 17.4\ cm^2$, axial moment of inertia of the section $J = 572\ cm^4$

Consider the problem of forced fixed oscillations at harmonic perturbations presented by uniformly distributed load $f_1(t)$ (dynamic perturbation) and right end movements $f_2(t)$ (kinematic disturbance) (Fig. 2).

The corresponding mathematical model of the problem is

$$(b(x)u'')'' - Pu''m(x)\ddot{u} + \eta m(x)\dot{u} = f_1(t), \quad x \in (0,l), \quad t > -\infty. \tag{2}$$

$$u(0,t) = 0,\ u''(0,t) = 0,\ u(l,t) = f_2(t),\ \ u''(l,t) = 0,\ t > -\infty \tag{3}$$

Here $b(x)$ – is the specified, continuous double differentiable positive function corresponding to the rod physico-mechanical and geometric characteristics, $m(x)$- continuous

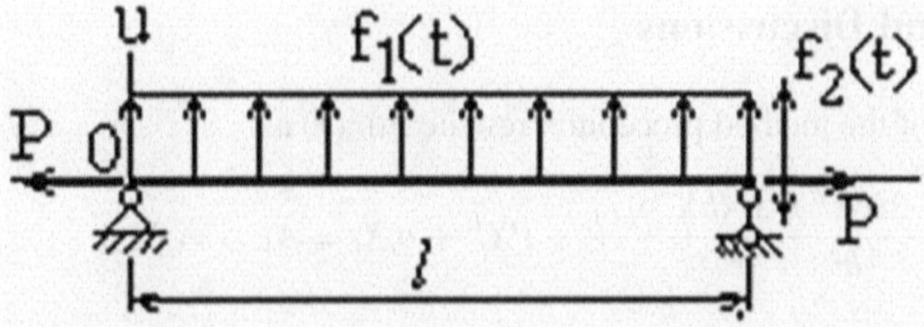

Fig. 2 Design diagram with transverse uniformly distributed load

function, mass intensity, η is the specific coefficient of linear-viscous resistance forces, $u(x, t)$ is the function of rod deflections when oscillation occurs.

Two disturbances form a vector harmonic process $f(t) = \{f_1(t), f_2(t)\}$, with frequencies Ω_k and initial phases Ψ_k. They are conveniently presented as complex valued functions

$$f_k(t) = a_k e^{i(\Omega_k t + \Psi_k)}, \quad \mathrm{k} = 1, \ldots, 5 \tag{4}$$

Enter a symbol for the complex amplitude:

$$A_k = a_k e^{i\Omega_k t}.$$

Here a_k – is the real harmonic perturbation amplitude. Simplify the description of disturbances

$$f_k(t) = A_k e^{i\Omega_k t}, \quad \mathrm{k} = 1, 2. \tag{5}$$

Ω_k The output process $u(x, t)$ is the sum of two harmonics with different frequencies. Such oscillations will be periodic only if the relations Ω_j / Ω_k turn out to be rational numbers. If both perturbation frequencies i.e. $\Omega_1 = \Omega_2 = \Omega$ are the same, the output process will be harmonic. The amplitude can also be determined for this problem.

The solution to the problem (2), (3) is sought by the method of partitioning variables as a product

$$u(x, t) = X(x)e^{\lambda t} \tag{6}$$

where $\lambda = j\Omega$ is the characteristic indicator, $X(x)$– is the amplitude function of forced oscillations to be determined. Denote

$$a = \lambda m(\lambda + \eta),$$

substitute (6) in (2), (3) and get

$$(bX'')'' - PX'' + aX = A_1, \quad x \in (0, l), \tag{7}$$

$$X(0) = 0, X''(0) = 0, X(l) = A_2, X''(l) = 0. \tag{8}$$

3 Results and Discussions

In the first step of the method procedure, rewrite Eq. (7) as

$$\frac{(bX'')_{i-1} - 2(bX'')_i + (bX'')_{i+1}}{h^2} - PX_i'' + a_iX_i = A_1, \ \mathrm{i} = 3,\ 4,\ \ldots,\ \mathrm{n} - 2$$

Multiply by h^2, bring similar terms and get

$$(bX'')_{i-1} - (2b''i + Ph^2)X_i'' + a_ih^2X_i = h^2A_1 \tag{9}$$

Thus.

$$b_i = b(x_i),\ \ a_i = b(x_i),\ \ X_i = X(x_i).$$

When moving to the finite difference method, the function will be represented by the vector $Y = \{y_1, y_2, ..., y_n\}^T$.

Having performed the necessary transformations of the finite differences method, instead of (9) write

$$b_{i-1}\frac{y_{i-2} - 2y_{i-1} + y_i}{h^2} - (2b_i + Ph^2)\frac{y_{i-1} - 2y_i + y_{i+1}}{h^2} + b_{i+1}\frac{y_i - 2y_{i+1} + y_{i+2}}{h^2} + c_ih^2y_i = h^2A_1 \tag{10}$$

Let us introduce the notation

$$\alpha_i = 2b_{i-1} - 2b_i - Ph^2, \beta_i = b_{i-1} + 4b_i + b_{i+1} + 2Ph^2 + c_ih^4,$$
$$\gamma_i = -2b_i - 2b_{i+1} - Ph^2.$$

Perform the simplest transformations and get instead (10).

$$b_{i-1}y_{i-2} + \alpha_iy_{i-1} + \beta_iy_i + \gamma_iy_{i+1} + b_{i+1}y_{i+2} = A_h,$$
$$A_h = h^4A_1, \quad \mathrm{i} = 3,\ 4,\ \ldots,\ \mathrm{n} - 2. \tag{11}$$

Adjust boundary conditions (8) to the method of finite differences and write

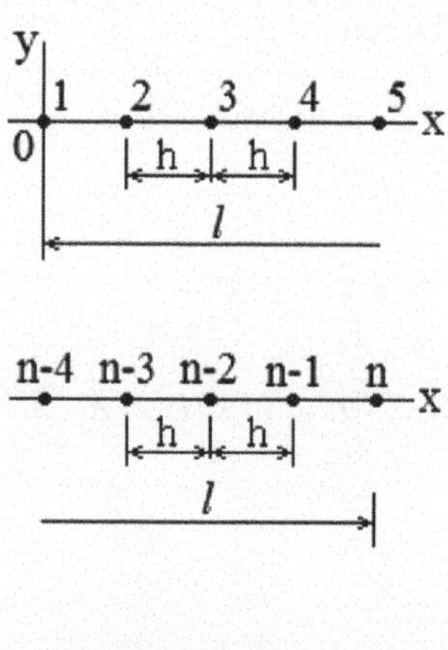

At the left end:

$$y_1 = 0,\ 2y_1 - 5y_2 + 4y_3 - y_4 = 0 \tag{12}$$

At the right end:

$$y_n = A_2, -y_{n-3} + 4_{n-2} - 5_{n-1} + 2y_n = 0. \tag{13}$$

Equations (11)–(13) create a homogeneous algebraic system that can be rewritten in matrix–vector form

$$G(\lambda)Y = Q, \tag{14}$$

where **G**(λ) – is the square matrix of order n. It can be easily shown that Eq. (14) as a coordinate takes the form

$$\begin{pmatrix} 1 & & & & & \vdots & & & & & \\ 2 & -5 & 4 & -1 & & \vdots & & & & & \\ b_2 & \alpha_3 & \beta_3 & \gamma_3 & b_4 & \vdots & & & & & \\ & b_3 & \alpha_4 & \beta_4 & \gamma_4 & \vdots & & & & & \\ \cdots & \cdots & \cdots & \cdots & \cdots & \cdots & \cdots & \cdots & \cdots & \cdots & \cdots \\ & & & & & \vdots & b_{n-3} & \alpha_{n-2} & \beta_{n-2} & \gamma_{n-2} & b_{n-1} \\ & & & & & \vdots & & -1 & 4 & -5 & \\ & & & & & \vdots & & & & & \end{pmatrix} \begin{pmatrix} y_1 \\ y_2 \\ y_3 \\ y_4 \\ \cdots \\ y_{n-2} \\ y_{n-1} \\ y_n \end{pmatrix} = \begin{pmatrix} 0 \\ 0 \\ A_h \\ A_h \\ \cdots \\ A_h \\ 0 \\ A_2 \end{pmatrix}. \tag{15}$$

The zero elements here are not written out. The solution to the system of Eq. (14) gives the vector $Y = G^{-1}Q$, and then an amplitude vector according to the formula.

$$a_u(x_i) = |Y|. \tag{16}$$

Example. Consider the hingedly operating rod system with parameters:

$$l = 3м, \quad n = 1001,\ m(x) = 10 + 0{,}5x^2 кг/м$$

$, P = 200H, \eta = 0,01c^{-1}, b(x) = 80 + 10x$ Нм^2.
Study the increasing perturbation frequency influence

$$\Omega = \{01, 82, 51429, 5\}\ c^{-1}$$

on the amplitude without phase shift, i.e.

$$\Psi_1 = \Psi_2 = 0.$$

According to the procedure described in [5] three elements of the natural frequency spectrum are defined, as shown in Table 1.

Table 1 Elements of the natural frequency spectrum

k	1	2	3
ω_k, c^{-1}	5,44	15,40	31,32

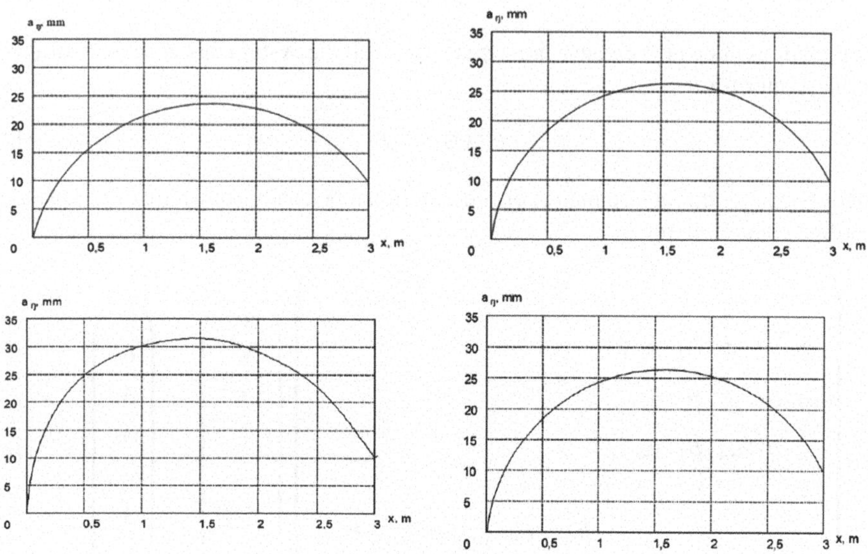

Fig. 3 Counting results

The counting results are shown by curves in Fig. 3. The curves are arranged in the order of the vector elements Ω in the example condition.

You can see that as the perturbation rate increases, the amplitude increases, and the shape of the rod length distribution coincides with the first natural function. When the first eigenvalue coincides with the perturbation frequency, the oscillations become resonant, and the amplitudes become very large.

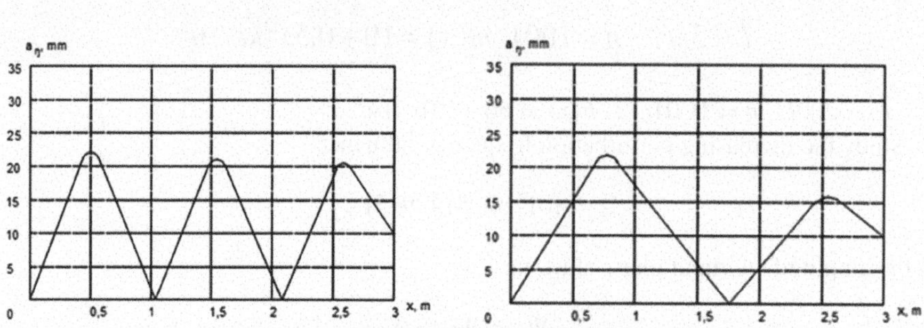

Fig. 4 Calculation results after resonance

After the first resonance when the perturbation frequency increases, oscillations occur in the second and third eigenforms (Fig. 4).

4 Conclusions

1. The finite difference method is used in solving the problem of determining the amplitude of forced vibrations of the rod.
2. It is shown that the size and shape of the amplitude distribution along the rod depend significantly on the spectra of eigenvalues and functions.
3. The influence of the phase shift of perturbations on the forced vibrations of rods is demonstrated by numerical example.
4. Problem models describing vibrational processes under dynamic and kinematic perturbations are developed,
5. Simple enough methods are proposed to determine the spectra of natural frequencies and eigenforms of vibrations.

References

1. Kulterbaev, Kh.P.: Fundamentals of the Theory of Oscillations, Fundamentals of the theory, Problems for Homework, Examples of Solutions. Nalchik: Publishing House KBRGU 130 p. (2003)
2. Ilyin, M.M., Kolesnikov, K.S., Saratov, Y.S.: Theory of Oscillations: Textbook for Universities Under the General Edition of K.S. Kolesnikov. 2nd edition is stereotypical 72 p. (2003)
3. Kulterbaev, Kh.P., Kaziev, A.M.: On harmonic vibrations of beams excited by vector perturbations Light construction structures. Rostov-on-Don: Rost. S. builds 146–154 (2003)
4. Kulterbaev, Kh.P., Chechenov, T.Y.: Forced oscillations of variable cross-section beams at vector harmonic perturbations Bulletin VolgGASU. Series: Construction and Architecture. Volgograd 5–12 (2007)
5. GOST P 60.3.4.2–2017 Robots and robotic devices. Industrial manipulation robots. Mechanical interfaces. Rods - Text: Electronic Electronic fund-normative documents [site – URL: https://docs.cntd.ru/document/1200157562 (reference date: 03.05.2023)
6. Sytin, A.B.: Design features mechatronic device for testing petal gas dynamic bearings Izvestia of the University. Techn. Sci. **12**, 2–5 (2017)
7. Tomilin, A.K., Prokopenko, E.V.: Control of longitudinal fluctuations of mechatronic system by means of capacitive parameter Izvestia TPU 6, (2015)
8. Tomilin, A.K.: Control of vibration frequencies of elastic electromechanical systems Tomsk state un-ty. Math. Mech. **3**(19), 20–25 (2012)
9. Homchenko, V.G.: Mechatronic and robotic systems: Manual V. G. Homchenko, V. Y. Solomin. Omsk: Izd-vo OmGTU 160 p. (2008)
10. Kulterbaev, Kh.P., Jankulayav, A.Y.: Introduction to Matlab Kabardino-Balkar State University Nalchik 57 p. (2006)
11. Verzhbitsky, V.M.: Basics of numerical methods M High School, 840 p. (2002)

Fibrofoam Concretes with Dispersed PolyARM Based on Volcanic Ash

Tolya Khezhev(✉), Muhamed Kokoev, Artur Zhurtov, and Adam Dzhankulaev

Kabardino-Balkarian State University Named After Kh.M. Berbekov, Nalchik 360004, Russia
hejev_tolya@mail.ru

Abstract. The article presents studies of the strength characteristics and shrinkage deformations of non-autoclaved foam concretes using local raw materials from the Kabardino-Balkarian Republic, in particular volcanic ash, as a result of their reinforcement with synthetic and basalt fibers. The technology of preparing a fiber-reinforced foam concrete mixture is presented. The research results showed that the use of reinforcing basalt fiber in foam concrete increases the tensile strength in bending of the original matrix by 4.6 times, and in comparison with foam concrete reinforced with nylon fibers, by 1.51 times. It was revealed that dispersed polyreinforcement of foam concrete on volcanic ash provides a significant decrease in shrinkage deformations from 3.97 mm/m to 2.39 mm/m compared to foam concrete and compared to fiber-reinforced foam concrete reinforced with basalt fibers, from 2.96 mm/m to 2.39 mm/m.

Keywords: Non-Autoclave Curing Foam Concrete · Fiberopen Concrete · Synthetic Fiber · Basalt Fiber · Volcanic Ash · Shrinkage Deformations

1 Introduction

New economic conditions in the country predetermine a new approach to the selection of effective building materials. When assessing their competitiveness, it is necessary to take into account that in recent years there has been a sharp increase in the cost of energy resources, an increase in the cost of transportation. First of all, this applies to the use of concrete and reinforced concrete structures, which are the basis of mass construction. Reducing the mass, and, consequently, improving the thermal characteristics of the enclosing structure can be achieved through rational design, the formation of voids in concrete due to the use of mixtures without fine aggregate, the use of natural and artificial porous aggregates or the creation of a large number of small, evenly distributed air bubbles in the body of concrete. Building products and structures made of cellular concrete are increasingly used in construction [1, 2]. The fine-pored structure is the main feature of cellular concrete. It provides such positive properties as light weight, low thermal conductivity, relatively low sorption moisture, and high frost resistance. Possessing a low average density, cellular concretes are characterized by sufficient strength and good thermal insulation properties, so products made from them can simultaneously perform both load-bearing and thermal insulation functions in building structures.

S. V. Klyuev et al. (Eds.): ICRAAC 2024, LNCE 627, pp. 50–56, 2025.
https://doi.org/10.1007/978-3-031-82938-3_7

Portland cement and finely dispersed silica aggregates are used in non-autoclave curing foam concrete [3]. At present, as aggregates for the production of cellular concrete is used mainly quartz sand, which does not always provide a material of a given density and strength at the permissible cement consumption, and their use, as a rule, provides the need for additional grinding.

Increasing the efficiency of foam concrete application is possible due to the use of local construction materials, for example, volcanic rocks [4–6]. The use of local raw materials for the production of foam concrete is an urgent task. The Kabardino-Balkarian Republic of the Russian Federation has large reserves of volcanic rocks. They have found application as aggregates in the production of lightweight concrete for various purposes. Fine dispersibility and high content of amorphous silica in volcanic rocks determine the efficiency of their use in foamed concrete [5].

2 Methods and Materials

Studies [7] have shown that the use of tuff sawing waste as a foam concrete aggregate is more effective than traditionally used quartz sand and other aggregates. In order to reduce shrinkage deformations of foamed concrete and increase its strength characteristics, they were reinforced with capron fibers. Raw material mixture for the preparation of cellular concrete contains, wt.%: cement 16.1–34.9; tuff sand with content of fractions, wt.%: less than 0,63 mm 70–85, 0.63–1.25 mm 15–30 32.2–34.9; glutinous foaming agent 1.25–1.32; synthetic kapron fiber 0.9; quicklime 0–16.1; building gypsum 0–0.9; water - the rest [7]. The researches obtained fibropenotufoeton with an average density of 500 kg/m^3, which meets the requirements of the state standard.

The disadvantages of these compositions are the lack of sufficient raw material base of volcanic tuff sawing waste, the need for their additional grinding, relatively low tensile and flexural strength of foam concrete.

One of the materials that are an effective substitute for tuff sand and other silica components can be volcanic ash. The use of volcanic ash for cellular concrete will make up for the existing huge deficit of aggregates for lightweight concrete and reduce the cost of production. Dust-like fractions of volcanic ash have good adhesion with cement stone, hydraulic activity, which can provide a reduction in the cost of raw mix and increase the strength of foam concrete [8, 9].

Raw material mixture for the production of cellular concrete contains portland cement, quick-lime, construction gypsum, volcanic ash, basalt fibers, water and as a powdering agent - foaming agent PB-2000.

Binders were used in the experiments: Portland cement PC400-DO produced by JSC "Belgorod Cement", gypsum binder of Ust-Dzheguta gypsum plant of G-5 BII grade, powdered quick-lime calcium lime produced by JSC "Izvestnyak", Ust-Dzheguta. Volcanic ash of Zayukovskoye deposit with maximum grain size of 1.25 mm was used as an active mineral additive and foam concrete aggregate. Foaming agent PB-2000 produced by PJSC "Ivkhimprom" was used. For disperse reinforcement of foam concrete was used basalt fiber produced by PJSC "Ivotsteklo" of RNB-9-1200-4c grade, the ratio of fiber length to diameter on the basis of preliminary experiments was accepted, the fiber length was 13 mm.

Production of cellular concrete samples from raw mix includes the following operations: preparation of cellular concrete mix, molding and heat and humidity treatment.

The mixture was prepared in a batch mixer. According to the usual (traditional) technology, cement, lime, gypsum and volcanic ash were initially mixed with water until a homogeneous mass was obtained, then basalt fibers and foam were added, after which mixing of all components was continued until the mixture of the specified density was obtained.

The samples with the size of 4x4x16 cm were molded by casting method.

Thermal treatment of samples was carried out after preliminary exposure for 16 h at $t = 20 \pm 2$ °C in a steaming chamber at $t = 80$ °C under the regime of $2 + 6 +$ natural cooling.

Before the tests the samples were dried to constant weight at $t = 105$ °C in a desiccator.

3 Results and Discussions

The compositions of the initial raw material mixtures of foamed concrete and their basic physical and mechanical properties are given in Tables 1 and 2.

Table 1. Ratio of components in the mix.

Compounds	Ratio of components in the mixture, wt. %						
	cement	ash	lime	gypsum	basalt fiber	blowing agent PB-2000	water
1	33.8	33.8	–	–	–	0.25	32.15
2	16.1	32.2	16.1	0.90		0.25	34.45
3	33.5	33.5	–	–	0.90	0.25	31.85

Table 2. Concrete property values.

Compounds	Concrete property values			
	average density, kg/m3	tensile strength at bending, MPa	compressive strength, MPa	softening factor
1	506	0.27	1.38	0.84
2	500	0.28	1.49	0.51
3	504	1.24	1.58	0.87

The results of the research showed that the developed composition on volcanic ash has higher flexural and compressive strength compared to foam concrete on tuff sand. The use of quicklime and gypsum as a stimulant of latent hydraulic activity of ash allows to reduce cement consumption by 2 times without reducing the strength of foam concrete and has increased strength properties compared to foam concrete on tuff sand. This is explained by the greater dispersibility and hydraulic activity of ash. At the same time, the use of additives leads to a decrease in the softening coefficient of foam concrete from 0.84 to 0.51. The use of reinforcing basalt fiber in foam concrete increases the tensile strength in bending of the original matrix by 4.6 times, and compared to foam concrete reinforced with capron fibers, by 1.51 times.

The disadvantages of these compositions are relatively high shrinkage deformations, relatively low tensile and flexural strength of foam concrete.

The main task in developing poly-reinforced fiber concrete is to determine the optimal combination and composition of two or more types of fibers with characteristics that ensure the production of a composite that fully meets the nature and conditions of its operation in the structure. Certain experience in reinforcing concrete with several types of fibers simultaneously for the purpose of comprehensive management of its properties is available [10–13].

At the same time, it is known that individual types of fibers, while selectively improving individual characteristics of concrete, have little effect on others, so further development of dispersed reinforcement methods should be aimed at combining different types of fiber, allowing their positive qualities to be combined in one composite. However, little attention has been paid to poly-reinforcement issues so far, and the available information sometimes indicates inconsistency in the research results, which reduces the volume of application of fiber-reinforced concrete. In connection with the above, the relevance of further development of promising options for dispersed reinforcement of fiber concrete and the study of their physical, mechanical, energy and deformation characteristics to improve the technical and economic efficiency, reliability and operational safety of building and structure structures becomes obvious [14, 15].

To reduce shrinkage deformations, improve the quality and strength of foam concrete, dispersed polyarming with basalt and kapron fibers was used.

Raw material mixture for the manufacture of cellular concrete contains portland cement, quicklime, construction gypsum, volcanic ash, basalt fibers, kapron fibers, water and as a pore-forming agent - foaming agent PB-2000.

The mixture was prepared in a batch mixer. According to the usual (traditional) technology, cement, lime, gypsum and volcanic ash were initially mixed with water until a homogeneous mass was obtained, then basalt fibers, capron fibers and foam were added, after which mixing of all components was continued until the mixture of the specified density was obtained [16–18].

The compositions of initial raw material mixtures of fibrofoam concrete and their basic physical and mechanical properties are given in Tables 3 and 4.

Tables 3 and 4 show that a significant increase in flexural tensile strength is observed both when reinforced with individual fibers and when reinforced simultaneously with two types of fibers, the maximum increase in strength is 4.7 times. The compressive strength of foam concrete reinforced with capron or basalt fibers, as well as with polydisperse fiber reinforcement increases insignificantly, the maximum strength increase is 17%.

Table 3. Ratio of components in the mix.

Compounds	Ratio of components in the mixture, wt. %							
	cement	ash	lime	gypsum	synthetic fiber	basalt fiber	blowing agent PB-2000	water
1	33.8	33.8	–	–	–	–	0.25	32.15
2	16.1	32.2	16.1	0.90	–		0.25	34.45
3	33.5	33.5	–	–	–	0.90	0.25	31.85
4	33.5	33.5	–	–	0.90	–	0.25	31.85
5	16.1	32.2	16.1	0.90	0.90	–	0.25	33.55
6	16.1	32.2	16.1	0.90	–	0.90	0.25	33.55
7	33.5	33.5	–	–	0.25	0.65	0.25	31.85
8	16.1	32.2	16.1	0.90	0.25	0.65	0.25	33.55

Table 4. Concrete property values.

Compounds	Concrete property values			
	average density, kg/m3	tensile strength at bending, MPa	compressive strength, MPa	softening factor
1	506	0.27	1.38	3.97
2	500	0.28	1.49	3.96
3	504	1.24	1.58	2.96
4	506	0.87	1.54	3.05
5	500	0.89	1.56	3.07
6	500	1.28	1.61	2.99
7	504	1.20	1.51	2.40
8	502	1.23	1.55	2.39

4 Conclusions

Thus, the results of the studies showed that the developed concrete using volcanic ash has greater bending and compression strength compared to foam concrete using tuff sand. The use of quicklime and gypsum as a stimulant of the latent hydraulic activity of ash allows to reduce cement consumption by 2 times without reducing the strength of foam concrete and has increased strength indicators compared to foam concrete on tuff sand.

The introduction of nylon or basalt fiber into foam concrete reduces shrinkage by up to 25%. Compositions with nylon fiber have a shrinkage of 3.05 mm/m. Samples of fiber-reinforced foam concrete reinforced with basalt fibers have a shrinkage of 2.96 mm/m.

Fiber-reinforced foam concretes, reinforced simultaneously with nylon and basalt fibers in the amount of 0.25% and 0.65% by weight, respectively, have the lowest shrinkage - 2.39 mm/m.

Thus, dispersed polyreinforcement of foam concrete on volcanic ash provides a significant reduction in shrinkage deformations from 3.97 mm/m to 2.39 mm/m compared to foam concrete and compared to fiber-reinforced foam concrete reinforced with basalt fibers, from 2.96 mm/m to 2.39 mm/m.

Studies have also shown that the compressive strength of foam concrete reinforced with nylon or basalt fibers, as well as with polydisperse fiber reinforcement, increases insignificantly, with the maximum increase in strength being 17% (Fig. 1).

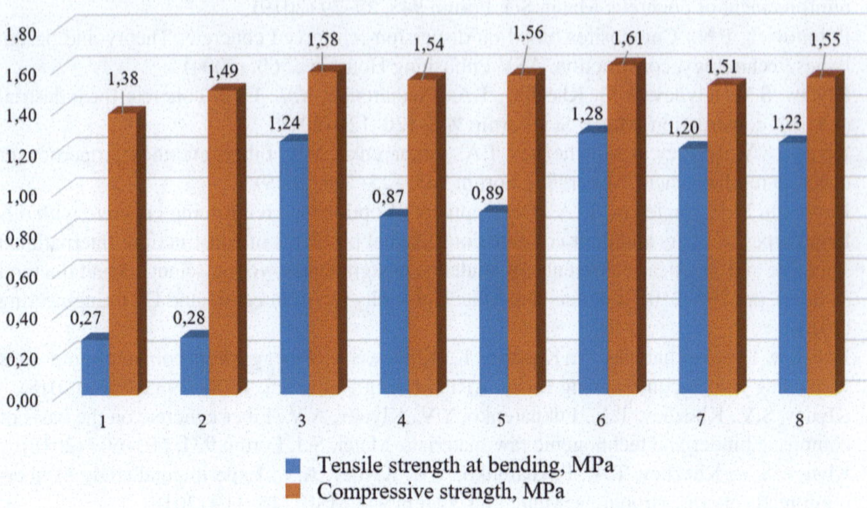

Fig. 1. Concrete strength depending on composition.

References

1. Morgun, L.V., Morgun, V.N., Pimenova, E.V., Smirnova, P.V., Nabokova, Ya.S.: Possibility of using non-autoclaved fibropenobeton in large-panel house building. Constr. Mater. **3**, 19–21 (2011)
2. Klyuev, S.V., Klyuev, A.V., Khezhev, T.A., Pucharenko, Yu.V.: High-strength fine-grained fiber concrete with combined reinforcement by fiber. J. Eng. Appl. Sci. **13**(8 SI), 6407–6412 (2018)
3. Morgun, V.N., Pushenko, O.V.: On the structure of fibrofoam concrete. Eng. Bull. Don **3**, 1–4 (2012)
4. Klyuev, S.V., Khezhev, T.A., Pukharenko, Y., Klyuev, A.V.: The fiber-reinforced concrete constructions experimental research. Mater. Sci. Forum **931**, 598–602 (2018)
5. Akhmatov, M.A.: Efficiency of application of lightweight concrete, products and structures made of them. Constr. Mater. **4**, 9–13 (1998)

6. Khezhev, T.A., Zhurtov, A.V., Kokoev, M.N., Hadzhyhalapov, G.N., Yazyev, B.M.: Foam containers based on volcanic tuff with disperse polyarmiation. Mater. Sci. Forum **974**, 424–428 (2019)
7. Khezhev, T.A., Pukharenko, Y.V., Khashukaev, M.N.: Russian Federation Patent 239600. Raw material mixture and method of manufacturing of products from foam concrete. Bulletin **33** (2008)
8. Khezhev, T., Ansokov, A., Kerefov, I., Malkarov, R.: Fiber foam tuffcrete structure and properties formation on the basis of volcanic tuff dusting waste. Mater. Sci. Forum **1011**, 130–135 (2020)
9. Khezhev, T, Kardanov, A, Bolotokov, E., Dottuyev, A, Mashfezh, I.: Fine-grained fibre concrete of run-of-crusher stone using volcanic ash. Mater. Sci. Forum **1043**, 61–65 (2021)
10. Klyuev, S.V., Klyuev, A.V., Khezhev, T.A., Pukharenko, Y.V.: To the question of fiber reinforcement of concrete. Mater. Sci. Forum **945**, 25–29 (2019)
11. Rabinovich, F.N.: Composites based on dispersion-reinforced concrete. Theory and design issues, technology, construction. ASV Publishing House, p. 560, (2004)
12. Klyuev, S.V., Klyuev, A.V., Khezhev, T.A., Pukharenko, Y.V.: Fiber concrete for industrial and civil construction. Mater. Sci. Forum **945**, 120–124 (2019)
13. Klyuev, S.V., Klyuev, A.V., Khezhev, T.A., Pukharenko, Y.V.: Fibers and their properties for concrete reinforcement. Mater. Sci. Forum **945**, 125–130 (2019)
14. Kostrikin, M.P., Panteleev, D.A.: Development of options for reinforcing concrete with different types of fibers and their combinations. Actual problems of construction: International scientific and practical conference of students, postgraduates, young scientists and doctoral students, pp. 206–210. St. Petersburg State University of Architecture and Civil Engineering (2014)
15. Khezhev, T.A., Pukharenko, Y., Khezhev, K., Klyuev, S.V.: Fiber gypsum concrete composites with using volcanic tuff sawing waste. ARPN J. Eng. Appl. Sci. **13**(8), 2935–2946 (2018)
16. Klyuev, S.V., Khezhev, T.A., Pukharenko, Y.V., Klyuev, A.V.: Fiber concrete on the basis of composite binder and technogenic raw materials. Mater. Sci. Forum **931**, 603–607 (2018)
17. Klyuev, S.V., Khezhev, T.A., Pukharenko, Y.V., Klyuev, A.V.: Experimental study of fiber-reinforced concrete structures. Mater. Sci. Forum 945 MSF, 115–119 (2018)
18. Klyuev, S.V., Klyuev, A.V., Khezhev, T.A., Pucharenko, Y.: Technogenic sands as effective filler for fine-grained fibre concrete. J. Phys. Conf. Ser. **1118**, 012020 (2018)

Evaluation of the Possibility of Using Slag from Out-Of-Furnace Steel Treatment as an Active Mineral Additive for Concrete and Mortars

L. N. Lomakina(✉), F. A. Agzamov, R. T. Yulberdin, E. I. Ganieva, and D. A. Sinitsin

Building Structures Department, Ufa State Petroleum Technological University, Ufa, Russia
lomakinaln@mail.ru

Abstract. Research has been carried out on multicomponent slag from out-of-furnace steel processing with the purpose of its use as an active mineral additive replacing a part of cement. The tricalcium form of C_3S silicate is represented in the form of three polymorphic structures: triclinic, monoclinic and rhombohedric in the total amount of 10.28%. The high-temperature form of α-C_2S is represented in the form of "traces". Most likely, it is the β + α form located within β-C_2S. The total β-C_2S content was 13.26%. One third of the composition is γ-C_2S, it is considered inert in natural conditions as it has a very stable crystalline structure that does not react with water. The total calcium silicate content was about 23–24%. Aluminate phases Mayenite $C_{12}A_7$ and tricalcium aluminate C_3A cubic were detected. Tricalcium aluminate hydrates with water to form calcium hydroaluminates, which harden rapidly, but the hardening products have low strength. Mayenite plays an important role in retarding the early hydration of cement with steel slag because it retards the precipitation of portlandite (CH) and C-S-H [8, 14]. Wollastonite 2M $Ca_3(Si_3O_9)$, Gehlenite C_2AS, merwinite $Ca_3Mg(SiO_4)_2$ and microcline maximum $KAlSi_3O_8$ are found in small amounts and have no binding activity at normal concrete curing temperatures. Periclase MgO content in the mixture is not more than 5%. Dolomite $Ca,MgCO_3$ is an inert substance, so it does not participate in the hydration process. Iron alpha and Iron Silicide cubic are residues from magnetic separation of iron by iron separators of initial slag. Hydration activity of calcium ferrite C_2F is insignificant, increases when their structure is modified with sulfate ion and basicity increases. When studying the strength of cement-sand compositions with different content of slag it was shown that the compositions with pure slag at the age of 28 days have 9.5MPa compressive strength, which meets the requirement of GOST R 56592-2015 on the strength for active mineral additives with binding properties. Replacing 25% of cement by weight with slag gives strength at the age of 28 days 41MPa compared to 42.2MPa of the composition without slag. When obtaining low-grade binders it is possible to save up to 50% of cement with compressive strength not less than 25MPa. A sharp increase in the strength of compositions "cement + slag" in the early age gives reason to believe that the surface of slag grains is covered with silicon-oxygen particles of amorphous structure, which interact with cement, playing the role of alkaline activator, with the formation of hydrosilicates. At a later age slag begins to show hydraulic properties.

S. V. Klyuev et al. (Eds.): ICRAAC 2024, LNCE 627, pp. 57–64, 2025.
https://doi.org/10.1007/978-3-031-82938-3_8

Keywords: Slag of Out-of-Furnace Steel Processing · Binding Ability of Slag · Active Mineral Additive · Metallurgical Wastes · Slag Hydration · Mineralogical Composition of Slag · Hydraulic Activity · Calcium Silicates · Calcium Aluminates

1 Introduction

Cement concrete is the most popular construction material in the building industry. However, the production of cement for these concretes accounts for up to 8% of global anthropogenic carbon dioxide emissions [1]. An alternative to cement concrete are concretes based on composite binders, in which part of cement clinker is replaced by inorganic materials - industrial by-products (ashes, slags) with binding capacity. Their use in the production of construction materials is not only an effective method of reducing carbon dioxide emissions into the atmosphere, but also solves the problem of utilization of these wastes.

Among industrial wastes sludge [10, 11, 12], as well as slags - wastes of metallurgical production are of interest. The use of these wastes as aggregates, finely ground fillers or pigment [13] in concrete and mortars is the most widespread direction of their application. Slags obtained from metallurgical production are subjected to magnetic separation from iron-containing compounds, which makes it possible to consider their use in both cement production and concrete production.

2 Methods and Materials

Grinded to a specific surface of 4000–4500 cm^2/g and purified from iron compounds slags are expediently used in the manufacture of concrete as a mineral additive. Although there are examples of use of slags enriched with iron up to 70 wt.% [2]. The authors [2] investigated the hydration of cements mixed with such slag and showed their reactivity similar to that of siliceous fly ash. The compressive strength of cement specimens decreased markedly with increasing slag content, especially at early ages. However, the authors [2] observed that after 28 days of curing, the compressive strengths of specimens with slag increased markedly compared to the strength of the control specimen, suggesting that slag improves later strengths. In [3] it was shown that open-hearth slag in the amount of 5–15% can replace the same amount of cement while maintaining the basic properties of concrete. The authors [4] investigated cement concrete with the addition of finely ground blast furnace granulated slag in an amount of up to 30% to obtain the required 28-day compressive strength of concrete. At the same time, the authors noted a slight decrease in strength gain in the first 1–7 days, however, by 28 days the rate of strength gain equalized compared to the control compositions, which apparently indicates the ability to gain strength at a later time, compared to cement clinker. These facts indicate that despite the higher reactivity, finely ground steel slag, unlike other active mineral additives, exhibit low hydration activity in cement-steel-slag composite mixture at a late age, compared to coarsely ground slag. Therefore, the hydration activity of steel slag at late age can be improved by changing its particle size distribution [15].

According to the studies of some authors, the activity index of cement binder with steel slag at 28 days increased from 45% to 80% when the specific surface area of steel slag increased from 475 m^2/kg to 768 m^2/kg [16].

Thus, many authors note the delayed effect of the positive impact of slag on the strength gain of cement-slag compositions. At early stages of curing slag can be considered only as an inert filler [9], although even at later stages of curing, unlike cement, most slag grains (especially large ones) can have a fairly small degree of hydration [9]. Since studies with slags are carried out for a duration of not more than a few years, the authors of [9] decided to investigate the chemical composition of partially and fully hydrated blast furnace slag grains in cement concrete of about 40 years of age in order to find out the degree of slag hydration depending on the size of slag particles. They argue that not only the fine fraction of slag (up to 8 microns) has good hydration activity, but also the medium sized grains (8–17 microns) are able to hydrate well when added to cement and hence have great potential, so their proportion in cement production can be increased.

Metallurgical slags have a multi-component composition depending on the production conditions and the type of metal produced in production. Depending on what ability fine ground slags have, their principle of action in concrete depends on, i.e. as pozzolans or as a binding component (GOST R 56592-2015).

3 Results and Discussions

We investigated slag, which is a product of out-of-furnace treatment of steelmaking production. The presented slag after grinding was a crushed mixture of light gray color (Fig. 1) with a bulk density of 1061 kg/m^3. To increase the specific surface area of the particles, the slag was additionally pulverized to the residue on the sieve 0.05, not exceeding 10%.

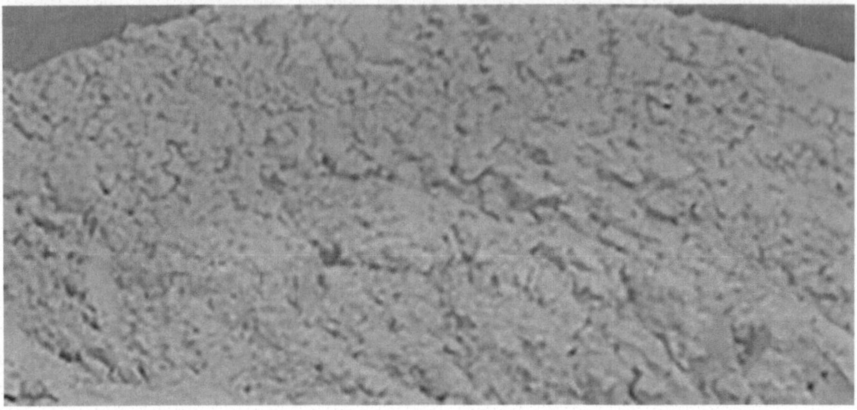

Fig. 1. Physical configuration of ground slag sifted through a 0.05 sieve.

X-ray phase analysis of two samples allowed to find out the mineralogical composition of slag (Figs. 2, 3, Table 1).

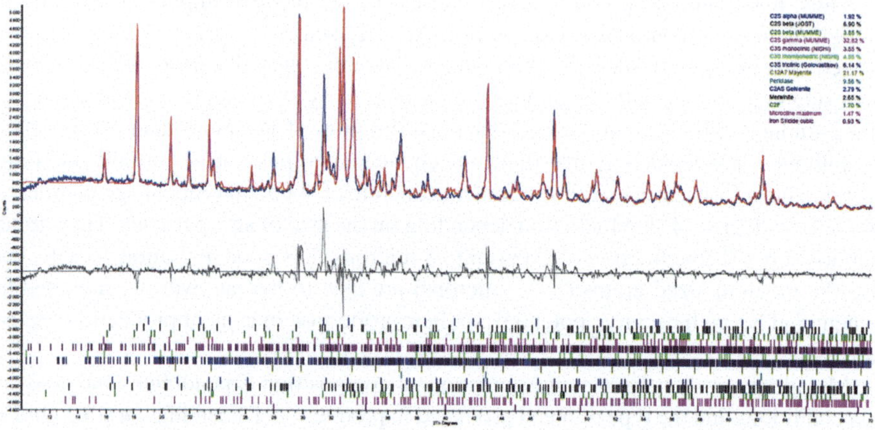

Fig. 2. Mineralogical composition of slag sample 1 according to XRPA results.

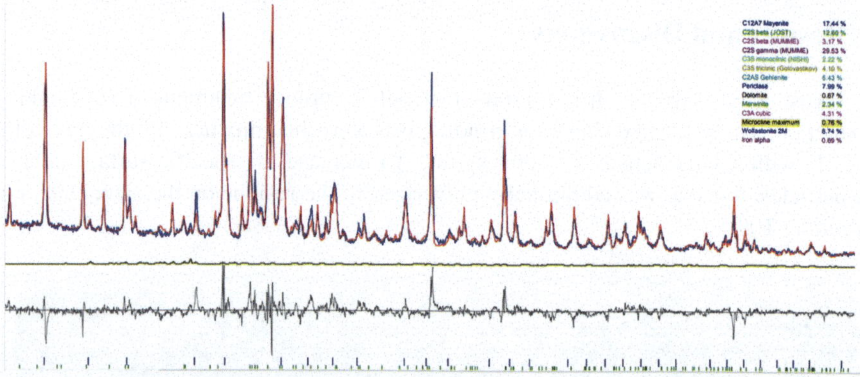

Fig. 3. Mineralogical composition of slag sample 2 according to XRPA results.

Calcium silicates are of the greatest interest from the hydration point of view. The three-calcium form of C_3S silicate is represented in the form of three polymorphic structures: triclinic, monoclinic and rhombohedric in a total amount of 10.28%. These structures arise as a result of sequential and reversible high-temperature phase transitions in a sequence starting from triclinic, then monoclinic, ending with rhombohedric one [7]. The high-temperature form of α-C_2S is represented as "traces". Most likely, this is the $\beta + \alpha$ form located within β-C_2S [6] - it is not of interest from the hydration point of view. The total β-C_2S content was, on average, 13.26%. Much of the β-C_2S hydration reaction occurs after 28 days and later within 1 year [6], meaning that the kinetics of strength gain is likely to increase after 28 days of curing for these formulations. One third of the composition (31.18%) is γ-C_2S, which is considered inert in natural conditions, as it has a very stable crystalline structure that does not react with water at all [6]. The activity of steelmaking slags, where the main minerals are also β-C_2S and γ-C_2S, was studied by the authors of [5], where the hardening of the compositions was observed at a later curing time due to β-C_2S; the activity of γ-C_2S was increased by autoclave treatment of the compositions. The total content of calcium silicates, which presumably have binding properties, was about 23–24%.

Table 1. Summary mineralogical analysis of two slag samples.

№	Name and formula of the mineral	Sample 1	Sample 2	Average content in both samples
1	$C_{12}A_7$ Mayenite	21.17	17.44	19.305
2	C_2S beta (JOST)	6.9	12.6	9.75
3	C_2S beta (MUMME)	3.85	3.17	3.51
4	C_2S gamma (MUMME)	32.82	29.53	31.18
5	C_3S monoclinic (NISHI)	3.55	2.22	2.885
6	C_3S triclinic (Golovastikov)	6.14	4.1	5.12
7	C_2AS Gehlenite	2.79	5.43	4.11
8	Periclase	9.56	7.99	8.775
9	Dolomite	0	0.67	0.335
10	Merwinite $Ca_3Mg(SiO_4)_2$	2.65	2.34	2.5
11	C_3A cubic	0	4.31	2.155
12	Microcline maximum $KAlSi_3O_8$	1.47	0.76	1.115
13	Wollastonite 2M $Ca_3(Si_3O_9)$	0	8.74	4.37
14	Iron alpha	0	0.69	0.345
15	C_2S alpha (MUMME)	1.92	0	0.96
16	C_3S rhombohedric (NISHI)	4.55	0	2.275
17	C_2F	1,7	0	0.85
18	Iron Silicide cubic	0.93	0	0.465

Aluminate phases of Mayenite $C_{12}A_7$ and tricalcium aluminate C_3A cubic were found. Three-calcium aluminate hydrate with water to form calcium hydroaluminates, harden rapidly, but the hardening products have low strength. Mayenite plays an important role in retarding the early hydration of cement with steel slag because it retards the precipitation of portlandite (CH) and C-S-H [8, 14].

Wollastonite 2M $Ca_3(Si_3O_9)$, Gehlenite C_2AS, Merwinite $Ca_3Mg(SiO_4)_2$ and Microcline maximum $KAlSi_3O_8$ are found in small amounts and at normal concrete curing temperatures do not have binding activity. The content of periclase MgO is regulated in GOST 31108-2020 - not more than 5%. Periclase slows down hydration of other available phases. Dolomite Ca,$MgCO_3$ is an inert substance, so it does not participate in the hydration process. Iron alpha and Iron Silicide cubic are residues from magnetic separation of iron by iron separators of initial slag. Hydration activity of calcium ferrite C_2F is insignificant, increases when their structure is modified with sulfate ion and basicity increases.

Thus, it was found out that in the initial slag there are substances with binding properties.

3.1 Experimental Studies of Binder Composition Selection

Cement CEM II B-K (SH-I) 42,5 N produced by "HeidelbergCement" (Sterlitamak, Republic of Bashkortostan) and sand from Kulushevsky deposit of the Republic of Bashkortostan were used for preparation of the compositions.

The plasticizing additive PFM-NLK produced by Polyplast Group of Companies (Russia, Ufa) was added to compositions 5 and 6. According to the quality passport, PFM-NLK is a complex additive with plasticizing, water-reducing, increasing frost resistance and increasing air content abilities.

Cement-sand compositions were selected for testing: composition 1 (control), containing no slag; composition 2 contained 100% slag, i.e. was without cement; composition 3 contained 75% cement by weight and 25% slag by weight; composition 4 - cement 50% and slag 50%; composition 5 - cement 75%, slag 25%, additives 2.3% by weight of slag + cement; composition 6 - cement 50%, slag 50%, additives 2.3% by weight of slag + cement. Sand was added at a ratio of 3:1 compared to cement. The introduction of plasticizer in compositions 5 and 6 allowed to reduce the water-cement ratio, it varied from 0.34 to 0.38, allowing obtaining workability of different values.

The compositions were tested for compressive and flexural tensile strength at the age of 3, 14 and 28 days to control the dynamics of strength change in the chamber of normal curing (Table 2). The results of strength evaluation of the composition with pure slag 2 (Table 2) show the dynamics of strength growth in time. This allows us to conclude that hydration is going on despite the presence of a large number of different components in the slag - at 28 days the compressive strength is 9.5MPa, which meets the requirement of GOST R 56592-2015 for strength of at least 5MPa for active mineral additives with binding properties.

At introduction of slag into the mixture with cement there was a sharp increase in the strength of the compositions compared to the composition with pure slag. This suggests that cement acts as an alkaline activator of slag, reacting with silicon-oxygen particles of amorphous structure present on the surface of slag grains, forming additional calcium hydrosilicates.

Thus, the introduction of the investigated slag with the replacement of 25% cement (composition 3) reduces the compressive strength of the composition at 28 days only by 3%, the flexural tensile strength - by 16%. Replacing 25% of slag with cement will not significantly affect the strength of the binder compared to cement. However, when increasing the amount of slag up to half of the binder (composition 4), the strength of the composition is significantly reduced (in compression - by 30%, in flexural tension - by 22.5%). However, it can be noted that the loss of flexural tensile strength is much slower than in compression with increasing slag consumption in the system. With the introduction of compositions 5 and 6 additive PFM-NLK observed an increase in strength at 28 days compared with no additive compositions 3 and 4. In particular, the composition with the addition of slag 25% has an increase in tensile strength in bending by 6.6% and in compression - by 2.6%. The composition with addition of slag 50% has an increase in flexural tensile strength by 3.7% and compressive strength by 7.7%.

Table 2. Average strength of concretes from different mix compositions.

№ of compositions	Composition: C – cement; Sl – slag; Ad – additive % by weight	Flexural tensile strength MPa, at age,			Compressive strength at bending MPa, at age		
		3 days	14 days	28 days	3 days	14 days	28 days
1	C + S	5.06	5.39	6.35	33.9	41	42.2
2	Sl + S	1.52	2.18	2.60	7.7	9.05	9.5
3	Sl(25%) + C(75%) + S	4.80	5.04	5.32	27.03	40.6	41
4	Sl(50%) + C(50%) + S	3.66	5.02	4.92	22.8	28.54	29.7
5	Sl(25%) + C(75%) + Ad + S	Not carried out		5.67	Not carried out		44.2
6	Sl(50%) + C(50%) + Ad + S			5.1			32

Thus, the results of the research show that if slag is introduced into the binder after out-of-furnace treatment of steel, instead of cement in an amount of 25% by weight, then cement savings are possible without much damage in strength at the age of 28 days (41 MPa with 25% slag and 42.2 MPa without slag). For obtaining low quality binders it is possible to save up to 50% of cement to ensure the strength of at least 25MPa (29.7MPa), which is more than 70% of the strength of the binder without slag (42.2MPa).

Introduction of plasticizer addition increases expectedly the strength of slag binder by reducing W/C. Thus, with the introduction of 25% slag, the compressive strength increases from 41 to 44 MPa; flexural tensile strength from 5.32MPa to 5.67MPa. With the introduction of 50% slag, the compressive strength increases from 29.7 to 32 MPa; flexural tensile strength from 4.92 MPa to 5.1 MPa.

4 Conclusions

The primary reason for the introduction of slag from out-of-furnace steel processing was to utilize the basic properties of slag and to obtain a composite binder with reduced cement content due to the hydraulic properties of slag. However, a significant increase in the strength of composite binders "cement + slag" in the early age gives reason to believe that the surface of slag grains is covered with silicon-oxygen particles of amorphous structure, which interacting with cement, playing the role of alkaline activator, form hydrosilicates of different composition. At a later age slag starts to show its hydraulic properties more and more.

References

1. Fediuk, R., Amran, M., Klyuev, S., Klyuev, A.: Increasing the performance of a fiber-reinforced concrete for protective facilities. Fibers **9**(11), 64 (2021)
2. Klyuev, S., et al.: Fresh and mechanical properties of low-cement mortars for 3D printing. Constr. Build. Mater. **338**, 127644 (2022). https://doi.org/10.1016/j.conbuildmat.2022.127644
3. Akhverdieva, T.A., Dzhafarov, R.: The effect of finely ground mineral additives on the properties of concrete. Constr. Mater. **3**, 73–76 (2019). https://doi.org/10.31659/0585-430X-2019-768-3-73-76
4. Klyuev, S., et al.: Phase formation of mortar using technogenic fibrous materials. Case Stud. Constr. Mater. **16**, e01099 (2022)
5. Kudeyarova, N.P., Bushueva, N.P., Starikova, E.D.: Activity of belite phases of steelmaking phases in autoclave conditions. Bulletin of BSTU named after V.G. Shukhov **9**, 106–111 (2017). https://doi.org/10.12737/article_59a93b12242ff2.20855442
6. Klyuev, S., et al.: Technogenic fiber wastes for optimizing concrete. Materials **15**(14), 5058 (2022)
7. Amran, M., Fediuk, R., Klyuev, S., Qader, D.N.: Sustainable development of basalt fiber-reinforced high-strength eco-friendly concrete with a modified composite binder. Case Stud. Constr. Mater. **17**, e01550 (2022)
8. Valuev, D.V., Gizatulin, R.A.: Technologies of metallurgical waste processing: a textbook. D.V. Valuev; Yurginsky Technological Institute. Tomsk: Publishing House of Tomsk Polytechnic University, p. 196 (2012)
9. Kulikova, E., Balovtsev, S.V., Skopintseva, O.V.: Complex estimation of geotechnical risks in mine and underground construction. Sustain. Dev. Mt. Territ. **15**(1), 7–16 (2023). https://doi.org/10.21177/1998-4502-2023-15-1-7-16
10. Kolesnikova, O., Vasilyeva, N., Kolesnikov, A., Zolkin, A.: Optimization of raw mix using technogenic waste to produce cement clinker. MIAB **10–1**, 103–115 (2022). https://doi.org/10.25018/0236_1493_2022_101_0_103
11. Kulikova, E.Y., Balovtsev, S.V.: Risk control system for the construction of urban underground structures. IOP Conf. Ser. Mater. Sci. Eng. **962**(4), 042020 (2020)
12. Kulikova, E: Safety and risk management in underground construction as a complex information process. MIAB **2–1**, 134–143 (2021). https://doi.org/10.25018/0236-1493-2021-21-0-134-143
13. Lomakina, L.N., Sinitsyn, D.A., Nedoseko, I.V., Rakhimova, O.N., Ryabykh, D.M.: Investigation of the possibility of obtaining decorative concrete using iron oxide slag. Constr. Geotech. **13**(4), 116–125 (2022). https://doi.org/10.15593/2224-9826/2022.4.09
14. Yessenov, M.K., Ramatullaeva, L.I., Kolesnikov, A.S., Ivakhniyuk, G.K.: Aspects of ecological modernization of technological equipment to reduce the level of dust from mining and processing production. MIAB **10**, 136–148 (2023). https://doi.org/10.25018/0236_1493_2023_10_0_136
15. Kulikova, E.Y.: Methodical principles for improving the ecological and technological reliability of urban underground structures. MIAB **6–1**, 176–185 (2020). https://doi.org/10.25018/0236-1493-2020-61-0-176-185
16. Wang, Q., Yang, J., Yan, P.: Cementitious properties of super-fine steel slag. Powder Tech. **245**, 35–39 (2013)

Analysis of the Space-Planning Solution of the School in Nalchik for Renovation

Valery Khuranov(✉), Khazret Guketlov, Fatima Shogenova, and Amirbek Mazloev

Kabardino-Balkarian State University Named After Kh.M. Berbekov, Nalchik 360004, Russia
valeriy-archi@mail.ru

Abstract. Within the framework of the implementation of the federal program for the overhaul of schools, the public and educational project "Dream School" is a platform for students of architectural and design universities of the region to obtain applied competencies. Modern schools need a fresh and modern look at the architecture of the building, which can be achieved with the help of students of architecture and design. For them, practice on real examples and work with experts in the development of this new direction for Russian schools is important. The main objectives of this project are to improve the architectural and artistic expressiveness of school buildings during their major repairs and to provide them with modern developing educational spaces. When developing a design project, special attention should be paid to the satisfaction of students, teachers and parents with the program, taking into account their wishes in the project, in accordance with the instructions of the president. Taking into account the above, in this article we considered an example in which, during the development of a design project, the need for additional project analysis was identified due to the fact that one of the schools did not meet modern requirements for its volumetric planning parameters. The only possible way to bring the building into a standard condition is to increase the area of the building. This can be solved in two ways: increasing the number of floors or adding an additional block. During the discussion of the issue and analysis of the experience of solving such problems, we came to the conclusion that it is necessary to develop a project for the reconstruction of this school with the addition of a separate block, into which the premises of the canteen and assembly hall are transferred.

Keywords: Public Education Project "Dream School" · Design Project of an Educational Institution · Major Repairs · Reconstruction

1 Introduction

Currently, thanks to the program of modernization of school education systems in our country, a lot of work is being done to repair and reconstruct schools. Their external appearance and internal condition are improving, which contributes to the creation of a favorable educational environment and improving the quality of education. In support of the federal program of capital repairs of schools, the public education project "Dream

S. V. Klyuev et al. (Eds.): ICRAAC 2024, LNCE 627, pp. 65–70, 2025.
https://doi.org/10.1007/978-3-031-82938-3_9

School" is also being implemented - a public education development project. Students of architectural and design schools of the regions are involved in the implementation of this project. This practice makes it possible to simultaneously acquire applied competencies in the learning process.

In Russia, there is a need to create scientific schools for the design of modern educational spaces, and here the "Dream School" project is the first step to solving this. It organizes educational student expeditions, during which students of the architectural and design direction get acquainted with the experience of the best schools in Russia, listen to lectures from experts and independently develop design projects, cover the expedition in social networks and the media. Since September 2023, the project has been systematically involving students in the Dream School project as a solution to important state issues in the format of service learning practice [1].

2 Methods and Materials

As part of the modernization of the school education system, the Department of Architecture and Design of KBSU was tasked with involving the most creatively developed part of students in the creation of design projects. This is achieved, including through the creation of an educational space that meets modern requirements.

The main objectives in the implementation of this project are to improve the architectural and artistic expressiveness of the building during its major repairs and provide modern developing educational spaces. During the creation of the design project, special attention must be paid to the satisfaction of students, teachers and parents with the program, taking into account their wishes in the project, in accordance with the requirements for work within the framework of the Dream School project.

Project objectives:

- conduct an analysis of existing interiors and planning solutions to identify strengths and weaknesses;
- develop a concept for a preliminary renovation project based on the identified problems.

3 Results and Discussions

The two-story building was built in 1990. In plan it is five rectangles connected to each other in a checkerboard pattern (see Fig. 1). The educational institution is designed for teaching primary grades. The number of students is 800 students.

An analysis of the space-planning solution of the school was conducted and a survey of teachers and students was conducted using the "participatory design" method [2–5]. The following problems were identified (see Fig. 2), including those that do not comply with the main regulatory framework SP 251.1325800.2016 "Buildings of general education organizations". Small capacity - the school operates in two shifts, the canteen does not accommodate the required number of students, the sports and assembly halls are very small and do not meet the area standards, the entrance group does not meet modern standards.

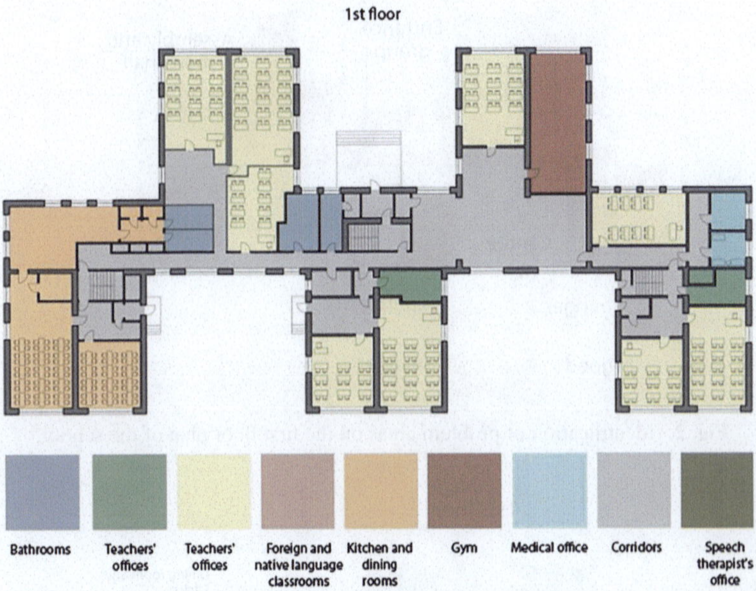

Fig. 1. School plan.

The design project provides for an interior renovation, which includes not only the decoration of the walls and the replacement of old furniture with new, but also a partial redevelopment of the premises for more efficient use of space during the educational process. The school space is an integral part of the educational environment, which is a historically established cultural form of meeting children, young people and adults [6, 7].

But to correct all the identified shortcomings, it turned out that design alone would not be enough. The design team decided to come up with a proposal on the need to reconstruct this facility to bring it up to standard.

The only way to eliminate the comments may be to increase the area of the building, which can be solved in two ways: increasing the number of floors or adding an additional block [8–10]. There are many methods for increasing the capacity of public buildings, including abroad [11–17].

In our case, the most preferable option is to install an extension block, since the canteen and assembly hall cannot be increased to the required parameters in the existing volume.

The checkerboard shape of the plan complicates the placement of the passage to the new building. The passage may limit access to natural light and insolation of the classrooms. We decided to preserve the symmetrical shape of the plan. The block-extension will be located along the central axis of the plan (see Fig. 3). Access to it is provided by two glass passages adjacent to the walls of the central projection. The upper and side glazing of the passage will allow preserving the lighting of the classrooms. An internal courtyard space - a recreation area - is formed between the extension, the existing building and the two passages.

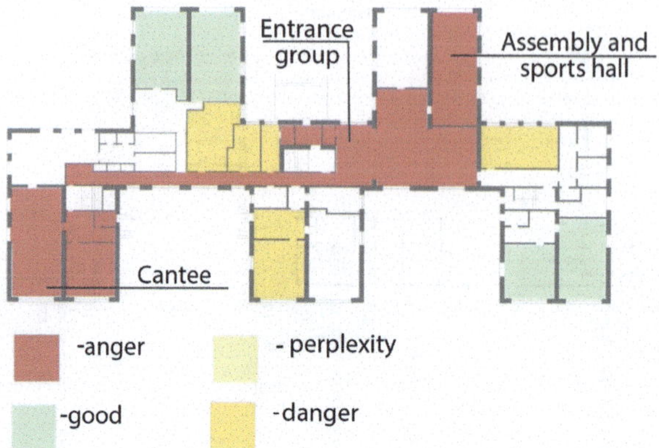

Fig. 2. Identification of problem areas on the first floor plan of the school.

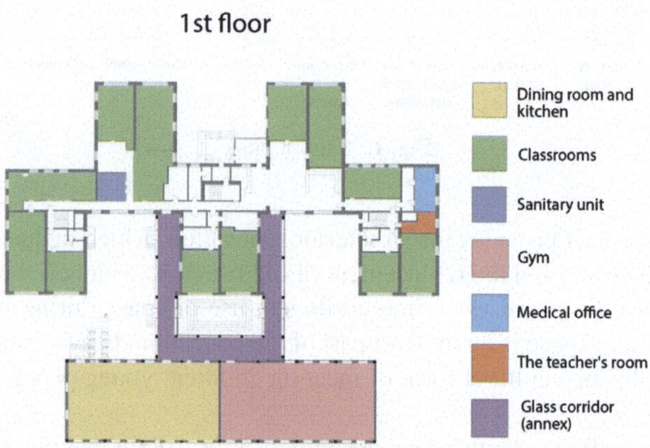

Fig. 3. The plan of the first floor of the school after reconstruction.

Technical and economic indicators of the increase in the area of the building due to the extension indicate a significant improvement in the capacity of the building. The difference in area before and after the reconstruction of the canteen is 39%, and the sports hall is 64%. The number of classrooms for classes has increased by 17% (Table 1).

To eliminate the comments on the entrance group, a redevelopment of the vestibule and the recreation hall was developed (see Fig. 4). The vestibule is moved outside to increase its size, and a security post is provided in the vacated space.

4 Conclusions

The development of the design project discussed above showed the need to expand design projects in some cases to a reconstruction project within the framework of the same public education project "Dream School". It is necessary to understand that the involvement of student architects in their first project must first of all begin with the

Table 1. Technical and economic indicators before and after reconstruction.

№	Designation	Unit of measurement	Before reconstruction	After reconstruction
1	Building area	m^2	1283	2160
2	Total area	m^2	2048	2917
3	Canteen area	m^2	142	230
4	Gym area	m^2	142	230
5	Number of classrooms	pcs	24	28

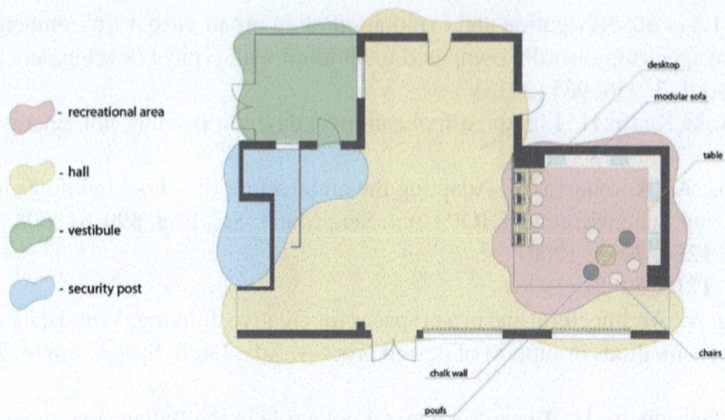

Fig. 4. Entrance group with recreation.

requirement for strict compliance with construction regulations. In our case, one of the five schools does not require major repairs, but reconstruction. Further work within the framework of this project will show whether other schools have such problems and the development of standard projects of block extensions to school buildings in this case can become unification and, accordingly, reduce the cost of construction. In the Kabardino-Balkarian Republic, there is experience in the development and further use of standard projects of extensions to kindergartens [18]. And the transition described in our case during renovation from a design project to a reconstruction project with an extension shows that similar work must be carried out for school buildings.

References

1. Reshetnikov, O.V.: Professionalization—the problem of primary inclusion of modern youth in labor relations. Bull. Pract. Psychol. Educ. **4**(41), 82–86 (2014)
2. Kudryavtseva, S.P., Dolotkazina, N.S.: Modern directions of creation of children's educational institutions. Arch. Mod. Inf. Technol. **3**(36) (2016)
3. Menconi, M.E., Grohmann, D.: Participatory retrofitting of school playgrounds: collaboration between children and university students to develop a vision. Think. Ski. Creat. **29**, 71–86 (2018)

4. Martovitskaya, A.: Architecture for children. Speech **14**(8), 16–37 (2015)
5. Zagvozdkin, V.A.: Waldorf Kindergarten in Harmony with the Child's Nature. Demetra, St. Petersburg – Moscow (2005)
6. Slobodchikova, V.I.: Anthropological Perspective of Domestic Education. Publ. department of the Ekaterinburg diocese, Ekaterinburg (2009)
7. Bolshunova, N.Ya.: Subjectivity as a Socio-Cultural Phenomenon. Publishing House NSPU, Novosibirsk (2005)
8. Sedegova, L.N.: Features of construction of civil buildings in the existing urban development. Eng. Bull. Don **2** (2013)
9. Maslovskaia, O., Kopeva, A., Srikauskas, L., Ivanova, O., Khrapko, O.: Humanization of the urban environment for children (on example of the residential yards in city of Vladivostok). IOP Conf. Series: Mater. Sci. Eng. **890**, 012001 (2020)
10. Siu, K.W.M., Wong, Y.L., Lam, M.S.: Inclusive play in urban cities: a pilot study of the inclusive playgrounds in Hong Kong. Procedia Eng. **198**, 169–175 (2017)
11. Fornasari, L., et al.: Navigation and exploration of an urban virtual environment by children with autism spectrum disorder compared to children with typical development. Res. Autism Spectr. Disord. **7**, 956–965 (2013)
12. Rosemann, A., Kaase, H.: Lightpipe application for daylight systems. Sol. Energy **78**, 772–780 (2005)
13. Galimullina, A., Korotkova, S.: Adapting the architecture of school buildings in the context of humanizing the environment. IOP Conf. Ser.: Mater. Sci. Eng. **890**, 012008 (2020)
14. Arch. Rev. **1256**, 70–71 (2001)
15. Arch. Rev. **1215**, 50–52 (1998)
16. Bonenberg, A.: Architectural and urban spaces for creative thinking. Visual data and questionnaire analysis methods in support of design process. Adv. Intell. Syst. Comput. **966**, 313–322 (2020)
17. Nucci, L., Galimberti, L.: Towards a new school's role in the Italian contemporary city. City Territ. Arch. **7**(2) (2020)
18. Khuranov, V.Kh., Bzhakhov, M.I., Shamparov, A.L.: Variants of planning structures of nursery group blocks for the conditions of the KBR. Eng. Bull. Don 3 (2019)

Investigation of the Relationship Between Early Shrinkage Cracking and the Mechanical Properties of Medium Strength Concrete for Cement Floors Based on Fractal Theory

O. V. Radaikin[1], A. D. Makarov[1,2](✉), D. M. Lobov[2], and I. V. Smagin[2]

[1] Kazan State Power Engineering University, Kazan, Russia
ad.makarov@mail.ru
[2] Nizhny Novgorod State University of Architecture and Civil Engineering, Nizhny Novgorod, Russia

Abstract. The paper presents the results of an experimental study of medium-strength concrete for cement floors, performed according to the ASTM C 1579–06 standard test method for evaluating shrinkage cracks. In the process of experimental study the growth of shrinkage cracks development was recorded using a digital camera. Processing of the obtained images of shrinkage cracks was performed, using the cube counting method, and the fractal dimension of cracks was determined. Approbation of the methodology for determining the Minkowski fractal dimension for cracks formed on the concrete surface by processing digital photographs has been carried out. Graphs of dependence of strength change on time according to the results of standard tests of specimens of concrete cubes, graphs of dependence of change of fractal dimension of a crack on time in the process of concrete strength gain are constructed. According to the results of the experimental study the dependence between concrete strength and fractal dimension of cracks was established.

A nondestructive method of express control of concrete strength at intermediate age is proposed on the basis of the use of fundamental provisions of fractal mechanics of concrete fracture.

Keywords: Concrete · Fractal Dimension · Fractal Mechanics of Concrete Destruction · Concrete Shrinkage

1 Introduction

The study and further development of concrete strength criteria is an urgent task and is given in studies [1–3]. The results of research indicate that composite materials (modern concretes) are complex fractal systems, which are similar at each scale level. The similarity of structures at different scale levels of the structure of modern concretes is emphasized in the works of V.P. Selyaev [4–6].

S. V. Klyuev et al. (Eds.): ICRAAC 2024, LNCE 627, pp. 71–78, 2025.
https://doi.org/10.1007/978-3-031-82938-3_10

Fractal dimension is a quantitative characteristic of self-similarity of a fractal object. In [7–9], the assumption of the fractal structure of cement concrete structure was experimentally substantiated, and dependencies for determining the strength of concrete with regard to the presence of defects (cracks), fractal dimensionality, and the scale level of the structure were proposed.

This paper studies the development of concrete shrinkage cracks in order to establish the relationship between the change of concrete strength in time and the fractal dimension of the cracks. The study of the relationship between early shrinkage cracking and mechanical properties of composite materials based on fractal theory is presented in [9–12].

2 Methods and Materials

Experimental study of medium strength concrete for industrial cement floors is carried out in accordance with the standard test method for shrinkage crack assessment according to ASTM C 1579–06. The application of this standard for the evaluation of shrinkage cracks is presented in the works [12, 13]. In the course of the experimental study, concrete rectangular slabs with the presence of stress concentrators at the bottom of the formwork were produced. The slab material used was medium strength topping concrete with aggregate in the form of crushed stone with a fraction of 5–10 mm. Technonikol Ready-Mix chemical additive was added to the concrete in the amount of 1% of the cement weight. These slabs were placed in a test chamber where constant environmental conditions were maintained. A scheme of the environmental chamber is shown in Fig. 1. A scheme of the specimen and the geometry of the stress concentrator is shown in Fig. 2.

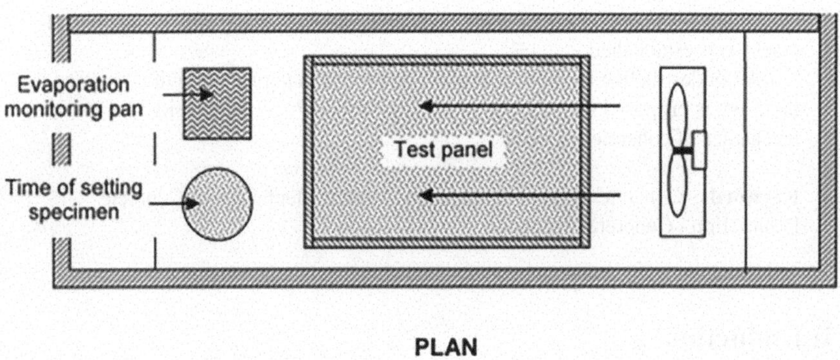

Fig. 1. The chamber scheme for maintaining environmental conditions.

To record the development of concrete shrinkage cracks, an experimental setup was assembled as shown in Fig. 3. The development of cracks was carried out using a digital camera with a given frequency of photo fixation.

Photographs of concrete slab specimens of different ages during the testing process are shown in Fig. 4. The photos show the development of shrinkage cracks.

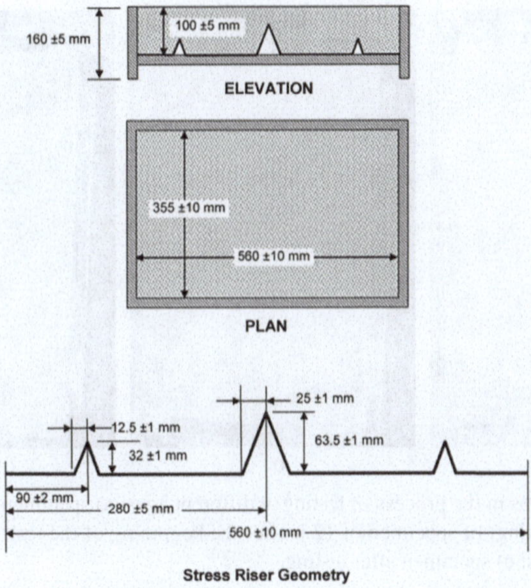

Fig. 2. Specimen scheme and stress concentrator geometry.

Fig. 3. Test setup scheme.

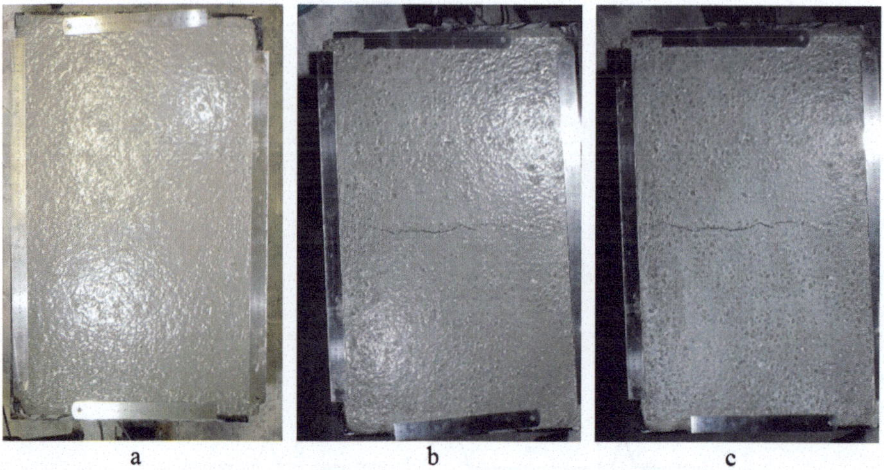

Fig. 4. Concrete slabs in the process of testing at different ages; a) specimen at the beginning of the test, before cracking; b) specimen of 12 h after the beginning of the test, shrinkage cracking development process; c) specimen after testing.

3 Results and Discussions

Photographs of shrinkage cracks obtained during the experimental study were processed. The Minkowski fractal dimension was calculated to determine the quantitative characterization of the crack. The Minkowski fractal dimension, which is one of the ways to determine the fractal dimension of a bounded set in a metric space, is calculated as follows:

$$D = \lim_{\epsilon \to 0} \frac{\log N(\epsilon)}{\log \frac{1}{\epsilon}} = \lim_{\epsilon \to 0} \frac{\log N(\epsilon)}{-\log \epsilon} \tag{1}$$

there D – Minkowski fractal dimension;

N(ϵ) – the minimum number of sets of size ϵ that can cover the original set;

ϵ - the size of the set.

To calculate the fractal dimension of a crack, which represents some bounded set in metric space, a uniform grid of cubes with a step ϵ was applied to the crack photo with subsequent fixation of those grid cells that contain at least one element of the desired set. The dimensionality determination was carried out by an iterative method, in which at the next iteration step a downward change of the mesh size was performed. At each iteration step, the cell size ϵ and the number of fixed grid cells N(ϵ) containing fracture fragments were fixed.

An example of processing the experimental data obtained is shown in Fig. 5.

For the obtained data set, a graph in the form of a regression line was plotted as an approximation of the Minkowski fractal dimension. Figure 6 shows the graph of dependence on the determination of the Minkowski fractal dimension for concrete of different ages.

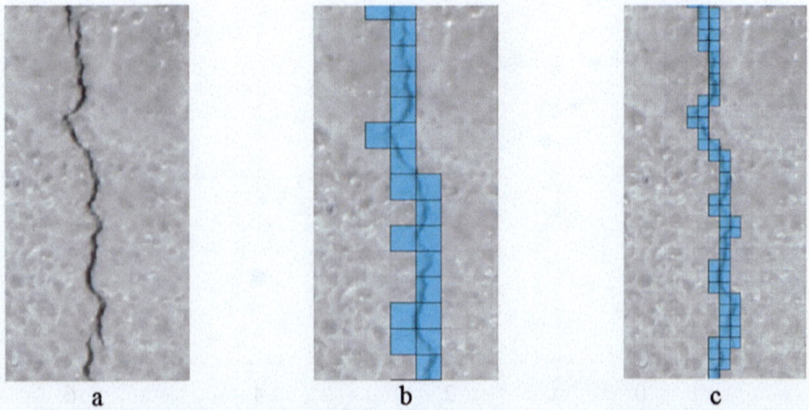

Fig. 5. Example of fracture photo processing for determining fractal dimensionality: a) photograph of the fracture under study; b) applying a dimensional grid to the fracture photo; c) scaling down the dimensional grid.

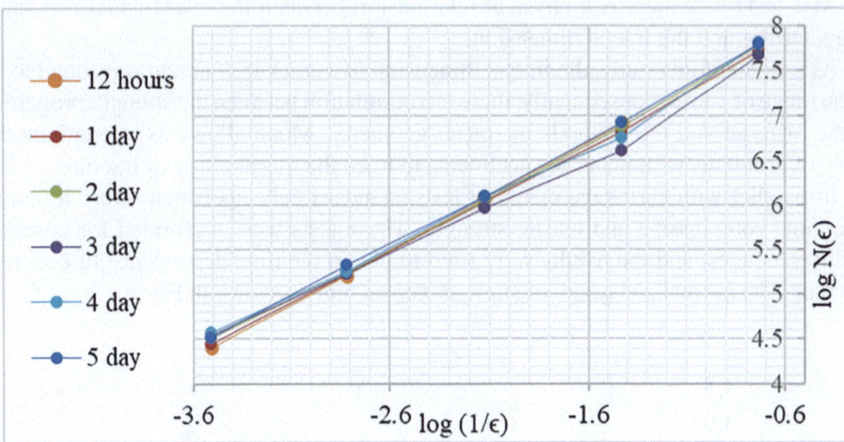

Fig. 6. Graph of dependence on the definition of Minkowski fractal dimension for concrete of different ages.

These graphs were plotted for concrete at different ages and the dependence of fractal dimension on time was obtained. The graph of dependence of fractal dimension on time is presented in Fig. 7.

The fractal dimension can characterize quantitatively the morphological characteristics of the crack and is an important constant calculated after processing the crack image.

At $D < 1$ the range goes from fractal geometry to Euclidean geometry, the crack is treated as a line.

At $D > 2$ the number of cubes needed to cover the cracks is equal to the area of the whole image.

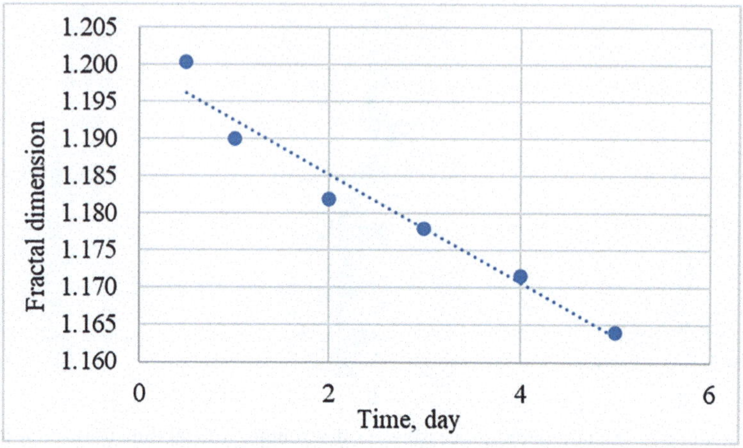

Fig. 7. Graph of dependence of fractal dimension on time.

At 1 < D < 2 there is a physical relationship between the investigated mechanic characteristic and the fractal dimension.

As a result of this study, the fractal dimension of a crack in concrete was found to be in the range of 1 to 2. Consequently, there is a correlation between the strength properties of the material and the methods of fractal geometry, which allows us to continue the study of problems with material specimens, to trace the morphology of fracture.

In parallel with shrinkage cracking tests, standard cube specimens were tested to determine compressive and tensile strength. These tests were performed for concrete of different ages, and the results were used to obtain the time dependence of concrete strength. The normalized graph of strength versus time is shown in Fig. 8.

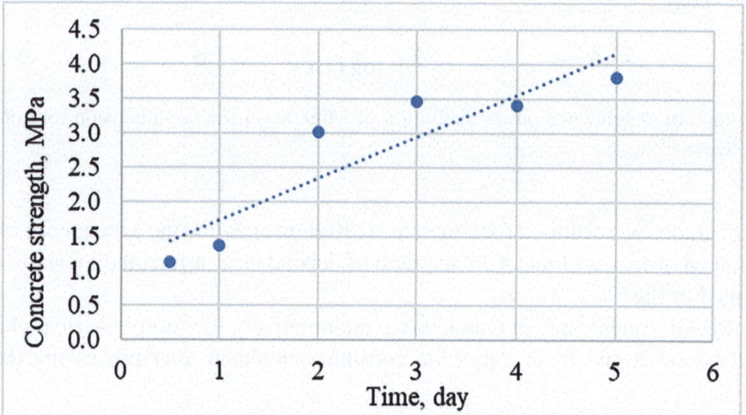

Fig. 8. Graph of time dependence of strength.

According to the results of superimposition of two dependences of strength on time and fractal dimension on time, the dependence of concrete strength on fractal dimension was obtained. The graph of the dependence of strength on fractal dimensionality is presented in Fig. 9.

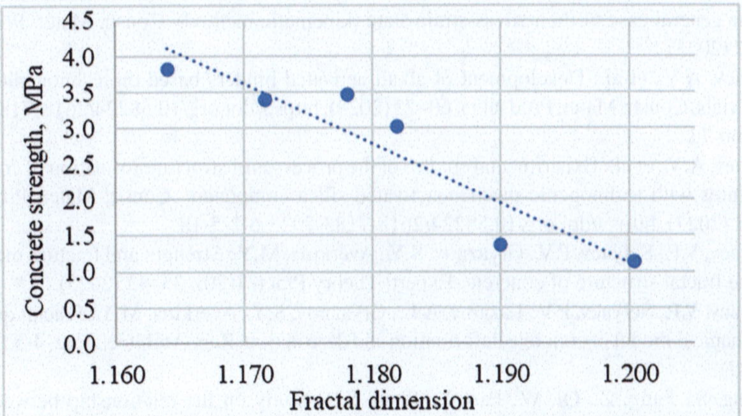

Fig. 9. Graph of strength dependence on fractal dimension.

This dependence can be approximated as a decreasing linear regression. The physical dependence of the parameters under consideration is as follows:

$$R = 100 \cdot (1.15 - 0.95 \cdot D) \tag{2}$$

R – concrete strength, MPa.
D – fractal dimension.

4 Conclusions

A nondestructive method of express control of concrete strength at intermediate age based on the use of fundamental provisions of fractal mechanics of concrete fracture is proposed.

The technique of quantitative characterization of the crack in the form of Minkowski fractal dimension by processing of digital photographs for concrete slab surfaces has been tested.

According to the results of the experimental study, the dependence of the change of the fractal dimension of the crack on time in the process of concrete strength gain is established. According to the results of the experimental study of standard cube specimens, the dependence of concrete strength change on time is obtained.

According to the results of comparison of experimental data of concrete for different types of tests, the physical dependence of concrete strength on Minkowski fractal dimension was established.

References

1. Novoselov, O.G., et al.: Method for calculating the strength of massive structural elements in the general case of their stress-strain state (parametric equations of the strength surface). Constr. Mater. Prod. **6**(2), 104–120 (2023)
2. Novoselov, O.G., et al.: Method for calculating the strength of massive structural elements in the general case of their stress-strain state (kinematic method). Constr. Mater. Prod. **6**(3), 5–17 (2023)
3. Klyuev, A.V., et al.: Development of alkali-activated binders based on technogenic fibrous materials. Constr. Mater. Prod. **6**(1), 60–73 (2023). https://doi.org/10.58224/2618-7183-2023-6-1-60-73
4. Klyuev, A.V., et al.: Experimental studies of the processes of structure formation of composite mixtures with technogenic mechanoactivated silica component. Constr. Mater. Prod. **6**(2), 5–18 (2023). https://doi.org/10.58224/2618-7183-2023-6-2-5-18
5. Selyaev, V.P., Selyaev, P.V., Gryaznov, S.Y., Averkina, M.Y.: Strength and fracture mechanics of the fractal structure of concrete. Expert: Theory Pract. **1**(20), 35–43 (2023)
6. Selyaev, V.P., Selyaev, P.V., Lazarev, A.L., Gryaznov, S.Y., Averkina, M.Y.: Fractal quantum-mechanical model of concrete deformation and destruction. Reg. Architect. Eng. **4**(53), 31–40 (2022)
7. Zhang, S., Fang, Z., Qi, W., Fan, Y., Shah, S.P.: Study on the relationship between early shrinkage cracking and mechanical properties of nano-clay cement mortar based on fractal theory. Buildings **13**, 123 (2023)
8. Szeląg, M.: Fractal characterization of thermal cracking patterns and fracture zone in low-alkali cement matrix modified with microsilica. Cement Concr. Compos. **114**, 103732 (2020)
9. Li, Y., Shen, A., Wu, H.: Fractal dimension of basalt fiber reinforced concrete (BFRC) and its correlations to pore structure, strength and shrinkage. Materials **13**, 3238 (2020)
10. Wang, L., et al.: Influence of MgO on the hy-dration and shrinkage behavior of low heat portland cement-based materials via pore structural and fractal analysis. Fractal Fract **6**, 40 (2022)
11. Zhang, L., Gong, H., Liu, J., Li, H.: Mechanical properties and chloride penetration resistance of copper slag aggregate concrete. Fractal Fract **6**, 427 (2022)
12. Wang, L., Yu, Z., Liu, B., Zhao, F., Tang, S., Jin, M.: Effects of fly ash dosage on shrink-age, crack resistance and fractal characteristics of face slab concrete. Fractal Fract **6**, 335 (2022)
13. Sayahi, F., Emborg, M., Hedlund, H., Cwirzen, A.: Effect of steel fibres extracted from recycled tyres on plastic shrinkage cracking in self-compacting concrete. Mag. Concr. Res.Concr. Res. **73**(24), 1270–1282 (2020)

Method of Taking into Account the Light Climate in the Calculation and Design of Upper Natural Lighting of Buildings

Khazret Guketlov(✉), Valery Khuranov, Alima Azhoeva, and Rasul Selyaev

Kabardino-Balkarian State University Named After Kh.M. Berbekov, Nalchik 360004, Russia
123guket@mail.ru

Abstract. An algorithm has been developed for calculating the area of light apertures in buildings with various systems of overhead natural lighting of premises, allowing for a transition to differentiated accounting of light climate resources in any geographical point on the globe. A mathematical model of the room lighting regime has been improved, taking into account the features of redistribution of light fluxes in the structures of point and strip type skylights, rectangular light-aeration lanterns and in inclined lighting devices. Based on calculations of the amount of lighting in the annual cycle (this criterion has a physiological justification and ensures average annual labor productivity) using a mathematical model and its software implementation on a computer, light climate coefficients for the city of Nalchik were obtained. For light apertures located in an inclined plane, the values of these coefficients change depending not only on the orientation of the light apertures to the sides of the horizon, but also on the angle of their inclination. The difference in coefficients can exceed 40%. Using the obtained coefficients when calculating natural lighting of premises will reduce the area of light openings by 8–15% while maintaining a full-fledged light environment and improving the microclimate in the premises, and thereby increase the technical and economic efficiency of overhead natural lighting systems.

Keywords: Mathematical Model · Amount of Illumination · Natural Illumination on Inclined Surface · Luminous Flux · Clear Sky Cie · Lanterns · Light Climate Coefficient · Direct Illumination from the Sun · Probability of Sunshine · Lighting Devices

1 Introduction

The currently existing system of methods for standardization and design of natural lighting [1] is based on integral criteria for assessing light climate resources (light climate coefficient m_N), the numerical values of which are obtained by averaging the light climate indicators within each administrative district.

Given the wide variety of light and climate characteristics in different geographical locations of an administrative district, averaging the light climate indicators in some cases leads to an underestimation of the area of light openings and, as a consequence, to an

S. V. Klyuev et al. (Eds.): ICRAAC 2024, LNCE 627, pp. 79–86, 2025.
https://doi.org/10.1007/978-3-031-82938-3_11

overspending of electricity on artificial lighting, and in other cases to an overestimation of the area of light openings and, consequently, to an increase in one-time costs and heat losses in natural lighting systems.

In this regard, a very urgent task is to divide administrative districts into local ones and move from average indicators to regional parameters of the light climate, which more fully reflect the features and real resources of natural lighting at the location of the building.

To solve this problem, we have developed a differentiated method for accounting for light climate indicators based on an improved mathematical model of the passage of light flux through various types of overhead natural lighting devices.

2 Methods and Materials

The main idea of the algorithm of the proposed method is to determine the required area of light openings based on a comparison of the lighting conditions of a room located in Moscow (this area is taken as a reference) and in the construction area under consideration (1).

$$S_p = m_N S_M = \frac{H_M}{H_p} S_M \tag{1}$$

where S_p,—the required area of light openings for the light and climate conditions of the construction area under consideration; m_N—the light climate coefficient; S_M—the area of light openings required for the light and climate conditions of Moscow, determined according to [1]; H_M—the amount of lighting in a room located in Moscow with an area of light openings equal to S_M; H_p—the amount of lighting in a room located in the area under consideration with an area of light openings equal to S_M.

In accordance with [2], the average annual amount of lighting in the room was taken as the criterion for assessing the lighting conditions of the room. This criterion has a physiological basis [2] and allows us to resolve the issue of how to change the area of the light opening in order to maintain the same average annual level of visual performance in different light and climate conditions.

Based on the results of theoretical studies [2, 5], a mathematical model of the lighting regime of premises with software implementation on a computer was developed, which in general form is described by the Eq. (2).

$$H = \sum_{i=1}^{12} N_i \left[S_i \sum_{j=t_1}^{t_2} E_{i,j}^{c.s.} + (1 - S_i) \sum_{j=t_3}^{t_4} E_{i,j}^{C} \right] \cdot \Delta t \tag{2}$$

where H—the amount of illumination in the room per year, lx·h; N_i—the number of days in a month; S_i—the probability of sunshine by month of the year, determined according to [2]; t_1, t_2, t_3, t_4—the time of switching on and off of artificial lighting in the room under clear and cloudy skies; $E_{i,j}^{c.s.}$, $E_{i,j}^{C}$—the values of average illumination in a room under clear and cloudy skies, respectively, determined taking into account the type of lantern and the presence of sun protection devices (SPD).

3 Results and Discussion

The redistribution of light fluxes when passing through different types of lighting devices, as can be seen from Fig. 1, is different, therefore the mathematical model is described by different equations [3]:

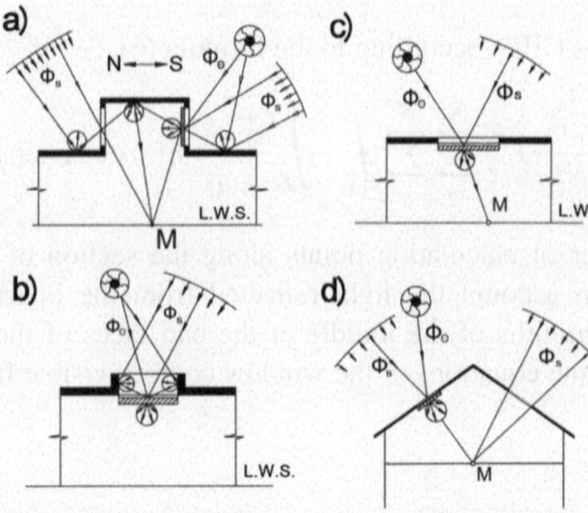

Fig. 1. Scheme of light flow through U-shaped (a), zenithal strip (b), zenithal spot (c) lanterns and inclined lighting devices (d).

Natural illumination in the room from point-type skylights (with dimensions in plan no more than 2.7 × 2.7 m) is determined under clear skies CIE according to the formula (3).

$$E_{i,j}^{c.s.} = \frac{nS_l}{K_f S_{fl}} \tau_o \tau_c r_2 G\left(\tau_{tr}^{c.s.} E_h^{c.s.} + \tau_{tr}^{\Theta} E_h^{\Theta}\right) \tag{3}$$

Under cloudy skies CIE—according to the formula (4).

$$E_{i,j}^{C} = \frac{nS_l}{K_f S_{fl}} \tau_o r_2 G \tau_{tr} E_h^{C} \tag{4}$$

where n—the number of skylights; S_l—the area of one skylight (in the clear); τ_o—the total light transmittance coefficient of the skylight filling: S_{fl}—the illuminated area of the room; G—the luminous flux utilization coefficient, determined according to [3]; $\tau_{tr}^{c.s.}$—the light transmittance coefficient of the skylight for diffuse illumination from the clear sky, determined according to [7]; τ_{tr}^{Θ}—the light transmittance coefficient of the skylight for direct illumination from the Sun, determined according to [6]; $E_h^{c.s.}$—the external diffuse illumination from the clear sky CIE on a horizontal surface, determined according to [2]; E_h^{Θ}—the external horizontal illumination from the Sun, determined according to [2]; E_h^{C}—the external horizontal illumination from the cloudy sky CIE, determined according to [5]; K_f—the safety factor.

For large-sized (more than 2.7 × 2.7 m) strip-type skylights, the illumination is determined under clear skies and the presence of a SPD in the skylights using the formula (5).

$$E_{i,j}^{c.s.} = \frac{E_h^{c.s.} + E_h^{\Theta}}{\pi N K_f} \tau_o \tau_c r_2 \sum_m^n \sum_k^N \int_{\alpha_1}^{\alpha_2} \int_{Z_1(\theta_1,\alpha)}^{Z_2(\theta_2,\alpha)} \cos z \sin z dz d\alpha \quad (5)$$

Under cloudy skies CIE—according to the formula (6).

$$E_{i,j}^{C} = \frac{\tau_o r_2}{\pi N K_f} L_{i,j}^{C} \sum_m^n \sum_k^N \int_{\alpha_1}^{\alpha_2} \int_{Z_1(\theta_1,\alpha)}^{Z_2(\theta_2,\alpha)} \eta(z) \cos z \sin z dz d\alpha \quad (6)$$

where N—the number of calculation points along the section of the room; r_2—the coefficient taking into account the light reflected from the interior surfaces of the room; α_1, α_2—the azimuths of the middle of the end faces of the skylight opening; $Z_1(\theta_1, \alpha)$,$Z_2(\theta_2, \alpha)$—the equations of the window contour visible from the calculation point (see Fig. 2).

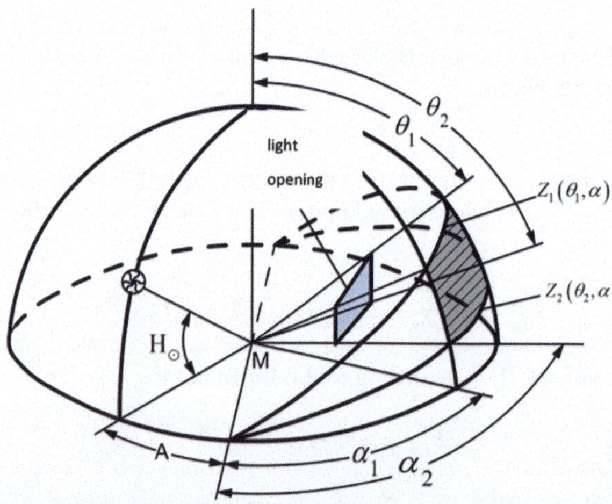

Fig. 2. Scheme of the arrangement of the light opening, the calculated point M and the Sun in spherical coordinates.

θ_1, θ_2—the zenith distance of the middle of the left and right edges of the light aperture, respectively (Fig. 2).; $\eta(z)$—the brightness distribution function over the cloudy sky CIE; $L_{i,j}^{C}$—the brightness of the cloudy sky at the zenith is determined by the formula (7).

$$L_{i,j}^{C} = 300 + 17117,4\sin H_{\Theta} + 4031,7 H_{\Theta} \sin^2 H_{\Theta} \quad (7)$$

where H_{Θ}—the height of the Sun in radians

For rectangular aeration lanterns, the illumination is determined in clear skies and the presence of SZU—according to the formula formula (8).

$$E_{i,j}^{c.s.} = \frac{\tau_o r_2 \kappa_l}{\pi N K_f} \sum_m^n \sum_k^N E_{i,j}^{tot} \int_{\alpha_1}^{\alpha_2} \int_{Z_1(\theta_1,\alpha)}^{Z_2(\theta_2,\alpha)} \cos z \sin z dz d\alpha \tag{8}$$

In clear skies and no SPD—according to the formula (9).

$$E_{i,j}^{c.s.} = \frac{\tau_o r_2 K_l}{\pi N K_f} L_{i,j}^{c.s.} \sum_m^n \sum_k^N \int_{\alpha_1}^{\alpha_2} \int_{Z_1(\theta_1,\alpha)}^{Z_2(\theta_2,\alpha)} S(z,\alpha) \cos z \sin z dz d\alpha \tag{9}$$

Under cloudy skies – according to the formula (10).

$$E_{i,j}^{C} = \frac{\tau_o r_2 K_l}{\pi N K_f} L_{i,j}^{C} \sum_m^n \sum_k^N \int_{\alpha_1}^{\alpha_2} \int_{Z_1(\theta_1,\alpha)}^{Z_2(\theta_2,\alpha)} \eta(Z) \cos z \sin z dz d\alpha \tag{10}$$

where $E_{i,j}^{tot}$—the total illumination on the outer vertical surface of the lantern, determined by [8]; $S(z,\alpha)$—the brightness distribution function over the clear sky CIE; K_l—the coefficient taking into account the type of lantern; $L_{i,j}^{c.s.}$—the brightness of the clear sky CIE at the zenith, determined by the formula (11).

$$L_{i,j}^{c.s.} = 60 + 600\cos^2(2H_\Theta) + 4800 H_\Theta^{1,55} \tag{11}$$

For inclined lighting devices, the illumination in the room is determined in clear skies CIE from only the diffuse component of the luminous flux falling on the light, determined by the formula (12).

$$E_{i,j}^{c.s.} = \tau_o r_2 \int_{\alpha_1}^{\alpha_2} \int_{z_1(\theta_1,\alpha)}^{z_2(\theta_2,\alpha)} L(z,\alpha) \cos z \sin z dz d\alpha \tag{12}$$

From the total diffuse and direct component of the luminous flux falling on the light aperture—according to the formula (13).

$$E_{i,j}^{c.s.} = \frac{\left(E_{i.p.}^{c.s.} + E_{i.p.}^{\Theta}\right)}{\pi} \tau_c \tau_o r_2 \int_{\alpha_1}^{\alpha_2} \int_{z_1(\theta_1,\alpha)}^{z_2(\theta_2,\alpha)} \cos z \sin z dz d\alpha \tag{13}$$

$E_{i.p.}^{c.s.}$—Diffuse illumination on a sloping roof surface under clear skies CIE, determined by the formula (14) [4].

$$E_{i.p.}^{c.s.} = \int_0^{\pi} \int_0^{Z_2(\theta_4,\alpha)} L(z,\alpha) \cos(z-\beta) \sin z dz d\alpha + \int_{\pi}^{2\pi} \int_0^{Z_2(\theta_3,\alpha)} L(z,\alpha) \cos(z+\beta) \sin z dz d\alpha \tag{14}$$

$E_{i.p.}^{\Theta}$—direct illumination from the Sun on the inclined surface of the roof, determined by the formula (15).

$$E_{i.p.}^{\Theta} = \frac{C_{\Theta}}{\Delta^2} p \cos\theta \tag{15}$$

Under cloudy skies—according to the formula (16).

$$E_{i,j}^{C} = \tau_o r_2 \int_{\alpha_1}^{\alpha_2} \int_{z_1(\theta_1,\alpha)}^{z_2(\theta_2,\alpha)} L(z) \cos z \sin z dz d\alpha \tag{16}$$

where C_{Θ}—the solar luminous constant; Δ—the distance from the Earth to the Sun for the year in question in astronomical units; θ_3, θ_4—angular values of the integration limits. $L(z, \alpha)$, $L(z)$—the brightness distribution function of clear and cloudy skies of the CIE; M—the optical mass of the atmosphere; θ—the angle formed by the normal of the illuminated plane with the direction to the Sun, determined by the formula (17).

$$\theta = \arccos[\cos z \cos z_{\Theta} + \sin z \sin z_{\Theta} \cos(A - A_{\Theta})] \tag{17}$$

where Z, Z_{Θ}, A, A_{Θ}—the zenith distance and azimuths, respectively, of the normal and the Sun to the surface of the light opening.

Simpson's cubature formula was used to solve double integrals [9].

Based on calculations using the considered mathematical model of the lighting regime of premises on a computer, the coefficients of the lighting climate were obtained (Tables 1, 2).

Table 1. Values of light and climate coefficients for zenith and rectangular lanterns in the city of Nalchik m_N..

Type of lantern					
anti-aircraft		rectangular, oriented to			
SNiP	new	N–S		W–E	
		SNiP	new	SNiP	new
0.9	0.708	0.9	0.733	0.9	0.697

The analysis of the calculation results showed that in the majority of regions of the Russian Federation there are unaccounted resources of natural light energy, which, in particular, is expressed in a decrease in light and climate coefficients by an average of 8–15%. This means that in the specified regions, when designing natural lighting for buildings, it is possible to reduce the area of light openings by 8–15% while maintaining a full-fledged light environment and improving the microclimate in the premises, as well as to reduce the reduced costs of lighting. Light and climate coefficients for inclined lighting devices (Table 2) differ slightly at small angles of inclination and differ significantly at large angles of inclination.

Table 2. Values of light and climate coefficients for inclined lighting devices in the city of Nalchik m_N.

Orientation of light openings to the cardinal directions	Light climate coefficient m_N					
	Angle of inclination of light openings, degrees					
	10	20	30	40	50	60
N	0.792	0.743	0.733	0.721	0.641	0.539
NE, NW	0.795	0.747	0.748	0.748	0.684	0.656
SE, SW	0.811	0.776	0.794	0.805	0.714	0.386
S	0.815	0.783	0.802	0.817	0.727	0.496
E, W	0.867	0.83	0.846	0.861	0.813	0.71

4 Conclusions

The mathematical model of redistribution of light fluxes in the structures of point and strip type skylights, rectangular light-aeration lanterns and in inclined lighting devices under clear and cloudy skies of the CIE, improved by the authors, will allow for a more complete consideration of light climate resources.

Based on the results of calculating the amount of lighting in the annual cycle for the city of Nalchik, the light climate coefficients were adjusted. It was found that for light openings located in an inclined plane, the values of these coefficients change not only depending on the orientation of the light openings to the sides of the horizon, but also on the angle of their inclination. The difference in coefficients can be more than 40%.

Using the obtained coefficients in calculating natural lighting of premises will reduce the area of light openings by 8–15% while maintaining a full-fledged light environment and improving the microclimate in the premises, and thereby increase the technical and economic efficiency of overhead natural lighting system.

References

1. SNiP 23–05–95* Natural and artificial lighting. Moscow (2004)
2. Guketlov, K.M., Bjakhov, M.I., Khuranov, V.K., Likhov, Z.R.: Improving the technical and economic ffectiveness of the system of natural lighting attic space. Proceedings of the 2017 International Conference Quality Management, Transport and Information Security, Information Technologies. IT and QM and IS 2017, pp. 728–729 (2017)
3. Guketlov, K.M., Shogenova, F.M., Fitsev, A.R.: Mathematical model of passing the light flow to the space in a clear sky through various types of lighting devices. Materials Sci. Forum **931**, 914–919 (2018)
4. Guketlov, K.M., Khachetlov, H.A., Kibishev, K.A.: Method for calculating the garret window light transmittance coefficient at clear sky CIE. IOP Conf. Series: Materials Science and Engineering **913**, 042001 (2020)
5. Guketlov, K.M., Gukov, A.H.: Accounting for the light climate of the city of Nalchik when calculating natural light. Published under licence by IOP Publishing Ltd.

6. Guketlov, K.M., Tatarov, I.A.: Method for determining the transmission of a garret window from a direct sun component. E3S Web of Conferences **281**, 04004 (2021)
7. Guketlov, H.M., Shogenov, O.M., Mashukova, M., Makushev, I.Z.: On the issue of determining the coefficient of light transmission of a point skylight under clear skies. Sci. Tech. Bull. Volga Region **1**, 79–84 (2022)
8. Guketlov, K.M., Khachetlov, H.A., Kibishev, K.A.: Accounting the photoclimate when calculating the reflected light from the operating buildings' facades. IOP Conf. Series Mat. Sci. Eng. **913**, 042002 (2020)
9. Demidovich, B.P., Maron, I.A.: Fundamentals of Computational Mathematics. Lan Publishing House, Moscow (2011)

Method of Experimentally Obtaining a Complete Stress-Strain Curve of Masonry Structures

M. L. Pozdeev, I. V. Smagin, D. M. Lobov(✉), and S. Yu. Likhacheva

Nizhny Novgorod State University of Architecture and Civil Engineering, Nizhny Novgorod, Russia

maksim.leon.pz@yandex.ru

Abstract. The paper describes the methodology of constructing the complete deformation diagram (stress-strain curve) of masonry using hydraulic presses without a tracking system (displacement control) and fast feedback. The technique of equilibrium tests is used, when during deformation of the tested specimen an additional element of increased stiffness is introduced, which allows redistributing the forces when realizing the downward branch of the deformation diagram of the masonry specimen. The paper presents a methodology for selecting the required stiffness taking into account the convexity of the deformation diagram on the downward section. This allows selecting optimal parameters of the loading system, which is especially important in case of a large nomenclature of stone products and variation of their deformation characteristics. The method is tested on thickened silicate bricks with hollowness. According to the results of experimental studies with the developed installation, complete deformation diagrams of masonry under uniaxial compression are successfully obtained.

Keywords: Masonry · Stress-Strain Curve · Deformation Diagram · Softening

1 Introduction

The construction of full deformation diagrams (stress-strain curve) of structural materials is an important task to ensure the reliability of building structures. The use of physically nonlinear material models allows to establish strength reserves, as well as to determine the ability of a structure to dissipate energy under dynamic loads.

Orthotropic flow theory models realizing the plane stress state [1] are known among the continuum models for masonry as well as failure under load action from the masonry plane [2]. In the set of deformation models, we can distinguish the orthotropic model of masonry [3], as well as isotropic deformation models of concrete [4, 5], which can be successfully applied to masonry under certain types of stress states [6]. For all these models, one of the important parameters is the dependence of stresses on strains in uniaxial compression and tension.

The paper [7] shows the importance of obtaining the plastic characteristics of masonry to substantiate the ability of the material to dissipate energy under seismic

S. V. Klyuev et al. (Eds.): ICRAAC 2024, LNCE 627, pp. 87–93, 2025.
https://doi.org/10.1007/978-3-031-82938-3_12

loads, the author obtains the corresponding dependencies numerically on the basis of the discrete finite element model of masonry for plane stress state developed by him [8].

In view of the large range of stone products and mortars used for the construction of masonry walls, there is a problem of normalization of the full deformation diagram of masonry. This problem is reflected both in Russian and foreign norms. The deformation diagram is described only on the ascending section. Also at the moment there is no developed strategy for experimental obtaining of the complete deformation diagram of masonry, which also complicates the use of modern physically nonlinear material models.

One of the problems of experimental studies of masonry structures is the size of a representative volume of masonry whose macromechanical behavior can be used with the required reliability for continuum plasticity models. This imposes limitations on the working areas of the testing equipment and the required levels of failure loads. In the known works on obtaining the full deformation diagram of masonry [9, 10], tests were performed on relatively small cube or column specimens, using testing machines with a tracking system (displacement control) (Fig. 1) and fast feedback. The use of specialized testing machines is a problem to the spread of the experimental method of constructing the descending section of the masonry deformation diagram.

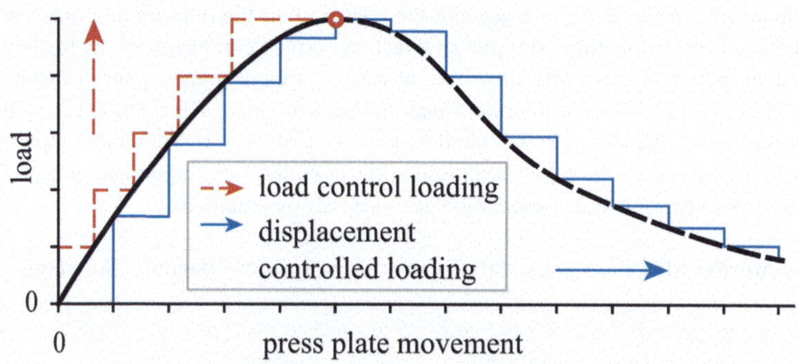

Fig. 1. Load-displacement relationship for different test strategies.

Tests with classical hydraulic presses are carried out with loading step control, i.e. the specimen is loaded in equal portions with continuously increasing load (Fig. 1). This leads to the fact that when the ultimate strength of the specimen is reached, the failure is dynamic, due to elastic energy stored in the specimen, which does not allow obtaining the required deformation diagram on the downward section under quasi-static loading.

In order to obtain a complete deformation diagram, there are known methods of equilibrium tests, set forth, in particular, in GOST 29167-2021 for determining the characteristics of crack resistance (fracture toughness) of concrete under static loading. The paper describes the use of high stiffness testing machines or standard testing machines equipped with an additional redistributing device. It is the principle of the redistributing device that is described in this paper for determining the complete deformation diagram of masonry.

In [11], a review of known designs of redistributing devices for concrete testing is given, and the design of an experimental installation in the form of a duralumin alloy pipe, which also acts as a strain gauge force sensor when selecting the gap between the loading plate and the pipe, is proposed. In work [12] as elastic inserts, which perceive forces on the downward section of deformation, calibrated struts are used, on the lower ends of which a stop thread is provided, allowing to regulate the gap between them and the loading plate and include them in operation at load levels of 0.5…0.6 of the destructive ones. In [13], an installation for calculating the modulus of slump was developed, which allows the specimen to be loaded with both tensile and compressive loads using a reverser. The use of this installation is limited to low-grade concrete, because when testing high-strength concrete, more energy is accumulated in the elastic elements of the installation than is required to fracture the concrete. In [14], this principle is used to determine the dynamic modulus of unstrengthening, where it is noted that when the stiffness of the test setup exceeds the unstrengthening modulus, it is possible to obtain a descending section of the deformation diagram in the quasi-static regime up to complete failure of the specimen.

In the case of masonry specimens, the disadvantages of the above-described installations are the required specimen size. Thus, when using a closed tube structure, the specimen size is limited by the size of the tube itself, and the use of compressed steel struts is ineffective under high compressive loads due to loss of stability.

The paper describes the method of selecting the parameters of the experimental installation for determining the full deformation diagram of masonry under uniaxial compression using a universal testing machine. In this case, the principle of operation is similar to that described earlier and allows obtaining the descending part of the deformation diagram due to the redistribution of forces between the specimen and the additionally introduced steel element.

2 Methods and Materials

The full deformation diagram was constructed for a masonry specimen of 510 × 510 × 120 mm in size, manufactured according to GOST 32047-2012. The thickened silicate brick with a void space of 9% of M150 grade according to GOST 379-2015 and mortar with an average compressive strength of 28.4 MPa were used. Characteristics of masonry materials are given in Table 1.

Table 1. Strength characteristics of brick and mortar.

Material	Test method	Compressive tensile strength, MPa	Bending tensile strength, MPa
Brick	GOST R 58527-2019	15.5 (11%)	3.1 (17%)
Mortar	GOST 5802-86 GOST 310.4-81	28.4 (2%)	7.8 (16%)

The value of the variation coefficient is given in parentheses.

The tests were carried out on the universal testing machine CD100. The scheme of the experimental installation is shown in Fig. 2.

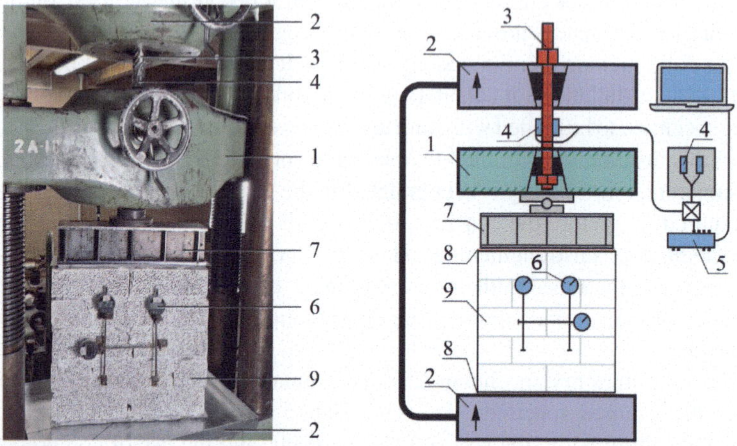

Fig. 2. Scheme of the experimental installation: 1 – elevating beam, 2 – fixed beams, 3 – reinforcing bar, 4 – active (on reinforcement) and passive strain gauges, 5 – strain-gauge station, 6 – clock type sensors, 7 – load-bearing girder, 8 – teflon (PTFE) gaskets, 9 – masonry specimen.

A masonry specimen is installed in the compression zone of the testing machine, the relative deformations of the specimen are recorded by six hour-type sensors installed on both sides of the specimen according to GOST 32047-2012. The compressive load is transferred from the fixed crosshead through a hinge to the steel girder and then to the specimen through a teflon (PTFE) gasket. In [15], a good effect of such spacers was shown to minimize the effect of strain constriction in the restrained zones of the specimen and to obtain a uniaxial stress state throughout its volume.

A reinforcing bar of A500C class according to GOST 34028-2016 with nominal diameter of 36 mm with threads and nuts cut on both sides is installed in the tensile zone of the testing machine. The presence of wedges of the testing machine is insufficient to fix the bar, which will be indicated in the results section. The rebar parameters (diameter, strength class and length) were selected based on the experimental methodology described below.

Before the load reaches a value of 0.8…0.9 of the destructive load, the upper nut of the reinforcing bar is released and the masonry specimen is deformed independently. This makes it possible to establish the initial elasticity characteristics (elastic modulus and Poisson's ratio) exactly as required by the normative documents. Then the nut is tightened and compression of the specimen is accompanied by tension of the reinforcing bar. When the masonry strength limit is reached, its stiffness decreases and part of the load is redistributed to the reinforcing bar, the acting forces in which are recorded by means of strain gauges installed on it, the bar itself must be pre-calibrated.

In order to determine the required theoretical stiffness of the reinforcing bar, the complete deformation diagram of the masonry specimen must be given. The diagram can be approximated by the broken *ABCD* in Fig. 3. As a basic dependence on the ascending section *ABC* (point *B* defines the moment of the reinforcement bar inclusion in the work, 0.8...0.9 of the strength of the masonry specimen) can be used the logarithmic dependence presented in SP 15.13330-2020, which in the coordinates of the force-displacement dependence *P-Δl* (Fig. 3) will take the form:

$$\frac{\Delta l}{l_0} = -\frac{1,1}{\alpha}\ln\left(1 - \frac{P}{1,1R_uA}\right), \tag{1}$$

In [9] it is shown that for masonry on cement mortars without the addition of plasticizing additives, for example, lime, the limiting relative strain is two times higher than the relative strain at reaching the tensile strength, i.e., the unstrengthening modulus can be equal to the secant strain modulus at reaching the tensile strength.

3 Results and Discussions

For a stable deformation process, the load-displacement dependence of the press plate (broken *ABKL* in Fig. 3) should increase throughout the deformation process, up to the complete failure of the specimen. Then the required theoretical stiffness of the bar k_m should not be less than the stiffness of the masonry specimen k_s on the downward section of the deformation diagram (Fig. 3b, c):

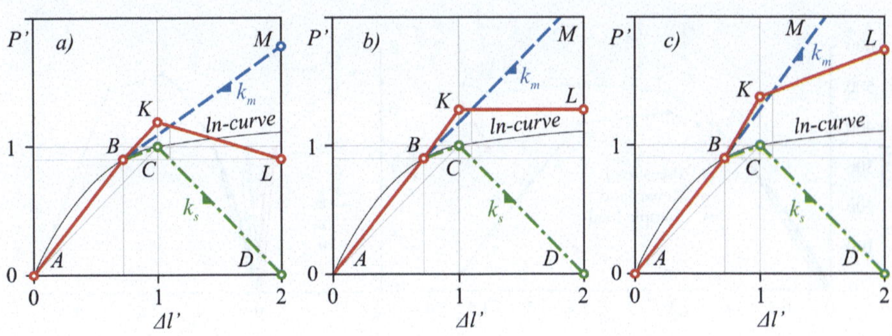

Fig. 3. Dependences of press plate displacement on *ABKL* load at different reinforcement bar stiffness: a – insufficient stiffness; b – ultimate stiffness; c – optimal stiffness.

It is important to note that the assumption of $k_m = k_s$ is incorrect due to the upward convexity of the typical diagram of masonry deformation in compression. In general, the first derivative of the graph of the load-displacement relationship of the press plate (curve *ABKL* in Fig. 4) should be positive, in which case the value of the required stiffness will be $k_m \geq 1{,}1\ k_s$.

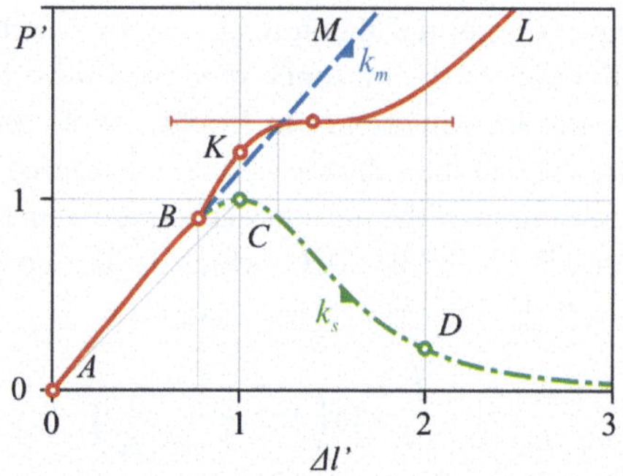

Fig. 4. Dependences of press plate displacement on load *ABKL* at curvilinear diagram *ABCD* of specimen deformation.

An experimental study was carried out on the installation shown in Fig. 2. The reinforcing bar was put into operation at stresses of about 0.8 of the ultimate ones. The forces in the specimen were obtained by subtracting the force in the reinforcement bar from the total force produced by the testing machine. As a result, the desired full deformation diagrams of the masonry specimen were constructed and are shown in Fig. 5. Since the main cracks open on the downward section of the diagram, it is difficult to determine the strains using clock-type sensors, so it is either necessary to record the movements of the press plate or to use the calibrated dependencies for the reinforcing bar. This is possible because after sampling the compression strains, the strains of the reinforcement and the specimen are combined.

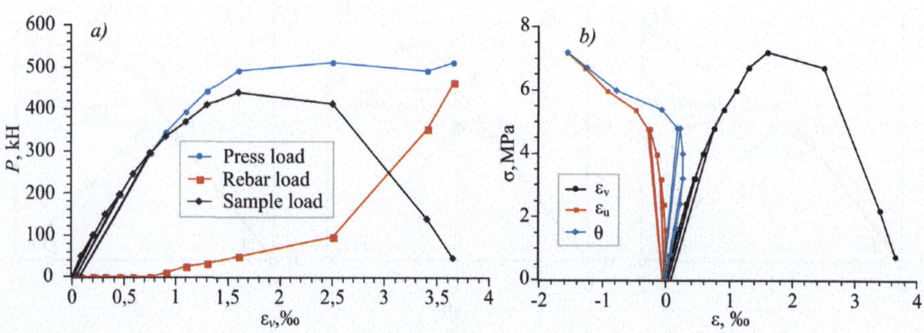

Fig. 5. Dependence of forces in the elements of the experimental installation on relative deformations of the masonry specimen (a) and obtained deformation diagrams (b)

4 Conclusions

As a result of the conducted research, a methodology for constructing a complete deformation diagram of masonry has been developed and tested. The experimental installation allows equilibrium tests to be performed using universal testing machines not equipped with a tracking system (displacement control) and fast feedback. The analytical method of steel bar parameters selection allows determining the required stiffness for different types of masonry units based on the assumed deformation diagram or on the basis of the first non-equilibrium test to determine the primary elastic characteristics and the values of ultimate deformations when the material strength limit is reached.

Acknowledgments. The research was carried out under the grant of the Government of Nizhny Novgorod region for young scientists from the regional budget in the form of a subsidy (Contract No. 316-06-16-13-13a/24).

References

1. Lourénço, P.B., De Borst, R., Rots, J.G.: A plane stress softening plasticity model for orthotropic materials. Int. J. Numer. Meth. Eng. **40**(21), 4033–4057 (1997)
2. Lourenco, P.: Anisotropic softening model for masonry plates and shells. J. Struct. Eng. **126**(9), 1008–1016 (2000)
3. Tyupin, G.A.: Deformation theory of plasticity of masonry. Constr. Mech. Calc. Struct. **6**, 28–30 (1980)
4. Geniev, G.A.: Variant of the deformation theory of plasticity of concrete. Concr. Reinf. Concr. **2**, 18 (1969)
5. Kruglov, V.M., Yerofeev, V.T., Vatin, N.I.,:Al Dulaimi Salman Dawood Salman: A variant of the deformation theory of plasticity of concrete in a flat stressed state Online journal "Transport facilities" **4** (2019)
6. Pozdeev, M.L., Likhacheva, S., Smagin, I.V., Radaykin, O.V.: Calculation of stone walls using the deformation theory of plasticity. Bull. Kazan State Energy Univ. **3**(59), 163–174 (2023)
7. Kabantsev, O.V., Tonkikh, G.P.: On deformability and seismic resistance of masonry structures. Ind. Civil Constr. **9**, 51–58 (2019)
8. Kabantsev, O.V.: Discrete model of masonry in conditions of biaxial stress state. Bull. Tomsk State Univ. Arch. Civil Eng. **4**(51), 113–134 (2015)
9. Kaushik, H., Rai, D., Jain, S., Asce, M.: Stress-strain characteristics of clay brick masonry under uniaxial compression. J. Mater. Civil Eng. **19** (2007)
10. Yang, K.-H., Lee, Y., Hwang, Y.-H.: A stress-strain model for brick prism under uniaxial compression. Adv. Civil Eng. 1–10 (2019)
11. Bezgodov, I.M., Levchenko, P.: On the question of the methodology for obtaining complete diagrams of concrete deformation. Technol. Concr. **8**(97), 27–29 (2014)
12. Gvozdev, A.A., Shubik, A.V., Matkov, I.G.: On the complete compression diagram of concrete reinforced with transverse grids. Concr. Reinf. Concr. **4**, 37–39 (1988)
13. Bich, P.M.: Experimental and theoretical studies of the supercritical characteristics of concrete. Concr. Reinf. Concr. **3**, 26–27 (1987)
14. Stavrogin, A.N., Protosenya, A.G.: Mechanics of Deformation and Destruction of Rocks. Nedra, Moscow 224 p. (1992)
15. van Mier, J.G.M., Shah, S.P., Arnaud, M.: Strain-softening of concrete in uniaxial compression. Mater. Struct. **30**(4), 195–209 (1997)

Application of Cement Soils in the Construction of Pile Anchoring of Ice-Resistant Gravity Platforms

N. G. Seregin(✉) and N. V. Shunko

Candidate of Engineering Sciences, Moscow State University of Civil Engineering, Yaroslavskoe Shosse, 26, Moscow 129337, Russia
SereginNG@mgsu.ru

Abstract. In the severe conditions of the Arctic shelf, ice-protective pile-gravity platforms have found application, which significantly improve the conditions of interaction with the soil base. When designing offshore ice-resistant gravity-type platforms, regardless of the design and material of the foundation block, it is usually envisaged to install elements buried below the base of the slab of one design or another. This allows deepening the calculated shear surface to the value required by the stability conditions to stronger soil layers, increasing the foundation stability due to additional support of the foundation soil, and protecting the foundation part along the perimeter from unwanted erosion of the foundation soil. Based on the theoretical studies presented in the article, a method for increasing the stability of an ice-resistant gravity platform in the form of a pile anchorage device in cement soils is presented. The physical and mechanical characteristics of cement soils are considered: brittleness measure, brittleness coefficient, compressive strength, splitting strength, strength variation coefficient. Based on the conducted research, practical recommendations are given for improving the composition of cement soils of pile foundations. A sequence of actions is proposed for installing an ice-resistant gravity platform at the point of its operation, with the parallel process of drilling production wells.

Keywords: Ice-Resistant Gravity Platform · Pile Anchorage · Cement Soils · Drilling and Mixing Method · Mechanical Activation

1 Introduction

Currently, various oil and gas projects are being implemented in our country in freezing seas. The development of oil and gas resources on the Arctic shelf requires solving complex engineering problems related to the design, construction and operation of facilities for exploration, production, processing and transportation of hydrocarbons. The particular complexity of these problems in the North is due to the combination of increased severity and high vulnerability of the region's natural conditions [1, 2].

According to Rosstat, the oil and gas complex accounts for 17–20% of the country's GDP. Therefore, research in the oil and gas sector is especially relevant, since it is aimed

S. V. Klyuev et al. (Eds.): ICRAAC 2024, LNCE 627, pp. 94–102, 2025.
https://doi.org/10.1007/978-3-031-82938-3_13

at achieving technological independence in the context of external sanctions restrictions from high-tech countries [3]. To develop promising fields, ice-resistant platforms are needed that can ensure year-round continuous drilling and operational operations. In such conditions, the most effective are gravity-type platforms, which can be manufactured and equipped with a production complex far from the construction area and then delivered to the field by water. It is known that when assessing the labor intensity of the construction of buildings and structures, the construction of foundations is no less than 15–20% of the total amount [4, 5].

In the case of gravity platforms designed for difficult natural conditions, a significant part depends on the design of their foundation, interacting with a water-saturated soil base; the cost of the foundation can reach up to 30% of the total cost of the platform.

A pile foundation can be considered a foundation that combines good performance Ensuring the stability of an ice-resistant gravity platform in northern latitudes is impossible without special measures to improve it. One of these measures is the installation of pile anchorage under the slab of the foundation part of the platform. The stability of an ice-resistant gravity platform in northern latitudes cannot be ensured without special measures to improve it. One of these measures is the installation of pile anchorage under the slab of the foundation part of the platform.

Properties and low cost [6, 7]. This type of foundation is less sensitive to dynamic changes in the state of the soil, including fluctuations in the groundwater level.

Ensuring the stability of an ice-resistant gravity platform in northern latitudes is impossible without special measures to improve it. One of these measures is the installation of pile anchorage under the slab of the foundation part of the platform. The stability of an ice-resistant gravity platform in northern latitudes cannot be ensured without special measures to improve it. One of these measures is the installation of pile anchorage under the slab of the foundation part of the platform.

When developing the continental shelf of freezing seas, the design of ice-resistant platforms of the gravity-pile type is often used.

The use of this type of combined foundations allows:

- to erect ice-resistant platforms of the gravity type on any soils;
- to improve the work of the foundation by transferring loads to stronger foundation soils;
- to reduce the size of the support block of ice-resistant platforms, ensuring a decrease in the overall material intensity of construction;
- to carry out the necessary adjustment of the foundation part design for any soil conditions and ice conditions by varying the number and design of piles;
- to speed up the process of installing the platform in the design position as much as possible [8, 9].

Given the shallow water conditions, it is possible to recommend free fastening of piles in the slab, allowing independent settlement of the slab under the action of vertical loads.

This is due to the following reasons:

- the upper soil layers are weak and of very low density, which under load will lead to significant settlements and displacements of the foundation slab;

- cyclic and repeated effects of ice and wave loads will worsen the characteristics of the foundation soil, which are difficult to predict with high accuracy;
- this method is environmentally friendly and does not harm the environment;
- driving short piles takes little time and can be carried out in parallel with drilling production wells.

One of the promising areas in the design of pile foundations is their installation in compacted soils [10, 11]. In this case, inexpensive building materials are used that maintain high operational reliability for a long time under increased loads. The aim of the research considered in this article is to develop a comprehensive method for improving the properties of cement soils for a pile foundation of a gravity platform using a drilling and mixing method with the use of mechanical activation of the soil-cement mixture in filler soils. The object of the study in the process of designing a pile foundation in filler soils is the technological system of interaction of the components of the soil-cement binder with the pile field. In this paper, it is proposed to use the method of physicochemical soil stabilization to increase the stability of an ice-resistant gravity platform using pile anchoring [12, 13]. The advantages of this method are that it imparts high strength and water resistance to soils, and creates new structural bonds; the disadvantages include the need for special equipment and the complexity of monitoring the soil stabilization process. Nevertheless, despite the indicated disadvantages, the methods of physicochemical soil stabilization are very promising. This is especially relevant when carrying out construction work in the harsh conditions of northern latitudes and the need for additional strengthening of weak soil foundations.

2 Methods and Materials

In the studies presented in this article, the probabilistic-deterministic method was chosen to create the experimental model [14]. This method considers a fractional replica $1/m^{k-1}$ from the full factorial experimental plan m^k, where m is the number of levels, and k is the number of factors. It can be represented as a polynomial with the corresponding regression equation. Let us consider the order of work on such plans. Each line of the plan characterizes certain conditions of the experiment, and the result of the study is recorded in the same line. Then a sample is made from the obtained results, after which the effects of all other factors are averaged. Each factor in the course of the study takes on the values of its levels, which allows compensating for their influence through weak and strong values. In the algebraic description of point dependencies, this method shows its advantages, since it does not impose restrictions on the type of private algebraic dependence. The adequacy of the model is assessed using the coefficient of nonlinear multiple correlation, and the verification does not require duplicating a series of experimental studies. The proposed method of planning experimental studies is called probabilistic-deterministic, since it combines elements of the probabilistic approach and deterministic description. The combined model in various conditions allows obtaining reliable output parameters and identifying internal cause-and-effect relationships.

3 Results and Discussions

Taking into account the above, in order to determine the optimal compositions of cement soils for installation of piles by the bored mixing method, studies were conducted to identify patterns of the main strength characteristics of soils. An important physical and mechanical characteristic of soils is their brittleness. The degree of brittleness is determined by studying the dependence of stress on deformation $\sigma = f(\varepsilon)$ (Fig. 1).

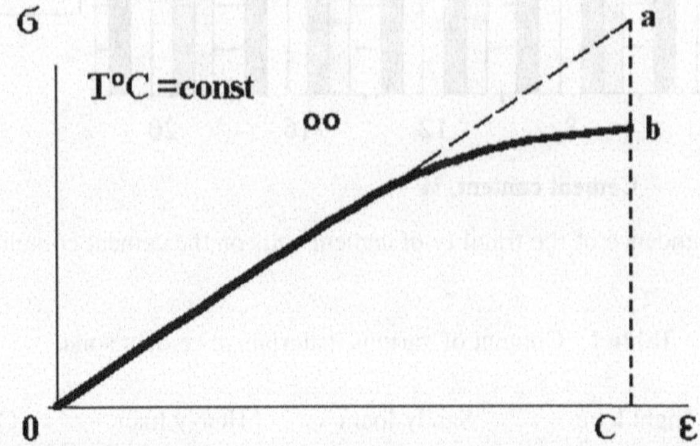

Fig. 1. Dependence of stress σ on deformation ε.

It is assumed that the curvature of the function $\sigma = f(\varepsilon)$ can serve as a fragility criterion. When developing this criterion for soils, it is necessary to take into account the nature and change of the curve $\sigma = f(\varepsilon)$ from the beginning of loading to the moment of failure of the soil sample. This condition is realized by calculating the area under the curve $\sigma = f(\varepsilon)$ and including this value in the criterion. In Fig. 1, the loading section of curve *0b* is extrapolated to straight line *0a*. The ratio of the actual area under curve *0bc* to the theoretical area under curve *0ac* is taken as a criterion or measure of soil fragility *X*. The lower the value of the fragility measure *X*, the less brittle the soil. For further research, it is advisable to consider the fragility measure *X* as the fragility coefficient *Xp*. The fragility coefficient *Xp* allows one to evaluate the structure of cement soils. Since the durability of building structures depends on the ability of soil materials to elastic deformation, the coefficient *Xp* helps to explain the physical nature of the processes occurring in cement soils made of different materials depending on the cement content (Fig. 2 and in Table 1). Based on the results of the conducted studies, presented in Fig. 2, it can be concluded that the fragility coefficient *Xp* of cement soils increases with the growth of their cement content.

This is explained by the fact that with an increased cement content in cement soils, crystalline bonds are formed, preventing the development of plastic deformations. In addition to the fragility coefficient *Xp*, the qualitative characteristics of cement soils were studied by testing soil samples for compression *Rcom* (Fig. 3 and in Table 2) and for splitting *Rspl* (Fig. 4 and in Table 3).

The results of the studies presented in Figs. 3 and 4 show that with an increase in the cement content in soils, the compressive strength (Fig. 3) and splitting strength (Fig. 4) of

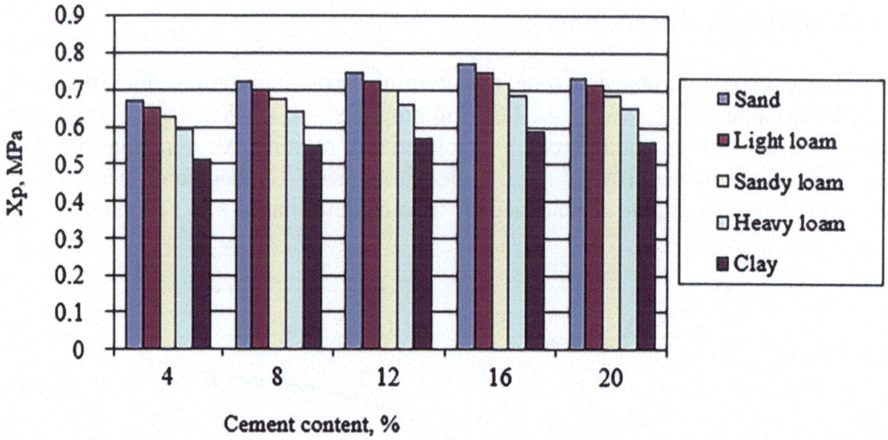

Fig. 2. Dependence of the fragility of cement soils on the cement content in them.

Table 1. Content of various materials in cement soils.

Sand	Light loam	Sandy loam	Heavy loam	Clay
0,6688938	0,648885	0,625275	0,594261	0,5109244
0,7218995	0,700305	0,674824	0,6413525	0,551412
0,7461153	0,723797	0,69746	0,6628664	0,5699089
0,7704656	0,747418	0,720223	0,6844998	0,5885085
0,7336038	0,711659	0,685765	0,6517509	0,5603522

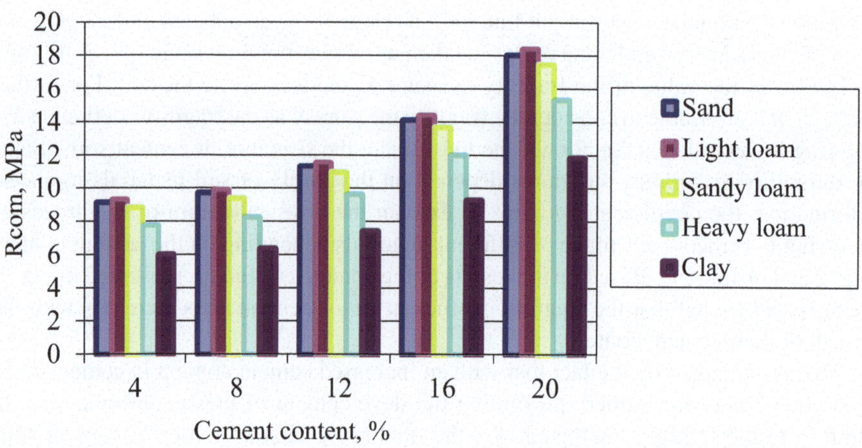

Fig. 3. Dependence of the compressive strength of cement soils on the cement content in them.

Table 2. Content of various materials in cement soils.

Sand	Light loam	Sandy loam	Heavy loam	Clay
9,1544043	9,3456857	8,888555543	7,7830138	6,0289308
9,7313411	9,9346776	9,448737794	8,2735217	6,4088913
11,356771	11,594071	11,02696453	9,6554515	7,4793712
14,14404	14,43958	13,73328926	12,025169	9,3150175
18,054956	18,432215	17,5306309	15,350204	11,890679

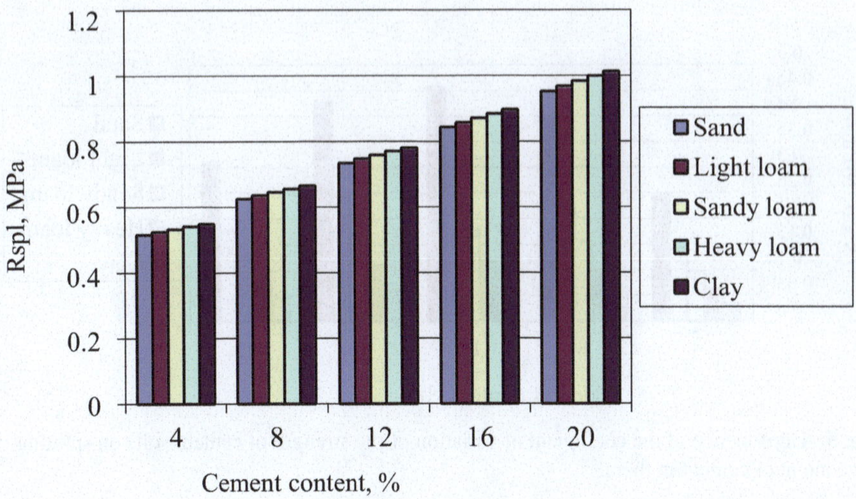

Fig. 4. Dependence of the splitting strength of cement soils on the cement content in them.

Table 3. Content of various materials in cement soils.

Sand	Light loam	Sandy loam	Heavy loam	Clay
0,5151371	0,523764799	0,5321971	0,5402876	0,54802
0,6243201	0,63477644	0,644996	0,6548012	0,6641725
0,7335031	0,74578808	0,7577949	0,7693149	0,780325
0,8426862	0,856799721	0,8705937	0,8838285	0,8964775
0,9518692	0,967811362	0,9833926	0,9983422	1,01263

cement soils increase. This is due to the fact that an increase in the proportion of cement changes the soil microstructure and increases the number of bonds formed during the fusion of individual crystals. The results of the studies presented in Figs. 3 and 4 show that with an increase in the cement content in soils, the compressive strength (Fig. 3) and splitting strength (Fig. 4) of cement soils increase. This is due to the fact that an increase in the proportion of cement changes the soil microstructure and increases the number of bonds formed during the fusion of individual crystals.

The most effective increase in the strength of cement soils is achieved due to the uniform distribution of cement throughout the entire volume of the soil, which is especially noticeable when testing samples for compression (Fig. 3). It should also be noted that with an increase in the proportion of mechanically activated cement soils, their compressive and splitting strength increases significantly due to an increase in the number of activated grains and an increase in the dispersion of the soil.

The degree of homogeneity of cement soils is characterized by the strength variation coefficient *Vvar* (Fig. 5).

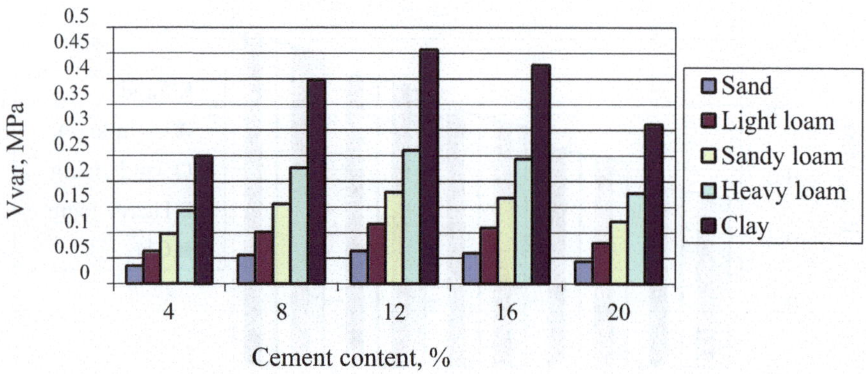

Fig. 5. Dependence of the coefficient of variation of the strength of cement soils on splitting on the content of cement in them.

With an increase in the cement content in cement soils to 12%, the strength variation coefficient first increases due to a more uniform distribution of cement, and then begins to decrease. According to the research results, the maximum strength was noted in loam-sandy soils, which are optimal for cement reinforcement. Sandy soils also showed high strength indicators. This is due to the fact that with an increase in the content of sand particles, the structure of the mixture changes. With a small amount of sand, the soil structure is coagulation, where large grains do not form mutual contacts. As the soil is saturated with large sand grains, the structure first passes into coagulation-condensation, and then into condensation-coagulation. As a result, a dense framework is formed in which sand grains are glued into a monolith by a thin layer of cement, which leads to an increase in the strength of cement soils.

Thus, we can formulate the main provisions:

1. Sandy loam is the most optimal soil for cement reinforcement.
2. With an increase in the cement content in the soil, the strength of cement soils increases until 12% cement is reached. A further increase in the proportion of cement over 12% does not lead to a significant increase in strength and is economically impractical.
3. Increasing the proportion of ground soil fraction over 30–35% does not provide a significant increase in the strength of cement soils.

4. Additives are effective when introduced in an amount of about 0.08% of the volume of cement soil.
5. Ground soil components must be added to the mixture without delay; the maximum time of their exposure to air should not exceed 3 h.

4 Conclusions

Based on the conducted research, the following sequence of actions is proposed for installing an ice-resistant gravity platform at the point of its operation:, the following sequence of actions is proposed for installing an ice-resistant gravity platform at the point of its operation:

- the gravity platform block with piles fixed in the guides is delivered to the installation site using floating craft, possibly using pontoons.
- then the block is installed on the guides, ballasted with water or sand and lowered into the design position.
- when installing piles using the bored mixing method, the weak upper soil layer of the foundation is subjected to physical and chemical consolidation with an optimized cement composition, which ensures tight contact of the sole with the surface of the foundation.
- the piles are driven to the design marks, after which drilling of production wells begins.
- after stabilization of deformations, work continues in the design mode, and the platform foundation is capable of withstanding significant ice loads (exceeding the calculated ones) due to the reserves of the bearing capacity of the piles within their permissible deformations.

References

1. Rogachko, S.I., Shunko N.V.: Scientific support of projects of Offshore hydraulic structures. Pow. Technol. Eng. **56**(1) (2022). https://doi.org/10.1007/s10749-023-01461-8
2. Kachelin, A.S.: Ensuring technological sovereignty and structural modernization in the oil and gas complex of Russia. Ener. Poli. **3**(194), 20–29 (2024)
3. James, B., Ang, Per, G. Fredriksson: Culture, legal heritage and the regulation of labor. J. Comparat. Econ. **46**(2), 616–633 (2018)
4. Hong, Z.: Executive labor market segmentation: How local market density affects incentives and performance. J. Corp. Finan. **50**, 1–21 (2018)
5. Gang-qiang. Kong, Te. Cao, Yao-hu. Hao, Yang. Zhou, Lian-wei. Ren: Thermomechanical properties of an energy micro pile – raft foundation in silty clay. Underground Space, pp. 1–9 (2019). https://doi.org/10.1016/j.undsp.2019.09.00528
6. Jiale, L., Xuefei, W., Yuan, G., Xiong (Bill), Y.: Vertical bearing capacity of the pile foundation with restriction plate via centrifuge modeling. Ocean Eng. **181**, 109–120 (2019).https://doi.org/10.1016/j.oceaneng.2019.04.026
7. Mirzoev, D.A., Arkhipova, O.L.: Development of deep-water oil and gas fields in the Arctic seas using a unified design of a mobile ice-resistant stationary platform. Neftegaz.RU 1, (2022)

8. Roberto, A.d.S., Edmundo, R.E.: Saturated anisotropic hydraulic conductivity of a compacted lateritic soil. J. Rock Mech. Geotech. Eng. **10**(5), 986–991 (2018). https://doi.org/10.1016/j.jrmge.2018.04.005
9. Zheng, L., Shaohua, X., Hailing, Y., Ren, F., Jian, S.: Influence of freeze-thaw cycles in the presence of a supplementary water supply on mechanical properties of compacted soil. Cold Reg. Sci. Technol. **157**, 42–52 (2019). https://doi.org/10.1016/j.coldregions.2018.09.009
10. Ruiming, Z., Rong, H., Lichao, L., Min, X., Like, A.: Effects of snowfall depth on soil physical–chemical properties and soil microbial biomass in moss–dominated crusts in the Gurbantunggut Desert, Northern China. Catena **169**, 175–182 (2018)
11. Qingyin, Z., Minguan, S., Xiaoxu, J., Xiaorong, W.: Changes in soil physical and chemical properties after short drought stress in semi-humid forests. Geoderma **338**, 170–177 (2019). https://doi.org/10.1016/j.geoderma.2018.11.051
12. Stephenson, C.L., Harris, C.A.: An assessment of dietary exposure to glyphosate using refined deterministic and probabilistic methods. Food Chem. Toxicol. **95**, 28–41 (2016). https://doi.org/10.1016/j.fct.2016.06.026

The Concept of Improvement of Apartment Building Survey Works with the Use of Artificial Intelligence Systems in the Arctic Zone

Azariy Lapidus, Tembot Bidov(✉), Alan Khubaev, and Andar Kagazezhev

National Research Moscow State University of Civil Engineering, 129337 Moscow, Russia
BidovTH@mgsu.ru

Abstract. The article discusses the possibilities of using artificial intelligence (AI) methods to improve the inspection process of apartment buildings (AB) in the Arctic zone. Given the specific conditions of the region, such as extreme temperatures, wind loads and high humidity, traditional inspection methods are often insufficiently effective. The research aims to integrate computer vision (CV), natural language processing (NLP) and machine learning (ML) technologies to optimise the main stages of inspection: analysis of technical documentation, visual and instrumental inspection, measurement, verification calculations and recommendation generation. As a result, key combinations of AI methods have been identified that automate routine processes, increase the accuracy of defect diagnosis, improve the quality of decision-making and reduce the time required to conduct inspections. This helps to improve the reliability and durability of buildings, reduce operational risks and cut repair costs in the Arctic zone.

Keywords: Survey Works · Arctic Zone · Artificial Intelligence Systems · Computer Vision · Natural Language Processing · Machine Learning

1 Introduction

The inspection of apartment buildings in the Arctic zone is a complex and multifaceted process that requires consideration of extreme climatic conditions and increased requirements to the condition of building structures. The cold climate, frequent temperature changes, strong winds and permafrost create significant loads on buildings and structures, which leads to accelerated deterioration of structural elements. In these conditions, traditional inspection methods often prove to be insufficiently effective, as they require significant time and material costs, and cannot always provide an accurate and complete assessment of the building's condition.

Modern artificial intelligence (AI) technologies offer new opportunities to optimise survey processes. The use of computer vision (CV), natural language processing (NLP), and machine learning (ML) methods can automate the analysis of documentation, speed up visual inspections, increase the accuracy of instrumental measurements, and improve the quality of predicting the technical condition of buildings. This is especially relevant in the Arctic zone, where field work is difficult due to harsh climatic conditions.

S. V. Klyuev et al. (Eds.): ICRAAC 2024, LNCE 627, pp. 103–115, 2025.
https://doi.org/10.1007/978-3-031-82938-3_14

The purpose of this paper is to develop a concept for improving the design and survey work for the inspection of apartment buildings in the Arctic zone using neural modelling systems. For this purpose, the main stages of the survey, including the analysis of technical documentation, visual and instrumental inspection, measurement works and verification calculations will be considered. At each stage, it is proposed to use combinations of AI methods to solve specific tasks, which will increase the efficiency and accuracy of the survey, as well as minimise the influence of the human factor.

The paper focuses on the consideration of key factors affecting the condition of buildings in Arctic conditions [1], and how AI technologies can help to solve the problems that arise. The proposed model of AI application at different stages of the survey will be presented in the form of functions reflecting the optimisation of each stage through automation and intelligent data processing.

2 Materials and Methods

This paper proposes to use artificial intelligence (AI) methods to optimise the survey process of apartment buildings (AH) in the Arctic zone. The Arctic climate is characterised by extreme conditions - low temperatures, frequent temperature changes, strong winds and the presence of permafrost [2]. These factors significantly affect the performance of buildings and structures, causing accelerated ageing of materials and the development of hidden defects, which complicates their detection using traditional inspection methods [3].

2.1 Key Survey Factors in the Arctic Zone

1. **Temperature fluctuations** - can cause cracking and deformation in structures.
2. **Low temperatures** - increase the likelihood of brittle failures, especially in steel structures and concrete.
3. **Permafrost** - causes uneven deformation of foundations and supporting structures.
4. **Effect of wind and snow loads** - causes additional stresses in structures, which requires careful analysis.
5. **Limited access and time windows for surveys** - field work can only be carried out during certain periods of the year, which limits the possibility of conducting long-term surveys.

The main problems of the traditional survey method:

- **Low responsiveness** - the survey takes a significant amount of time, especially in hard-to-reach Arctic environments.
- **Human** error - errors can occur during visual inspection or manual data processing.
- **Lack of accuracy of instrumental measurements** - in low temperatures and difficult geographical conditions, data inaccuracy can occur.
- **Difficulties in analysing large amounts of data** - project documentation, archive reports and results of multiple surveys are difficult to analyse manually.

2.2 AI Methods Used for ICD Surveying

Three key AI techniques are proposed to address the above challenges: computer vision (CV), natural language processing (NLP) and machine learning (ML). They can be applied at different stages of the survey to automate processes and improve their efficiency. Each AI technique optimises the relevant stage of the survey (Fig. 1) [4].

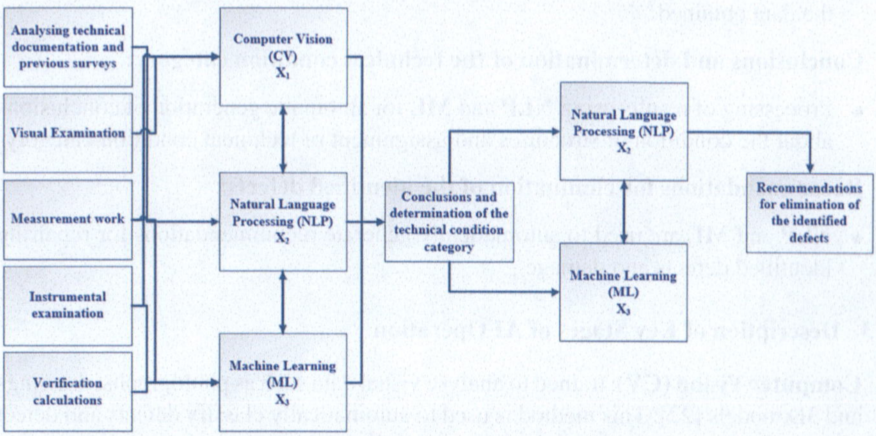

Fig. 1. Block diagram 1. AI application for ICD survey in the Arctic zone.

The presented flowchart shows the main stages of a survey of apartment buildings (AH) in the Arctic zone using different combinations of artificial intelligence (AI) methods. Each of the stages has its own input, applied AI methods [5] that optimise the process, and output results. The diagram reflects the following stages:

1. **Analysing technical documentation and previous surveys:**
   - The AI techniques used are computer vision (CV) and natural language processing (NLP).
   - **CV** analyses drawings and schematics, identifying key assemblies and potential defects.
   - **NLP** processes textual documentation, extracting important information and speeding up analysis.
2. **Visual Examination:**
   - Computer Vision (CV) is used to automatically classify and identify defects in photographs and videos.
3. **Measurement work:**
   - Used **CV** to build accurate 3D models of the building and **ML** to analyse the data and identify deviations from design parameters.

4. **Instrumental examination:**
    - The combination of **CV** and **ML** helps to analyse data from thermal imaging cameras and other tools, automatically identifying hidden defects.
5. **Verification calculations:**
    - **ML** is used to carry out strength and stability calculations of structures, based on the data obtained.
6. **Conclusions and determination of the technical condition category:**
    - Processing of results using **NLP** and **ML** for automatic generation of conclusions about the condition of structures and assignment of technical condition category.
7. **Recommendations for elimination of the identified defects:**
    - **NLP** and **ML** are used to automatically generate recommendations for repairing identified defects and damage.

2.3 Description of Key Stages of AI Operation

- **Computer Vision (CV)**: trained to analyse visual data such as photographs, drawings and 3D models [22]. This method is used to automatically classify defects and detect problems in designs.
- **Natural Language Processing (NLP)**: Is used to analyse textual data, including archival documentation and technical reports, which helps to speed up the processing and retrieval of important information [21].
- **Machine Learning (ML)**: is used to analyse complex data, perform verification calculations and predict the condition of structures, and generate recommendations to correct defects [23].

Each stage of the survey can be described as a function F(x) = y:

- F(x_CV, x_NLP) = y_1: automatic analysis of drawings and technical documentation, speeding up the analysis process and identifying key defects.
- F(x_CV) = y_2: automatic detection of defects in photos and videos, damage classification.
- F(x_CV, x_ML) = y_3: constructing 3D models and analysing outliers.
- F(x_CV, x_ML) = y_4: processing of instrumental inspection data, detection of hidden defects.
- F(x_ML) = y_5: carrying out strength and stability calculations of structures.
- F(x_NLP, x_ML) = y_6: forming inferences about the state of the structures.
- F(x_NLP, x_ML) = y_7: generation of defect recommendations.

Thus, the presented AI approach can significantly improve the accuracy and speed of ICD survey in the Arctic zone [6], minimise the influence of human factor and improve the processing of large volumes of data.

2.4 Description of Methods

- **Computer Vision (CV)**: is used to analyse visual data such as photographs, videos, schematics and drawings [20]. During the analysis phase of drawings and assemblies, CV is trained to recognise key elements and typical defects, which speeds up the inspection process.
- **Natural Language Processing (NLP)**: allows processing textual information, automating the analysis of technical documentation [19], archive reports, and generating conclusions and recommendations for defect remediation.
- **Machine Learning (ML)**: is used to analyse complex sensor data, perform verification calculations and predict the condition of structures. ML [18] is used to make recommendations to eliminate defects and predict the consequences of ignoring them (Table 1).

This methodology of AI application at each stage of ICD inspection allows to improve the accuracy of technical condition assessment of buildings in the Arctic zone, increase the speed of work performance and reduce the risk of human errors [7].

Functions of the survey stages:

1. **Analysing technical documentation and previous surveys:**
   - Input data: Technical documentation, drawings, archive data.
   - Applicable AIs:
     - x1x_1x1 (CV - Computer Vision) to analyse drawings and schematics.
     - x2x_2x2 (NLP - Natural Language Processing) to analyse text documents.
   - Function: F(x1,x2) = y1, where y1 - automatic analysis and classification of documentation, identification of key parameters and features. F(x_1, x_2) = y_1, where\ y_1\ - automatic analysis and classification of documentation, identification of key parameters and features. F(x1,x2) = y1, where y1 - automatic analysis and classification of documentation, identification of key parameters and features.
2. **Visual Examination:**
   - Input data: Photographs, video footage, visual data.
   - Applicable AIs:
     - x1x_1x1 (CV - Computer Vision) to recognise defects in images.
   - Function: F(x1) = y2, where y2 - automatic detection and classification of defects (cracks, corrosion, etc.). F(x_1) = y_2, where\ y_2\ - automatic detection and classification of defects (cracks, corrosion, etc.). F(x1) = y2, where y2 - automatic detection and classification of defects (cracks, corrosion, etc.).
3. **Measurement work:**
   - Input data: Laser measurement data, 3D scans, photographs.
   - Applicable AIs:

Table 1. Use of AI at different stages of the survey.

Survey phase	Input data for the AI	AI methods used (CV, NLP, ML)	Results of AI data processing	Function F(x) = y
1. Analysing technical documentation and previously conducted surveys	Technical documentation, archive reports, drawings, assemblies, schematics	NLP (Natural Language Processing), CV (Computer Vision)	Automated analysis of texts, drawings and diagrams, extraction of key data, faster information retrieval and classification	F(x_NLP, x_CV) = y_1: analyse documentation and drawings, identify key data
2. Visual inspection	Photos, drone and camera footage, visual data about the building	CV (Computer Vision)	Automatic defect detection (cracks, corrosion), damage classification and localisation	F(x_CV) = y_2: automatic defect recognition
3. Measurement works	3D scans, photographs, laser rangefinder data	CV (Computer Vision), ML (Machine Learning)	Creating accurate 3D models of the building, analysing geometry and deviations from design data	F(x_CV, x_ML) = y_3: building 3D models and analysing geometry
4. Instrumental examination	Data from thermal imaging, laser measurement, sonicometry and other instruments	CV (Computer Vision), ML (Machine Learning)	Detection of hidden defects (heat leaks, structural problems), automatic data processing	F(x_CV, x_ML) = y_4: analysing data from instruments, identifying hidden defects
5. Verification calculations	Load data, physical and mechanical properties of materials	ML (Machine Learning)	Verification calculations of strength and stability of structures, prediction of possible deformations	F(x_ML) = y_5: strength and stability calculations
6. Conclusions and determination of the technical condition category	Survey data, results of instrumental measurements and calculations	NLP (Natural Language Processing), ML (Machine Learning)	Automatic generation of reports, classification of structural condition (normative, limited-operational, etc.)	F(x_NLP, x_ML) = y_6: automatic pin formation
7. Recommendations for elimination of the identified defects	Final inspection data, recommendations on defect elimination	NLP (Natural Language Processing), ML (Machine Learning)	Formation of recommendations on defect elimination, prediction of possible consequences without repairs	F(x_NLP, x_ML) = y_7: automatic defect recommendations

    - x1x_1x1 (CV - Computer Vision) to create 3D models.
    - x3x_3x3 (ML - Machine Learning) to analyse geometry and outliers.

- Function: F(x1, x3) = y3, where y3 - construction of accurate 3D models and detection of deviations from design parameters. F(x_1, x_3) = y_3, where\ y_3\ - construction of accurate 3D models and detection of deviations from design parameters. F(x1,x3) = y3, where y3 - construction of accurate 3D models and detection of deviations from design parameters.

4. **Instrumental examination:**
    - Input data: Data from thermal imagers, sound meters, laser meters.
    - Applicable AIs:
        - x1x_1x1 (CV - Computer Vision) for image processing.
        - x3x_3x3 (ML - Machine Learning) to analyse data and identify hidden defects.
    - Function: F(x1,x3) = y4, where y4 is the detection of hidden defects such as heat leaks or cracks. F(x_1, x_3) = y_4, where\ y_4\ is the detection of hidden defects such as heat leaks or cracks. F(x1,x3) = y4, where y4 is the detection of hidden defects such as heat leaks or cracks.
5. **Verification calculations:**
    - Input data: Load data, physical and mechanical properties of materials.
    - Applicable AIs:
        - x3x_3x3 (ML - Machine Learning) for strength calculations and strain prediction.
    - Function: F(x3) = y5, where y5 - calculations of strength, stability and deformations of structures. F(x_3) = y_5, where\ y_5\ - calculations of strength, stability and deformations of structures. F(x3) = y5, where y5 - calculations of strength, stability and deformations of structures.
6. **Conclusions and determination of the technical condition category:**
    - Input data: Results of all previous stages of the survey.
    - Applicable AIs:
        - x2x_2x2 (NLP - Natural Language Processing) to generate a report.
        - x3x_3x3 (ML - Machine Learning) to classify the state of the structures.
    - Function: F(x2,x3) = y6, where y6 - automatic formation of conclusions and determination of technical condition category. F(x_2, x_3) = y_6, where\ y_6\ - automatic formation of conclusions and determination of technical condition category. F(x2,x3) = y6, where y6 - automatic formation of conclusions and determination of technical condition category.

7. **Recommendations for elimination of the identified defects:**
   - Input data: Survey totals.
   - Applicable AIs:
     – x2x_2x2 (NLP - Natural Language Processing) to generate recommendations.
     – x3x_3x3 (ML - Machine Learning) to predict consequences without repair.
   - Function: F(x2,x3) = y7, where y7 - automatic recommendations on elimination of defects and prediction of consequences. F(x_2, x_3) = y_7, where\ y_7\ - automatic recommendations on elimination of defects and prediction of consequences. F(x2,x3) = y7, where y7 - automatic recommendations on elimination of defects and prediction of consequences.

2.5 General Function of the Survey Process

F(x1,x2,x3) = y, where y is a comprehensive conclusion on the technical condition of the building with recommendations for defect elimination. F(x_1, x_2, x_3) = y, where\ y\ is a comprehensive conclusion on the technical condition of the building with recommendations for defect elimination. F(x1,x2,x3) = y, where y is a complex conclusion about the technical condition of the building with recommendations on defects elimination [8].

In this way, each step of the examination is optimised using a combination of AI techniques to speed up the process, improve accuracy and reduce the likelihood of errors.

3 Results and Discussions

An analysis of the application of Artificial Intelligence (AI) techniques to the survey of multi-family dwellings (MFDs) [9] in the Arctic zone identified the main stages that can be optimised using AI technologies, and identified the combinations of AI techniques that are most appropriate for each stage of the survey.

1. **Analysis of technical documentation and previous surveys**: The use of computer vision (CV) and natural language processing (NLP) has automated the process of analysing drawings, schematics, technical reports and conclusions. This significantly reduces the time required to analyse documentation and avoids human errors [10] associated with inattention or incomplete data analysis.
2. **Visual inspection**: The use of computer vision for automatic defect classification based on photographic and video data has improved the accuracy and speed of defect detection. CV is trained to recognise typical damages such as cracks [11], corrosion, deformations of structures, which greatly facilitates the task of engineers and reduces the time for visual inspection of objects.
3. **Measurement works**: Using a combination of CV and machine learning (ML) allowed to automate the construction of 3D models of buildings, which made it possible to accurately compare the current state of the object with the design parameters. This is especially important in Arctic conditions [12], where temperature fluctuations can cause deformation of structures.

4. **Instrumental survey**: The use of CV and ML to analyse data from instruments such as thermal imaging cameras and acoustic sensors has made it possible to detect hidden defects that cannot be seen visually. This helps to more accurately assess the technical condition of structures, especially in the low temperatures and harsh climatic conditions of the Arctic [13].
5. **Verification calculations**: Machine learning has proven useful in performing complex strength and stability calculations of structures [14]. Automating these processes using AI has reduced the probability of errors and significantly reduced the time spent on the calculations.
6. **Conclusions and determination of technical condition category**: Application of NLP and ML for processing of survey results allowed to automate formation of conclusions about technical condition [15] of the object and assignment of condition category of structures.
7. **Recommendations to address the identified defects**: NLP and ML were used to generate recommendations to address the identified defects, which not only identifies problems [16], but also suggests solutions based on analysing data on typical defects and methods to address them [17].

3.1 Comparative Analysis of AI-Enhanced Inspection Process

This document presents a graphical comparison of the time efficiency and error reduction achieved by integrating artificial intelligence (AI) into the inspection process of apartment buildings in the Arctic zone. The following figures illustrate the time savings and the decrease in diagnostic errors at various stages of the inspection when using AI methods compared to traditional methods.

3.2 Time Comparison for Inspection Stages

The line graph below compares the time required for each inspection stage when using traditional methods versus AI-enhanced methods. Each stage benefits from AI-driven automation, resulting in significant time savings, which is especially valuable under the challenging conditions of the Arctic zone (Fig. 2).

3.3 Error Comparison in Diagnostic Stages

The bar chart below shows the percentage of errors in three main types of diagnostics: visual inspection, data processing, and documentation analysis. The AI-enhanced method significantly reduces error rates, highlighting the potential for AI to improve accuracy in defect detection and analysis (Fig. 3).

4 Conclusions

The study showed that the introduction of artificial intelligence methods into the process of surveying apartment buildings in the Arctic zone can significantly improve the efficiency and accuracy of work, as well as reduce time and resource costs [24–32, 33]. The main conclusions that can be drawn on the basis of the analysis:

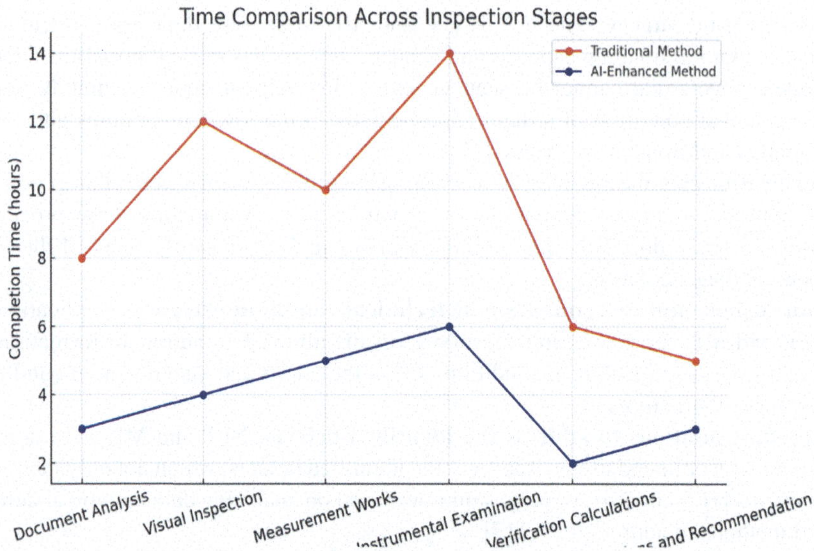

Fig. 2. Time Comparison for Inspection Stages.

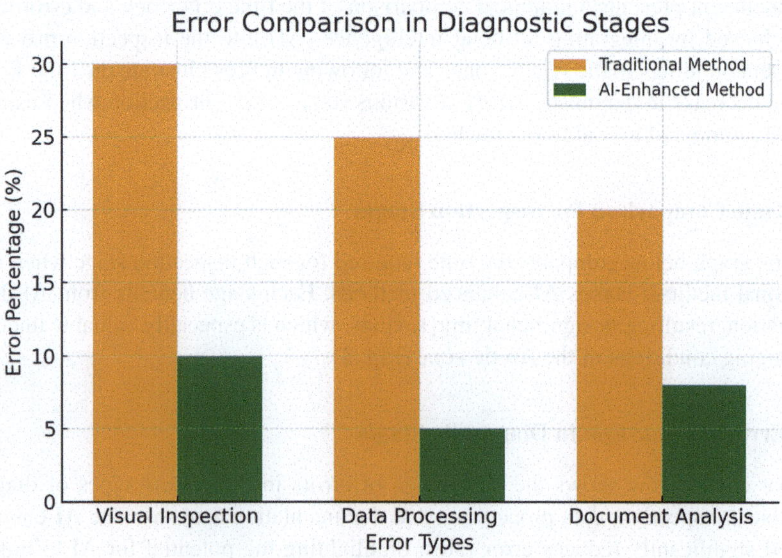

Fig. 3. Error Comparison in Diagnostic Stages.

1. **Improving survey accuracy**: The application of AI, such as computer vision and machine learning, can improve the accuracy of defect detection, eliminating human error and providing a more objective assessment of the condition of facilities.

2. **Automating routine processes**: Using AI techniques to analyse technical documentation, visual inspection and instrumental inspection reduces the workload of engineers and allows them to focus on more complex decision-making tasks.
3. **Optimising data analysis**: Natural language processing and machine learning techniques help to efficiently process large volumes of data, such as reports and survey results, which significantly speeds up decision-making and recommendation-making.
4. **Prediction and recommendation**: AI can not only analyse current data, but also predict potential problems, suggesting measures to prevent them at an early stage, which is especially important in the harsh climate of the Arctic zone.

Thus, the integration of AI into the process of ICD inspection in the Arctic is a promising direction that will improve the reliability and durability of structures, reduce repair costs and minimise the risks of building operation in extreme conditions.

References

1. Simakov, O.A.: Analysis of the factors causing the need for reinforcement of reinforced concrete structures. Constr. Reconstr. **1**(81), 76–84 (2019). https://doi.org/10.33979/2073-7416-2019-81-1-76-84
2. Linkov, N.: Calculation for the progressive collapse of the steel frame based on the shutdown of an element of the structural scheme. E3S Web Conf. **389**, 06023 (2023). https://doi.org/10.1051/e3sconf/202338906023
3. Brovko, I., Brovko, K., Dosaliyev, D., et al.: Confirmation of the operational reliability of building structures at the place of their installation in the building. In: X International Annual Conference "Industrial Technologies and Engineering" (ICITE 2023), Shymkent, Kazakhstan, 09–10 November 2023. vol. 474, 01039 (2023). https://doi.org/10.1051/e3sconf/202447401039
4. Bidov, T.Kh., Khubaev, A.O., Pomytko, E.A., Kotelnikova, A.D.: Methodology of evaluating of factors affecting the efficiency of construction project implementation by general contracting organization in Fit-out direction. Vestnik MGSU **19**(9), 1562–1569 (2024). https://doi.org/10.22227/1997-0935.2024.9.1562-156
5. Fatullaev, R.S.S., Bidov, T.Kh., Shesterikova, Ya.V., Nikolenko, D.M.: Analytical study of the actual cost of the completed overhaul of multi-apartment residential buildings in Russia. E3S Web Conf. **389**, 06029 (2023). https://doi.org/10.1051/e3sconf/202338906029
6. Bratukhina, E., Nagovitsyna, E., Tusin, D., et al.: Assessment of the development of innovative and ecological potential of agriculture. E3S Web Conf. **462**, 01038 (2023). https://doi.org/10.1051/e3sconf/202346201038
7. Bidov, T.H., Khubaev, A.O., Shabanova, A.A.: Organisational and technical modelling of the complex system of concrete works production in winter period during the construction of residential buildings. Constr. Geotech. **12**(2), 15–25 (2021). https://doi.org/10.15593/2224-9826/2021.2.02
8. Lapidus, A.A., Kuzmina, T.K., Bolshakova, P.V.: Developing administrative and process-related solutions for the benefit of residential construction projects. Constr. Mater. Prod. **7**(6), 2 (2024). https://doi.org/10.58224/2618-7183-2024-7-6-2
9. Fatullaev, R., et.al, Analysis of the labour of roof repair in different countries. AIP Conf. Proc. 050019 (2023). https://doi.org/10.1063/5.0178176
10. Lapidus, A., Bidov, T., Khubaev, A.: The study of the calibration dependences used when testing the concrete strength by nondestructive methods. MATEC Web Conf. **117**, 00094 (2017). https://doi.org/10.1051/matecconf/20171170009

11. Fatullaev, R.S., Bidov, T.: The method of mutual use of the labour intensity of finishing works of different countries. Networked Control Systems for Connected and Automated Vehicles: Conference proceedings 510–512 (2023). https://doi.org/10.1007/978-3-031-11051-1_125
12. Fatullaev, R.S., Bidov, T.: Evaluation of biotic damage to structures as a risk factor for environmental pollution during a comprehensive survey of the cultural heritage site of regional significance "The Building of the Izvestia Newspaper". In: XV International Scientific Conference "INTERAGROMASH 2022": Collection of materials of the 15th International Conference "INTERAGROMASH 2022", Rostov-on-Don, 02–04 March 2022, vol. 575–2, pp. 150–160 (2023). https://doi.org/10.1007/978-3-031-21219-2_15
13. Shesterikova, Ya.V., Bidov, T.Kh., Fatullaev, R. S., Elnikova, D.: Evaluating the Impact of a Comprehensive Technical Building Survey using a Cultural Heritage Site as an Example. Int. J. Eng. Trends Technol. **71**(6), 1–7 (2023). https://doi.org/10.14445/22315381/IJETT-V71I6P201
14. Fatullaev, R.S., Shesterikova, Ya.V., Bidov, T.Kh., Koroteeva, M.: Influence of the results of verification calculations of the building on the comprehensive examination and determination of the technical condition of the bearing structures of the building. Int. J. Eng. . and Technol. **71**(7), 134–140 (2023). https://doi.org/10.14445/22315381/IJETT-V71I6P216
15. Korobeinikova, A.E.: Design methods of perimetral residential development of slope areas of settlements of the Arctic zone of the Russian Federation taking into account wind comfort. Izvestiya vysshee obrazovaniya vysshee obrazovaniya. Construction **2**(746), 92–102 (2021). https://doi.org/10.32683/0536-1052-2021-746-2-92-102
16. Melnikov, V.P., Osipov, V.I., Brushkov, A.V., et al.: Adaptation of the arctic and subarctic infrastructure to changes in the temperature of frozen ground. Earth Cryosphere **25**(6), 3–15 (2021). https://doi.org/10.15372/KZ20210601
17. Fedosov, S.V., Lapidus, A.A., Narmaniya, B.E., Ayzatullin, M.M.: Cauchy problem for modeling of unsteady mass transfer processes in an unbounded plate by the integral Laplace transform method. Constr. Mater. Prod. **7**(5), 4 (2024). https://doi.org/10.58224/2618-7183-2024-7-5-4
18. Falaleeva, A.A., et al: Estimation of changes in the bearing capacity of frozen soils under possible changes in temperature and thawing of permafrost in the Arctic zone of the Russian Federation by the middle of the XXI century. Bull. Moscow Univ. Series 4: Geology **5**, 105–116 (2023). https://doi.org/10.55959/MSU0579-9406-4-2023-63-5-105-116
19. Zilberova, I., Bobkina, V., Sukhinin, S., Novoselova, I.: Organizational and technological aspects of major repairs in the maintenance and operation of the housing fund. E3S Web Conf. **531**, 05002 (2024). https://doi.org/10.1051/e3sconf/202453105002
20. Khotamov, A.T.: The value of the intensity of wear to ensure the durability of multi- apartment housing in the renovation of residential areas. IOP Conf. Ser. Mater. Sci. Eng. 022052 (2020). https://doi.org/10.1088/1757-899X/962/2/022052
21. Ivanova, S., Kuznetsov, A., Zverev, R., Rada, A.: Artificial intelligence methods for the construction and management of buildings. Sensors **23**(21), 8740 (2023). https://doi.org/10.3390/s23218740
22. Garyaev, A., Garyaev, N.: Integration of artificial intelligence and video surveillance technology to monitor construction equipment. E3S Web Conf. **410**, 04002 (2023). https://doi.org/10.1051/e3sconf/202341004002
23. Rosłon, J.: Materials and technology selection for construction projects supported with the use of artificial intelligence. Materials. Rosłon **15**(4), 1282 (2022). https://doi.org/10.3390/ma15041282
24. Lapidus, A., Khubaev, A., Bidov, T.: Organizational and technological solutions justifying use of non-destructive methods of control when building monolithic constructions of civil buildings and structures. MATEC Web Conf. **251**, 05014 (2018). https://doi.org/10.1051/matecconf/201825105014

25. Khubaev, A., Bidov, T., Rybakova, A.: Analysis of physical and mechanical properties of vacuum treated claydite-concrete. MATEC Web Conf. **196**, 04071 (2018). https://doi.org/10.1051/matecconf/201819604071
26. Khubaev, A., Bidov, T., Bzhienikov, A., Nesterova, V.: Modern Russian high-tech construction materials and their application in domestic construction industry (on example of metal-ceramic panels Hardwall). IOP Conf. Ser. Mater. Sci. Eng. **365**, 032005 (2018). https://doi.org/10.1088/1757-899X/365/3/032005
27. Lapidus, A., Khubaev, A., Bidov, T.: Development of a three-tier system of parameters in the formation of the organizational and technological potential of using non-destructive testing methods. E3S Web Conf. **97**, 06037 (2019). https://doi.org/10.1051/e3sconf/20199706037
28. Lapidus, A.A., Kuzmina, T.K., Bolshakova, P.V.: Improving the procedure for interaction between participants in the construction control system in order to increase labor productivity. Constr. Mater. Prod. **7**(4), 1 (2024). https://doi.org/10.58224/2618-7183-2024-7-4-1
29. Lapidus, A. et al.: Development software for the non-destructive control of monolithic structures in housing construction. E3S Web Conf. **258**, 09003 (2021). https://doi.org/10.1051/e3sconf/202125809003
30. Khubaev, A.O.: Determination of the efficiency of the production process of winter concreting based on field studies. E3S Web Conf. **389**, 06012 (2023). https://doi.org/10.1051/e3sconf/202338906012
31. Zholtikov, A., Zinovkin, A., Khubaev, A.: Improving the efficiency of applying the method of slab and floor lifting. E3S Web Conf. **389**, 01067 (2023). https://doi.org/10.1051/e3sconf/202338901067
32. Bidov, T.H., Khubaev, A.O., Shabanova, A.A.: Organizational and technical modeling of the complex system of concrete works production in winter period during the construction of residential buildings. Constr. Geotech. **12**(2), 15–25 (2021). https://doi.org/10.15593/2224-9826/2021.2.02

Theoretical Substantiation of Strength of Joints of Composite Multi-Component Structures on Steel Plate Dowels

S. V. Klyuev[1(✉)] and A. S. Toropov[2]

[1] Belgorod State Technological University Named After V.G. Shukhov, Belgorod, Russia
Klyuyev@yandex.ru

[2] Nizhny Novgorod State University of Architecture and Civil Engineering, Nizhny Novgorod, Russia

Abstract. Multi-component structures include load-bearing elements made of different materials joined together by means of bonds. One of the types of mechanical bonds used to connect individual monoconstructs into a single load-bearing structure is steel plate dowels. The method of calculation of connections on plate dowels is based on the assumption of elastic-plastic work of the material of monoconstructs on buckling in the dowel sockets and full utilization of the bearing capacity of the steel dowel to bending. The strength calculation of such connections is based on the following assumptions: the strength limit state of the connection is taken as the state when the load-bearing capacity of the connection by buckling of the dowel socket material or bending of the dowel is completely exhausted; at construction of calculation formulas conditional stress-strain diagrams of buckling of one sign at the moment of failure of joints from buckling of material on the model of rigid-plastic body are accepted; actual curvilinear diagram of material work on buckling and steel at bending is approximated by straightened diagram of deformation of ideal elastic-plastic material; the dowel axis is assumed to be straight until plastic hinges are formed in it. According to the theory of calculating dowel joints, the design scheme for plate dowel joints in all cases is a single-slotted scheme. Formulas for calculating the ultimate force are obtained: at failure of the joint from complete buckling of the material in the dowel sockets, at bending of the dowel, the general case; at failure of connection with the formation of two plastic hinges in the dowel. The obtained dependences can be used in the design of multi-component structures of reconstructed historical buildings.

Keywords: Multicomponent Composite Structure · Plate Dowel · Timber · Concrete

1 Introduction

The study of composite section elements consisting of a timber beam and a concrete slab is an urgent task. The first works on the study of joint operation of timber and concrete elements are presented in the works [1, 2].

S. V. Klyuev et al. (Eds.): ICRAAC 2024, LNCE 627, pp. 116–125, 2025.
https://doi.org/10.1007/978-3-031-82938-3_15

An overview of applied solutions to ensure joint operation of concrete slab and timber beam is given in [3–5].

The research in the field of studying the joint performance of timber and concrete elements is widely developed in the structures of multi-storeyed timber construction. In [6–9], the joint performance of CLT panels and reinforced concrete slabs is studied.

Various connection elements are used to make the wood and concrete work together, in [10] screw elements are used as connection elements and in [11] plate elements are used. Special attention is paid to the study of shear deformation at the fusion boundary in composite elements, paper [12] is devoted to the theoretical study, paper [13] is devoted to the experimental study of shear bearing capacity. The study [14] considers the design of a panelized timber floor with wood-composite beams. Wood and concrete have creep properties; work [15] is devoted to the study of the load-bearing capacity of composite sections.

The present study deals with the method of strengthening beam structures by joint operation of a timber beam and a reinforced concrete floor slab. Modernization of coatings of historical buildings with the addition of useful functions to the premises allows for a tangible economic and social result at minimal cost. Due to the increase in operational loads, it is necessary to reinforce the existing timber beams, which are quite suitable for further operation and have not required repair yet.

2 Methods and Materials

One of the types of mechanical bonds used to connect individual monoconstructs into a single load-bearing structure are steel plate dowels (Fig. 1a and 1c). The plate dowel with thickness t_H, length 1_H and width b resembles a key, but by the nature of work is a typical dowel, because, pinched in the socket, works on bending and, unlike key joints, does not give spacing. The method of calculation of connections on plate dowels is based on the assumption of elastic-plastic work of material of monoconstructives on buckling in dowel sockets and full utilization of bearing capacity of steel dowel to bending (Fig. 1b, 1d,1e, and 1f).

The strength calculation of such connections is based on the assumptions of the general theory of calculating the dowel joints of P.A. Dmitriev [16, 17], namely:

1. The strength limit state of a connection is the state when the load-bearing capacity of the connection by buckling of the dowel socket material or bending of the dowel is completely exhausted. The buckling load-bearing capacity of the material is considered to be fully utilized when the buckling stresses under the dowel reach the limit values Ra and Rc. The full utilization of the buckling bearing capacity of the dowel occurs with the formation of one or two plastic joints along the length of the dowel.
2. At construction of calculation formulas conditional stress-strain diagrams of buckling of one sign at the moment of failure of joints from buckling of material on the model of rigid-plastic body are accepted. When the joint fails from dowel bending, the ultimate buckling stresses are assumed to be uniformly distributed in the areas between the sections where plastic hinges occur and eventually the dowel fracture occurs.
3. The actual curvilinear diagram of material buckling and steel in bending is approximated by the straightened deformation diagram of an ideal elastic-plastic material (Fig. 2).

4. The dowel axis is assumed to be straight until plastic joints are formed in the dowel. At the moment of formation of plastic joints, the dowel axis is assumed to be rectilinear in the areas between the sections where the bending moments reach their limit value:

$$M_{np} = W_{n\pi} \cdot R_u = 0.25 \cdot b \cdot t_H \cdot R_u \quad (1)$$

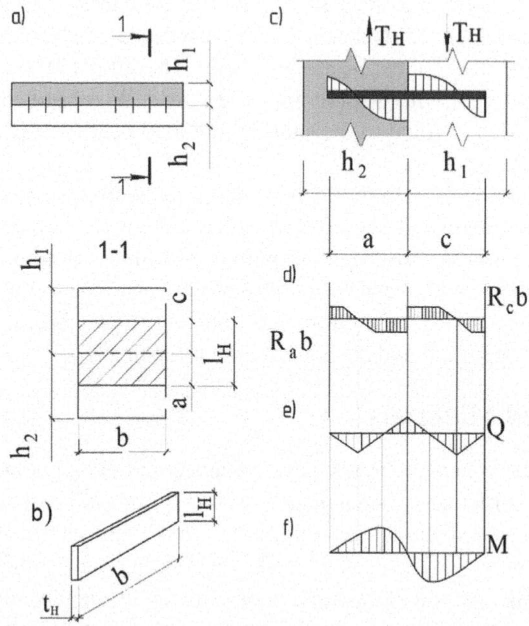

Fig. 1. Fusion of mono-structural elements on plate dowels (a), plate dowel (b), general case of single-cut element mating (c), limit load diagrams for dowel (d), transverse force diagram (e), bending moment diagram (f).

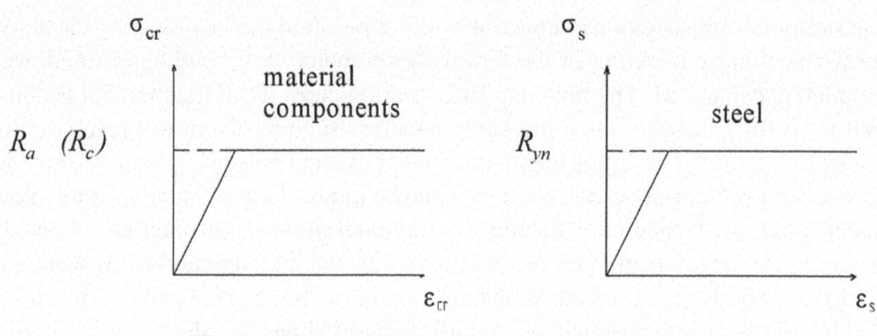

Fig. 2. Bilinear diagrams of material deformation.

3 Results and Discussions

Figure 3 shows a diagram of the failure of the connection from complete buckling of the material in the dowel sockets, taking into account the assumption of straightness of the dowel axis.

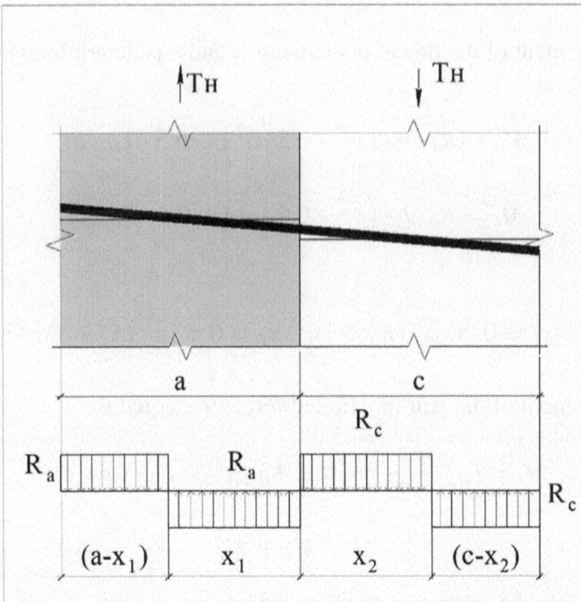

Fig. 3. Scheme of joint failure from complete buckling of the material in the dowel sockets.

According to the theory of calculating dowel joints [1], the design scheme for plate dowel joints in all cases is a single-cut scheme. Let us consider the general case of a connection with unequal values of the ultimate force for the buckling connection of materials:

$$T_a = K_a \cdot a \cdot b \cdot R_a; \tag{2}$$

$$T_c = K_c \cdot a \cdot b \cdot R_c; \tag{3}$$

there K_a, K_c - numerical coefficients characterizing the degree of “useful” use of the dowel socket.

For connections with equal strength in terms of buckling of the element material, the condition must be fulfilled:

$$T_a = T_c = T_{H,} \tag{4}$$

Consequently:

$$K_a = K_c \cdot \frac{c}{a} \cdot \frac{R_c}{R_a}; K_c = K_a \cdot \frac{a}{c} \cdot \frac{R_a}{R_c} \tag{5}$$

Taking into account the notations:

$$n = \frac{a}{c};\ m = \frac{R_a}{R_c}; \tag{6}$$

$$\text{К}_a = \text{К}_c \cdot \frac{1}{n} \cdot \frac{1}{m};\ \text{К}_c = \text{К}_a \cdot n \cdot m; \tag{7}$$

The pinching moment of the dowel in elements *a* and *c* is determined by the following formula:

$$M_a = R_a \cdot b \cdot \left(x_1^2 - 0.5 \cdot a^2\right) - 0.5 \cdot T_H \cdot a; \tag{8}$$

$$M_c = R_c \cdot b \cdot \left(x_2^2 - 0.5 \cdot c^2\right) - 0.5 \cdot T_H \cdot c; \tag{9}$$

there:

$$x_1 = 0,5 \cdot a \cdot (K_a + 1); \quad x_1 = 0,5 \cdot c \cdot (K_c + 1); \tag{10}$$

Equating the moment of the pair of external forces *Tн*, equal to

$$M = T_H \cdot \frac{a+c}{2} = 0{,}5 \cdot K_a \cdot a \cdot b \cdot R_a \cdot (a+c); \tag{11}$$

$$M = M_a + M_c \tag{12}$$

from (12) we obtain the following quadratic equation

$$K_a^2 \cdot (1+m) + 2 \cdot K_a \cdot (1 + \frac{1}{n}) - \left(1 + \frac{1}{mn^2}\right) = 0; \tag{13}$$

from where

$$K_a^2 = \frac{-(1+\frac{1}{n}) + \sqrt{(1+\frac{1}{n})^2 + (1+m)\left(1+\frac{1}{mn^2}\right)}}{1+m} = 0; \tag{14}$$

From the equation

$$K_c^2 \cdot \left(1 + \frac{1}{m}\right) + 2 \cdot K_c \cdot (1+n) - \left(1 + mn^2\right) = 0; \tag{15}$$

we get

$$K_a^2 = \frac{-(1+n) + \sqrt{(1+n)^2 + (1+\frac{1}{m})(1+mn^2)}}{1+\frac{1}{m}} = 0; \tag{16}$$

In a particular case with n = 1 and m = 1, elements of equal thickness, of equal strength materials we have coefficients Ka = Kc = 0.414.

Failure of the connection can also occur due to bending of the dowel. The maximum permissible load on the connection under the condition of dowel bending is determined by the following formulas [1]:

$$T_u = K_u \cdot b \cdot t_H \cdot \sqrt{R_u \cdot R_a}; \quad (17)$$

$$T_u = K_u \cdot b \cdot t_H \cdot \sqrt{R_u \cdot R_c}, \quad (18)$$

there R_u- the bending resistance of the dowel, assumed to be equal to the yield strength of the steel.

For connections with equal strength in terms of buckling of the material of the elements to be joined and bending of the dowels, the following condition must be fulfilled:

$$T_u = Ta = T_c.; \quad (19)$$

Whence, taking into account dependences (2) and (3):

$$K_u = K_a \cdot \frac{a}{t_H} \cdot \sqrt{\frac{R_a}{R_u}} = K_c \cdot \frac{c}{t_H} \cdot \sqrt{\frac{R_c}{R_u}}; \quad (20)$$

The minimum value of the coefficient K_u is determined from the condition of the bending strength of the plate dowel with the formation of one plate joint according to the schemes shown in Fig. 4.

$$T_a = K_a \cdot a \cdot b \cdot R_a = x_1 \cdot b \cdot Ra; \quad (21)$$

$$x_1 = K_a \cdot a; \quad (22)$$

The coefficient Ka can be determined from the equation of the sums of the moments of all forces with respect to the element fusion joint:

$$M_{\Pi\text{p}} - 0.5 \cdot x_2^2 \cdot b \cdot R_c - 0.5 \cdot x_1^2 \cdot b \cdot R_a = 0; \quad (23)$$

Taking into account the adopted notations (6), (7) and

$$x_2 = K_c \cdot c; \quad (24)$$

from (22) we obtain;

$$M_{\Pi\text{p}} = 0.5 \cdot R_a \cdot b \cdot K_a^2 \cdot a^2 + 0.5 \cdot R_a \cdot b \cdot K_a^2 \cdot a^2 \cdot m; \quad (25)$$

or

$$M_{\Pi p} = 0.5 \cdot R_a \cdot b \cdot K_a^2 \cdot a^2 \cdot (1+m); \quad (26)$$

from where

$$K_a = \frac{1}{a} \cdot \sqrt{\frac{2 \cdot M_{\text{пр}}}{R_a` \cdot b \cdot (1+m)}}; \quad (27)$$

by substituting (27) into (20) we find:

$$K_u = \frac{1}{t_H} \cdot \sqrt{\frac{2 \cdot M_{np}}{R_a` \cdot b \cdot (1+m)}}; \tag{28}$$

The theoretical value of the breaking load for the connection shown in Fig. 4, which is equal in material on buckling and bending of the plate dowel, can be determined by one of the following formulas:

$$T_a = a \cdot b \cdot R_a; \tag{29}$$

$$T_u = K_u \cdot b \cdot t_H \cdot \sqrt{R_u \cdot R_a}; \tag{30}$$

$$T_c = x_2 \cdot b \cdot R_a = a \cdot m \cdot b \cdot R_a; \tag{31}$$

The maximum value of the numerical coefficient in formulas (17) and (18) is determined from the condition of the plate dowel operation on bending at its excessive pinching in each of the connected elements, characterized by the formation of two plastic hinges on Fig. 5.

In this case, the equality is valid

$$T = K_a \cdot a \cdot b \cdot R_a = x_1 \cdot b \cdot R_a; \tag{32}$$

$$T_c = K_c \cdot c \cdot b \cdot R_c = x_2 \cdot b \cdot R_c; \tag{33}$$

$$x_1 = K_a \cdot a \quad x_2 = x_1 \cdot m = K_a \cdot a \cdot m; \tag{34}$$

$$M_{np} - 0,5 \cdot x_1^2 \cdot b \cdot R_a - 0,5 \cdot x_2^2 \cdot b \cdot R_c + M_{np} = 0; \tag{35}$$

Considering (34) from (35) we obtain

$$2M_{np} - 0,5 \cdot b \cdot R_a (K_a \cdot a)^2 - 0,5 \cdot b \cdot R_a \cdot \frac{1}{m} \cdot (K_a \cdot a \cdot m)^2 = 0; \tag{36}$$

or

$$4M_{np} = R_a \cdot b \cdot K_a^2 \cdot a^2 \cdot (1+m) = 0; \tag{37}$$

from where

$$K_a = \frac{2}{a} \cdot \sqrt{\frac{M_{np}}{R_a \cdot b \cdot (1+m)}}.; \tag{38}$$

Substituting K_a into (20) we obtain:

$$K_u = \frac{2}{t_H} \cdot \sqrt{\frac{M_{np}}{R_a \cdot b \cdot (1+m)}}, \tag{39}$$

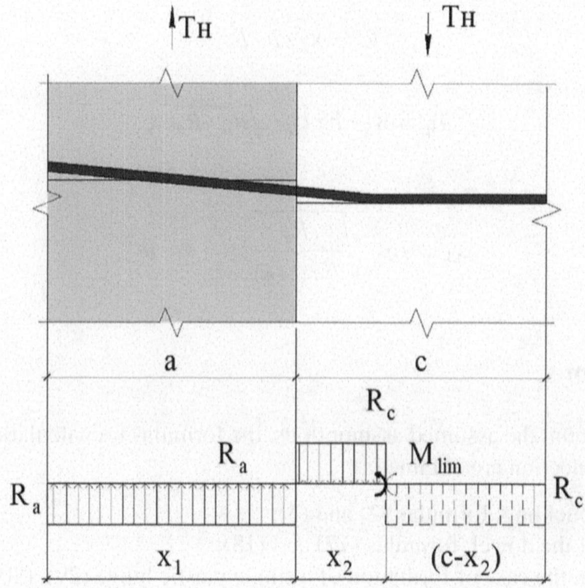

Fig. 4. The scheme of destruction of the joint with the formation of one plastic hinge in the dowel.

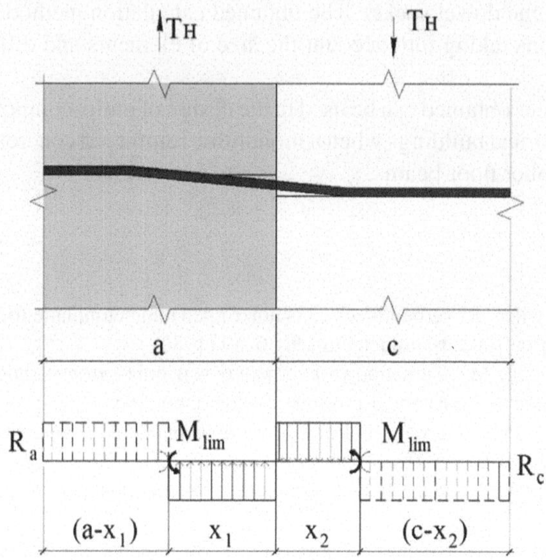

Fig. 5. The scheme of destruction of the joint with the formation of two plastic hinges in the dowel.

Then for the scheme of destruction of Fig. 5 the theoretical value of the breaking load can be determined by one of the formulas:

$$T_a = x_1 \cdot b \cdot R_a; \tag{40}$$

$$T_c = x_2 \cdot b \cdot R_c; \tag{41}$$

$$T_u = K_u \cdot b \cdot t_H \cdot \sqrt{R_u \cdot R_a}\ ; \tag{42}$$

there

$$x_1 = t_H \cdot \sqrt{\frac{R_u}{R_{a\cdot(1+m)}}};2 = x_1 \cdot m. \tag{43}$$

4 Conclusions

Taking into account the assumed assumptions, the formulas for calculating the ultimate force in the connection are obtained:

- on material buckling, formulas (2) and (3);
- at bending of the dowel, formulas (17) and (18);
- at bending, in the case of formation of a single plastic hinge (29), (30), (31);
- at bending, in the case of formation of two plastic hinges (40), (41), (42).

The formulas include numerical coefficients K_a, K_c, which characterize the degree of "useful" use of the dowel socket. The obtained calculation methodology allows calculating connections taking into account the size of elements and different strength of materials.

The dependences obtained can be used in the design of multi-component structures of reconstructed historical buildings when a monolithic reinforced concrete slab is installed on the existing timber floor beams.

References

1. Godycki, T., Pawlica, J.: Verbunddecke aus holzrippen und betonplatte forschunqserqebnisse und ausfuhrunqsbeispile. Bauinqenieur **59**, 56–59 (1984)
2. Natterrer, J., Hoeft, M.: Zum traqwerkverhalten von holz-beton-verbund- konstruktionen ecole Politehnique de Lousanne, Lousanne, Switzerland (1987)
3. Klyuev, S., et al.: Phase formation of mortar using technogenic fibrous materials. Case Stud. Constr. Mater. **16**, e01099 (2022)
4. Klyuev, S., et al.: Technogenic fiber wastes for optimizing concrete. Materials **15**(14), 5058 (2022)
5. Amran, M., Fediuk, R., Klyuev, S., Qader, D.N.: Sustainable development of basalt fiber-reinforced high-strength eco-friendly concrete with a modified composite binder. Case Stud. Constr. Mater. **17**, e01550 (2022)
6. Est´evez-Cimadevila, J., Martín-Guti´errez, E., Su´arez-Riestra, F., Otero-Chans, D., V´azquez-Rodríguez, J. A.: Timber-concrete composite structural flooring system. J. Build. Eng. **49**, 104078 (2022)
7. Martín-Gutiérrez, E., Estévez-Cimadevila, J., Suárez-Riestra, F., Otero-Chans, D.: Flexural behaviour of a new timber-concrete composite structural flooring system. Full scale testing. J. Build. Eng. **64**, 105606 (2023)

8. Forsberg, A., Farbäck, F.: Timber concrete composite floors with cross laminated timber-structural Behaviour & Design, Division of Structural Engineering, Faculty of Engineering; Report TVBK-20/5279. Lund University, Lund, Sweden (2020)
9. Kromoser, B., Holzhaider, P.: An innovative resource-efficient timber-concrete-composite ceiling system: Feasibility and environmental performance. Civ. Eng. Des. **3**, 179–191 (2021)
10. Woschitz, R., Deix, K., Huber, C., Kampitsch, T.: Entwicklung neuartiger Holz-Betonverbunddecken in Fertigteilbauweise (2021)
11. Fediuk, R., Amran, M., Klyuev, S., Klyuev, A.: Increasing the performance of a fiber-reinforced concrete for protective facilities. Fibers **9**(11), 64 (2021)
12. Huber, C., Deix, K.: Comparison of calculation methods of elastic bonding: limits of the gamma method using an example of a wood–concrete composite floor with single loads. Materials **14**(23), 7211 (2021)
13. Shephard, A.B., Fischer, E.C., Barbosa, A.R., Sinha, A.: Fundamental behavior of timber concrete-composite floors in fire. J. Struct. Eng. **147**, 2 (2021)
14. Sergeev, M., Lukina, A., Shunqi, M., Glebova, T., Kryukov, A.: Work of wood-composite beams in panel floors of prefabricated buildings. In: Vatin, N., Roshchina, S., Serdjuks, D. (eds) Proceedings of MPCPE 2021. LNCE 182, 493–499. Springer, Cham (2022)
15. Ferrier, E., Michel, L.: High performance timber concrete floor pannel connected by bonding. In: Ha-Minh, C., Pham, C.H., Vu, H.T.H., Huynh, D.V.K. (eds) Proceedings of the 7th International Conference on Geotechnics, Civil Engineering and Structures, CIGOS 2024, 4–5 April, Ho Chi Minh City, Vietnam. CIGOS 2024. LNCE **482**, 443–450 (2024)
16. Dmitriev, P.: Experimental studies of joints of elements of wooden structures on metal and plastic nails and the theory of their calculation taking into account elastic-viscous and plastic deformations. Doctor of Engineering Sciences. NISI named after Kuybyshev, Novosibirsk, RF (1975)
17. Dmitriev, P., Orlovich, R.: On strengthening wooden floors of reconstructed buildings by including beams in joint work with a reinforced concrete slab. Proceedings of Universities. Construction **1**, 6–10 (2000)

Reconstruction of Kabardinskaya Street in Nalchik

Marita Mashukova(✉), Alim Dugulubgov, Aminat Sultanova, and Zalina Unezheva

Kabardino-Balkarian State University Named After Kh.M. Berbekov, Nalchik 360004, Russia
mashukovamarita@yandex.ru

Abstract. This article presents a study of Kabardinskaya Street in Nalchik, one of the longest streets in the city, with a history dating back to the 19th century. The street, once called Vorontsovskaya in honor of Prince Mikhail Vorontsov, was central and played an important role in the life of the city. Particular attention is paid to the surviving historical buildings, such as the Pobeda Cinema, the house of merchant Zipalov P.G. and the Shuisky Hotel I.E., which are valuable monuments of architecture and cultural heritage, as well as buildings constructed in the modern period and how they fit into the architectural ensemble of that time. The work provides historical background on these objects, architectural, artistic and planning solutions, and their significance for the city. The main purpose of preserving the street. Kabardinskaya, is the preservation of the buildings of that time for the citizens and residents of the republic, as a historical and cultural heritage of the people, which must be recreated for future generations and preserve the recognizable unique image of the city. The article emphasizes the importance of research and preservation of the cultural heritage of the urban environment, pointing out the need not only to carefully preserve historical objects, but also to update their structure to create a viable urban space. The main goal is that the preservation and development of the historical urban environment requires a balanced approach, including restoration, reconstruction, improvement and landscaping in order to preserve the historical appearance of the street and at the same time adapt it to the modern needs of citizens.

Keywords: Architectural Analysis · Historical Monument · Arbat · Urban Environment · Restructuring · Reconstruction

1 Introduction

The study of modern buildings and structures, historical monuments of cultural heritage, including architectural monuments, plays a huge role in the development of the urban environment.

The harmonious city design has to become not only esthetically subject expression of spiritual and material human life, but also the latent instrument of management and coordination of life of society by means of formation of a valuable picture of the world, support of traditions, formation of new trends of style and a way of life [1].

S. V. Klyuev et al. (Eds.): ICRAAC 2024, LNCE 627, pp. 126–134, 2025.
https://doi.org/10.1007/978-3-031-82938-3_16

The image of the city is only one of translators of symbolical information on the city which forms city identity. Understanding of by the citizen and understanding difference the settlement from another, to formation of pride for the territory of accommodation are promoted as well other translators which participate in process of city identification: local myths and legends, ceremonial traditions, components of a cultural landscape (monuments, etc.), holidays and festivals, city signs and symbols. All of them serve in order that represent and keep city uniqueness which is a kernel of city identity [2].

The purpose of the city is accumulation and transformation of its own experience to create a comfortable and safe environment for people [3].

And R. Rogers, one of the classics of the humanistic approach to the study of cities, directly points out: the city is primarily a public space convenient for humans [4].

Urban (city) environment is one of the key resources of city development in social-economic and territorial aspects [5].

The modern urban environment is determined by the peculiarities of architecture, heritage of the past and the natural environment. When an architect takes on a project, he should take into account all these features, not forgetting about the architectural value of adjacent buildings and their strength. And sometimes after the new modern structures' construction, it is necessary to deal with the reconstruction and strengthening the old buildings. It is very important to build not a stand-alone facility, but an architectural ensemble that harmoniously fit into the environment [6].

Studying the history of any region and even a small settlement on the basis of existing monuments of architectural heritage spiritually enriches and significantly raises the cultural level of the modern generation and is of great importance in the spiritual and moral formation of the future. The preservation of this heritage is impossible without due respect, love and care for it. It should be emphasized that over a long period of society development, the objects of historical and cultural heritage, in particular, some of the types of this heritage—the monuments and ensembles are the most vulnerable and poorly protected in comparison with the objects from other spheres, and the measures taken to maintain them did not always lead to the desired result, and sometimes they were wrong and further aggravated their situation [7].

The strict linear layout of the planning structure, along with the enlarged residential districts, has become the most important testing ground for the accelerated construction of mass housing by industrial methods [8].

The concept of "city design". The modern theory and design methodology put forward new principles and methods for organizing the spatial environment of the city, responding to the ongoing changes in society and its scientific and technological development [9].

Among the urban reconstruction urgent tasks we should note the problem of transforming the centers of small cities that have preserved historically valuable buildings, the renewal of which should focus not only on improving the urban environment quality, but also on preserving and adapting the historical environment to new requirements [10].

The object of this study is st. Kabardinskaya in Nalchik, one of the longest in the city, its length is about 1700 m. It is located between the street. Keshokov and Prokhladnenskoe highway. A century and a half ago, the central street was called Vorontsovskaya (Fig. 1), in honor of the royal governor in the Caucasus, Prince Mikhail Vorontsov.

Before the revolution, there were merchant houses on the street, and on the site of the Pobeda cinema there was a bazaar [11].

Fig. 1. St. Vorontsovskaya.

2 Methods and Materials

The Soviet government renamed Vorontsovskaya Street to Kabardinskaya Street; these days it has not undergone major restructuring and continues to quietly preserve its history. At the end of the 20th century and the beginning of the 21st, projects for the reconstruction of the street were repeatedly developed, where it was supposed to become a pedestrian zone. So, after the restoration of the section from the street. Keshokov to st. Nogmov in 2016, the street became pedestrian and was named "Arbat" [12] (Fig. 2).

Fig. 2. "Arbat" in our time.

On Kabardinskaya Street, the vast majority of buildings are private homes built between 1910 and 1950.

Among the surviving historical buildings, we can note the Pobeda cinema, the house of the merchant P. G. Zipalov and the hotel of I. E. Shuysky. The main street of Nalchik, Vorontsovskaya (Kabardinskaya), was built up mainly with adobe houses. At the beginning of it stood a church, and at the end—a brick and tile factory of the village elder Krivchik. There were only three stone two-story houses in the village: a non-classical secondary school, the hotel of I. E. Shuysky and the house of the merchant P. G. Zipalov. If you go down from the non-classical secondary school towards Kabardinskaya Street, you will see an unremarkable building of a dance hall. In its place once stood the Church of Simeon the Stylite. It was a stone temple built in the middle of the 19th century with funds bequeathed by General Khlyupin; the donor himself is buried in one of the aisles of the temple. In 1929, by order of the KBAO leadership, the church was demolished and Khlyupin's grave was plundered. This was the period of the Soviet government's struggle with religious prejudices and the inculcation of atheism among the population.

The house of the merchant Zipalov P.G. is an old building with balconies, with figured masonry, attracting the close attention of the townspeople, the date of construction of the house is clearly visible: "1903", and next to it are the initials of the owner of the house: "P.3.", that is, "Pyotr Zipalov" (Fig. 3).

A two-story brick house with the inscription: "1913" is closely adjacent to it. This is the house of Pyotr Zipalov's brother Afanasy Grigorievich [13]. The house of the merchant Ilya Yemelyanovich Shuisky was located opposite the house of the Zipalov merchants—Pyotr and Afanasy, that is, on the corner of Vorontsovskaya (Kabardinskaya) Street and Pochtovaya (Nogmova) Street. It was one-story, unlike the two-story neighboring one, built in the middle of the 19th century. A garden and numerous outbuildings, including a stable, adjoined the house.

Fig. 3. Balcony of the house of merchant Zipalov P.G.

Alexey Ilyich married the daughter of the 1st guild merchant Grigory Zipalov—Olga. Thus, two of the largest entrepreneurs of Nalchik—Pyotr and Grigory Zipalov and Alexey Shuysky—became related. Mikhail Ilyich Shuysky lived in his own house, located on the corner of Kabardinskaya and Kommunisticheskaya streets. He was not warned about the upcoming arrest; he was captured, sent on a prisoner transfer and, apparently, died on the way, since his family did not receive any news from him. The history of the descendants of Ilya Yemelyanovich Shuysky is recreated from the words of his daughter Olga Mikhailovna, who lived to a ripe old age.

When Soviet power was established in Nalchik, the first thing it did was turn its gaze on the "bloodsuckers and oppressors", taking away their property. Alexey, Shuisky's eldest son, was warned by someone and managed to escape from Nalchik and settled for a time in Novocherkassk. But his hotel, built at the beginning of Kabardinskaya Street, Vorontsovskaya at that time, was nationalized. It was a solid two-story mansion with a large balcony with twisted openwork lattices running along the entire façade. During the occupation, Alexey Shuisky's hotel was destroyed; after the war it was restored, but not in its original form—in particular, the famous balcony from which the Bolsheviks greeted the masses during the revolutionary holidays disappeared [14] (Figs. 4 and 5).

Fig. 4. Balcony of the house of merchant Shuisky I.E.

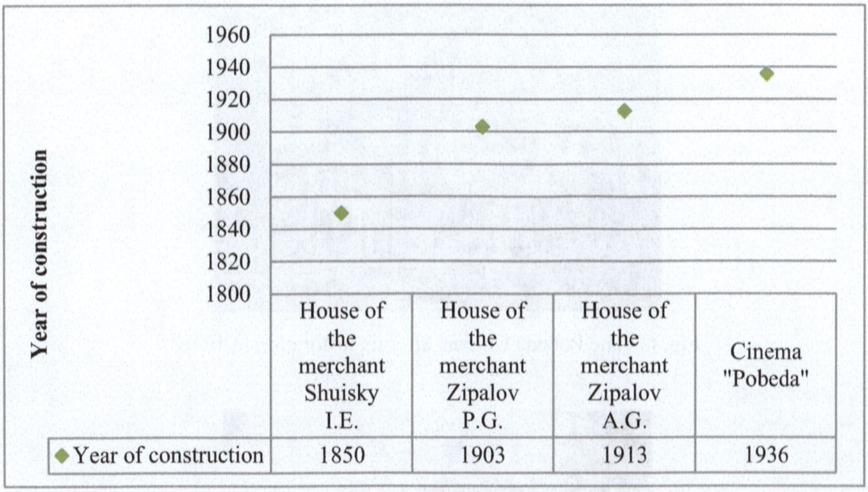

Fig. 5. Chronology of historical buildings on Vorontsovskaya (Kabardinskaya) Street.

3 Results and Discussion

Unfortunately, pre-revolutionary houses are a rarity for Nalchik, and in addition to the above-mentioned surviving buildings, the Pobeda cinema (before 1945—“Winter”) is one of the oldest cinemas in the republic, built in 1932, destroyed by the Germans in 1943 and one of the first to be restored after the war.

It was decided to build a cinema on the site of the so-called small Nalchik bazaar, for which purpose the shopping arcades were demolished and construction began in May 1936—the foundation of the building was laid. In the spring of 1938, the main work was completed (architect—Kalmykov V.P.).

There is very little reliable information about the construction of the Pobeda cinema (Fig. 6). The author of the dancing couple installed on the facade of the cinema is Leonid Grigorievich Molodzhanin (Leo Mol).

The cinema, destroyed during military operations during the occupation, was one of the first to be restored. During the war, it was hit by an air bomb, which almost completely destroyed the building, but miraculously the façade with two sculptures survived. By 1946, the cinema was restored.

In 2016, a major fire broke out in it, after which the façade with two sculptures suffered the least damage. It should also be added that by 2016 (Fig. 7), the cinema had been abandoned for decades.

For 6 years now, the cinema has been delighting city residents with its restored facade. In 2022, a proposal was accepted for consideration to carry out an internal reconstruction of the building (at the moment, only some load-bearing structures, in places ceilings, stairs and columns have been preserved there) in order to create an open lecture space there. However, the fate of “Pobeda” is still undetermined [15, 16]. In order to preserve and develop the existing urban space, it is necessary to make it truly viable, not only to carefully maintain, but also to renew its structure. This evolutionary process includes

Fig. 6. The Pobeda Cinema after its restoration in 1946.

Fig. 7. Cinema "Pobeda" before restoration.

the spheres of urban construction, restoration of buildings, major repairs, improvement and landscaping of the territory (Fig. 8).

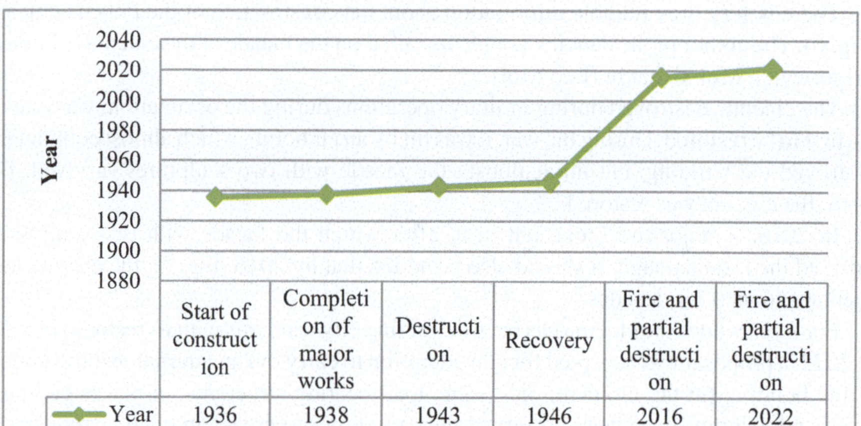

Fig. 8. Chronology of the development of the "Pobeda" cinema.

4 Conclusions

In order to preserve and develop the existing urban space, it is necessary to make it truly viable, not only carefully maintain, but also update its structure. This evolutionary process includes the areas of urban development, building restoration, major repairs, improvement and landscaping of the territory.

Kabardinskaya Street in Nalchik is not just a busy artery of the city, but a keeper of centuries-old history, capable of telling about the destinies of people and times. Each house, each building on this street is imbued with the spirit of the past, which is intertwined with modernity. Epochal architecture, preserved since the beginning of the 20th century, attracts the attention of not only tourists, but also city residents, filling their hearts with nostalgia (Table 1).

Table 1. Chronology of historical buildings on Vorontsovskaya (Kabardinskaya) Street.

Name of buildings	Year of construction
House of the merchant Shuisky I.E.	1850
House of the merchant Zipalov P.G.	1903
House of the merchant Zipalov A.G.	1913
Cinema "Pobeda"	1936

Reconstruction and creation of a pedestrian zone in recent years have contributed to the revival of interest in Kabardinskaya Street. New projects and restoration of old buildings actively contribute to the integration of cultural heritage into modern urban realities. It is important to carefully preserve history so that future generations can not only admire the architectural heritage, but also understand its significance. Thus, Kabardinskaya Street continues to be a center of attraction for all who value the symbiosis of the past and the present. It inspires to comprehend the history that is important for the cultural identity of Nalchik and the entire republic as a whole.

References

1. Hokhrin, E.V., Smolkov, S.A.: Regeneration of architectural style of the historical environment of the cultural and public downtown. IOP Conf. Ser. Mater. Sci. Eng. **880**, 012062 (2020). https://doi.org/10.1088/1757-899X/880/1/012062
2. Hokhrin, E.V., Smolkov, S.A.: City identity: mechanisms of formation of design of the urbanized territories. IOP Conf. Ser. Mater. Sci. Eng. **880**, 012061 (2020). https://doi.org/10.1088/1757-899X/880/1/012061
3. Kudryavtseva, V.A.: Modern city: areas of modernization of the urban environment. IOP Conf. Ser. Mater. Sci. Eng. **880**, 012064 (2020)
4. Polyakova, N.V., Polyakov, V.V., Zaleshin, V.E.: Public spaces as a factor of the quality of the urban environment. IOP Conf. Ser. Mater. Sci. Eng. **880**, 012070 (2020)
5. Khusnutdinova, S., Faizrakhmanova, G., Khusnutdinov, R.: Tangible and intangible assets of city environment: architecture, city lighting and art aspects. IOP Conf. Ser. Mater. Sci. Eng. **890**, 012020 (2020). https://doi.org/10.1088/1757-899X/890/1/012020

6. Yazyeva, S.B., Mayatskaya, I.A.: Eco-sustainable architecture and comfortable living environment. IOP Conf. Ser. Mater. Sci. Eng. **1083**, 012018 (2021). https://doi.org/10.1088/1757-899X/1083/1/012018
7. Subbotin, O.S.: Updating the architectural heritage of the Krasnodar Territory, in the context of its preservation and restoration. IOP Conf. Ser. Mater. Sci. Eng. **1083**, 012026 (2021). https://doi.org/10.1088/1757-899X/1083/1/012026
8. Zakirova, Y., Dembich, A., Latypova, M.: The concept of transforming Naberezhnye Chelny urban planning structure. IOP Conf. Ser. Mater. Sci. Eng. **890**, 012007 (2020). https://doi.org/10.1088/1757-899X/890/1/012007
9. Mikhailov, S., Mikhailova, A., Nadyrshine, N.: Multilinear approach to representing the historical evolution of urban design (industrial and post-industrial periods). IOP Conf. Ser. Mater. Sci. Eng. **890**, 012006 (2020). https://doi.org/10.1088/1757-899X/890/1/012006
10. Arsenteva, Y., Zaletova, E., Ismagilova, S.: Structural transformation of small cities. IOP Conf. Ser. Mater. Sci. Eng. **890**, 012011 (2020). https://doi.org/10.1088/1757-899X/890/1/012011
11. Walk along Kabardinskaya Street [Electronic resource]. https://etokavkaz.ru/gorod/progulka-po-kabardinskoi
12. Reconstruction of the historical center of Nalchik - Kabardinskaya street [Electronic resource]. http://sk-news.ru
13. Introduction to Nalchik and Kabardino-Balkaria. https://nalchik.ru.livejournal.com/38779.html?ysclid=ldirvi7iul594078565
14. Houses of the Shuisky merchants. https://viktorkotl.livejournal.com/212304.html?ysclid=ldivdp3yzv624171879
15. Nalchik cinema "POBEDA". https://viktorkotl.livejournal.com/210232.html?ysclid=ldivyy9to959451589
16. Kotlyarov, M.: Unique Nalchik: an entertaining journey with the authors and heroes of books around the capital of Kabardino-Balkaria. Publishing House of M. and V, Kotlyarov (Poligrafservis i T) (2006)

To the Calculation of Reinforced Concrete Flexible Elements with Combined Prestress, Taking into Account Complete Diagrams of Material Deformation

Zalimkhan Likhov(✉), Amerkhan Dzhankulaev, Aslan Kaziev, and Marita Mashukova

Kabardino-Balkarian State University Named After H.M. Berbekov, Nalchik 360004, Russia
zalimhan@mail.ru

Abstract. Formulas are given for determining losses in pre-compressed high-strength reinforcement, as well as the condition for closing the initial technological cracks in bent elements with pre-compressed reinforcement necessary for calculating reinforced concrete beams with pre-compressed and stretched high-strength reinforcement. Analytical descriptions of concrete and steel deformation diagrams: provided specific equations for both materials, taking into account their non-linear behavior under stress. Pre-stress effects on steel deformation: explained how pre-stressing affects the yield point and elastic limit of the steel, and how this is incorporated into the calculations. Concrete deformation under compression: discussed the non-linear behavior of concrete under compression and how to model it using a combination of linear and non-linear equations. The importance of using realistic stress plots: highlighted the need to use a curved stress plot for the compressed zone of the concrete element instead of a Here are some potential improvements: It would be helpful to explicitly state the goal of this method. What specific types of reinforced concrete elements is it intended for? Further explanation of the limitations: mention some inaccuracies associated with the method. Providing more detail about these limitations and their potential impact on the results would be valuable.

Keywords: Reinforced Concrete Beams · Combined Prestress

1 Introduction

To develop a method for calculating reinforced concrete elements with combined prestress, taking into account the complete deformation diagrams of reinforcing steel and concrete, it is necessary to have an analytical description of them.

Among the many expressions for the analytical description of concrete deformation diagrams, we have adopted the Eq. (1) and (2) proposed by M. Sargin [1–6] and recommended by ECB-FIP [3–6]

$$\sigma_B = R_B \frac{K\frac{\varepsilon_B}{\varepsilon_{BR}} - \left(\frac{\varepsilon_B}{\varepsilon_{BR}}\right)^2}{1 + (K-2)\frac{\varepsilon_B}{\varepsilon_{BR}}} \tag{1}$$

S. V. Klyuev et al. (Eds.): ICRAAC 2024, LNCE 627, pp. 135–141, 2025.
https://doi.org/10.1007/978-3-031-82938-3_17

where

$$K = E_B \varepsilon_{BR} / R_B \tag{2}$$

R_B – prismatic strength of concrete;

ε_{BR} – the corresponding deformation.

The acceptance of this equation is justified not only by its comparative simplicity and accuracy, but also by the possibility of its application to the analytical description of the deformation diagram of high-strength reinforcing steel. This is confirmed by a number of previous studies [1–6].

2 Methods and Materials

To describe the curved part of the reinforcement steel deformation diagram that does not have a physical yield point, we use the following expression (3) and (4) [1–5]

$$\sigma_s = \sigma_{el} + (\sigma_u - \sigma_{el}) \frac{K_s\left(\frac{\varepsilon_s - \varepsilon_{el}}{\varepsilon_u - \varepsilon_{el}}\right) - \left(\frac{\varepsilon_s - \varepsilon_{el}}{\varepsilon_u - \varepsilon_{el}}\right)^2}{1 + (K_s - 2)\frac{\varepsilon_s - \varepsilon_{el}}{\varepsilon_u - \varepsilon_{el}}} \tag{3}$$

where

$$K_s = \frac{E_s(\varepsilon_u - \varepsilon_{el})}{\sigma_u - \sigma_{el}} \tag{4}$$

σ_{el} and ε_{el} - apparent yield stress and the corresponding deformation;

σ_u and ε_u – the tensile strength and the corresponding deformation;

E_s – modulus of elasticity of steel.

By $\varepsilon_s \le \varepsilon_{el}$ or $\sigma_s \le \sigma_{el}$ chart $\sigma_s \le \varepsilon_s$ linear (straight 1 in Fig. 1,a), therefore, instead of the expression (3), Hooke's law is accepted $\sigma_s = E_s\varepsilon_s$.

When the armature is prestressed to stresses above the conditional yield point $\sigma_{sp} > \sigma_{0,2}$, prestress losses occur from stress relaxation σ_1 without changes in deformations (the vertical segment on Fig. 1,a) and the remaining losses occur with changes in deformations. The sum of all losses is equal to σ_{los}, and the set pre-voltages are σ_{sp2}.

As a result of prestressing and removing part of the plastic deformations, the conditional limits of elasticity from σel to σ'el and yield from $\sigma_{0.2}$ to $\sigma'_{0.2}$ (see Fig. 1,a).

In contrast to the Eq. (3) describing the initial diagram of deformation of high-strength steel, after prestressing it will take the form (5) and (6):

$$\sigma_s = \sigma'_{el} + (\sigma_u - \sigma'_{el}) K'_s \frac{\left(\frac{\varepsilon_s - \varepsilon'_{el}}{\varepsilon_u - \varepsilon'_{el}}\right) - \left(\frac{\varepsilon_s - \varepsilon'_{el}}{\varepsilon_u - \varepsilon'_{el}}\right)^2}{1 + (K'_s - 2)\left(\frac{\varepsilon_s - \varepsilon'_{el}}{\varepsilon_u - \varepsilon'_{el}}\right)} \tag{5}$$

where

$$K'_s = \frac{E_s(\varepsilon_u - \varepsilon'_{el})}{\sigma_u - \sigma'_{el}} \tag{6}$$

σ'_{el} and ε'_{el} - new values of the conditional limit of elasticity and corresponding deformation;

ε_{si} and $\varepsilon'_{el} = \sigma'_{el}/E_{sp}$ - measured from the new origin.

When $\varepsilon_s \leq \varepsilon'_{el}$ or $\sigma_s \leq \sigma'_{el}$ the dependence $\sigma_s - \varepsilon_s$ is linear (line 2 in Fig. 1,a), it obeys Hooke's law.

The pressed strain at the pre-stress σ_{sp} will be $\varepsilon_{sp,pl} + \sigma_1 / E_s$, where $\varepsilon_{sp,pl}$ – is the plastic strain at the stress $\sigma_{sp} > \sigma_{el}$, and σ_1 / E_s - is the same, with the loss of pre-stress from stress relaxation σ_1.

3 Results and Discussion

Experimental data show that it is possible to accept the following relationship between $\varepsilon_{sp,pl}$ and σ_{sp} (7):

$$\varepsilon_{sp,pl} = 0{,}25\left(\frac{\sigma_{sp}}{R_{sp}} - 0{,}8\right)^3 \tag{7}$$

Change chart σ_s – ε_s high-strength steel, in particular the increase in yield point caused by the prestressing reinforcement in norms, is taken into account only if pre-stress exceeding the apparent yield stress of steel in the delivery state [3–12].

It should be noted, however, that this phenomenon is also observed at $\sigma_{sp} < \sigma_{el}$, i.e. at a relatively low level of pre-stresses, which is a consequence of the development of creep deformations of steel with a constant total deformation, which leads to the loss of pre-stresses from stress relaxation equal to σ_1 (Fig. 1,b).

When the armature is prestressed within the limits of its elastic operation, the conditional elastic limit changes slightly – only due to residual deformations caused by prestress relaxation.

In this regard the conditional elastic limit is assumed to be the same as for non-stressed reinforcement, i.e. $\sigma_{el} = 0.8\sigma_{0,2}$.

At a higher level of prestressing norms recommend that the conditional elastic limit of high strength rod reinforcement be determined by the formula

$$\sigma'_{\mathrm{el}} = \beta\sigma_{0.2} \tag{8}$$

where

$$\beta = 0.5\frac{\sigma_{sp}}{\sigma_{0,2}} + 0.4 \geq 0.8$$

Pre-compression of the rebar of the compressed zone of the beam above the conditional elastic limit is impractical, since the stress increment in the rebar S'_p under external loading it will lead to excess of the calculated resistance by the total tension ηR_s. So, for a class of valve A1000 $\sigma_{el} = 0.8 \bullet 815 = 652$ MPa, and the increment of stress in the long-term of the current external load $R_{scp} = 500$ MPa. The total tension $R_{scp} + \sigma_{el} = 1152$ MPa, what's higher $\eta R_s = 1.1 \bullet 815 = 896.5$ MPa.

Therefore, the pre-compression of the S'_p rebar must not exceed the value $\sigma'_{sp} = \eta R_s - R_{scp}$, which is lower than the conditional elastic limit.

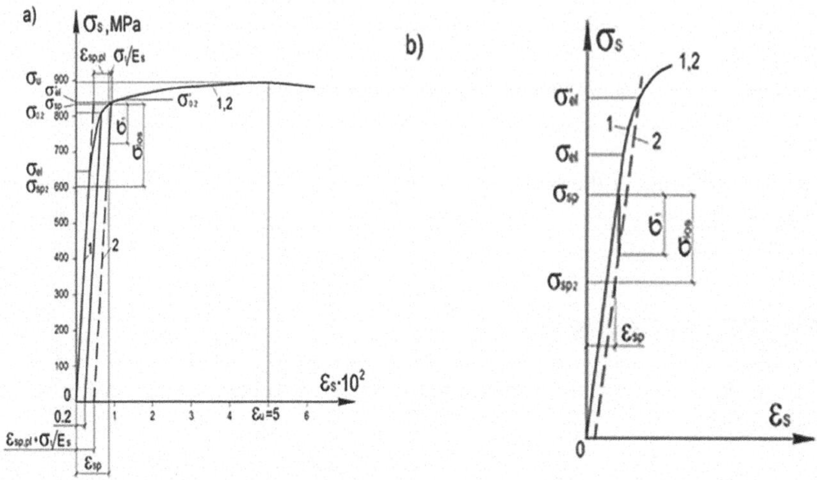

Fig. 1. Deformation Diagrams for rebar steel of class A1000: 1-initial, 2 - at a pre-stress higher than the elastic limit (a); the same, at a pre-stress not higher than the elastic limit (b).

This condition does not apply to a preliminary compression reinforcement S_p, as stress from external loads have opposite sign, so the total prestressing is equal to the difference of these tensions.

For class A1000 rebar, which is used in reinforced concrete structures proposed by the author in Chapter 2, the deformation diagrams before and after prestressing have the form shown in Fig. 1,a. The diagram is based on the design characteristics of the specified valve, equal to $R_s = 815$ MPa; $\eta = 1{,}1$; $\sigma_u = R_s\eta = 815\bullet 1{,}1 = 896{,}5$ MPa; $\sigma_{el} = 652$ MPa; $E_s = 19\bullet 10^4$ MPa; $\varepsilon_s = 652/19\bullet 10^4 = 0{,}34\bullet 10^{-2}$; $\varepsilon_u = 5\bullet 10^{-2}$.

After substituting the specified values in formulas (3) and (4), we get the following expression describing the deformation diagram of A1000 class reinforcing steel in the delivery state (9) (curve 1 in Fig. 1):

$$\sigma_s = 652 + \frac{1.907 \cdot 10^5 \Delta\varepsilon_s - 1.127 \cdot 10^5 \Delta\varepsilon_s^2 - 645.8}{133.9\Delta\varepsilon_s - 1.495} \tag{9}$$

where $\Delta\varepsilon s = \varepsilon s - \varepsilon el$.

In the developed method of calculation, instead of a rectangular stress plot in the compressed zone of the bent reinforced concrete element, a real curvilinear one is adopted, which in the case when the relative height of the compressed zone of the section ξ is close to the boundary ξR, significantly clarifies the calculation. For this purpose, the dependence curve $\sigma_B - \varepsilon_B$ obtained by central compression of concrete prisms is used. It should be noted that when you transfer the curve to the compressed area of the flexible element is allowed some inaccuracy associated with the fact that in contrast to the Central grip, when all "fibers" of the cross section given the same longitudinal strain in the compressed area of the flexible element in the "fibers" located at different distances from the neutral axis, the deformation will be different. Less loaded "fibers" will exert a restraining influence on neighboring more loaded ones. This certain inadequacy significantly less affects the accuracy of the calculation than replacing the actual curved plot of compressive stresses in the concrete of the bent element with a rectangular one. When constructing a concrete deformation diagram, we divide it into two sections. On the site 0 – 0.3Rʙ concrete practically works elastically dependence σ_B – ε_B can be assumed linear (Fig. 2) [12–22].

In this case, the beginning of the curve described by formulas (1) and (2) should be moved to a point with coordinates 0.3Rʙ – εel The analytical description of the complete dependence σ_B – ε_B will take the form (10) and (11)

$$\sigma_B = 0{,}3R_B + \frac{K_1 \frac{\varepsilon_B - \varepsilon_{B,el}}{\varepsilon_{BR} - \varepsilon_{B,el}} - \left(\frac{\varepsilon_B - \varepsilon_{B,el}}{\varepsilon_{BR} - \varepsilon_{B,el}}\right)^2}{1 + (K_1 - 2)\left(\frac{\varepsilon_B - \varepsilon_{B,el}}{\varepsilon_{BR} - \varepsilon_{B,el}}\right)} \tag{10}$$

where

$$K_1 = E_B \frac{\varepsilon_{BR} - \varepsilon_{B,el}}{0{,}7R_B} \tag{11}$$

Characteristics of ordinary heavy concrete of class B30 equal $R_B = 17$ MPa; $\varepsilon_{BR} = 2 \cdot 10^{-3}$; $E_B = 32.5 \cdot 10^3$ MPa; $\varepsilon_{BRt} = 0.2 \cdot 10^{-3}$; $R_{Bt} = 1.2$ MPa; $\sigma_{B,el} = 0.3R_B = 5.1$ MPa; $\varepsilon_{B,el} = 0.16 \cdot 10^{-3}$, dependency (1) will take the form (Fig. 2) (12):

$$\sigma_B = 5.1 + \frac{3.37 \cdot 10^4 \Delta\varepsilon_B - 3.5 \cdot 10^6 \Delta\varepsilon_B^2 - 5.3}{1.65 \cdot 10^3 \Delta\varepsilon_B + 0.737} \tag{12}$$

where $\Delta\varepsilon_B = \varepsilon_B - \varepsilon_{el}$.

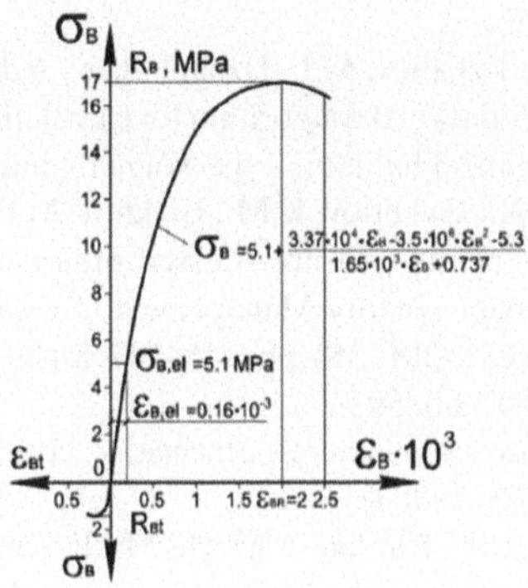

Fig. 2. Diagram of deformation of heavy concrete of class B30 under axial compression and its analytical description.

4 Conclusions

In conclusion, the development of an analytical method for calculating reinforced concrete elements under combined pre-stress has been adequately addressed through the equations and theories presented. The acceptance of M. Sargin's deformation diagrams for both concrete and high-strength reinforcing steel demonstrates a balanced approach, accounting for the complexities involved in the behavior of materials at various stress levels.

The delineation of the elastic and plastic behaviors of high-strength steel under pre-stressing loads reinforces the need for accuracy in design calculations. The outlined relationships between stresses, strains, and the resulting modifications to the elastic limits provide essential guidance for engineers in achieving optimal safety and performance in reinforced concrete structures.

Furthermore, acknowledging the nuances of prestress relaxation and the creep deformations of steel enriches our understanding of long-term durability and performance. Thus, the meticulous consideration of these factors not only enhances the structural integrity of concrete elements but also facilitates a more reliable engineering practice. Continued research and practical application of these methods will undoubtedly yield further advancements in the field, ensuring the progression of reinforced concrete technology.

References

1. Likhov, Z.R.: On the calculation of reinforced concrete bending elements with combined pre-stressing taking into account complete deformation diagrams of materials. Collection of reports of the International Conference "Construction - 2003", pp. 62–63 (RSUCE, Rostov-on-Don, 2003)
2. Analytical description of the deformation diagrams of concrete and pre-stressed high-strength reinforcement steel. https://doi.org/10.1109/ITMQIS.2017.8085933 eid: 2-s2.0–85040113757 isbn: 9781538607039
3. Likhov, Z.R.: Prestressed concrete structural beams and their design methods (Rostov-on-Don, RGSU, 2007)
4. Likhov, Z.R., Khuranov, V., Bzhahov, M.I., Dzhankulaev, A.J.: Taking into account full diagrams of deformation of materials in the algorithm for calculating bending reinforced concrete elements. Scientific and technical bulletin of the Volga region **6**, 213–218 (2014)
5. Likhov, Z.R., Khuranov, K.K., Guketlov, K.M., Bjakhov, M.I.: Analytical description of the deformation diagrams of concrete and pre-stressed high-strength reinforcement steel. In: 2017 International Conference "Quality Management, Transport and Information Security, Information Technologies" (IT&QM&IS), pp. 745–747. St. Petersburg, Russia (2017). https://doi.org/10.1109/ITMQIS.2017.8085933
6. Dzhankulaev, A.Y.: Diagram of concrete deformation. Bulletin of the KBSU. Series of Technical Sciences **5**, 101–103 (2003)
7. Dzhankulaev, A.Y., Gutov, A.A., Kardanov, M.M., Mashukov, I.K., Kablo, M.: Modeling of the concrete deformation diagram in the MATLAB system. University Scientific Collection **3**, 103–105 (2019)
8. Mailyan, D.R., Kurgin, K.V.: On the need to transform the basic analytical dependence "σb-eb" of concrete. Engineering Bulletin of the Don **4**, 1–5 (2011)
9. Mailjan, R.L., Likhov, Z.R.: Methodology for calculating reinforced concrete elements with combined prestressing taking into account full deformation diagrams of materials. Development of science and practice of reinforced concrete structures (Rostov-on-Don, SevkavNIPIagroprom RGSU, 2003)
10. Mailian, R.L., Likhov, Z.R.: Improving the efficiency of reinforced concrete rafter beams. Collection of reports of the International Conference "Concrete and Reinforced Concrete in the Third Millennium" (Rostov-on-Don, RGSU, 2002)
11. Khuranov, V.K., Likhov, Z.R., Kaziev, A.M., Tsipinov, A.S., Mailyan, V.D.: Bending resistance of reinforced concrete elements under various classes of concrete and ratios of reinforcement. In: 2016 IEEE Conference on Quality Management, Transport and Information

Security, Information Technologies (IT&MQ&IS), pp. 86–89. Nalchik, Russia (2016). https://doi.org/10.1109/ITMQIS.2016.7751930. eid: 2-s2.0-85006925455 isbn: 9781509036806
12. Dzhankulaev, A.Ia., Likhov, Z.R., Khuranov, V.Kh., Shogenov, O.M.: Effect of sigma-epsilon concrete diagram on the safety performance of reinforced concrete slabs. In: Information technology, quality management, information safety. Global conferences, pp. 305–308. Abrau-Durso (2015)
13. SP 63.13330.2018 Betonnye i zhelezobetonnye konstruktsii
14. SP 52-102-2004. Predvaritel'no napriazhennye zhelezobetonnye konstruktsii
15. Mailian, R.L., Likhov, Z.R.: Reinforced concrete beams of a rational section with a combined prefabricated structure. Sbornik dokladov International conference "Construction - 2002" (Rostov-on-Don, RGSU, 2002)
16. Likhov, Z.R., Dadova, M.Kh., Tokhaev, G.Kh, Khussein, A.Sh.: To the calculation of reinforced concrete flexible elements with combined pre-stress, taking into account the complete diagrams of material deformation. IOP Conference Series Materials Science and Engineering 913(2), 022005 (2020). https://doi.org/10.1088/1757-899X/913/2/022005 eid: 2-s2.0-85091992148 issn: 1757899X 17578981
17. Likhov, Z.R., Dzhankulaev, A.J., Tokhaev, G.Kh, Khussein, A.Sh.: Reinforced concrete beams with console ribs along their side surfaces. IOP Conference Series Materials Science and Engineering 913(2), 022004 (2020). https://doi.org/10.1088/1757-899X/913/2/022004 eid: 2-s2.0-85091963566 issn: 1757899X 17578981
18. Dzhankulayev, A.Y., Likhov, Z.R., Dzhankulayev, A.A., Kaziyev, I.A.: Modeling of nonlinear rebuff grounds by finite element methods. In: 2016 IEEE Conference on Quality Management, Transport and Information Security, Information Technologies (IT&MQ&IS), pp. 44–45. Nalchik, Russia (2016). https://doi.org/10.1109/ITMQIS.2016.7751897 eid: 2-s2.0-85006934239 isbn: 9781509036806
19. Dzhankulayev, A.Y., Likhov, Z.R., Shogenov, O.M.: The Finite Element of the Plate with the Account of the Transformation of the Cross Section and the Nonlinear Foundation. In: 2018 IEEE International Conference "Quality Management, Transport and Information Security, Information Technologies" (IT&QM&IS), pp. 432–434. St. Petersburg, Russia (2018). https://doi.org/10.1109/ITMQIS.2018.8525048 eid: 2-s2.0-85058044693 isbn: 9781538667576
20. Efimenko, E.A., Chepurnenko, A.S., Mailyan, D.R., Saibel, A.V.: The industrial buildings reinforced concrete floor slabs with rational choice of the column pitch. IOP Conf. Series: Materials Science and Engineering 896, 012004 (2020). https://doi.org/10.1088/1757-899X/896/1/012004
21. Mailyan, D.R.: Valerio Arreola Lucia Del Socorro: Effective reinforced concrete structures of monolithic frame buildings and structures. IOP Conf. Ser.: Mater. Sci. Eng. 913, 032049 (2020). https://doi.org/10.1088/1757-899X/913/3/032049
22. Muradyan, V., Mailyan, D., Lyapin, A., Chubarov, V.: Energy-efficiency increase of reinforced concrete columns with recessed working fittings. IOP Conf. Ser.: Earth Environ. Sci. 90, 012032 (2017). https://doi.org/10.1088/1755-1315/90/1/012032

1A Strategy for Managing the Safety of Work in Construction Based on BIM Solutions

S. V. Shilkina(✉)

National Research Moscow State University of Civil Engineering, Moscow, Russia
shilkinasv@mgsu.ru

Abstract. The construction industry is one of the most traumatic industries. Despite the existence of strict safety regulations for all stages of construction work, violation of safety and labor protection requirements regularly leads to tragedies on construction sites. The high rates of injury to workers are also influenced by the fact that a large proportion of construction projects are largely individual. Therefore, in almost every project, non-standard situations arise when the implementation of some operations had no analogues in previous projects and, accordingly, was not sufficiently studied for the risk of injury to personnel. In addition, there are often cases when two relatively safe operations, but carried out at the same time in the same territory, repeatedly increase the risk of injury. The introduction of modern innovative technologies allows to avoid such problems, with the help of which, at the early stages of the design of construction facilities, it is possible to assess all risks related to the safety of processes and develop practical recommendations to minimize the likelihood of situations that may cause harm to the life and health of workers. Today, one of the main tools for improving the situation with ensuring industrial safety throughout the entire process of building construction and operation is building information modeling (BIM – Building Information Model). The article analyzes the current situation with occupational injuries in the world and in Russia, highlights the main causes of occupational injuries. To improve the situation, the main tasks have been identified, the solution of which will positively affect the situation with occupational injuries, and the need for the introduction of BIM technologies in the field of occupational safety and health to solve the tasks set has been justified. Recommendations for improving construction safety and reducing injuries based on BIM modeling at various stages of the project are proposed, which will be useful for both designers and construction specialists.

Keywords: Safety · Labor protection in construction · Injury reduction · Information modeling technologies · Safety rules · Risk assessment · Hazard prevention

1 Introduction

The problems of occupational health and safety of work on construction sites are quite acute, are relevant and urgent. Since in modern conditions of development of the construction industry, work safety management is becoming the most important key factor in

S. V. Klyuev et al. (Eds.): ICRAAC 2024, LNCE 627, pp. 142–153, 2025.
https://doi.org/10.1007/978-3-031-82938-3_18

creating jobs where there will be no accidents [1–3]. Of course, this situation, which goes beyond simple altruism, is a reaction to significant human, material and financial losses associated with high mortality in the workplace. An analysis of the current situation with occupational injuries, conducted according to the UN, showed that 6.5 thousand people die daily from occupational diseases and more than 1 thousand people from injuries sustained at work, which is more than 2.6 million people per year. It is significant to compare the fact that workplace mortality exceeds the total average annual mortality as a result of: Road accidents (999 thousand people), wars (502 thousand people), crime (563 thousand people) and HIV (312 thousand people). But there are still more than 370 million. People are injured without fatal outcome, the disability of which reduces global GDP from 4% to 6%. Mortality and morbidity in production are extremely unevenly distributed across regions and industries. About two thirds are in Asia and 11% -12% each in Africa, Europe and the Americas. Among the industries in the top three of the anti-rating are industrial production, construction and logistics (transportation and warehousing). Of course, there are many factors "inside" industries that directly or indirectly increase the level of occupational injuries. The most significant among them is the low level of application of industrial safety measures in the work of temporary, seasonal workers and, especially, in attracting informally employed migrant workers.

Despite the fact that the relative rate of occupational injuries (the number of industrial accidents per 100,000 employees) in Europe is several times lower than in Asia or Africa, the total number of such incidents, which led to at least four calendar days of absence from work according to data for 2021 in the EU, was about Three million. At the same time, 3,347 fatal accidents occurred in the same period. Despite all the efforts made by the EU, this indicator has been stable for many years (Fig. 1) [4].

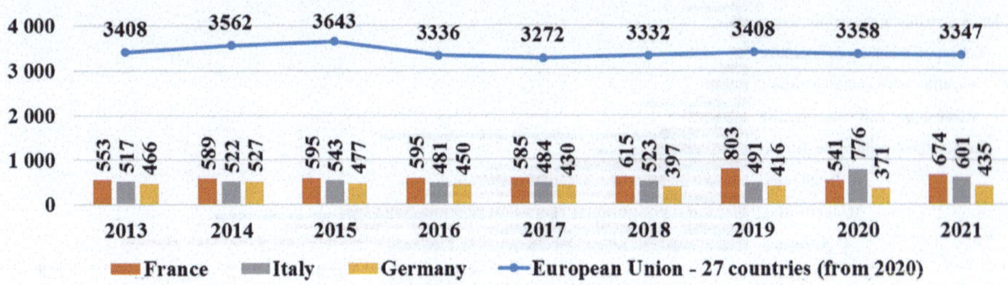

Fig. 1. The dynamics of deaths from industrial accidents in the EU as a whole and in the three countries with the highest absolute mortality rates, people per year (built by the author according to Eurostat data).

2 Methods and Materials

France is consistently leading in terms of the total number of fatal accidents, and in terms of relative indicators, calculated as the number of fatal accidents at work per 100,000 workers in 2021, the Baltic countries are in the first place (Fig. 2) [4]. It is noteworthy that within the EU, the relative mortality rates at work differ significantly by country.

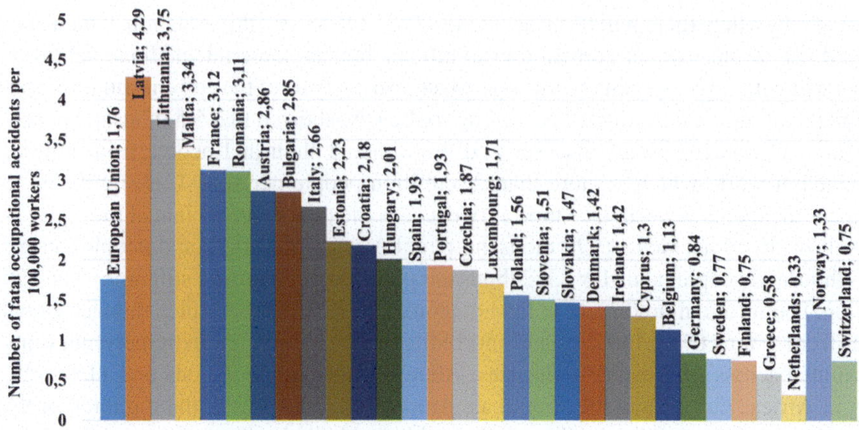

Fig. 2. The number of fatal industrial accidents per 100,000 workers in the EU in 2021, (built by the author according to Eurostat data).

When analyzing statistical data on the death rate from accidents by industry in the EU, the same three above-mentioned "leaders" are determined: the construction sector, logistics and industrial production. At the same time, it should be noted that the construction industry accounts for between one quarter and one fifth of the total number of fatal accidents in the EU in 2021 (Fig. 3) [4].

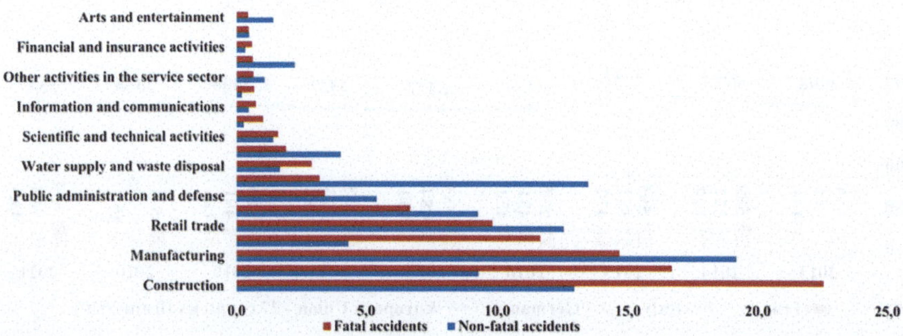

Fig. 3. Fatal and non-fatal accidents by industry in the EU in 2021, as % of the total (built by the author according to Eurostat data).

Analysis of the situation with occupational injuries in the Russian Federation. In Russia, information about industrial accidents is received, analyzed and published by several government agencies, in particular the Federal State Statistics Service (Rosstat), the Social Fund of Russia (SFR) and the Federal Service for Labor and Employment (Rostrud). Each department has its own assessment methodology, so the registration and processing of information is different, the final results that they publish often differ, although not significantly. In Russia, according to Rosstat, there is a large proportion of harmful and dangerous industries (more than 36%), as well as heavy work (more than

20%). This naturally carries an increased risk of serious and fatal industrial accidents (Fig. 4) [5].

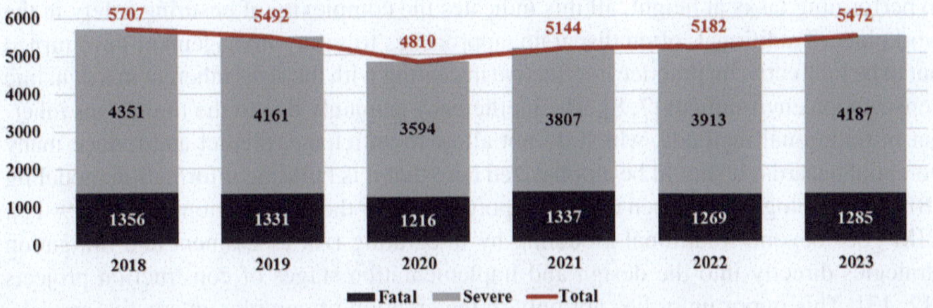

Fig. 4. The dynamics of industrial accidents with severe and fatal outcomes, people/100,000 workers (built by the author according to Rostrud and FNP).

According to the territorial distribution, Moscow is the anti–leader in industrial deaths - 12.65% of the total number in the country of the Russian Federation, which is more than the total of the three regions following Moscow (Sverdlovsk Region – 4.15%, Krasnodar Territory – 3.64%, St. Petersburg – 3.19%).

An analysis of Rosstat data for 2021 showed that, by industry, the most workers died in construction - 339, in manufacturing – 306, in transportation and storage – 232, in mining – 177. The graphs of the ratio of the number of injured and killed in accidents to the total average number of employees by industry in 2022 are indicative (Fig. 5) [6].

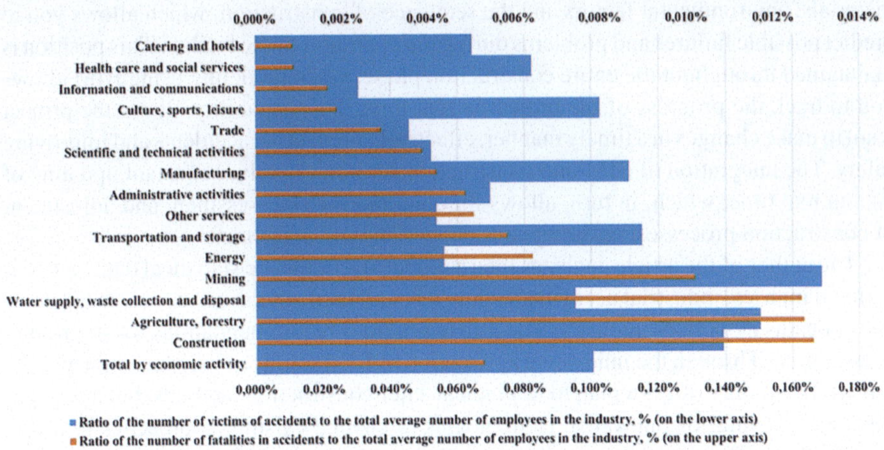

Fig. 5. The ratio of the number of injured (on the lower axis) and dead (on the upper axis) in accidents to the total average number of employees in the industry in 2022, % (built by the author according to Rosstat data).

Thus, the analysis of statistical information clearly shows that the problem of occupational safety exists and requires urgent solution. The construction sector ranks first in terms of the number of deaths compared to other industries, which underscores the urgent need for comprehensive safety measures [2, 7]. It should be noted that the numerous risks inherent in construction projects [8], ranging from working with heavy machinery to performing tasks at height, all this indicates the complexity of ensuring safety in the workplace. Traditional, often disparate, approaches to safety management have turned out to be ineffective in practice, ineffective in dealing with the risks inherent in a dynamic construction environment [7, 8]. The inefficiency is mainly due to the limitations inherent in traditional methods, which do not allow to anticipate, predict and reduce many potential hazards. It should be emphasized here that it is building information modeling (BIM) technologies that open up new opportunities for the construction industry [9–11]. BIM goes beyond traditional modeling by integrating risk assessment and mitigation strategies directly into the design and implementation stages of construction projects [12, 13]. This opportunity lies not only in the creation of complex interactive models, but also in a significant transition to security management at a new qualitative level. BIM technologies that take into account security issues at each stage of the project lifecycle ensure the identification and elimination of risks in advance, increasing the overall level of security [14, 15]. By integrating risk assessment and mitigation strategies directly into the design and implementation phases, BIM provides an integrated approach to safety, with particular attention to prevention and timely intervention [16–18].

This article proposes a new BIM solution-based security management strategy that expands the capabilities of BIM models beyond simple design and planning. BIM models serve as dynamic platforms for integrated risk management. At the design stage, models are systematically supplemented with parameters that describe potential risks in detail, turning BIM into a tool that promotes early identification and mitigation of risks before construction begins. The addition includes the integration of data on the properties of materials, environmental factors and the sequence of construction, which allows you to predict possible failures and problems during the operation of the facility. This position is maintained throughout the entire construction phase, when modeling using BIM allows you to track the progress of the project in real time. This capability allows the project team to make changes in a timely manner, effectively preventing accidents and improving safety. The integration of BIM into construction activities involves constant updating of data in real time, which, in turn, allows for continuous risk assessment and adjustment of construction processes, ensuring continuous safety management.

The author of the article analyzes international and domestic statistical data on occupational injuries, industrial accidents with severe and fatal outcomes, as well as the number of deaths at work by industry. The article provides recommendations for improving the situation. Through the innovative application of BIM solutions to reduce injury rates, this study not only offers a plan to implement safer construction methods, but also identifies the potential of innovative technologies to change existing approaches to safety management. The research contributes to the development of knowledge bases, contributes to the growth of joint efforts of specialists in improving the safety standards of the construction industry.

3 Results and Discussions

To develop recommendations for reducing the number of accidents at work, we will analyze the type (Fig. 6) [5] and the causes of occupational injuries (Fig. 7) [5].

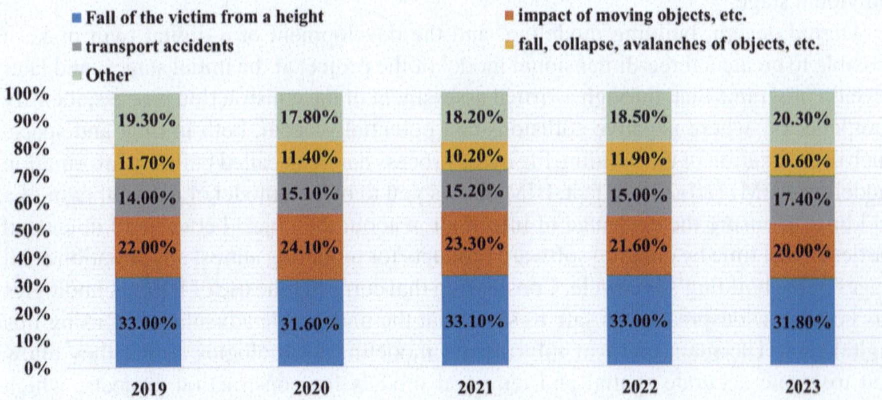

Fig. 6. Types of accidents with severe consequences, in% of the total number (built by the author according to Rostrud data).

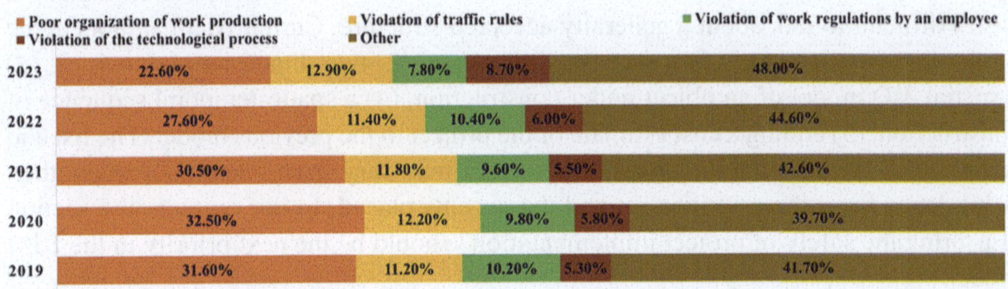

Fig. 7. Causes of accidents with severe consequences, in% of the total number (built by the author according to Rostrud data).

An analysis of the structure of the causes of accidents allows us to draw conclusions about the need for more thorough development of safety measures at the earliest stages of the project. This is especially true in construction projects, as the most traumatic. After all, most negative situations can be avoided if, at the design stage, all possible personnel movements, their potential intersections with the places of implementation of dangerous operations are taken into account, all temporary transport corridors are provided with the necessary margin of safety in case of emergency situations, and much more. Of course, it is impossible in advance, at the design stage, to take into account all the possible consequences of each construction operation when implementing large projects, where many contractors usually work, sometimes they do not know what the "neighbors" are doing. Therefore, the only real possibility is to use computer—aided design methods to calculate in advance, if not all, then most dangerous situations. For several decades now, this area has been developing rapidly, which has led to the appearance of many software products on the software market [19]. It became possible at the initial stages to analyze all operations for their intersection with each other in time and space in order to avoid negative consequences. This has a positive effect on both the cost and time of the project, and helps with choosing the safest and most effective ways to perform each individual stage.

Digital design, building modeling, and the development of a digital twin make it possible to create a three-dimensional model of the project at the initial stages, and later develop programs that, through a virtual assessment of the construction process, identify "bottlenecks" where negative collisions can potentially occur, both in time and space. Such digitalization of the building life cycle process has been called building information modeling (BIM) [20–22]. In fact, BIM allows you to build a model of a digital twin of a building, to ensure the exchange of information about the object between all interested parties in real time by creating software products for modeling almost any situation at all stages of the building's life cycle. Considering that currently the use of BIM technologies has become widespread, it is safe to say about the undeniable advantages of using this digital tool. The main merit of information modeling technologies is that they allow you to create accurate spatial and temporal models for construction projects, which simplify the unhindered exchange of information and integration of processes between stakeholders [20, 21, 23].

The development of BIM modeling is described in detail in the literature [14, 16, 18, 21, 23, 24]. The number of sections of the BIM model is constantly increasing and it is still difficult to talk about a generally accepted structure. Currently, all authors agree on the first five sections: (1) a structured database, (2) two-dimensional drawings, (3) a spatial 3-D model of an object under construction, (4) a spatio-temporal sequence of construction, (5) adding a cost estimate of the project to the previous model. The list and sequence of further sections of the BIM model may vary from one author to another. A lot depends on the priorities set and the accents placed. In our opinion, the section "Ensuring the safety of project implementation" should be the next priority in the BIM model.

After the analysis, the analysis of data on the causes and types of accidents in the implementation of projects in the construction industry, it becomes necessary to develop recommendations on the directions and tasks of using BIM technologies in order to improve safety in construction to reduce the level of injuries, especially with severe and fatal consequences. The developed recommendations are proposed to be structured as follows (Fig. 8).

The introduction of algorithms for solving the above tasks in the design of large construction projects will certainly lead to a significant reduction in the level of injuries.

The moral and ethical component of the need to introduce advanced methods to reduce the level of injuries at work is beyond doubt, because saving people's lives and health should be a priority, both at the state level and for managers and executors of each construction project. But if concrete steps in this direction are increasingly visible on the part of the state, then efforts on the part of "business" are clearly not enough.

After all, despite dozens of adopted legislative and regulatory legal acts containing state regulatory requirements for labor protection, until then, owners and managers of

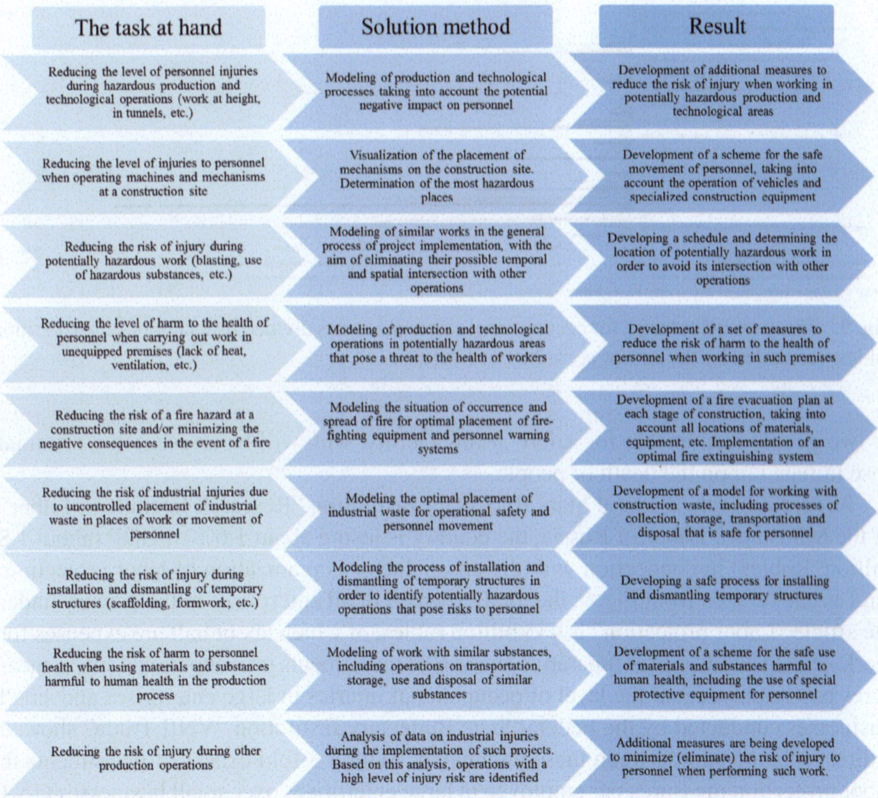

Fig. 8. The algorithm of using BIM technologies to increase the level of safety in construction (developed by the author).

enterprises do not realize the need to constantly protect the life and health of their employees, significant improvements in this area cannot be expected. Realizing this, enterprises spend quite a lot of money on labor protection. However, if we analyze the structure of these expenses more closely, for example, based on Rosstat data for 2022, we can see that about 44% of all expenses are the purchase of workwear, safety shoes and other personal protective equipment. And the implementation of organizational and technical and technological measures, which should bring the greatest effect, accounts for only 16–18% of total expenses. At the same time, these figures are even lower in the construction industry – 13.6% and 7.4%, respectively (Fig. 9) [6].

In a specially conducted study published in December 2023, scientists from the Central Economic and Mathematical Institute of the Russian Academy of Sciences (CEMI RAS) calculated losses for the Russian economy from premature deaths [25, 26]. Their calculations showed that the economic damage to the country from the premature death of an employee is 226–513 thousand rubles for each year that is not lived. And the losses from surviving but disabled workers are even higher due to the cost of their social security. As an indirect cost, you can add the retirement from the production process of one

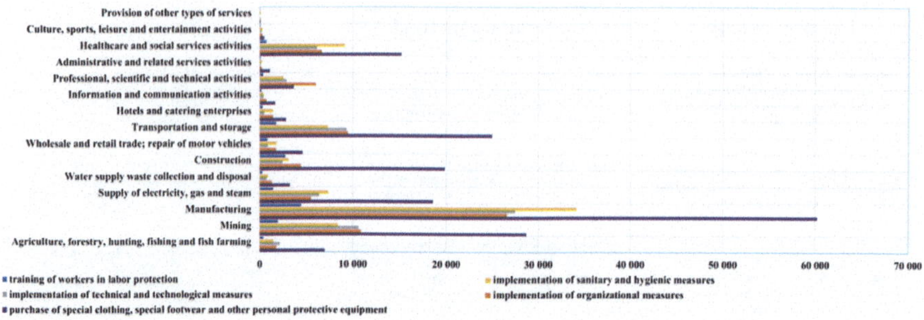

Fig. 9. Expenses by industry for occupational safety measures divided by categories in 2022, mln rubles. (compiled by the author according to Rosstat).

or two people responsible for safety at this facility, who will be arrested in a criminal case initiated upon the death of people.

According to the Federal State Budgetary Institution "Research Institute of Labor" of the Ministry of Labor of Russia, the country is losing about 1.6% of GDP (about 1.8 trillion. Rubles) in connection with the unsatisfactory organization of labor protection. And as can be seen from Fig. 9, the total expenses of all enterprises in the country under the article "Labor protection" is 435 billion rubles, of which about half are expenses for workwear. In addition, a comparative analysis of investments in occupational safety measures per employee and the level of occupational injuries in large enterprises and small businesses conducted by the Federal State Budgetary Institution "VNII Truda" showed a direct relationship between these parameters. With a 3-fold excess of investments in labor protection measures per employee in large enterprises over small businesses (15.1 and 4.8 thousand rubles, respectively), the number of deaths per thousand employees in the whole country in large businesses is 1.5 times lower than in small businesses (0.053 and 0.081, respectively). The level of fatal occupational injuries in small businesses is also much higher than that of large and medium-sized enterprises (from 1.2 times to 6.6 times, depending on the type of activity).

It turns out that the losses from the poor state of working conditions and labor protection in Russia are many times greater than investments in labor protection. The cost of attracting a team of BIM engineers focused on creating a section "Organization of safe working conditions" in a large construction project is estimated at 7–10 million rubles in the first year, including the organization of two workplaces. This includes the purchase of high–performance computers with high network access bandwidth, the organization of cloud storage, an annual subscription to one of the most popular software products - Autodesk BIM Collaborate Pro, as well as the salaries of two developers.

Based on the above data, the author of this article has built a financial and economic model, as a result of calculations it was obtained that the payback of such investments, capable of preventing at least one accident during the construction of an object, is no more than one year. This is more than an acceptable result.

The main problem is that due to the lack of a strict regulatory framework for the implementation of BIM technologies in the development of safety sections in construction projects, many managers do not want to incur additional costs, saving on such obviously necessary resources. But it must be borne in mind that in 2022, according to Rosstat, out of 10750 organizations providing services in construction, registered accidents occurred at 869 enterprises and, according to experts, at least as many incidents were hidden.

In recent years, the Government of the country has taken a number of significant steps to develop BIM modeling in the implementation of large state construction projects.

4 Conclusions

As can be seen from the above, the government of the country is aware of the need to introduce BIM technologies in the development of construction projects. But now it is necessary at the legislative level to oblige designers to use information technologies that reduce the risk of occupational injuries.

The research presented in the article clearly proves that underestimating the need for investments in measures to improve working conditions and labor protection leads not only to economic losses, but also to a deterioration in the demographic situation in the country. At the same time, given the rapid development of BIM technologies and the emergence of a large number of specialists who are able to adapt the visualization of a construction project into the development of a model for the safest process of organizing construction work at all stages of the project, increasing the level of safety in construction when building a BIM model of an object ceases to be something exotic.

The conducted research, the tasks formulated in this article, as well as the proposed algorithms for their solution will significantly improve the situation with industrial injuries in construction. Because BIM modeling, by providing the opportunity to virtually study each stage of a construction project, allows you to optimize the planning process by improving interaction between all its participants. The proposed organizational and technological solutions based on the analysis of the rational organization and coordination of all work, taking into account the "historical" data on incidents that occurred in similar projects earlier, will avoid many situations that can lead to irreparable mistakes.

References

1. Wang, Y., Griffis, F.: The theory of zero incident safety management. J. Civ., Constr. Environ. Eng. **3**(3), 83–98. (2018). https://doi.org/10.11648/j.jccee.20180303.15
2. Akinlolu, M., Haupt, T.C., Edwards, D.J., Simpeh, F.: A Bibliometric review of the status and emerging research trends in construction safety management technologies. Int. J. Constr. Manag. **22**(5), 2699–2711 (2020). https://doi.org/10.1080/15623599.2020.1819584
3. Sherratt, F.: Exploring 'zero target' safety programmes in the UK construction industry. Constr. Manag. Econ. **32**(7–8), 737–748 (2014). https://doi.org/10.1080/01446193.2014.894248
4. https://ec.europa.eu/eurostat/statistics-explained/index.php?title=Accidents_at_work_statistics
5. https://rostrud.gov.ru/
6. https://rosstat.gov.ru/working_conditions?print=1

7. Konda, S., Tiesman, H.M., Reichard, A.A.: Fatal traumatic brain injuries in the construction industry, 2003–2010. Am. J. Ind. Med. **59**, 212–220 (2016). https://doi.org/10.1002/ajim.22557
8. Sousa, V., Almeida, N.M., Dias, L.A.: Risk-based management of occupational safety and health in the construction industry—Part 1: Background knowledge. Saf. Sci. **66**, 75–86 (2014)
9. Sawhney, A., Riley, M., Irizarry, J.: Construction 4.0: An Innovation Platform for the Built Environment, 1st ed.; Routledge: London, UK, (2020). https://doi.org/10.1201/9780429398100
10. Dasovic, B., Galic, M., Klansek, U.: Active BIM approach to optimize work facilities and tower crane locations on construction sites with repetitive operations. Buildings **9**, 21 (2019)
11. Asadzadeh, A., Arashpour, M., Li, H., Ngo, T., Bab-Hadiashar, A., Rashidi, A.: Sensor-based safety management. Autom. Constr. **113**, 103128 (2020)
12. Zhang, J., Seet, B.-C., Lie, T.: Building information modelling for smart built environments. Buildings **5**, 100–115 (2015)
13. Cascone, S.: Digital technologies and sustainability assessment: a critical review on the integration methods between BIM and LEED. Sustainability **15**, 5548 (2023). https://doi.org/10.3390/su15065548
14. Ganbat, T., Chong, H.-Y., Liao, P.-C.: Mapping BIM uses for risk mitigation in international construction projects. Adv. Civ. Eng. **1**/ 5143879 (2020)
15. Eleftheriadis, S., Mumovic, D., Greening, P.: Life cycle energy efficiency in building structures: A review of current developments and future outlooks based on BIM capabilities. Renew. Sustain. Energy Rev. Elsevier, **67**(C), 811–825 (2017)
16. Akram, R., Thaheem, M.J., Khan, S., Nasir, A.R., Maqsoom, A.: Exploring the Role of BIM in construction safety in developing countries: Toward automated hazard analysis. Sustainability **14**, 12905 (2022)
17. Díaz-Curbelo, A., Gento, Á.M., Redondo, A., Aqlan, F.: A fuzzy-based holistic approach for supply chain risk assessment and aggregation considering risk interdependencies. Appl. Sci. **9**(24), 5329 (2019)
18. Farghaly, K., Collinge, W., Hadi Mosleh, M., Manu, P., Cheung, C.: Digital information technologies for prevention through design (PtD): A literature review and directions for future research. Constr. Innov. **22**(4), 1036–1058 (2022). https://doi.org/10.1108/CI-02-2021-0027
19. Shilkina, S.V., Ivanova, O.V.: The choice of software for the implementation of projects based on information modeling technologies. Constr. Arch. **2**, 13–13 (2023). https://doi.org/10.29039/2308-0191-2023-11-2-13-13
20. Golizadeh, H., Hon, C.K.H., Drogemuller, R., Reza Hosseini, M.: Digital engineering potential in addressing causes of construction accidents. Automation in Construction 95, 284–295. (2018)
21. Borngrund, C., Bodin, U., Sandin, F.: Machine vision for construction equipment by transfer learning with scale models. In Proceedings of the 2020 International Joint Conference on Neural Networks (IJCNN), IEEE, Glasgow, UK, 19–24 July, pp. 1–8. (2020)
22. Hoeft, M., Trask, C.: Safety built right in: exploring the occupational health and safety potential of BIM-Based platforms throughout the building lifecycle. Sustainability **14**, 6104 (2022)
23. Shilkina, S.V.: Construction project management with the implementation of information modeling technology. In the book: Ginzburg A., Galina K. (ed.) Life Cycle Management of Buildings. Information Systems and Technologies. Lecture Notes on Civil Engineering, volume 231. Springer, Cham. https://doi.org/10.1007/978-3-030-96206-7_24 April, 10 (2022)

24. Shilkina, S.V.: Conceptual and informational approach to the design of buildings, taking into account technical solutions at the stages of their life cycle. Mod. Sci. Innov. **3**(19), 68–76 (2017)
25. Varshavsky, A.E., Kuznetsova, M.S.: Assessment of economic damage in reducing human life expectancy as a result of major types of diseases. National interests: priorities and security **19**(12), 2206–2236 (2023)
26. Zubets, A.N., Novikov, A.V.: Numerical assessment of the cost of human life in Russia and in the world. Financ.: Theory Pract. **22**(4), 52–75 (2018)

Assessment of the Load-Bearing Capacity of the Braces of Structures of Energy Facilities and Methods for Selecting Their Sections

Liya Tokareva[1(✉)], Renat Khusainov[1], Marat Ayzatullin[1], Linar Sabitov[2,3], and Edvard Abdullazyanov[1]

[1] Kazan State Energy University, Kazan 420066, Russia
la_tokareva@mail.ru
[2] Moscow State University of Civil Engineering, Moscow 129337, Russia
[3] Kazan (Volga region) Federal University, Kazan 420008, Russia

Abstract. Guy ropes are a structural bearing element that is widely used in mast structures. The task of this element is to ensure the overall stability of the main element - the mast due to the perception of horizontal loads, reducing its estimated length, and ensuring the required deformability of mast structures. An assessment of the values of the reserve coefficients of cable ties of mast structures of energy facilities using the reliability index revealed a significant and possibly excessive safety margin for the ties, when calculating them in accordance with modern building regulations. In this article, the form of the methodology for assessing the bearing capacity of cable ties for cable ties of mast structures is considered. Based on the analysis and calculation, it was concluded that the use of the reserve coefficient to assess the condition of cable ties of mast structures significantly simplifies the assessment of their reliability and defects during operation. The values of the reserve coefficients of cable ties of mast structures for the values of forces from standard loads using the safety index known in reliability theory, depending on the planned service life of the structures, are determined.

Keywords: Guy Ropes · Mast Structures · Construction · Methodology · Efficiency Assessment · Load-Bearing Capacity

1 Introduction

Guy ropes are a structural bearing element, widely used in mast structures. The element's task is to ensure the overall stability of the main element that is the mast, by perceiving horizontal loads, reducing its estimated length, and ensuring the required deformability of mast structures. Mast structures with guy ropes are all-purpose and are used in power transmission lines' supports, starting from 10 to 500 kW of power, as well as in aerial towers and cellular communication supports [1, 2]. Figures 1 and 2 show the designs of 10 kW to 330 kW power transmission lines' supports with guy ropes.

The designs of guy ropes are typically made up of several structural elements made of different materials and having different nodal solutions and design.

S. V. Klyuev et al. (Eds.): ICRAAC 2024, LNCE 627, pp. 154–161, 2025.
https://doi.org/10.1007/978-3-031-82938-3_19

Fig. 1 Guys on power transmission poles.

Fig. 2 Designs of mast structures for power lines over 100 kW.

2 Methods and Materials

The main methods of scientific research used in writing this work include:

- collection and analysis of literary sources;
- analysis of previously developed solutions;
- carrying out calculations, determining the value of the reserve coefficients of cable ties of mast structures for the values of forces from standard loads using the safety index known in the theory of reliability, depending on the planned life of the structures

The method of calculating mast structures with the determination of the forces in guy ropes, described in paper [3], is a multi-stage task. The calculation of the mast and its guys is carried out in several stages. First comes the setting of the preliminary tensionof the guy ropes σ_0 and their sections A, followed by determining of bending moments once all the necessary calculations have been performed, as well as a system of equations has been solved. The distribution of moments is likely to be unsuccessful, thus all the operations need to be performed several times to achieve optimization.

3 Results and Discussion

The process of calculating moment diagrams for the trunk of a mast is a less laborious and more convenient way from the technical implementation point of view. First, the moment diagrams are set to describe the distribution of moments along the length of the mast trunk, and then the deflections and support reactions in anchorage nodes are sequentially determined. Such an approach allows effectively taking into account the mechanical behavior of the mast trunk under the influence of loads and provides more accurate calculation results. Shearing force diagram is built in accordance with the moment diagram. To determine longitudinal forces in the structures, the mass of all the above elements is taken into account, since these elements create vertical components of tension. For each section of the structure, it is necessary that both the load from the structure itself and the vertical components of the forces from the mass of all elements above this section are taken into account. Such an approach allows calculating the longitudinal forces in each section of the structure with the necessary accuracy.

The verification of the accepted section of the rope of the guy ropes is carried out in the form of a comparison of the maximum calculated force in the guy rope with the load bearing capacity of the rope. $S^{\sim} = R^{\sim} - F^{\sim}$

$$N_{\max} = \sigma_{\max} \cdot A \leq N_p \cdot Y_c \cdot Y_{bt} / Y_m \tag{1}$$

where A – sectional area of the guy rope,

N_p – overall rope capacity,

γ_m – reliability factor in relation to materials, taken equal for ropes,

$\gamma_c = 0.8$ with the number of guy ropes in one tier less than 5,

γ_{bt} – coefficient of the rope attachment in the bushing, equal 0.95.

The tasks, solved at the stage of operation of mast structures to assess their load bearing capacity and make decisions on their repair or replacement, require simpler engineering techniques.

The technique used for lifting machine ropes [4] is one of the most effective methods for assessing load-bearing capacity with rejection criteria at the stage of operation in relation to rope elements. In this technique, the calculation method based on permissible stresses is used in the form of applying a margin factor instead of the calculation method based on limit conditions. It requires that the ropes – prior to the design and installment – are to be checked using the calculation with the below formula:

$$F_o = Z_p \cdot S \tag{2}$$

where F_o– overall rope capacity (H), equaling the value indicated in the certificate (test certification);

Z_p– minimum rope utilization factor (factor of safety),

S – the highest tension of the rope branch (H) specified in the crane passport.

The values of factors of safety for ropes in crane structures are accepted according to [4]. M1–M3 rope mechanisms are used in various boom-lifting machines, such as cranes, jib cranes on a special chassis and manipulator cranes. The establishment of the relationship between the factor of safety and the coefficients of critical load design is considered in paper [5]. As per paper [5], when designating the calculated values of loads and strength with values F and R respectively, the reliability of structures is characterized not only by controlled guideline values, but also by values of the factors of safety.

To estimate the values of factors of safety in guy ropes, it is necessary to determine the values of numerical characteristics describing the values of load bearing capacity and forces in guy ropes of mast structures, i.e. it is necessary to set the values of μ_R, v_R, μ_F, v_F.

The guideline value of the load bearing capacity of guy ropes is determined by the formula:

$$R_n = m_R(1 - \mu_R \cdot v_R) \quad (3)$$

The guideline value of the force from wind load in the guy rope equals:

$$T_n = m_f(1 + \mu_F \cdot v_F) \quad (4)$$

For the load-bearing capacity parameters described by the normal law of distribution with a known security of values of $\mu_R = 1.65$, $v_R = 0.1$.

The task of determining the coefficients of variation and the corresponding quantile values for the values of the rated load is more difficult.

For mast structures, the load in guy ropes typically arises because of the action of the wind load, which is described by asymmetric distribution laws. Papers [6, 7], when describing the wind load, recommend that Gumbel distribution for annual maximum wind loads is used.

Below we use Gumbel distribution and write down the equations for periods of non-exceeding values for the standard wind load with N = 5 years and for the calculated wind load with N = 50 years.

Rated wind load equals:

$$w_N^- = w_1^- + 0,78\, Sw\, ln5 \quad (5)$$

Designed wind load equals:

$$w_p^- = w_1^- + 0,78\ \mathrm{Sw\ ln50} \quad (6)$$

With that $w_p^- = 1.4 w_N^-$.

Solving a system of two equations allows determining the values of the unknown mathematical expectation of the wind load w_1^- and the standard deviation of the wind load S_w.

$$\mathrm{Sw} = 0.4\, w_N^-/1.784,\ \ w_1^- = 0.717 w_N^- \quad (7)$$

Then $v_F = \frac{S_W}{w_1^-} = 0.31$

μ_F is determined from the equation as $w_N^- = w_1^-(1 + \mu_F v_F)$

Then the quantile for standard wind load equals:

$$\mu_F = 1.27 \tag{8}$$

The rated factor of safety for rope cross-section braces is determined by the formula:

$$K_n = F_0 \cdot n \cdot sin\alpha / Tn = \gamma_n \gamma_f \gamma_m \gamma_d = 2.8 = 1 \cdot 1.4 \cdot 1.6/0.8 \tag{9}$$

where Tn – the horizontal total force from the rated values of the loads falling on the fastening unit of the guy ropes,

n – the number of braces in the node, perceiving Tn,

α –the angle of inclination of the guy ropes to the horizon, for example α = 570 for tightening masts and power transmission line supports according to series 3.407,

F_0 – overall rope capacity.

When comparing, the factor of safety of guy ropes determined according to building codes, corresponds to the operating mode of cranes A2-A3 (light operation mode). Most automobile and self-propelled cranes fall in this group of cranes.

To analyze the obtained value of the factor of safety, such a reliability indicator as the safety index β is used.

If the load variability and strength variation is known, then sometimes reliability can be assessed by the variability of the function, which is called the strength reserve:

$$S^{\sim} = R^{\sim} - F^{\sim} \tag{10}$$

where R˜and F˜– values of load bearing capacity and load (load effect) with the same dimensions.

Mathematical expectation $m_s = m_R - m_F$ and the variance $s^2 S = S^2 R + S^2 F$ are the main probabilistic characteristics of this (in the absence of a correlation between load and load bearing capacity).

The probability of failure (risk) or the probability of a negative value of the strength reserve S ˜ at a known value β is determined from the formula:

$$P_F = P\left(S^{\sim} < 0\right) = 1 - \Phi(\beta) \tag{11}$$

where $\Phi(\beta) = \frac{1}{\sqrt{2\pi_0}} \int^{\beta} \exp\left(-\frac{u^2}{2}\right) du$ – the probability integral or Laplace function of the variable $u = (S^{\sim} - m_s)/s_s$, corresponding to the standard function $\varphi(u)$.

Taking into account the property of the function φ(u) it can be also written down

$$P_F = P\left(S^{\sim} < 0\right) = 0.5 - \Phi * (\beta) \tag{12}$$

The probability of failure-free operation or reliability of the structure $P_R = 1 - P_F$.

The safety index β is a standardized probability of a random variable S˜. The method of evaluating reliability by means of the β coefficient is called the two-moment method, since two characteristics of random variables R˜and F˜ are used to determine it.

This safety index is used in the European standards EN1990.

Below the values of the safety index set for construction structures by our regulatory documents are determined. For n years of planned operation, the reliability index is determined by the formula:

$$\Phi(\beta_n) = (\Phi(\beta 1))^n \tag{13}$$

With n = 50 years $\beta_1 = 5.2$ from the equation $[\Phi(\beta_1)]^{50} = 0.999995$.
The values of the reliability index β will be determined by the formula:

$$\beta = \frac{m_s}{s_s} = \frac{m_R - m_F}{\sqrt{s_R^2 + s_F^2}} = \frac{m_k - 1}{\sqrt{m_k^2 v_R^2 + v_F^2}} \tag{14}$$

where $m_k = m_R/m_F$–the ratio of the average values of the rope breaking forces and the pulling force from the wind load.

Taking into account the defined parameters μ_R, v_R, μ_F, v_F

$$k_n = m_k \frac{1 - \mu_R v_R}{1 + \mu_F v_F} = m_k \cdot 0.6 \tag{15}$$

Then with $K_n = 2.8$ $m_k = 2.8/0.6 = 4.67$, which corresponds to the reliability index β = 6.55. When operating an object with guy ropes up to 50 years, the rated reliability index determined according to GOST 31937-2011 is β = 5.2. With the planned operation of the object up to 50 years, the obtained reliability index value for guy ropes of mast structures is higher than the rated one, which indicates suboptimal overestimated factors of safety used in calculating the guy ropes of mast structures.

Table 1 shows values of the safety factors for ropes of guy ropes for tightening forces from the standard load, determined using the reliability index β for various periods of planned operation of structures with guy ropes.

Table 1 Values of the safety factor for guy ropes for tightening forces from the standard load

Operation period	1 year	6 years	10 years	30 years	50 years
Value β	4.5	4.8	4.97	5.1	5.2
Value m_k	2.925	3.1	3.23	3.3	3.39
Value *Kn* the factor of safety for the ropes of the guy ropes for the force in the rope from the standard load	1.76	1.86	1.94	1.98	2.0

The use of the factor of safety to assess the condition of ropes of mast structures significantly simplifies the assessment of their reliability and defects during operation [8–12, 13]. To assess the state of the load bearing capacity of guy ropes, it is not necessary to perform step-by-step calculations according to a deformed scheme. It is enough to determine the horizontal rated total force Tn, which falls into the attachment point of the mast ropes and, using the following formula, to determine the sufficiency of the breaking force of the rope in the guy rope F_0:

$$F_0 \geq T_n \cdot K_n/(n \cdot sin\alpha) \tag{16}$$

where α - the angle of the guy ropes to the horizon, n – the number of guy ropes in the node, perceiving Tn,

$$T_n = w_{\bar{N}} \cdot C_X \cdot A_K \cdot K_Z$$

where $w_{\bar{N}}$ – standard wind force according to SP 20.13330.2016,

A_K –the area of the contour of the mast structure, which falls on the anchorage assembly,

Cx, Kz – the aerodynamic coefficient and the coefficient of increase in wind force in height according to SP 20.13330.2016 in the area of the anchorage assembly.

4 Conclusions

1. Guy ropes are a structural bearing element that is widely used in mast structures. The task of this element is to ensure the overall stability of the main element- the mast due to the perception of horizontal loads, reducing its estimated length, and ensuring the required deformability of mast structures. Mast structures with guy ropes are widely used and are used in power transmission line supports, ranging from 10 to 500 kW of power, in antenna structures and cellular communication supports.
2. The process of selecting cross sections and determining the forces in the braces is a complex and multi-stage task that is solved at the design stage. For tasks solved at the stage of operation of mast structures, simpler engineering techniques are required to assess their bearing capacity and make decisions on their repair or replacement.
3. An assessment of the values of the reserve coefficients of cable ties of mast structures of energy facilities using the reliability index revealed a significant and possibly excessive safety margin for the ties, when calculating them in accordance with modern building regulations. Thus, the coefficient of the reserve of the guy ropes of mast structures corresponds in magnitude to the reserve coefficients for ropes used in boom cranes of A1-A2 mode, which have more stringent operating conditions.
4. The form of the methodology for assessing the bearing capacity of rope ties for rope ties of mast structures in the form used to diagnose the condition of ropes of lifting machines is convenient for engineering and technical workers involved in the diagnosis and operation of mast structures, and is simple and clear.
5. The values of factors of safety of guy ropes of mast structures for the values of forces from standard loads using the safety index known in reliability theory, depending on the planned life of the structures, are determined.

References

1. Tokareva, L.A.: Analysis of design solutions for tower structures in the energy sector. In the collection: XXV All-Russian Postgraduate and Master's Scientific Seminar dedicated to Power Engineer's Day. Conference materials. In 3 volumes. Under the general editorship of E.Yu. Abdullazyanova. Kazan, pp. 240–242 (2022)

2. Sabitov, L.S., Kashapov, N.F., Strelkov, Yu.M., Kuznetsov, I.L.: Method of the steel aluminum wires protection of air transmission lines from fast vibration damage at the output from the support clamp. IOP Conference Series: Materials Science and Engineering 012059 (2017)
3. Sabitov, L.S., Khamidulin, I.N., Kuznetsov, I.L., Khusanov, D.: Stress-strained state of supports for energy construction. Int. J. Appl. Eng. Res. **10**(24), 45438–45448 (2015)
4. Federal norms and rules in the field of industrial safety "Safety rules for hazardous production facilities where lifting structures are used." Approved by order of the Federal Service for Environmental, Technological and Nuclear Supervision dated November 12, 2013 N 533
5. Krasnoshchekov, Yu.V., Zapoleva, M.Yu.: Fundamentals of Designing Structures of Buildings and Structures. Infra-Engineering, Omsk (2019)
6. Reiser, V.D.: Theory of Reliability of Structures. ASV, Moscow (2010)
7. Sabitov L.S., Kashapov N.F., Strelkov Yu.M., Kuznetsov I.L.: Method of the steel aluminum wires protection of air transmission lines from fast vibration damage at the output from the support clamp. IOP Conference Series: Materials Science and Engineering. 2017. C. 012059
8. Sabitov, L.S., Ilyin, V.K., Ibragimova, A.V., Strelkov, Yu.M.: Method for protecting steel-aluminum wires of overhead power lines from fatigue oscillatory damage at the exit from the supporting clamp. Bulletin of the Irkutsk State Technical University 21, 5 (124), 122–130 (2017)
9. Strelkov, Yu.M., Sabitov, L.S., Klyuev, S.V., Klyuev, A.V., Radaykin, O.V., Tokareva, L.A.: Technological features of the construction of a demountable foundation for tower structures. Constr. Mater. Prod. **5**(3), 17–26 (2022). https://doi.org/10.58224/2618-7183-2022-5-3-17-26
10. Novoselov, O.G., Sabitov, L.S., Sibgatullin, K.E., Sibgatullin, E.S., Klyuev, A.V., Klyuev, S.V., Shorstova, E.S.: Method for calculating the strength of massive structural elements in the general case of their stress-strain state (parametric equations of the strength surface). Constr. Mater. Prod. **6**(2), 104–120 (2023). https://doi.org/10.58224/2618-7183-2023-6-2-104-120
11. Novoselov, O.G., Sabitov, L.S., Sibgatullin, K.E., Sibgatullin, E.S., Klyuev, A.S., Klyuev, S.V., Shorstova, E.S.: Method for calculating the strength of massive structural elements in the general case of their stress-strain state (kinematic method). Constr. Mater. Prod. **6**(3), 5–17 (2023). https://doi.org/10.58224/2618-7183-2023-6-3-5-17
12. Khamidullin, I.N., Sabitov, L.S., Kuznetsov, I.L.: development and research towers for wind turbines. Scientific Herald of the Voronezh State University of Architecture and Civil Engineering. Constr. Archite. **4**(28), 25–32 (2015).

Study of the Influence of Liquid Glass Density on the Properties of the Raw Mix and the Physical and Mechanical Characteristics of Geopolymer Stone

V. B. Petropavlovskaya(✉), Babaev D.D., K. S. Petropavlovskii, and T. B. Novichenkova

Tver State Technical University, 170026 emb. Af. Nikitina, 22, Tver, Russia
kspetropavlovsky@gmail.com

Abstract. This study presents the results of experiments on the development of formulations of geopolymer binders based on industrial waste - enriched ashes of thermal power plants and granulated metallurgical slag. The technology has great potential in solving the problems of resource and energy conservation, as well as reducing the impact of exhaust gases and heat on the environment. Reducing carbon dioxide emissions compared to traditional cement production technology increases interest in this study from low-carbon concrete manufacturers. Finding the optimal percentage of the alkaline activator used is one of the main factors in the formed geopolymer system based on enriched fuel ash and ground metallurgical slag, since the alkali activator will be decisive in the process of achieving maximum bending and compressive strength, as well as ensuring stable properties during operation of the geopolymer stone. The results showed that the concentration of the alkaline sealer determines the dissolution of the aluminosilicate phase in the ash and, at the same time, the active structuring of the geopolymer structure and its technical and operational properties.

Keywords: Bottom Ash · Metallurgical Slag · Alkaline Activator · Polymerization · Concentration · Strength · Microstructure

1 Introduction

Ash is a by-product of coal combustion that contains minerals. It can be effectively used in many construction composites, including geopolymers [1]. Such composites can contain a sufficient amount of ash and its components, which are obtained by enriching mixtures of ash and slag [2]. Geopolymers based on waste fuel ash and slag, hardened in the environment, usually do not have sufficient strength due to low reactivity. However, enrichment of fuel waste can help improve the properties of geopolymer stone [3].

Aluminosilicate microspheres, consisting of glass-ceramic hollow particles of spherical shape with a diameter of 50–200 μm and a density of 380–400 kg/m^3, are used as fillers due to their high dispersion, low density, increased thermal stability and resistance

S. V. Klyuev et al. (Eds.): ICRAAC 2024, LNCE 627, pp. 162–170, 2025.
https://doi.org/10.1007/978-3-031-82938-3_20

in aggressive environments [9, 10]. The process of alkaline activation or geopolymerization allows the transformation of such glassy structures into compact cementitious composites [4]. Hydroxides and silicates are most often used as activated alkaline materials, and potassium, sodium and sometimes calcium are usually used as alkaline elements. If geopolymer concrete is made from ash with a low calcium content, it requires thermal or steam curing. It is known, for example, that at a temperature of 75 °C for 24 h of thermal curing, the compressive strength increases by an average of 50% compared to samples that were cured in the environment [5].

However, the need for thermal curing makes this material less suitable for use in construction.

Industrial waste, such as slag, can increase the strength of geopolymer concrete. They can be used when mixed with ash without additional curing. In this case, the final strength may remain unchanged [6]. The addition of materials with high calcium oxide CaO content, such as slag, cement can accelerate the reaction kinetics in alkaline environment and increase the strength of ash waste. Granulated blast furnace slag contains about 35–50% CaO, 30–35% SiO_2 and 8–15% Al_2O_3, as well as FeO, MgO and TiO_2. It is used in various fields including road bases, asphalt, embankments, earthworks, concrete, insulation, glass, railway ballast, filters and Portland cement production. Blast furnace slag can undergo alkali activation due to its high $CaO/(SiO_2 + Al_2O_3)$ ratio, which makes it rich in calcium [7, 8]. Liquid glass as an activator plays a key role in composite materials. It binds mineral powders. A liquid glass-based sealer determines the rheological properties of molding mixtures and activates the participation of fillers in the processes of structure formation of artificial stone.

Studies have shown that a geopolymer based on ash, sodium hydroxide NaOH and liquid glass does not harden at 20 °C even after 3 days. Therefore, various techniques can be used to improve the characteristics of such a geopolymer composition, including the introduction of slag, changing the types and concentrations of alkali activators and high-temperature curing [11–15]. For example, the introduction of alkali-activated slag into the composition of such a material promotes an increase in compressive strength, but determines a short setting time and large shrinkage during drying [11].

Thus, it is of interest to obtain a geopolymer binder with high performance properties based on fuel ash and ground granulated slag. It is necessary to consider the issue of the influence of the hardening activator – liquid sodium glass – on the properties of geopolymer mixtures and the resulting stone [16, 17].

2 Methods and Materials

Granulated blast furnace ground slag. Blast furnace granulated ground slag is a waste product of ferrous metallurgy. This waste product was provided by a large Russian company, Mechel-Materials LLC (Moscow, Russia).

Samples of this waste product were examined using X-ray fluorescence analysis (XRF), scanning electron microscopy (SEM) and granulometric analysis.

SEM analysis of the slag sample showed that the material consists mainly of large, thickened, irregularly shaped particles and their aggregates, with diameters ranging from fractions of a micron to several microns. In this case, finer particles are subject to aggregation.

As a result of the studies, it was found that the sulfur oxide (SO_3) content for the slag sample used in this work is at the level of 1.53 wt. %. At the same time, the mass fraction of magnesium oxide content is at the level of 10.61%.

In addition, the value of the sum of alkali oxides ($Na_2O + 0.65\ K_2O$) was determined to be 0.86 wt. %. The mass fraction of the content of the main oxides CaO, MgO and SiO_2 was also determined. This value is presented as the sum of the main oxides (CaO + MgO + SiO_2) and was at the level of 81.15%. The ratio of the mass fractions of these oxides ((CaO + MgO)/ SiO_2) was at the level of 1.27.

As a result of the studies, it was found that the content of sulfur oxide (SO_3) for the slag sample is at the level of 1.53 wt. %. At the same time, the mass fraction of the magnesium oxide content is at the level of 10.61%.

In addition, the value of the sum of alkali oxides ($Na_2O + 0.65K_2O$) was determined. It is equal to 0.86 wt. %. The mass fraction of the main oxides CaO, MgO and SiO_2 was also determined. This value is presented as the sum of the main oxides (CaO + MgO + SiO_2) and is at the level of 81.15%. The ratio of the mass fractions of these oxides ((CaO + MgO)/ SiO_2) is at the level of 1.27. The crushed slag contained mainly fine particles with sizes from 2 μm to 45.022 μm (85.03%). The amount of particles with sizes less than 2 μm was 14.97%. The average particle size of the blast-furnace granulated ground slag was 11.073 μm.

Enriched bottom ash of Kashirskaya SDPP. Ash, a product of enrichment of ash and slag waste from the Kashirskaya State District Power Plant (Moscow Region), was used as a microdispersed modifier in the studies. It is represented by a fraction with a particle size of 0 to 0.1 μm.

The structure of the ash is represented by a glass phase, in some areas of which crystalline quartz formations are visible. Aluminosilicate ash is a light-gray powdery material containing 1.47% by weight of carbon and 4.95% of magnetite (Table 1).

A study of the granulometric composition of the enriched ash showed that the range of particle size dispersion is quite narrow, from 1 to 50 μm.

The properties of the initial powder mixtures were studied using standard methods. Differential and integral particle size distribution of the aluminosilicate product of the enrichment of ash and slag waste from thermal power plants was performed using a laser particle analyzer. In order to study the physical and mechanical characteristics reflecting the structural properties of the hydration hardening system, samples were formed by casting. The studies were carried out on beam samples measuring 40 × 40 × 160 mm. The samples were hardened under normal air-dry conditions in accordance with the research program.

The setting time, normal density of the raw mix, as well as the compressive strength, true and average density, humidity, water absorption and water resistance of the geopolymer modified samples were determined using standard methods in accordance with GOST and TC.

The density of the liquid glass solution was measured using water and a graduated cylinder. A certain amount of liquid glass was placed in a graduated cylinder, then 20 ml of water was measured out and added to the graduated cylinder. After each addition of water, the density of the resulting solution was checked using a hydrometer. The operation was carried out until the desired density of the solution was achieved. As

Table 1. Oxide composition of the component of the ash and slag waste of the Kashirskaya SDPP, Moscow region.

Oxide	Na_2O	MgO	Al_2O_3	SiO_2	K_2O	CaO	TiO_2	MnO	Fe_2O_3	P_2O_5	SO_3	C
Content, %												
Aluminosilicate component	0.6	1.6	21.5	57.8	2.1	2.3	0.9	0.1	4.6	0.4	0.1	3

a result, the optimal ratios of the density of the liquid glass solution and water were selected.

3 Results and Discussions

The study of the effect of liquid glass solution on the properties of raw mixes was carried out by conducting comparative experiments using mathematical experimental design. The variable factor was the density of the liquid glass additive, with variation levels: 1; 1.05; 1.1; 1.15; 1.2 g/cm^3. The definition of normal density and setting time was determined for each composition. Optimum values of normal density and liquid-solid ratio were selected for each composition, the results are presented in Table 2.

Table 2. Determination of normal density and setting time of geopolymer mixture.

№ of composition	Normal density	Start of setting, min	End of setting, min
1	0.4	135	205
2	0.4	150	225
3	0.45	165	255
4	0.45	180	285
5	0.48	195	315

In order to determine the effect of the density of the liquid glass solution, the compressive and flexural strength of the obtained geopolymer castings was estimated on the 28th day of hardening. A comparative analysis of geopolymer compositions with different densities of the alkaline activator is presented in Fig. 1. A comparison of compositions No. 2 – No. 4 with the control composition No. 1 (without an alkaline activator) is given in Table 3.

Table 3. Properties of geopolymer stone.

№ of composition	Average density, g/cm^3	Tensile strength, MPa	Compressive strength, MPa
1	1.25	1.75	10.3
2	1.31	3.21	15.15
3	1.35	3.57	17.01
4	1.4	2.19	12.33
5	1.41	1.83	11.25

The liquid glass solution contains hydroxide ions (OH^-) and sodium ions (Na^+), which initiate a reaction between the internal silicate (Si) and aluminate (Al) components, initiating the process of dissolution and strengthening. The density increases as

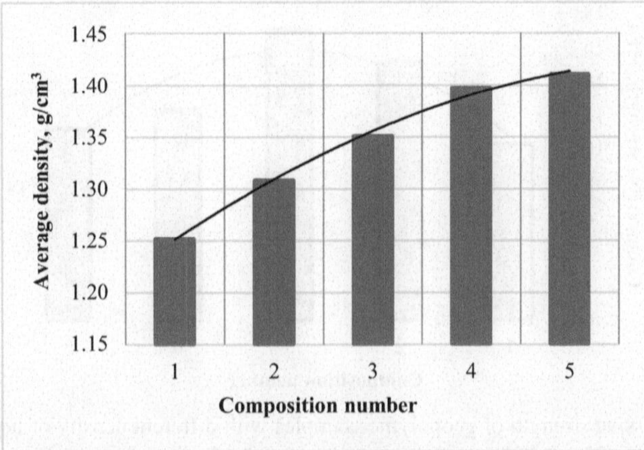

Fig. 1. Average density of geopolymer samples with different density of liquid glass solution, where composition 1 is 1 g/cm^3, composition 2 is 1.05 g/cm^3, composition 3 is 1.1 g/cm3, composition 4 is 1.15 g/cm^3, composition 5 is 1.2 g/cm^3.

the concentration of the alkaline activator increases (Fig. 2). Compared with the control composition, the density increased by 13% with an alkaline activator density of 1.2 g/cm^3. Higher strength corresponds to a higher content of sodium silicate in the solution. However, too much alkali also negatively affects the strength of geopolymer stone. The results of a comparative analysis of the strength properties of geopolymers (Figs. 2, 3) showed that the samples of the 3rd composition have the highest bending strength (Fig. 2).

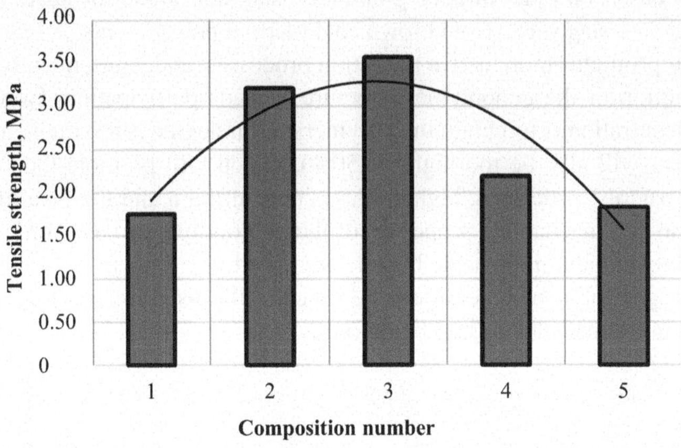

Fig. 2. Tensile strength of geopolymer samples with different density of liquid glass solution, where composition 1 is 1 g/cm^3, composition 2 is 1.05 g/cm^3, composition 3 is 1.1 g/cm^3, composition 4 is 1.15 g/cm^3, composition 5 is 1.2 g/cm^3.

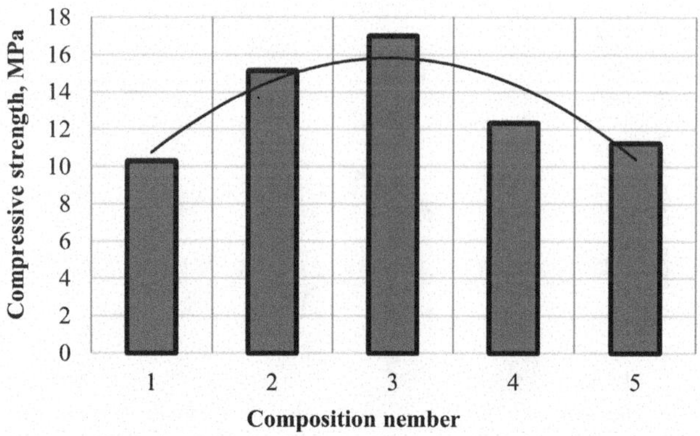

Fig. 3. Compressive strength of geopolymer samples with different density of liquid glass solution, where composition 1 is 1 g/cm^3, composition 2 is 1.05 g/cm^3, composition 3 is 1.1 g/cm3, composition 4 is 1.15 g/cm^3, composition 5 is 1.2 g/cm^3.

At a solution density of 1.1 g/cm^3, the compressive and bending strengths reached peaks of 3.57 and 17.01 MPa, respectively, which was 2 times higher than the similar indicator of the control composition. However, with a further increase in the solution density, both compressive and bending strength decreased. At a solution density of 1.2 g/cm^3, a noticeable decrease in strength indicators was observed.

4 Conclusions

Thus, the activator concentration is a key factor influencing the properties of the studied geopolymers based on blast-furnace granulated slag and aluminosilicate beneficiated product of ash and slag waste from brown coal thermal power plants. A higher activator concentration promotes more active hydration processes, and, consequently, the degree of polymerization of the geopolymer structure with its participation. However, if the activator concentration is too high, the polymerization degree is too high, and therefore the gel content will also be too high (> 50%), which will promote the development of strong destruction processes, higher stress concentration and the resulting decrease in the physical and mechanical properties of these geopolymers based on blast-furnace granulated slag and aluminosilicate beneficiated product of ash from brown coal combustion. Such geopolymer material can be successfully used in road foundations, soil stabilization and as part of structural materials.

Acknowledgements. The research was supported by Russian Science Foundation (project No. 21–79-30004).

References

1. Babaev, D.D., Petropavlovskaya, V.B., Novichenkova, T.B., Petropavlovskii, K.S., Zavadko, M.YU.: Overview: Geopolymer binders based on tpp ash. Vestnik of Tver State Technical University. Series Build. Electr. Eng. Chem. Technol. **3**(19), 5–19 (2023)
2. Babaev, D.D., Petropavlovskaya, V.B., Petropavlovskii, K.S., Novichenkova, T.B.: Geopolymers based on industrial waste. In: Third International Scientific and Practical Symposium on Materials Science and Technology (MST-III 2023), pp. 1298615. Spie-soc photo-optical instrumentation engineers, Washington (2024)
3. Babaev, D.D., Petropavlovskaya, V.B., Novichenkova, T.B., Petropavlovskii, K.S., Zavadko, M.YU.: Synthesis of geopolymer compositions based on alkaline technogenic products of fuel stations. Vestnik of Tver State Technical University. Series Build. Electr. Eng. Chem. Technol. **4**(20), 83–90 (2023)
4. Petropavlovskaya, V., Novichenkova, T., Sulman, M., Petropavlovskii, K.: Coal ash enrichment with its full use in various areas. Materials **15**(9), 6610 (2022)
5. Kalinkin, A.M., Kalinkina, E.V., Ivanova, A.G., Kruglyak, E.A.: Geopolymers based on mechanically activated mixtures of coal-fired power plant fly ash with natural calcium and magnesium carbonates. Proc. Fersman Sci. Session GI KSC RAS **19**, 144–148 (2022)
6. Khaidarov, B.B., et al.: Development of energy-efficient technology for the production and research of clinker-free binders based on blast furnace granular slags. In: Modern Methods and Technologies of Creation and Processing of Materials: A Collection of Scientific Papers, 1, pp. 243–252. Physical-Technical Institute of the National Academy of Sciences of Belarus (2017)
7. Salehi, H., Mazloom, M.: Opposite effects of ground granulated blast-furnace slag and silica fume on the fracture behavior of self-compacting lightweight concrete. Constr. Build. Mater. **222**, 622–632 (2019)
8. Mahantesh, N.B., Amarnath, K., Raghuprasad, B.K.: Ductility Of fly ash - slag based reinforced geopolymer concrete elements cured at room temperature. MATEC Web of Conferences **138**(9), 01003 (2017)
9. Miryuk, O., Fediuk, R., Amran, M.: Porous Fly Ash/Aluminosilicate microspheres-based composites containing lightweight granules using liquid glass as binder. Polymers **14**(17), 3461 (2022)
10. Salieva, M.G.: Some aspects of the use of hydraulic removal ash in the construction industry. Mater. Sci. **2**(30), 101–103 (2019)
11. Da Silva Rocha, T., Dias, D.P., França, F.C., De Salles Guerra, R.R., De Oliveira, L.R.: Metakaolin-based geopolymer mortars with different alkaline activators (Na+ and K+). Constr. Build. Mater. **178**, 453–461 (2018)
12. Shi, X., Zhang, C., Wang, X., Zhang, T., Wang, Q.: Response surface methodology for multi-objective optimization of fly ash-GGBS based geopolymer mortar. Constr. Build. Mater. **315**(1), 125644 (2021)
13. Singh, B., Rahman, M.R., Paswan, R., Bhattacharyya, S.K.: Effect of activator concentration on the strength, ITZ and drying shrinkage of fly ash/slag geopolymer concrete. Constr. Build. Mater. **118**(11), 171–179 (2016)
14. Jiao, Z., Li, X., Yu, Q., Yao, Q., Guan, X., Hu, P.: Chloride resistance of class C/class F fly ash-based geopolymer mortars with different strength grades. Case Stud. Constr. Mater. **18**(11), e01811 (2023)
15. Samantasinghar, S., Singh, S.P.: Fresh and hardened properties of fly ash–slag blended geopolymer paste and mortar. Int. J. Concr. Struct. Mater. **13**(1), 47 (2019)

16. Klyuev, A.V., et al.: Experimental studies of the processes of structure formation of composite mixtures with technogenic mechanoactivated silica component. Constr. Mater. Prod. **6**(2), 5–18 (2023). https://doi.org/10.58224/2618-7183-2023-6-2-5-18
17. Klyuev, A.V. et al.: Development of alkali-activated binders based on technogenic fibrous materials. Constr. Mater. Prod. **6**(1), 60–73 (2023). https://doi.org/10.58224/2618-7183-2023-6-1-60-73

A Critical Analysis of Constructions of Process Pipe Racks, Cabled Pipe Racks and Unstrutted Pipe Racks of Industrial Enterprises

Konstantin Adushkin[1](✉), Liya Tokareva[2], Marsel Zaripov[2], Linar Sabitov[2,3], and Edvard Abdullazyanov[2]

[1] Ural Federal University Named After the First President of Russia B.N. Yeltsin, Ekaterinburg 620062, Russia
79126251270@yandex.ru
[2] Kazan State Energy University, Kazan 420066, Russia
[3] Moscow State University of Civil Engineering, Moscow 129337, Russia

Abstract. The article describes the role of linear engineering structures in ensuring continuous technological process of industrial enterprises. This article provides a definition and description of unstrutted pipe racks and pipe racks industrial enterprises use, and key differences between them. Possible loads and actions affecting structures are listed. Examples of stain-stress state of a section of a pipe rack with an anchor structure and a section of unstrutted pipe racks without an anchor structure are provided. A detailed review of the structural above-ground part of such structures was carried out: anchor structures and intermediate supports, support beams and longitudinal struts. Examples of constructions made of steel and reinforced concrete are listed. Separate references to standard series developed in the Soviet Union and Russia are provided. Prospects for the development and improvement of the structural part of the pipe racks are noted, an analysis of patent base referring to the discussed constructions has been made. A growing use of constructions made of composite materials is noted. It is concluded that it is possible to conduct research in order to find more efficient forms of pipe racks that have lower steel consumption compared to existing structures.

Keywords: Process Pipe Rack · Unstrutted Pipe Rack · Cabled Pipe Rack · Steel Structures · Reinforced Concrete Structures

1 Introduction

Every industrial enterprise consists of numerous buildings and structures, connected with one another by a communication system, ensuring a continuous technological process. Service lines in production facilities are laid underground or aboveground; special engineering structures are used for aboveground laying [1].

Pipe racks and unstrutted pipe racks are linear structures used for laying utility networks aboveground. Pipelines and cable products are the typical network types laid. Pipe racks are divided into process pipe racks – they serve only for laying technological

S. V. Klyuev et al. (Eds.): ICRAAC 2024, LNCE 627, pp. 171–181, 2025.
https://doi.org/10.1007/978-3-031-82938-3_21

pipelines [2], cabled pipe racks, through which cable products are laid [3], and combined ones, which combine the laying of pipes and cables.

If needed, maintenance platforms are located on pipe racks. They are used to perform planned maintenance and they facilitate the use of utility networks, located above the ground.

Today, there is a wide variety of pipe racks and unstrutted pipe racks. They vary in size, material and structural concept. Engineering structures are made of steel and reinforced concrete, with and without anchor structures, with longitudinal struts (pipe racks) and without them (unstrutted pipe racks), etc.

The goal of this article is to analyze the structures of pipe racks and unstrutted pipe racks, to review design experience of the Soviet Union time and that of modern Russia, and to note the latest trends in design. One of the tasks is to perform an overview of the loads and actions that affect the mentioned above structures.

2 Methods and Materials

To achieve this goal, the author of the article has carried out an analysis of the work he has so far performed on the structural investigation of building structures of pipe racks and unstrutted pipe racks (more than 10 years experience). To perform the analysis, 59 completed works were used, with a total length of investigated structures of 71.2 km, where 6.7 km were made of reinforced concrete and the remaining 64.5 km were made of steel. The analyzed works were mainly carried out in the northern regions of Russia – about 95% of all the objects are located in hard-to-reach areas with severe climatic conditions. Most of the investigated pipe racks and unstrutted pipe racks are located in oil-and-gas fields or at enterprises occupied in oil and gas processing.

To analyze the history of development of structural concepts, all the standard series, created in the Soviet Union (starting from the year 1963) and modern Russia (a total of 157 units) have been reviewed, as well as other regulations devoted to design and use of linear engineering structures on the subject of the research (a total of 32 units). When working on the article, the patent base for the discussed structures (a total of 28 units) was thoroughly analyzed. This base consists of patents from different countries (Soviet Union/modern Russia and USA) dated from 1956 till today. The review of the recent patents and older ones allowed to develop modern trends and prospects for pipe racks and unstrutted pipe racks, as well as to indicate possibilities for improvement in their construction.

3 Results and Discussion

From the structural point of view, a pipe rack consists of a set of supports connected to each other by longitudinal struts [4]. Utility networks laid through support beams are placed on supports and longitudinal beams. As a rule, unstrutted pipe racks do not have longitudinal beams connecting the supports. Pipelines are located only on the supports, trying to assign the largest spaces between the supports, while maximizing the self-supporting capacity of the laid pipelines. Figure 1 below, taken from the "Manual on the design of unstrutted pipe racks and pipe racks for process pipelines (to SNiP 2.09.03–85)" shows the structure diagrams of the construction of unstrutted pipe racks and pipe racks.

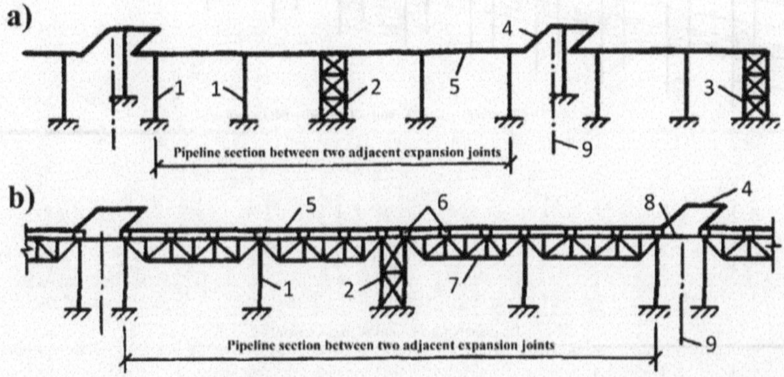

Fig. 1. Schematics of unstrutted pipe racks (subparagraph – a) and pipe racks (subparagraph – b) construction. 1 – intermediate support; 2 – anchoring intermediate structure; 3 – anchoring end structure; 4 – pipeline expansion joint; 5 – pipeline; 6 – support beam; 7 - longitudinal beam; 8 – insert of pipeline section between two adjacent expansion joints; 9 – the axis of the temperature gap.

Pipe racks and unstrutted pipe racks are divided into pipeline sections between two adjacent expansion joints to compensate for temperature deformations of building structures and pipelines. The classic pipeline section between two adjacent expansion joints includes one anchor structure and many intermediate supports. Each such section, as a rule, is limited by various types of pipeline expansion joints, whose task is to compensate for thermal expansions of pipelines [5]. Anchor structures mainly serve to perceive lateral longitudinal loads from pipelines [6]; they try to be place them in the middle of pipeline sections between adjacent expansion joints (see Fig. 2) and at the turning points or ends of the track.

Typically, when using reinforced concrete unstrutted pipe racks, anchor structures are not assigned, and in this case all lateral loads from pipelines are distributed among all supports of pipeline section between two adjacent expansion joints (see Fig. 3).

Special supports are used to place pipelines on building structures. The supports are divided into two types: movable and fixed [7]. Fixed supports reliably fix the pipeline from displacements. The task of movable supports is to ensure lateral movement of the pipeline in places where loads are transferred to load-bearing structures.

In the Soviet Union, many standard series dedicated to unstrutted pipe racks and pipe racks were developed. Back then, the vast majority of structures were made of reinforced concrete. Metallurgical industry success of the recent decades resulted in wider use of steel structures in construction [8]. In modern Russia, steel structures have almost completely replaced reinforced concrete during the construction of pipe racks and unstrutted pipe racks.

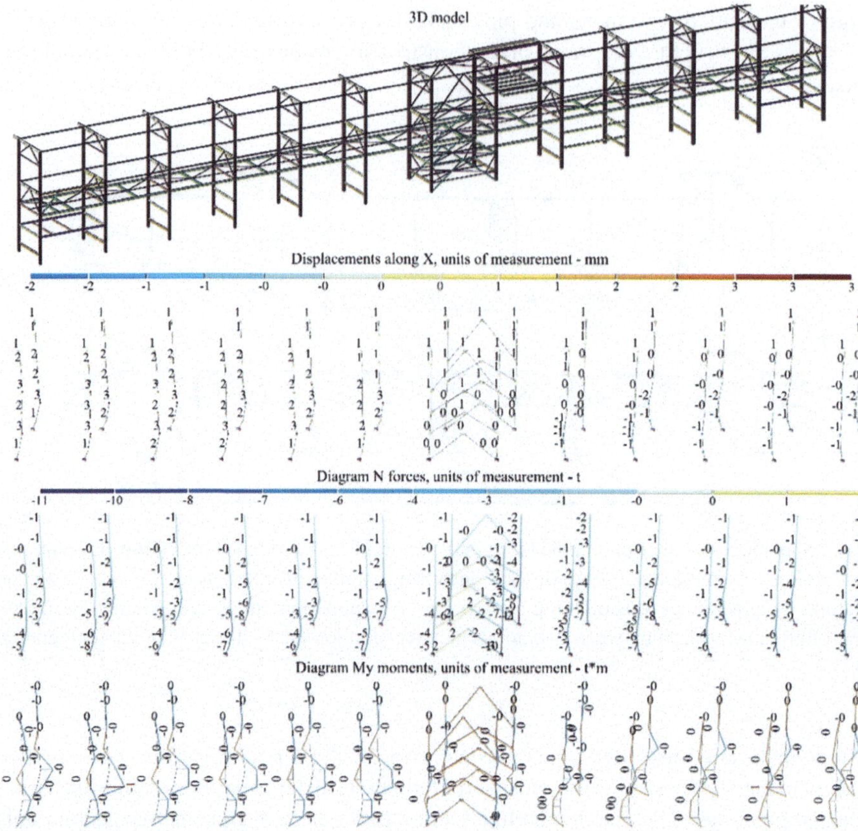

Fig. 2. Design model and the achieved displacements and forces for a section of a pipe rack with an anchor structure.

Intermediate supports are made single-column or transverse frames. Closed profiles are typical for steel single-column supports: pipes and paired channels, less often I-beams. There are also intermediate supports constructed as a steel space structure in which elements are made of angles or pipes, all elements work as centrally compressed or stretched. For reinforced concrete single-column supports, columns of square and rectangular cross-sections are used [9], including those with consoles for supporting longitudinal beams and support beams on them, as well as variable cross-section (stepped) columns. In 1979, along with the 3.015-1/77 and 3.015-3/77 standard series, centrifugal reinforced concrete ring-section columns were experimentally put into operation, but they did not become widely used.

Transverse frames combine several columns in a plane across the track (see Fig. 5). When made of steel, columns are typically interconnected by a system of vertical bracings that reduce the effective column length in the plane of the support. Columns in such supports are made in various cross-sections: I-beams, paired channels, pipes; and vertical bracings are constructed as single or paired angles. When made of reinforced concrete, columns in transverse frames are usually connected only by support beams, without

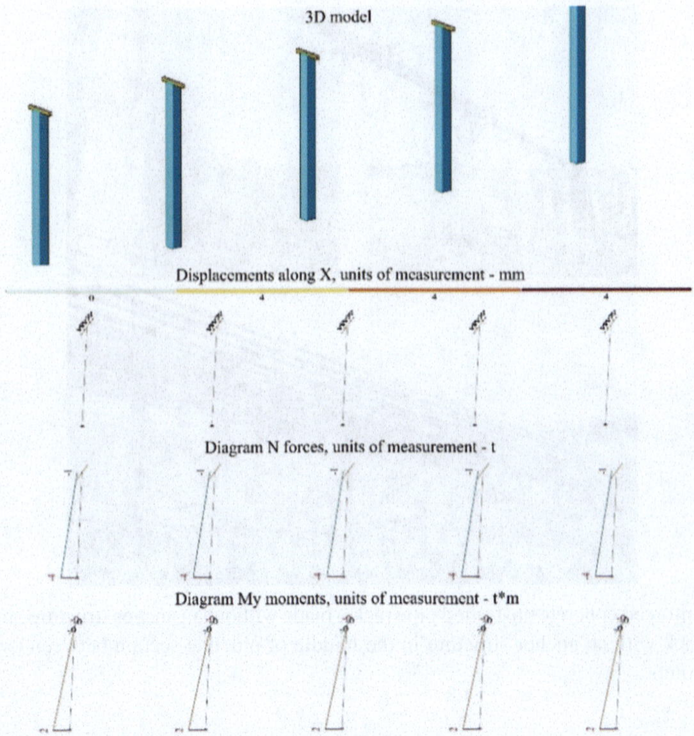

Fig. 3. Design model and the achieved displacements and forces for the section of a pipeline section between two adjacent expansion joints of unstrutted pipe racks without an anchor structure.

vertical bracings. In such supports, columns of same sections as in the single-column ones are used: square and rectangular sections and round centrifugal ones. The IS-01-3 and IS-01-06 standard series also provided reinforced concrete double-leg columns.

Pinned base columns can be used in columns of transverse frames, if the stability of geometrical shape of the transverse frames along the route is provided by a system consisting of longitudinal beams or struts and an anchor structure (see Fig. 2). In a plane across the route, the transverse frame column can pinned base columns when there are vertical bracings or fixed joints of columns with support beams (Fig. 4).

As mentioned earlier, anchor structures mainly serve to perceive longitudinal lateral loads from pipelines in the intermediate supports of their pipeline section between two adjacent expansion joints, often providing stability of geometrical shape of intermediate supports in them (see Fig. 2). Anchor structures are often space (see Fig. 6): support chords are made of angles, pipes, I-beams or paired channels; the elements connecting the chord is made of single or paired angles. Due to the triple lattice, the forces in the elements of such a support are predominantly tensile or compressive.

Flat anchor structures are also met; they are oriented in the plane of the route. These anchor structures can be made in both of steel and reinforced concrete (see Fig. 6). In both cases, connecting elements of hot-rolled section is used. The columns of such anchor structures are made of I-beams, pipes, paired channels and reinforced concrete

Fig. 4. Reinforced concrete unstrutted pipe racks, made without an anchor structure, and a classic steel pipe rack with an anchor structure in the middle of pipeline section between two adjacent expansion joints.

Fig. 5. Transverse frames made of reinforced concrete and steel.

sections. Same as in intermediate supports, it is possible to use flat two-leg reinforced concrete columns, standard series IS-01-3 and IS-01-06, oriented along the route.

The longitudinal beams connecting the supports of the pipeline section between two adjacent expansion joints are made of reinforced concrete and steel. In reinforced concrete longitudinal beams, classical beams of rectangular or I-shaped cross-section with a span of 12 m are most often found. Additionally, panel girders with a span of 12 m according to the 3.015-2 standard series and a span of 18 m according to the 3.015-2/82 standard series were developed. We can particularly mention the longitudinal beams

Fig. 6. Anchor space steel structure and anchor flat reinforced concrete structure.

made in the form of segmented reinforced concrete trusses with a span of 18 m according to the 3.015-3/77 standard series, but they have not become widely used.

Steel longitudinal beams are mainly constructed as I-beams, followed by sections of channels, paired channels and closed pipe profiles, as well as classic trusses with a span of 12 and 18 m [10]. When crossing significant obstacles, atypical shapes with large span sizes are used. When investigating one the objects, the author of the article met trusses with a span of 36 m (see Fig. 8). Support beams for pipelines, as well as lateral bracing and struts, increasing the stability of longitudinal beams from the deformation plane and ensuring the joint operation of building structures under the influence of lateral loads directed across the route (wind and earthquake loads, as well as the load from pipelines during turns of the route and branches [11]) are placed on longitudinal beams. Longitudinal beams also act as struts, connecting all the supports. Thus, longitudinal beams are involved in ensuring the stability of geometrical shape of the intermediate supports of the pipe track along the route, and in transferring longitudinal forces from the temperature deformations of pipelines and pipe rack structures to anchor structures (Fig. 7).

Fig. 7. Longitudinal beams made of reinforced concrete as prefabricated lattice girder and steel I-beams.

Fig. 8. Longitudinal beam made as a steel truss with a span of 36 m.

Support beams, serving for direct perception of loads from pipelines, platforms, cable lines are placed on supports and longitudinal beams. Support beams made of reinforced concrete are made of rectangular cross–section, and steel ones are made of standard hot-rolled sections: closed pipes, I-beams, channels [12]. During operation, support beams experience a complicated stress-strain state. Gravity loads from the weight of the laid engineering networks and lateral loads from pipelines are transmitted to the support beams. Thus, the support beams experience bending in two planes (skew bending) and torsional stresses caused by lateral loads from pipelines that are applied to the upper face of the support beams. As mentioned above, movable and fixed supports for pipelines are located on support beams. Fixed supports transmit significant lateral forces. Therefore, for these support beams, widely developed sections in the horizontal plane are provided, for example, paired profiles (double I-beams, channels, quadrant or rectangular pipes).

Moving now to cabled pipe racks, it is worth being noted that structurally they consist of a number of supports that are connected by longitudinal beams [13]. Due to the lack of lateral loads from pipelines, cabled pipe racks are often designed without anchor structures within a pipeline section between two adjacent expansion joints. Typically, the supports of steel cabled pipe racks are made of single-column pipes or other closed sections [14] (see Fig. 9). Longitudinal beams are made of I-beams, channels, paired channels, closed sections of pipes or trusses. The cable lines being laid are usually placed in special cable trays with lids that protect the cables from solar radiation and weather elements.

Speaking of the prospects for the development and improvement of the structural part of pipe racks, the growing use of structures made of various composite materials in construction should be noted [15]. The advantages of composite materials are well known: light weight, relatively high strength, minimal labor costs during installation, corrosion resistance and, consequently, durability. However, the materials also has their disadvantages: high cost, poor development of the regulatory framework in Russia, as well as low fire resistance. They have been trying to combat the latter disadvantage for a long time and now new types of materials with increased fire resistance are beginning to appear [16].

Fig. 9. Lightweight cabled pipe track with pipes as columns.

The use of modular structures in construction is worth mentioning on a separate note. In the past 15 years, at least three types of modular collapsible lightweight structures of cabled pipe racks have been patented in Russia [17–19]. Such designs have a number of advantages: high interchangeability of unified modular elements; the use of effective thin-walled profiles provides reduced steel consumption; light weight of structures and the absence of installation welding minimize costs at the stage of transportation and installation, and the factory protective coating applied to the elements by hot-dip galvanizing ensures high corrosion resistance of steel structures [20].

4 Conclusions

The article provides a description of unstrutted pipe racks and pipe racks; it indicates the importance of their role in process chain of industrial enterprises. A detailed review of structural part of these engineering structures was performed; key elements of pipe racks were described: anchor structures, intermediate supports, longitudinal struts and support beams. A review of steel and reinforced concrete structures of these engineering structures was made. Additionally, the article describes loads and actions, acting on unstrutted pipe racks and pipe racks with an estimate of stain-stress state of these structures.

When working on the article, the author used personal experience of more than 10 years in structural investigation of unstrutted pipe racks and pipe racks. Apart from examples of his own works used in the article, an additional analysis of all the standard series, created in the Soviet Union and modern Russia (summing up to a total of 157 units) was made, other regulatory framework and patent base on the research subject were thoroughly analyzed.

As a result of the conducted work, it can be stated that most design organizations use the developments of standard series, created back in the Soviet Union when designing unstrutted pipe racks and pipe racks.

However, technical capabilities of enterprises and construction and installation organizations for the production and installation of more advanced forms of building structures have increased significantly since then. The author of the article thinks that further research should be carried out in order to find new more effective forms with low steel consumption for unstrutted pipe racks and pipe racks.

References

1. Shebeko, A.Y., Shebeko, Y.N., Gordienko, D.M.: A settlement assessment of equivalent fire duration for steel structures of pipe rack of a refinery. Fire Safety **1**, 25–29 (2017)
2. Zhdanov, D.S., Shenkman, R.I.: Features of the calculation of free-standing pillar of technological pipelines, taking into account the joint operation of the "ground-foundation construction" system. Modern technologies in construction. Theory Pract. **1**, 148–151 (2022)
3. Novoselov, O.G., et al.: Method for calculating the strength of massive structural elements in the general case of their stress-strain state (parametric equations of the strength surface). Constr. Mater. Prod. **6**(2), 104–120 (2023). https://doi.org/10.58224/2618-7183-2023-6-2-104-120
4. Smirnov, V.V., Svitsov, M.A., Shileeva, A.: About the technical condition of the process pipe racks of the metallurgical enterprise. Sci. Technol. Educ. **10**(16), 119–121 (2015)
5. Novoselov, O.G., et al.: Method for calculating the strength of massive structural elements in the general case of their stress-strain state (kinematic method). Constr. Mater. Prod. **6**(3), 5–17 (2023). https://doi.org/10.58224/2618-7183-2023-6-3-5-17
6. Kablukov, A.V.: Numerical modeling of optimal design of linear pipeline system bearing structures. Review of research papers on the issue. Scholarly Notes of Komsomolsk-na-Amure State Technical University 5(45), 31–39 (2020)
7. Pavlyuchenko, A.S., Grigorenko, N.I.: Arrangement of fixed supports during aboveground installation of the water supply and sewage pipelines. Proceeding of the Donbas National Academy of Civil Engineering and Architecture **5**(133), 51–55 (2018)
8. Shereshevsky, I.A.: Construction of industrial buildings and structures: an album of drawings. Moscow (2010)
9. Mannapov, R.M.: Analysis of SSS of reinforced concrete panels on the basis of calculation in SP LIRA-SAPR with damage included. Sci. Res. XXI Century **2**(22), 56–58 (2023)
10. Nikolenko, S.D., Sazonova, S.A., Osipov, A.A., Akamsina, N.V.: The process of assessing the condition of structures of flasses in forestry. The modern forest complex of the country: problems and prospects of development: Materials of the All-Russian Scientific and Practical Conference, Voronezh, pp. 202–208 (2022)
11. Nikolenko, S.D., Sazonova, S.A., Osipov, A.A.: Control of the performance characteristics of building materials of supports of technological overheads. Aspects of modeling systems and processes: Materials of the All-Russian Scientific and Practical Conference, Voronezh, pp. 242–249 (2022)
12. Nikolenko, S.D., Sazonova, S.A., Akamsina, N.V.: Assessment of the bearing capacity of the structures of technological restas. Aspects of modeling systems and processes: Materials of the All-Russian Scientific and Practical Conference, Voronezh, pp. 220–229 (2022)
13. Garkin, I.N., Sabitov, L.S., Polyakov, L.G.: Strengthening the building structures of the cable rack—a factor in improving environmental safety. Eng. J. Don **5**(101), 404–411 (2023)
14. Kotkova, O.N.: The results of the survey of cabled pipe racks. Traditions and innovations in construction and architecture. Natural Sciences and Technosphere safety: A collection of articles based on the materials of the 72nd All-Russian Scientific and Technical Conference, Samara, pp. 414–417 (2015)
15. Ivlev, A.O.: The use of composite materials for the construction of cabled pipe racks. In: Science and Youth: Proceedings of the XX All-Russian Scientific and Technical Conference of Students, Postgraduates and Young Scientists, Barnaul, pp. 391–392 (2023)

16. Ermakova, E.O., Korkishko, A.N.: The possibility of introducing composite building structures in the field development. Eng. J. Don **11**(71), 46–55 (2020)
17. Farzaliev, V.A.: Patent No. 178233 Russian Federation, IPC E01D 18/00. Collapsible pipe rack for laying communications. Application 2017133720, publ. 03/28/2018
18. Strelkov, Y.M., Sabitov, L.S., Klyuev, S.V., Klyuev, A.V., Radaykin, O.V., Tokareva, L.A.: Technological features of the construction of a demountable foundation for tower structures. Constr. Mater. Prod. **5**(3), 17–26 (2022). https://doi.org/10.58224/2618-7183-2022-5-3-17-26
19. Kolpashnikov, D.N., Ogloblin, A.V.: Patent No. 2606730 Russian Federation, IPC E01D 18/00. Universal support structure (platform). Application 2013153871, publ. 01/10/2017
20. Kucha, G.T.: Application of the light cabling modular racks of domestic production in oil and gas industry branch. Oil Ind. **7**, 120–121 (2011)

Variational Analysis of Deformation of a Rod with Initial Bending Under Induced Anisotropy

Sergey Kalashnikov[1(✉)], Elena Gurova[1], Evgeny Shvedov[1], and Tatiana Ivanova[1,2]

[1] Volgograd State Technical University, 1 Akademicheskaya St., Volgograd 400005, Russia
kalashnikov@vstu.ru

[2] Volgograd Institute of Management - Branch of RANEPA, 8 Gagarin Str., Volgograd 400066, Russia

Abstract. Longitudinal bending of a flexible rectilinear rod with an initial imperfection in the form of an initial deflection is considered. The inhomogeneity of the stress state in the bent state leads to constrained deformations causing variability of the elastic characteristics of the material. To obtain the equation of the rod in the deflected state, the theory of deformation of bodies in inhomogeneous stress fields with induced anisotropy of properties proposed earlier by the authors is used. The solution is based on the variational analysis of the rod deflections using the Bubnov-Galerkin method. The growth of the maximum deflection at the steps of successive loading with increasing force is analyzed. Three variants of initial curvatures, including vanishing small curvature, are considered. Approximate integration is performed using simple basis functions. Regardless of the degree of initial curvature and the choice of approximating function, a marked increase in the compressive force corresponding to a significant increase in the deflections is found compared to the bifurcation approach.

Keywords: Initial Curvature · Stress Gradient · Critical Forces · Induced Anisotropy · Basis Functions

1 Introduction

A number of experimental works have established that isotropic bases [1], polycrystalline and multicomponent metals and alloys can exhibit anisotropy properties during deformation [2–8] due to the reorientation of crystallographic axes during fabrication and deformation. Elastomers and polymer composites can exhibit anisotropy depending on the modes of fabrication, deformation or operation [9–13]. Anisotropy of this type is called induced anisotropy.

Previously, the authors proposed and experimentally verified [14] a model of elastic material behavior and a group of physical equations, when it is considered that the stress distribution causes anisotropy of physical properties of the material, depending on the degree of inhomogeneity of the stress state in the vicinity of the point under consideration. Accounting for increments of the sought quantities realizes the incremental

S. V. Klyuev et al. (Eds.): ICRAAC 2024, LNCE 627, pp. 182–190, 2025.
https://doi.org/10.1007/978-3-031-82938-3_22

approach, which in recent decades has been used in deformable body mechanics [15–17], for rod systems, including nonlinear stability problems [18, 19]. In the proposed model, it is assumed that the inhomogeneous stress state causes strain constriction and local transversal isotropy takes place at each point of the body. By the initial linear solution, similarly to the elastic solution method, the elastic characteristics are adjusted for multiple realizations of the solving differential equations. The foregoing is a model and solving incremental relations of a body with curvilinear transversal anisotropy induced by the type of stress state. In [20, 21], this model was applied to the longitudinal bending of a compressible rod. In [22], the problem is solved by variational analysis of deflections of an idealized rod using the Bubnov-Galerkin method. In [23], the problem of longitudinal compression of a rod with initial deflection was solved. As a result, a noticeable increase in the compressive force corresponding to a significant increase in deflections was found.

It is reasonable to apply the Bubnov-Galerkin method, which is an approximate method of integration of differential equations of the boundary value problem. For an equation of the following form $\Phi(x, y, y', \ldots) = 0$, the solution is found in the form

$$y = \sum_{k=1}^{n} C_k \varphi_k(x), \tag{1}$$

where the basic functions $\varphi_k(x)$ satisfy all the boundary conditions of the problem at $x = x_1$ and $x = x_2$, and C_k are uncertain parameters, which are found from the condition of orthogonality of the non-convexity to the basic functions:

$$\int_{x_1}^{x_2} (x, y, y', \ldots)\varphi_k(x)dx = 0, \tag{2}$$

The latter decomposes into a system of C algebraic equations.

2 Methods and Materials

Let us consider a simple rod compressed by a force F applied at the center of gravity of a rectangular cross-section. The line of action of the force passes through the center of the supporting joints. Due to technological inaccuracies in manufacturing, it has an initial loss (Fig. 1) in the form of a half-wave of a sinusoid, where f_m is the highest value of the deflection in the middle of the rod.

Any increase in force will give the curvature an increment and increase the displacement of each point along the length of the rod by an additional deflection y.

The inhomogeneity of the stress state is created along the height of the cross-section "h". In each cross-section along the length of the rod, the degree of inhomogeneity is different due to different deflection magnitudes, hence, a transversal anisotropy induced by deflection "y" is created, functionally depending on the cross-sectional coordinate y^*. The location of the anisotropy axes is shown in Fig. 2. The direction of the gradient vector T is normal to the plane S, defining it as the elastic symmetry axis α. In the plane

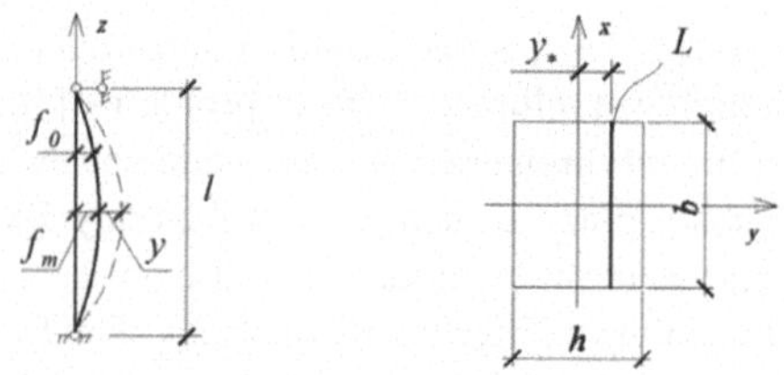

Fig. 1. Calculation diagram of a compressed rod with initial deflection.

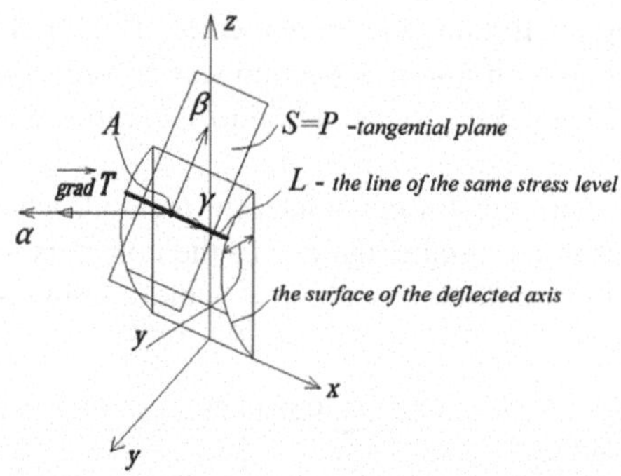

Fig. 2. Position of anisotropy axes in compression of a rod with initial deflection.

of isotropy, let us assume the direction of the γ axis coinciding with the line of the same stress level L, and the β axis - normal to the line L in the plane S. The material layers in the direction α are compressed, hence the increased modulus of elasticity is $E_\beta = E_{gr}$, a $E_\alpha = E_0$ [20, 21].

Modulus of elasticity of the material in the plane tangent to the curved axis [14] is

$$E_{gr} = E_0 \frac{\lambda_E + 1,5g}{\lambda_E + g}, \tag{3}$$

where the stress state inhomogeneity function for any section is

$$g = \frac{gradT}{T} = \frac{12y_{tot}}{h^2 + 12y_{tot} \cdot y_*}, \text{ a } gradT = \sqrt{\left(\frac{\partial T}{\partial y_*(h)}\right)^2}. \tag{4}$$

The known "moment-curvature" relation written for the rod axis at any stage of loading taking into account (3) - (4) leads to equation

$$\left[\frac{a + 1.5\left(y + f_m \sin\frac{\pi z}{l}\right)}{a + y + f_m \sin\frac{\pi z}{l}}\right] \cdot E_0 J_x y'' = -F\left(y + f_m \sin\frac{\pi z}{l}\right), \tag{5}$$

where $a = \lambda_E h^2/12$ denoted, and the material characteristic $\lambda_E = 20.1587\text{m}^{-1}$ was determined in [14] from experimental data of other authors.

Using the Bubnov-Galerkin method, we find an approximate solution of the second-order nonlinear Eq. (5) satisfying the boundary conditions, and a simple parabola can be chosen as the basis function:

$$y = C_1 z(l - z). \tag{6}$$

Substituting y and y'' by (6) into the left-hand side of (5), we obtain the incoherence:

$$R_1(z, C_1) = C_1^2\left[3z(z-l) + \frac{F}{E_0 J} z^2 (l-z)^2\right] + \frac{F}{E_0 J} f_m \sin\frac{\pi z}{l}\left(a + f_m \sin\frac{\pi z}{l}\right)$$
$$- C_1\left\{2a + 3f_m \sin\frac{\pi z}{l} - \frac{F}{E_0 J}\left[(l-z)\left(az + 2zf_m \sin\frac{\pi z}{l}\right)\right]\right\}. \tag{7}$$

The condition of orthogonality (2) of the function R_l to the basis function should be used to determine the integration constant

$$\int_0^l z(l-z) R_l(z, C_1) dz = 0. \tag{8}$$

When using a simple sinusoid as a basis function

$$y = C_1 \sin\frac{\pi z}{l}, \tag{9}$$

we have the incoherence and the orthogonality condition

$$R_2(C_1, z) = C_1^2 \sin^2\frac{\pi z}{l}\left(-1,5\frac{\pi^2}{l^2} + \frac{F}{E_0 J}\right) + \frac{F}{E_0 J} f_m \sin\frac{\pi z}{l}\left(a + f_m \sin\frac{\pi z}{l}\right)$$
$$- C_1 \sin\frac{\pi z}{l}\left[\frac{\pi^2}{l^2}\left(a + 1,5 f_m \sin\frac{\pi z}{l}\right) - \frac{F}{E_0 J}\left(a + 2f_m \sin\frac{\pi z}{l}\right)\right]; \tag{10}$$

$$\int_0^l R_2(C_1, z) \sin\frac{\pi z}{l} dz = 0. \tag{11}$$

For a given value of force F, the equations derived from (8) and (11) give solutions in the form of two roots of the integration constant C_I.

3 Results and Discussions

For a steel rod with length $l = 1$ m, dimensions $b = 0.05$ m; $h = 0.02$ m; $i_x = 5.77$ mm, Young's modulus $E_0 = 210 \cdot 10^6$ kPa, yield strength $\sigma_0 = 260$ MPa critical force $F_E = 68.4$ kN. The loading was carried out in steps starting from $0.25F_E$ to $1.4F_E$. The realization of Eqs. (8) and (11), integration and construction of the elastic line (6), (9) were performed in the author's software module running Wolfram Mathematica [24]. The solution is based on the use of the method of variable elasticity parameter at loading stages: the deformation process is numerically modeled and the load-deflection relationship is calculated. The roots of the equations have positive and negative values. The latter have no physical meaning according to (6) and (9).

The value of the initial loss $f_m = 1.0 \cdot 10^{-4}$ m is assumed to be vanishingly small. The numerical values of the roots and the value of the maximum deflection in the middle of the rod height for the loading stages are given in Table 1. Setting the rod to other values of initial curvature, not exceeding the standard allowable imperfections, we obtain for f_m = 1.0 mm (0.1% per meter) and f_m = 2.0 mm (limiting 0.2% per meter of length) results, not given in tabular form due to the economy of space, but in graphical form presented in Fig. 3. The deformation curves obtained using the parabolic basic function are shown with a solid line and the trigonometric function with a dashed line. The boundary of the Eulerian force is shown by a solid line.

The deflection growth curves remotely resemble the well-known "load-deflection" bifurcation diagram. The initial deflection of the rod leads the problem to the loss of stability of the 2nd kind: what value of the compressive force should be considered critical, especially for a vanishingly small deflection with $f_m = 0.1$ mm. Let us note that at load up to approximately $1.25F_E = 85.5$ kN the deflections do not exceed 1/500 of the span length, and up to load $1.4F_E = 95.7$ kN - 1/200 of the span (in Fig. 3 the area is shaded with ascending shading). It seems reasonable to assume that the critical force is $F_{cr} = 1.25F_E$.

At $f_m = 1.0$ mm, deflection exceeding 1/200 of the span is overcome at a load of approximately $1.15F_E$. At a load of $\approx 1.25F_E$, the displacements start to exceed the span (in Fig. 3, this area is shaded with a downward shading). Thus, according to the deformation criterion, the ultimate force can be assumed to be $1.15F_E < F_{cr} < 1.25F_E$. For a rod with $f_m = 2.0$ mm, the longitudinal force corresponding to a deflection of 1/200 of the span $< F_E$, a deflection of 1/100 of the span length is overcome at $1.15F_E$ (the area is shaded with dots in Fig. 3). Therefore, the critical force can be assumed to be $F_{cr} \approx 1.0F_E$.

The results of using the well-known formula from [25] in the classical approach are shown in Fig. 3 by a pale line (the area of allowable deflections is indicated by a filler). A distinctive characteristic of a particular compressible rod according to [25] is the ratio of the maximum loss to the radius of inertia of the cross-section in the plane of least stiffness. In this case there are $\zeta_{0,1} = 0,0173; \quad \zeta_{1,0} = 0,173; \quad \zeta_{2,0} = 0,347$. The larger this value is, the faster the additional deflections should increase with increasing compressive force, which means increasing deformability. The graphs shown in Fig. 4 are plotted in these semi-inverse coordinates.

A coordinate system is superimposed on them, which in semi-logarithmic coordinates shows the dependence of the magnitude of total deflection $|y_{tot}|$ on the ratio of the allowable total deflection to the magnitude of maximum deflection f_m. Dotted lines show curves for different values of f_m. The areas of realizable deflections for each case are shaded.

The joint analysis of Figs. 3 and 4 shows a qualitative difference in the behavior of the system when the incremental material deformation model is used, satisfactorily coinciding with [23]. The deflections increase at loads larger than the Euler force, in addition, the larger is the initial death related to the radii of inertia, the greater is the resistance. That is, the larger the value of the initial death f_m, the smaller the value of the additional deflection will be required to enhance the gradient effect.

Table 1. Values of integration constants by (8) and (11) and deflections by (6) and (9) for the rod with $f_m = 0.1$ mm.

F, kN	*F/ FE*	Parabola by (6)			Sine curve by (9)		
		C1	Additional deflection *y*, mm	Total deflection y_{tot}, mm	C_1	Additional deflection *y*, mm	Total deflection y_{tot}, mm
5	0.075	$28.7 \cdot 10^{-6}$	0.0071	0.107	$7.4 \cdot 10^{-6}$	0.074	0.11
10	0.15	$61.8 \cdot 10^{-6}$	0.0154	0.115	$15.9 \cdot 10^{-6}$	0.0159	0.12
17.1	0.25	$118.3 \cdot 10^{-6}$	0.03	0.13	$30.5 \cdot 10^{-6}$	0.0305	0.13
34.2	0.5	$329.7 \cdot 10^{-6}$	0.082	0.182	$84.1 \cdot 10^{-6}$	0.0844	0.18
51.3	0.75	$781 \cdot 10^{-6}$	0.195	0.295	$194.5 \cdot 10^{-6}$	0.1945	0.29
68.4	1.0	$20.5 \cdot 10^{-4}$	0.51	0.61	$4.75 \cdot 10^{-4}$	0.475	0.58
69.4	1.015	$21.8 \cdot 10^{-4}$	0.55	0.65	$5.03 \cdot 10^{-4}$	0.503	0.60
85.5	1.25	$81.8 \cdot 10^{-4}$	2.044	2.144	$14.91 \cdot 10^{-4}$	1.491	1.59
88.9	1.3	0.0127	3.18	3.28	$20.2 \cdot 10^{-4}$	2.02	2.12
90.6	1.325	0.0167	4.175	4.275	$24.01 \cdot 10^{-4}$	2.401	2.50
92.3	1.35	0.0237	5.93	6.03	$29.1 \cdot 10^{-4}$	2.91	3.01
94	1.375	0.0386	9.65	9.75	$36.2 \cdot 10^{-4}$	3.62	3.72
95.7	1.4	0.0919	22.97	23.07	$46.9 \cdot 10^{-4}$	4.69	4.49

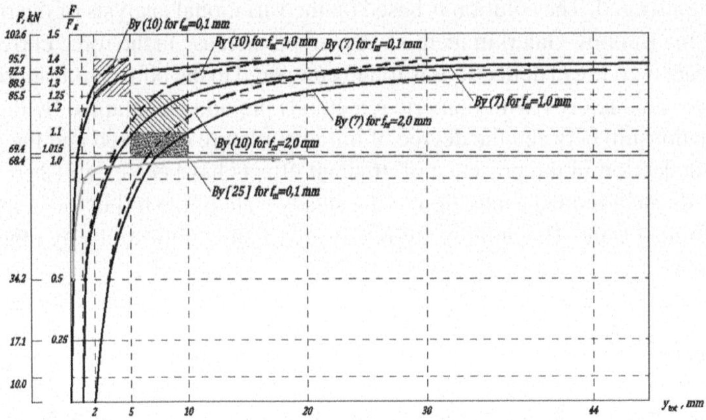

Fig. 3. Plots of the dependence of the deflection in the middle of the rod height on the compressive force at different values of f_m.

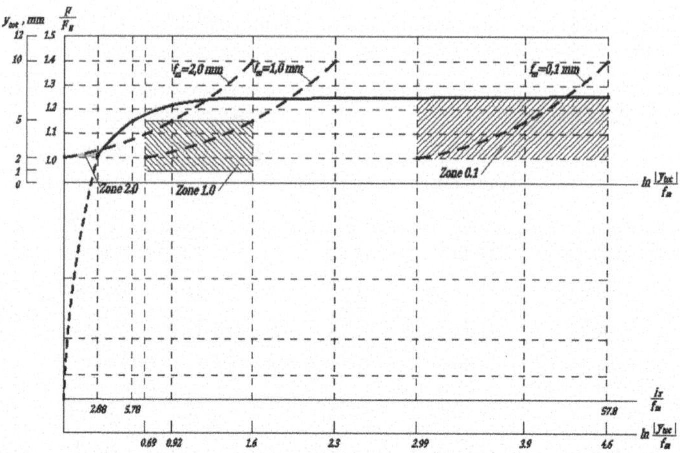

Fig. 4. Graphs of dependence of total relative additional deflection on compressive force at different initial deflections.

4 Conclusions

The theoretical constructions presented in the paper confirm the previously published results that in a compression-bent rod due to the inhomogeneity of the stress state in the cross-sections a curvilinear anisotropy induced by the type of stress state with complexly changing elastic characteristics is created.

The efficiency of solving the corresponding equation of a bent rod in the deflected state is demonstrated. The solution is based on the variational analysis of the rod deflections using the Bubnov-Galerkin method. A marked increase in the force corresponding to a significant increase in deflections is found compared to the bifurcation approach. The results agree satisfactorily with numerical methods previously obtained by the authors.

The relationship between the degree of initial curvature of the rod and the growth of additional deflection in the presence of gradient effects has been established. The new solutions indicate a refined evaluation of the steady state of the rod in the compression-curved deformed state. The authors are not aware of any publications by other authors in this setting.

References

1. Xiao, Y., Zhang, Z., Wang, J.: Granular hyperelasticity with inherent and stress-induced anisotropy. Acta Geotech. **15**, 671–680 (2020). https://doi.org/10.1007/s11440-019-00768-z
2. Busko, V.N., Osipov, A.A.: Application of the magneto-noise method for control of mechanical anisotropy of ferromagnetic materials. Instrum. Methods Meas. **10**(3), 281–292 (2019). https://doi.org/10.21122/2220-9506-2019-10-3-281-292
3. Popovich, A.A., Sufiyarov, V.Sh., Borisov, E.V., Polozov, I.A., Masailo, D.V., Grigoriev, A.V.: Anisotropy of mechanical properties of products made by selective laser melting of powder materials. Proc. Universities. Powder Metall. Funct. Coat. **3**, 4–11 (2016). https://doi.org/10.17073/1997-308X-2016-3-4-11

4. Novoselov, O.G., et al.: Method for calculating the strength of massive structural elements in the general case of their stress-strain state (parametric equations of the strength surface). Constr. Mater. Prod. **6**(2), 104–120 (2023). https://doi.org/10.58224/2618-7183-2023-6-2-104-120
5. Novoselov, O.G., et al.: Method for calculating the strength of massive structural elements in the general case of their stress-strain state (kinematic method). Constr. Mater. Prod. **6**(3), 5–17 (2023). https://doi.org/10.58224/2618-7183-2023-6-3-5-17
6. Muravyev, V.V., Muravyeva, O.V., Volkova, L.V.: Influence of anisotropy of mechanical properties of thin-sheet rolled steel on informative parameters of Lamb waves. Steel **10**, 75–79 (2016)
7. Finelli, A., Labanti, M.: Analysis of the influence of the anisotropy induced by cold rolling on duplex and super-austenitic stainless steels. Frattura ed Integrità Strutturales **13**, 24–30 (2010)
8. Loginov, Y.N., Puzanov, M.P.: Influence of properties anisotropy on stress-deformed state at rolling stripes from electrical steel. Chernye Metally **10**, 22–27 (2018)
9. Ustinov, K.B.: About induced anisotropy of mechanical properties of elastomers. Izv. Russ. Acad. Sciences. Mech. Solid Body **5**, 27–36 (2019). https://doi.org/10.1134/S05723299190 50167
10. Mokhireva, K.A., Svistkov, A., Solod'ko, Vladislav N., Komar, L., Stöckelhuber, K.: Experimental analysis of the effect of carbon nanoparticles with different geometry on the appearance of anisotropy of mechanical properties in elastomeric composites. Polym. Test. **59**, 46–54 (2017). https://doi.org/10.1016/j.polymertesting.2017.01.007
11. Shadrin, V.V., Mokhireva, K.A., Komar, L.A.: Anisotropy of mechanical properties of filled vulcanizers under the influence of external load. Bull. Perm Fed. Res. Cent. **1**, 93–98 (2017)
12. Korneev, S.A., Korneev, V.S., Romanyuk, D.A.: Mathematical modeling of the effect of induced deformation anisotropy of rubber-cord elastic element of flat coupling. Omsk. Sci. Bull. **3**(153), 10–15 (2017)
13. Komar, L.A., Mokhireva, K.A., Morozov, I.A.: Investigation of the appearance of anisotropic properties of polymer nanocomposites as a result of pre-deformation under biaxial loading conditions. Bull. Perm Fed. Res. Cent. **2**, 61–66 (2017)
14. Kalashnikov, S.Yu.: Experimental Verification of the Model of Material Deformation in Conditions of Inhomogeneous Stressed State: Monograph. VolgSTU, Volgograd 80 p. (2017)
15. Petrov, V.V.: Incremental equations of deformable body mechanics in complete functions. Bulletin of the Department of Building Sciences. 1 RAASN Ivan. State Institute of Archit. Anf Civil Eng. M. Ivanovo, pp. 159–166 (2010)
16. Zinoviev, A.S.: Stress-strain state of the system "slab - base layer" on the basis of incremental model of deformation. Bull. Saratov State Tech. Univ. **1**(37), 27–33 (2009)
17. Petrov, V.V.: Nonlinear Incremental Construction Mechanics. Infra – Engineering, 480p. (2014)
18. Kupavtsev, V.V.: Basic functions of the method of two-sided estimations in stability problems of elastic inhomogeneously compressed rods. Bull. MSCU **6**, 63–70 (20136)
19. Inozemtsev, V.K., Sineva, N.F., Inozemtseva, O.V.: Incremental model for investigating the stability of a high-rise structure on an inhomogeneous base. Constr. Mech. Eng. Struct. Constr. **2**, 41–46 (2008)
20. Kalashnikov, S.Yu., Gurova, E.V., Kuramshin, R.Kh., Yazyev, B.: About the distortion model of operational compressed-bent bars with induced anisotropy. In: Ginzburg, A., Kashevarova, G. (eds.) International Scientific Conference on Building Life-cycle Management. Information Systems and Technologies (Moscow 26 November 2021), pp. 95–102. Springer, Cham (2022). https://doi.org/10.1007/978-3-030-96206-7_10

21. Kalashnikov, S., Gurova, E., Bandurin, N.: Longitudinal compression of a rod with initial deflection acquiring induced anisotropy. Lect. Notes Civ. Eng. **436** (2024). https://doi.org/10.1007/978-3-031-44432-6_19/
22. Kalashnikov, S., Gurova, E., Shvedov, E.: Influence of the type of basis functions in the Bubnov-Galerkin method in the deformation analysis of a compressed-curved rod with induced anisotropy. Lect. Notes Civ. Eng. **436** (2024). https://doi.org/10.1007/978-3-031-44432-6_6
23. Kalashnikov, S., Gurova, E.V., Bandurin, N.G.: Longitudinal compression of a rod with initial perdition acquiring induced anisotropy. Constr. Reconstr. **2**(106), 25–38 (2023)
24. Kalashnikov, S.Yu., Gurova, E.V., Kuramshin, R.Kh., Ivanova, T.B., Shvedov, E.G., Knyazev, D.S.: Certificate of state registration of computer program 2023660982, dated May 25, 2023, Russian Federation. Analysis of the critical state of a rod with initial perdition at induced anisotropy. Volgograd State Technical University (2023)
25. Volmir, A.S.: Stability of Elastic Systems. M. Fizmatgiz, 880 p. (1963)

Experimental Studies of the Structures of Protective Piers

A. A. Bunov(✉) and N. V. Shunko

Candidate of Engineering Sciences, Moscow State University of Civil Engineering, Yaroslavskoe Shosse, 26, Moscow 129337, Russia
A_bunov@mail.ru

Abstract. Design and construction of sea port hydraulic structures is a complex and costly process. When developing a project for hydraulic structures, the question of optimizing design solutions always arises. One of the methods of such optimization is to clarify the accepted design loads on the structures. In particular, when calculating slope structures, loads and effects from waves should be determined based on the results of physical modeling. This article considers hydraulic structures of protective piers intended for construction in the southern region of our country. Their designs are presented in detail and the main types of figured concrete blocks often used in modern sea port construction are described. Experimental studies of wave action on structures of protective piers are presented as part of scientific support and construction of high-responsibility class structures. Based on the conducted experimental studies, data on the value of wave splash on the structures under study were obtained and the value of the wave reflection coefficient was measured. The absence of water overflow through the superstructure of the protective piers along the wave movement front was experimentally established, and the stability of the figured concrete blocks in the protective fill of tetrapods on the slope of the protective structures under study was experimentally substantiated.

Keywords: Protective Pier · Wave Parameters · Wave Splas · Tetrapod · Physical Modeling · Wave Reflection Coefficient

1 Introduction

To protect against the impact of destructive wind waves, various types of hydraulic structures (HS) were developed. They differ in their functional purpose and various design features. All types of HS can be considered as an obstacle to the propagation of waves. Based on this, all types of structures are divided into several classes. The first class includes structures in the form of a continuous vertical wall of long duration. This design is designed to protect the water area of ports from penetrating waves, mainly these are moles and breakwaters. The main feature of such structures is that the transverse dimensions of such protective structures exceed their length in the direction of movement of the calculated wave front. When waves approach such structures, they will be completely reflected, including with the formation of a new wave system. And if the

S. V. Klyuev et al. (Eds.): ICRAAC 2024, LNCE 627, pp. 191–198, 2025.
https://doi.org/10.1007/978-3-031-82938-3_23

reflection of waves occurs without damping, then due to the phenomenon of interference of approaching and reflected waves, so-called standing waves will be formed. A feature of this type of waves is their double height, with their initially small length, which does not change over time. To eliminate such a dangerous wave transformation for the GTS, a slope design of piers with a protective stone coating has been used in engineering practice. In the absence of stone quarries in the vicinity of the construction area, slope coverings with figured concrete blocks of various configurations are used. Some of the most common and applied figured concrete blocks in modern seaport construction are tetrapod and hexabit [1]. The tetrapod shape was developed in 1950 in France and a concrete block intended for an anti-tank barrier was taken as a prototype. The experience of conducting physical modeling of cast-on GTS, as well as observation of natural objects, allows us to outline the defining main characteristics of such structures. These are, first of all, the type of the figured concrete block itself, the method of its placement on the slope (laying or cast-on), its slope, and the height of this structure [2]. To protect the water area of the dry cargo seaport of Taman from penetrating wind waves, two protective piers are provided: the Northern pier and the Southern pier. The piers are designed as slopes with a protective covering made of tetrapods. Since these structures belong to class I, then according to the main regulatory document of our country [3–7], it is necessary to conduct physical modeling of the wave impact on the structures of the studied GS.

2 Methods and Materials

Physical modeling of the calculated wave impact of the SSW direction of the storm on the protective structures in the hydrowave flume was carried out with the formation of two characteristic sections of protective moles in the wave flume. After installing the model of the investigated section of the protective mole on the bottom, the flume was filled with water to the design mark, with further reproduction of the impact of regular waves of the calculated storm of the south-southwest direction, lasting 18 h (natural data) [4] or three hours in terms of model data. Physical models of the hydrotechnical structures were made in a wave flume at a scale of 1:50, ensuring kinematic and dynamic similarity. In the experiments, a standard technique was used, observing the similarity between the natural structure and the model according to the Froude number [4, 5]. The wave parameters were measured instrumentally during the experimental studies using wave recorder sensors, the signal from which was sent via a cable system to a personal computer, where the measurement results were processed using a special program included in the Wallingford measuring complex (UK) used in this work. The wave recorders were calibrated at the beginning of each series of experimental studies and consisted of successively changing the level of calm water and recording the corresponding signal value from the wave recorder. This procedure consisted of installing the wave recorders at a certain depth, then, with a selected number of steps, taking readings from them and calculating calibration coefficients using the HR DAQ program included in the kit of the above-mentioned measuring equipment.

HR DAQ analysis programs are based on a linear relationship between the measured parameter and the output signal of the device. The calibration program applies two coefficients and a model scale to the result data, which allows them to be converted into a format acceptable for calculation, which is required for the analysis program. In the process of collecting experimental data, the values of the initial data (usually expressed in volts) are converted to binary format. When analyzing the results, the initial data are first converted to calibrated values in the model scale.

Thus, the operation of the measuring system can be represented as follows: model value = model calibration factor * (the value of the obtained data (in volts) - the zero value (in volts)),

where the model value is the value in real units (meters, tons, etc.), and the model calibration factor is expressed in model units/volt (for example, in meters/volt or tons/volt).

If the analysis is selected based on the so-called prototype (a full-scale object under study), then the operation of the measuring system is as follows: prototype value = model value * scale factor of the prototype using one of the similarity laws used in research.

2.1 Linearity of Wave Recorders

For most measuring devices, the dependence of the output signal on the input signal differs from the ideal linearity. In addition, wave measuring instruments may have a calibration factor that varies with temperature or conductivity, so it is essential to calibrate all instruments regularly.

This is done by recording a number of output signal voltages (x) for different model input values (y). The calibration program collects these N ($N > 3$) records, calculates the calibration factor, and determines the approximation accuracy using linear regression analysis.

For most tests, an approximation accuracy parameter greater than 0.999 for at least 3 calibration points in the operating range is acceptable. If the approximation accuracy parameter is less than 0.999, additional calibration points should be used as needed, and the calibration should be extended or repeated. Table 1 shows the results of the wave sensor calibration.

Table 1. Table of calibration of wave recorders in experiments.

№ wave recorder	Calibration factor, K, m/V	R^2, %
1	0.0102411	0.9962
2	0.0123172	0.9930
3	0.0114235	0.9955
4	0.0121453	0.9961
5	0.0112011	0.9942
6	0.0123172	0.9934

3 Results and Discussions

The studied section of the Southern protective pier is located in the alignment of the head section of the pier, with a foundation on the seabed at the mark: −9.300 m.

The structure of the Southern protective pier is made of the following components:

- the pier core is filled with unsorted quarry stone, the slopes of which are covered with sorted stone weighing 50–100 kg and 800–1000 kg - from the open sea side (in-kind data), respectively, in model units (M 1:50) these data were: 0.4 - 0.8 g and 6.0 - 8.0 g;
- on the slope of the Southern protective pier from the open sea and water area side, tetrapods weighing 13 tons (in-kind data) were laid on the surface of sorted stone weighing 800–1000 kg (100.0 g model data). At the bottom of the slope, on a berm made of sorted stone weighing 800–1000 kg (in-kind data) (6 - 8 g model data), thrust tetrapods weighing 20 t (in-kind data) (160.0 g model data) were installed.

The concrete pavement of the pier was 20 cm thick. On the side of maximum wave action, a reinforced concrete parapet was installed with a top mark of + 7.500 m. On the inner side of the pier, the roadway was limited by a protective beam, 50 sm high.

The design of the Northern protective pier structure was made from the sea side in a similar way, but with some differences:

- seabed mark: −8.800 m;
- on the slope of the Northern protective pier from the side of the water area, tetrapods weighing 3 t (in-kind data) (0.9 g model data) were placed on the surface of sorted stone weighing 300–500 kg (in-kind data) (0.15 g model data). At the bottom of the slope, on a berm of sorted stone weighing 300–500 kg (in-kind data) (0.15 g model data), thrust tetrapods weighing 5 t (in-kind data) (1.5 g model data) were installed. A schematic representation of the experimental studies is shown in Fig. 1.

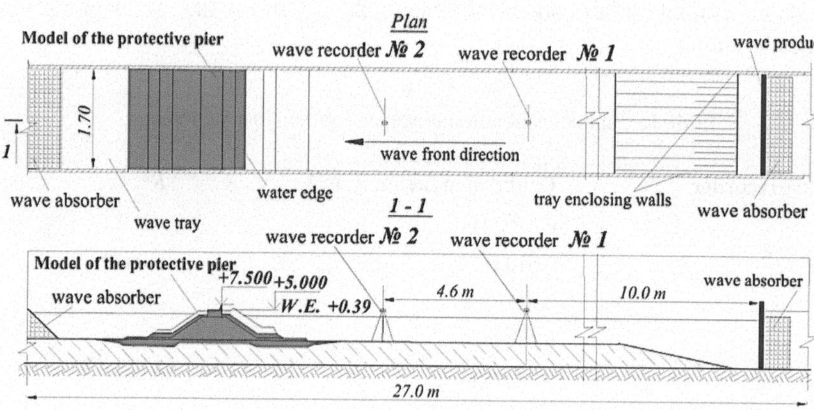

Fig. 1. Schematic representation of experimental studies.

The purpose of the experiment was to determine:

- the amount of wave splash on the structures under study;
- the wave reflection coefficient;
- the stability of the elements of the slope covering of the structures;
- the overall efficiency of the structures of the protective structures of the piers.

The water level of 1% probability was: $H_{1\%} = +\ 0.39$ m BS (in-kind data) [4]. The parameters of the storm waves in the SSW direction, with a repeatability of 1 time in 100 years: $h_{1\%} = 7.2$ m, $T_{av} = 10.0$ s (in-kind data) [4]. The experimental studies are shown in Fig. 2. The reflection coefficient values were measured in the experiments using a system of four wave sensors recording the oscillations of the wave surface during the experiment [2, 4, 5], Fig. 3.

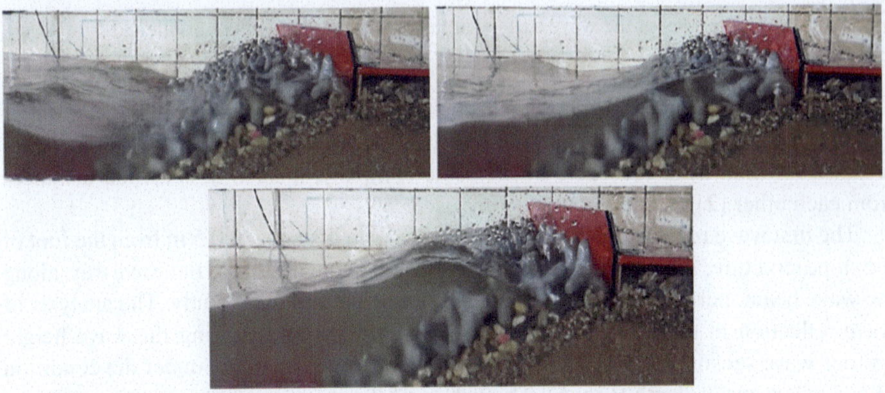

Fig. 2. Impact of waves on a protective tetrapod lining.

Fig. 3. Measuring the reflection coefficient.

Wave realizations from wave recorder No. 1 are shown in Fig. 4.

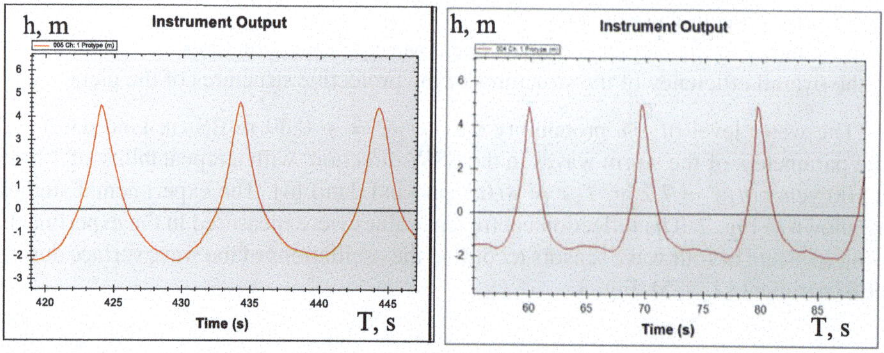

Fig. 4. Oscillations of the wave surface at the protective piers of the South and North.

To calculate the reflection coefficients using the HR DAQ program of the measuring system, in accordance with its operating instructions [2], the results of wave measurements from four wave recorders of the same company, located at specified distances from each other [2] on a flat bottom, were used.

The first wave recorder of the four was located at a distance of 0.5 m from the foot of the slope structure. All wave recorders were located in the middle of the wave tray, along the wave beam, perpendicular to the slope of the structure under study. The analysis of wave reflection in the HR DAQ program [2] is based on measuring the wave height on four wave sensors located at a known distance from each other, under the condition of a constant water depth (Fig. 5). The values of the wave reflection coefficients were automatically calculated in the HR DAQ program, using Fourier transforms, with the task of finding the Fourier coefficients of the incident and reflected waves. Testing and adjustment of the measuring equipment was carried out on the basis of the data set out in the regulatory document [2, 4].

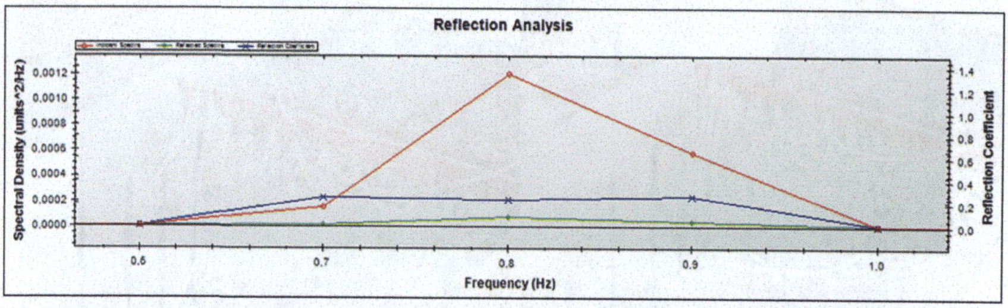

Fig. 5. Readings from the monitor screen when determining the wave reflection coefficient: red line – spectral energy of the reflected wave; green line – spectral energy of the incident wave; blue line – readings of the reflection coefficient distributed over the range of available frequencies

4 Conclusions

The slope cast-on protective structures of the piers bear the main wave and ice loads [8] from the most dangerous storms approaching the port water area from the open sea. As a rule, these are structures of I and II class of responsibility [9]. In this case, it is necessary to calculate the entire slope structure based on regulatory documentation, and then conduct physical modeling of wave effects in order to verify the adopted design solutions [3, 4].

Based on the analysis of the conducted experimental studies on the impact of a SSW storm on the sections of the Northern and Southern breakwaters, it was established that:

- the value of the wave splash on the structures under study reached +7.00 m;
- no water overflow through the superstructure of the protective breakwater structures along the wave front was recorded;
- there was no discharge of elements of the protective fill made of tetrapods from the slope;
- the wave reflection coefficient was $Cr = 0.24$.

Thus, it should be concluded that the overall efficiency of the protective breakwater structures is ensured. The structures are stable.

The application of the research results presented in this work is possible when making design and other decisions on the designs of cast slope protective structures at all stages of project development.

Based on the obtained experimental data, it should be concluded that all elements of the pier slope structures are stable, there is no damage to the structures, the structures under study are fully operational. Protective fastenings made of cast-on figured concrete tetrapod blocks ensure the stability of the entire structure from the impact of estimated storms of the SSW direction. Good wave-damping capacity of the structures, combined with the loss of wave energy during collapse on the slope, the absence of wave spillover over the top of the parapet and the passage of waves through the body of the piers, provide an improvement in the wave regime in the protected water area of the port. The overall efficiency of the protective piers is ensured. The structures are stable.

The application of the research results presented in this paper is possible when making design and other decisions on the designs of cast-on slope protective structures at all stages of project development, taking into account natural factors [10].

References

1. VSP 33-03-07/MO RF. Instructions for the design of slope and through protective structures and special underwater stands (developed by NIC 26 TsNII). Moscow (2008)
2. Zuev, N.D., Shunko, A.S., Shunko, N.V.: Investigation of coefficient of reflection of waves produced by a rock-fill inclined bank protection structure. Power Technol. Eng. **53**(1) (2019). https://doi.org/10.1007/s10749-019-01029-5
3. Rogachko, S.I., Shunko, N.V.: Scientific support of projects of Offshore hydraulic structures. Power Technol. Eng. **56**(1) (2022). https://doi.org/10.1007/s10749-023-01461-8
4. SP 38.13330.2018. Loads and impacts on hydraulic structures (wave, ice and from ships). - M., Ministry of Regional Development of the Russian Federation (2018)

5. GOST R 70023-2022. Physical modeling of wave impacts on port hydraulic structures. Requirements for model construction, experiments and processing of results. M. Rosstandart (2022)
6. Suvarna, P.S., Sathyanarayana, A.H., Umesh, P., Shirlal, K.G.: Laboratory investigation on hydraulic performance of enlarged pile head breakwater. Ocean Eng. **217**, 107989 (2020)
7. Shunko, N.V., Zuev, N.D., Shunko, A.A.: Physical modeling of wave action on the structures of the object Construction of a pedestrian embankment along Morskaya Street in Saki (Republic of Crimea). Power Technol. Eng. **57**(6) (2024). https://doi.org/10.1007/s10749-024-01745-7
8. Sharapov, D.A, Sumtsova, AS.: Stability of a stone outline to ice shear by finite element methods. Power Technol. Eng. **57**, 228–232 (2023). https://doi.org/10.1007/s10749-023-01646-1
9. Zuev, N.D., Shunko, A.A., Shunko, N.V.: Physical modeling of the projected marine terminal of the Nakhodka mineral fertilizer plant. Power Technol. Eng. **56**(2) (2022). https://doi.org/10.1007/s10749-023-01512-0
10. Korchagin, E.A: Technologies for taking into account natural factors in the construction of postal facilities. Power Technol. Eng. **56**, 862–865 (2023). https://doi.org/10.1007/s10749-023-01601-0

Beams Based on Plywood Sheets

V. I. Zhadanov(✉), S. V. Lisov, and M. A. Nesterenko

Orenburg State University, Orenburg, Russia
organ-2003@bk.ru

Abstract. Due to the increased strength and deformation properties of waterproof construction plywood in comparison with wood, it is advisable to design load-bearing beam elements consisting only of plywood. High technical and economic indicators distinguish such structures. In addition, plywood beams can be easily made in a collapsible version if their elements are manufactured on numerically controlled machines. The methods and results of experimental studies of a new type of collapsible beams developed entirely from plywood sheets without adhesive joints are considered. Experimental structures are beams 1.5 m long. The load was applied in the form of two concentrated forces in thirds of the span with the beams hinged to the supports. Experimental studies have confirmed a sufficient degree of strength and rigidity of the proposed design solutions. It is shown that the design methodology for plywood beams made on tenon ties should be based on the theoretical provisions for calculating composite wooden elements. The experiments carried out confirmed the correctness of this approach. In the engineering calculations of the proposed type of structures, it is recommended to determine the geometric characteristics of the design cross sections in the form of the moment of resistance and the moment of inertia, taking into account the compliance of wall connections with chords using the coefficients k_w and k_s.

Keywords: Beam · Plywood · Experiment · Stresses · Deformation

1 Introduction

Buildings and structures using wood and wood materials are widely used in construction practice [1–8]. Such objects are distinguished by environmental cleanliness, speed of construction, ease of processing of components. In load-bearing and enclosing structures, wood is often combined with plywood. Such wood elements include various types of roof slabs and wall panels, beams, trusses, arches, etc. Due to the use of rigid adhesive joints, wood and plywood together participate in the perception of external loads. In the known structural solutions, wood takes up the bulk of the load, and plywood performs a secondary function, acting as skins or beam walls for concentrating wood in belts. However, it seems advisable to develop purely plywood supporting beams. This is technically justified by the design resistances and moduli of elasticity of plywood that are quite high in comparison with wood under almost all types of stress-strain state. In addition, known solutions use adhesive joints, which must be made only in the factory.

S. V. Klyuev et al. (Eds.): ICRAAC 2024, LNCE 627, pp. 199–206, 2025.
https://doi.org/10.1007/978-3-031-82938-3_24

In this regard, large-sized elements are supplied to the construction site, which significantly increases transportation costs. The purpose of the study was to develop collapsible beam structures consisting only of plywood and made without the use of adhesive joints, which ensures the convenience of transportation to the place of erection of the object in convenient transport packages. To ensure the tightness of the joints, the manufacture of the elements of the proposed beams is provided on numerically controlled machines. Experimental studies were carried out to confirm the operability of the developed beams.

2 Methods and Materials

Experimental studies of the proposed types of beam structures based on plywood sheets were carried out on large-size models, which were, including in terms of nodes for connecting walls with belts, a complete analogue of full-scale elements. Three identical beams were tested, which made it possible to additionally assess the quality and stability of the proposed structures according to the technological regulations based on the use of numerically controlled machines. The effect without adhesive mechanical joints on the stress-strain state of the structure under the action of stepped increasing loads up to the moment of destruction was also evaluated. By comparison of actual and required safety factors determined according to recommendations of GOST R 57790–2017 "Wooden bearing structures. Strength and deformation test methods" assessment of the operational reliability of the tested beams was carried out taking into account their service life for 50 years. The main parameters of the experimental designs are shown in Table 1.

Table 1. Basic parameters of experimental beam structures.

Parameter	Value	Parameter	Value
Overall dimensions, $L x H x B$, mm	1500 × 300 × 150	Dimensions of spikes, $l x h x \delta$, mm	70 × 20 × 10
Distance between chords axes, mm	250	Type and thickness of plywood, mm	PRP, 10
Effective span, mm	1450	Type of supports	Hinged

To obtain an objective picture of the operation of experimental structures under load, during the tests were measured:

- the main deformations of the system are: deflection of the beams in the middle and thirds of the span, precipitation of the supports, deformation of the plywood sheet in the middle of the height of the cross section at a distance equal to half the height of the beam from the supports (assessment of the stability of the plywood wall), the amount of shear of the spikes relative to the sheets of the plywood walls of the beam in the middle of the span;
- fiber deformations of the beam wall and chords in the middle of the span with a symmetrical arrangement of redundant resistance strain gauges.

On each of the three tested beams, 6 deflection gauges, 8 indicators with a division value of 0.01 mm, 4 indicators with a division value of 0.001 mm, 12 resistance strain gauges were installed. To determine the stresses according to the readings of the resistance strain gauges, the actual values of the elastic modulus of plywood along and across the fibers of the outer layers (9000 MPa and 6000 MPa based on the test results of standard plywood samples) were taken into account. The diagram of the load application with the arrangement of measuring instruments is shown in Fig. 1, and the general view of the beam prepared for testing in Fig. 2. The design load P on the beam determined from the condition of operation for the transverse bending of plywood walls, taking into account the full inclusion in the overall work of the design of the belts, was $P = 30.0$ kN.

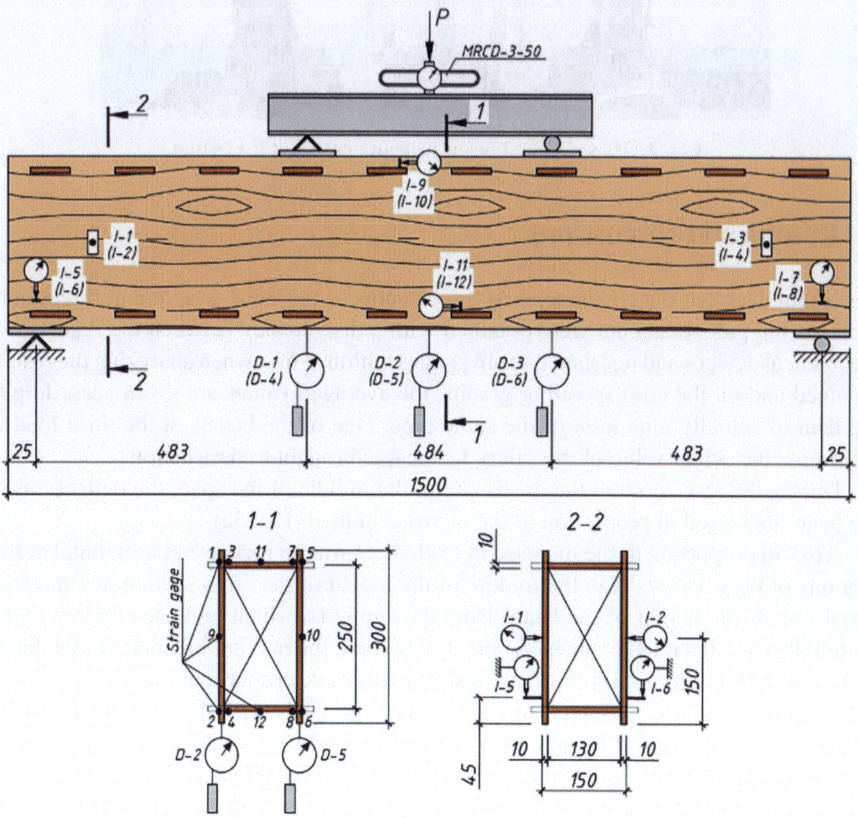

Fig. 1. Instrument arrangement and plywood beam layout for testing.

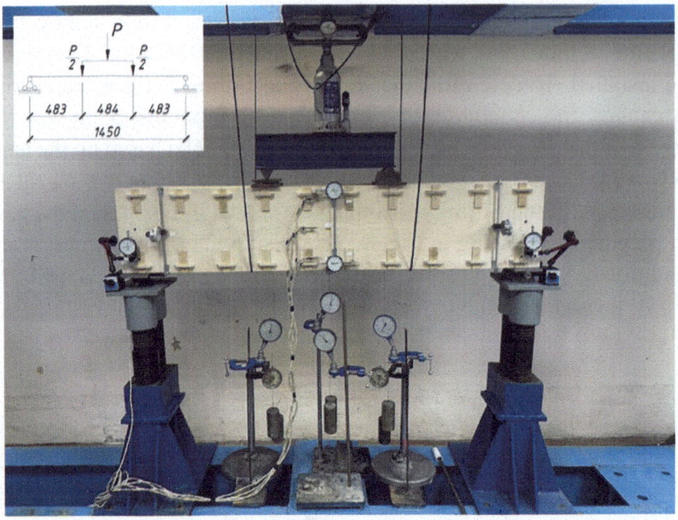

Fig. 2. Experimental beam structure prepared for testing.

3 Results and Discussions

Due to the fact that during the tests of three beams of the same type and at three times their loading, identical data were obtained with a discrepancy between the readings of mechanical devices and resistance strain gauges within + 6%, when analyzing the results obtained and on the corresponding graphs, the averaged values are given according to the data of actually nine tests of the same type. One of the beams at the third load to determine the actual value of the safety factor was brought to destruction.

During the tests, both in the thirds and in the middle of the span, the deflections of the beam increased in proportion to the increase in load (Fig. 3a).

Also, in proportion to the increase in load, there was an increase in horizontal deformations of plywood walls in the middle of the height of the cross section at a distance equal to half the height of the beam from the supports, which indicates their stability within the calculated load values. At the design load, the maximum value of this deformation was 0.08 mm or 1/125 of the wall thickness. A proportional increase in shear deformations was also observed in the joints of the walls with the studs of the belts, while their values were 0.12 mm and 0.15 mm in the lower stretched and upper compressed zones, respectively (at the calculated load value). When the load $P = 36.0\ kN$ was reached, plastic deformations began to appear during the deformation of the beam, which is most characteristically evident from the graph of the difference in deflections by loading stages (Fig. 3b). As the load increased, the proportion of plastic deformations also increased, which was an indirect sign of the exhaustion of the bearing capacity of the structure in the event of a prolonged operational load. At the same time, in the joints of the walls with the studs of the chords, a proportional increase in shear deformations was observed to the value of the load $P = 48.0\ kN$, which indicates the elastic nature of the proposed type of connection until the beam reaches the limit state.

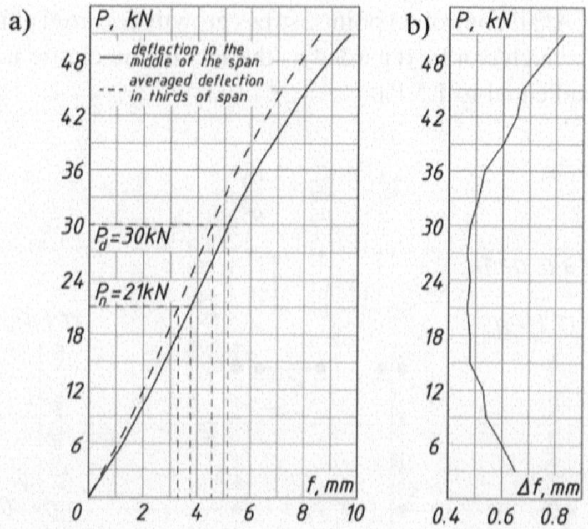

Fig. 3. Graphs of beam deflections in the middle and thirds of the span (a), and the difference in deflections along the loading stages (b).

During the experiments, stress values at characteristic points were also determined using glued resistance strain gauges. The most typical measured values of stresses are given in Table 2 (arrangement of resistance strain gauges by numbers corresponds to Fig. 1).

Table 2. Values of normal stresses by points of strain gauges location in the middle of beam span («–» – compression; « +» – stretching).

Load, P, kN	Stresses by points of strain gauges location, MPa							
	p.1	p.2	p.3	p.4	p.5	p.6	p.7	p.8
6	−2.31	2.19	−1.55	1.46	−2.28	2.16	−1.53	1.61
12	−4.66	4.36	−3.05	2.94	−4.71	4.32	−3.16	3.16
18	−6.91	6.48	−4.62	4.46	−6.81	6.38	−4.68	4.69
24	−8.94	8.55	−6.22	5.96	−8.97	8.34	−6.35	6.22
30	−11.44	10.47	−8.06	7.64	−11.26	10.32	−8.03	7.63
36	−13.56	12.53	−9.76	9.02	−13.61	12.46	−9.71	9.28
42	−15.99	14.64	−11.7	10.74	−16.06	14.52	−11.46	10.95
48	−18.35	17.07	−13.44	12.84	−18.40	16.78	−13.14	12.64

Diagrams of distribution of normal stresses in beam elements at a load of 30.0 kN in the middle of the span are shown in Fig. 4.

Analyzing the data given in Table 2 and Fig. 4, we note that the nature of the stress distribution in the left and right walls, as well as in the upper and lower chords of the beam, was completely identical and generally corresponded to the elementary theory of transverse bending. At all controlled points, stress growth occurred almost in proportion to the load growth, which can be regarded as the fact of the elastic nature of the beam deformation to a load equal to 1.5 P_d.

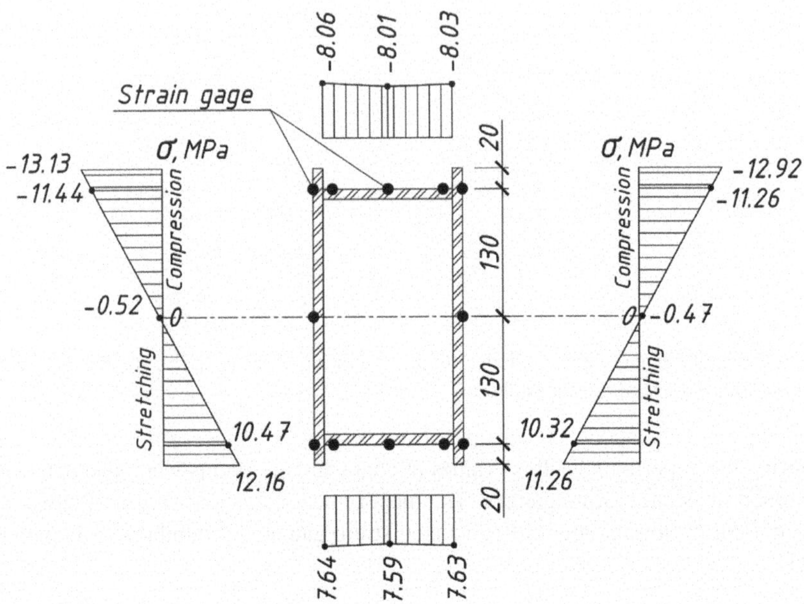

Fig. 4. The nature of the distribution of normal stresses in the walls and belts in the middle of the beam span at design load.

The value of maximum normal stresses along the line of installation of strain gauges at the design load ($P_d = 30$ kN) in the compressed and stretched zones was 11.44 MPa and 10.47 MPa, respectively. In the same sections, the maximum normal stresses in the shelves are equal: compression—8.06 MPa, tension—7.64 MPa.

The above results of the experiments show that plywood walls fully participate in the work of the beam, and the effectiveness of the participation of belts is reduced by 28%. This fact can be explained by the presence of partial compliance of the spike joints of the wall with the belts. Similarly, the actual deflections of experimental structures are higher than the theoretical values for the case of rigid connection of walls and chords (for example, on glue) by 23% when determining the calculated load value at the corrected moment of resistance (reduced by 28%).

Based on the theories of calculating composite rods, set forth, for example, in works [9–15], it is logical to take into account the share of belts in determining the geometric characteristics of the cross sections of the beam with the introduction of reducing coefficients that take into account this compliance. For example, performing the calculation of the beam by normal stresses according to the formula:

$$\sigma = \frac{M_x}{J_b} \cdot 0,5H \leq R_{p.b.}, \quad (1)$$

the moment of inertia of the cross section J_b should be determined as:

$$J_b = J_w + J_s \cdot k_w, \quad (2)$$

where

J_w— the moment of inertia of the walls;
J_s— the moment of inertia of the belts;
k_w— coefficient taking into account the value of compliance of connections of chords with walls.

For the beam in question, the coefficient k_w will be 0.72. When calculating by deformations, the moment of inertia should be determined with the coefficient k_s, the value of which for the tested beams will be equal to 0.77. Undoubtedly, when the beam span is changed, the compliance factors k_w and k_s will also change. To determine them, additional experimental work will be required when developing the engineering methodology for calculating the considered class of structures.

4 Conclusions

Experimental studies of the developed type of beams based on plywood sheets confirmed their performance and the possibility of application in construction practice. The absence of adhesive joints ensures their collapsibility. The inclusion of plywood belts of beams in the operation of the structure for the perception of external forces has been proven. The degree of their inclusion should be taken into account when calculating the moment of resistance and the moment of inertia of cross sections, taking into account the compliance of mechanical connections "belt-wall". It is advisable to evaluate the degree of compliance by the compliance coefficients k_w when calculating strength and k_s when calculating deformations. Tightness of joints is ensured by manufacture of beam elements on numerically controlled machines.

References

1. Premrov, M., Žegarac Leskovar, V.: Innovative structural systems for timber buildings: a comprehensive review of contemporary solutions. Buildings **13**(7), 1820 (2023)
2. Stefańska, A., Cygan, M., Batte, K., Pietrzak, J.: Applications of timber and wood-based materials in architectural design using multi-objective optimisation tools. Constr. Econ. Build. **21**(3), 105–121 (2021)
3. Roshchina, S., Gribanov, A., Lukin, M., Chibrikin, D., Shunqi, M.: Investigation of the stress-strain state of wooden beams with rational reinforcement with composite materials. Lecture Notes in Civil Engineering **182**, 475–483 (2022)
4. Preobrazhenskaya, I.P., Pogorel'Tsev, A.A., Turkovskij, S.B.: Development of design and construction of potassium chloride storehouse with framework from prefabricated wood frames with size of 63 m. Stroit Mater, 14–16 (2003)
5. Lukin, M., Prusov, E., Roshchina, S., Karelina, M., Vatin, N.: Multi-span composite timber beams with rational steel reinforcements. Buildings **11**(2), 46 (2021)

6. Brandner, R.: Cross laminated timber (CLT) in compression perpendicular to plane: testing, properties, design and recommendations for harmonizing design provisions for structural timber products. Eng. Struct. **171**, 944–960 (2018)
7. Ramage, M.H.: The wood from the trees: The use of timber in construction. Renew. Sustain. Energy Rev. **68**, 333–359 (2017)
8. Johannes, A., Huber, J., Ekevad, M., Berg, S.: Structural robustness and timber buildings–a review. Wood Mat. Sci. Eng. **14**(2), 107–128 (2019)
9. Ataei, A., et al.: Cyclic behaviour of bolt and screw shear connectors in steel-timber composite (STC) beams. J. Constr. Steel Res. **161**, 328–340 (2019)
10. Bedon, C., Fragiacomo, M.: Numerical analysis of timber-to-timber joints and composite beams with inclined self-tapping screws. Compos. Struct. **207**, 13–28 (2019)
11. Cavalli, A., et al.: Mechanical performance of full scale steel-timber epoxy joints after exposure to extreme environmental conditions. Int. J. Adhes. Adhes. **54**, 86–92 (2014)
12. Chiniforush, A.A., Valipour, H.R., Ataei, A.: Timber-timber composite (TTC) connections and beams: an experimental and numerical study. Constr. Build. Mater. **303**, 124493 (2021)
13. Hassanieh, A., Valipour, H.R., Bradford, M.A.: Experimental and numerical investigation of short-term behaviour of CLT-steel composite beams. Eng. Struct. **144**, 43–57 (2017)
14. Jurkiewiez, B., et al.: Experimental and analytical study of hybrid steel-timber beams in bending. Structures **39**, 1231–1248 (2022)
15. Miller, J.F.: Design and Analysis of Mechanically Laminated Timber Beams Using Shear Keys. PhD thesis in Civil Engineering. Michigan Technological University, 211 (2009)

Elimination of Historical Environmental Damage During the Comprehensive Development of Brown Fields and Other Urban Areas

Dmitry Spitsov and Igor Yazhlev(✉)

Moscow State Construction University, Yaroslavskoe Shosse, 26, Moscow 129337, Russia
hoved22@mail.ru

Abstract. The paper examines the practice of realization of projects for the comprehensive development of industrial zones and the reorganization of brownfields at the stage of developing plans to eliminate historical environmental damage and restore disturbed lands. The analysis of the content and quality of existing regulatory instruments for regulating the choice of reclamation technologies, assessing their economic and environmental effectiveness, and evaluating alternative options is made. Proposals have been developed to make changes to state standards, information technology reference books on best available technologies, construction legislation to ensure to create better conditions for citizens, environmental restoration in residential areas of urban and rural settlements.

Keywords: Comprehensive Development of Territories · Reclamation of Disturbed Lands · Historical Environmental Damage · Brownfields · The Best Available Technologies (Bat)

1 Introduction

The Russian urban development code contains mechanisms for comprehensive development of industrial zones and brown-fields. These events can be held in residential and non-residential areas, in undeveloped areas and at the initiative of land owners [1].

The legislative mechanisms allowed:

- the capabilities of regional authorities to include apartment building in projects of the comprehensive development of industrial zones and brown-fields;
- the development of transport, social, engineering infrastructure;
- create conditions for the elimination of dilapidated housing and the allocation of subsidies to reimburse the costs of relocating residents;
- attract extra-budgetary funds to finance the renovation of industrial zones and brownfields;
- create a mechanism for the seizure of inefficiently used buildings, structures and land plots for the purposes of the comprehensive development of industrial zones and brown-fields;

S. V. Klyuev et al. (Eds.): ICRAAC 2024, LNCE 627, pp. 207–216, 2025.
https://doi.org/10.1007/978-3-031-82938-3_25

- creation of jobs;
- and the others [3, 8]. However, on the other hand, this has increased the requirements for the quality of the condition of such territories.

2 Methods and Materials

In recent years, industrial zones have been reorganized in Moscow in non-residential development projects, and land plots used inefficiently and inappropriately have been seized. [9]. The total area of urban industrial zones is 18 thousand hectares, or 17% of the city's territory. There are also residential and public development areas within the boundaries of industrial territories. Such territories are often inefficiently used, and due to accumulated environmental damage, they have a high negative impact on the urban environment [6, 10].

The reorganization of production areas is an important area of projects for comprehensive development of industrial zones and brown-fields buildings in St. Petersburg, including the modernization of infrastructure and renovation of built-up areas. Industrial zones account for more than 6 thousand hectares out of 143.9 thousand hectares or 4.2% of the total area of the city. The production areas that changed their functional purpose during the reorganization account for about 30% of the total volume of housing put into operation in the city, about 5.6 million square meters, over the past five years [11, 12].

The total area of Yekaterinburg is 111.069 thousand hectares. The territories of the existing production facilities are 7222.31 hectares, which is 6.5% of the total area. In accordance with the City master plan, the area of territories changing their functional purpose will amount to 1177.84 hectares [13].

In Kazan, the territories of non-residential buildings make up 5,706.8 hectares, 8.3% of the total area of the city. The high level of environmental contamination is recorded in production areas and around busy highways. As a result of the integrated development of these territories, the area of negative environmental impact will decrease from 19 to 14% of the total area of the city [7].

Providing a favorable living environment in cities, the main issues of the comprehensive development of industrial zones and brown-fields can be considered:

- increased negative impact of territories on the urban environment due to significant historical harm to the environment;
- the need for high-quality engineering and environmental studies of land plots, land reclamation and elimination of accumulated environmental damage;
- inefficient use of territories;
- the presence of residential development sites and public use areas within the boundaries of production zones or in the immediate vicinity, along with production facilities [11].

In engineering and environmental studies for the development of a plan for the ecological restoration of contaminated areas, it's necessary:

- to establish the horizontal and vertical dimensions of the pollution area, concentrations, mass, properties of pollutants in all natural components and the possibility of their spread beyond the boundaries of the studied area;

- to establish the characteristics, features of the surface layers of the soil, properties of the underlying soils and site features, including topography and depth of groundwater;
- identify sources of pollution, routes of movement to actual or potential recipients of pollutants through air, soil, rock, precipitation, groundwater, surface water, infrastructure and buildings in the contaminated area;
- obtain data for assessing alternatives for ecological restoration, self-healing of the environment;
- to investigate the ways of pollutants;
- if necessary, evaluate the data on the degree of negative impact on fauna and flora.

If we imagine the process of developing a recovery plan and eliminating accumulated damage, the choice of recovery technologies, in the form of a set of variables, and the factors affecting the quality of its development and compliance of the selected technologies with the conditions at the site under study, in the form of results obtained during engineering and environmental studies, then:

$f$plan eff. = {$f$bound.,$f$charact.,$f$ways,$f$self heal.,$f$threats,$f$neg.impact,f restrict., ...,fn}K error.

where: f plan eff. - the quality, environmental and economic efficiency of the restoration plan;

$f$bound – the accuracy of determining the boundaries of the pollution area;

f charact. - characteristics of soil and soil properties;

f ways - the sources and ways of movement of pollutants;

$f$self heal. – the possibility of self-healing of the natural environment;

f threats - actual and potential threats to the people and fauna and flora;

$f$neg. impact – the degree of negative impact on fauna and flora;

$f$restrict. – possible restrictions on the conduct of preventive measures;

The selection order of technologies for the restoration of disturbed lands and the development of a reclamation plan is presented in the tables below. The development of models of the processes of transformation and distribution of pollutants in a polluted area is a part of the study of the behavior of pollutants in the environment (Table 1).

The exact questions for the study of contaminated site, given in Table 2, including:

- maximum permissible concentrations of contamination;
- sanitary and epidemiological requirements;
- data on the actual and potential harmful effects on human health and environmental components;
- the required level of purification from pollutants for the studied area [2, 4].

When choosing the necessary recovery technologies, the following criteria must be taken into account:

- the protection of the environment;
- compliance with legal requirements;
- duration of effective exposure after carrying out rehabilitation measures;
- reduction of toxicity, mobility and volume of pollutants after reclamation. When choosing recovery technologies, preference is given to measures, as a result of which the toxicity, mobility and volume of pollutants in the study area are significantly reduced;

- the possibility of a short-term negative impact during the implementation of restoration measures on the population, specialists, and the environment during reclamation activities;
- the technical and organizational possibility of implementing rehabilitation measures.

Table 1 The content of modeling the transformation processes, the spread of pollutants in the contaminated area

Elements	Possible content
1) Sources of pollution	Site plan indicating points of release, discharge, leaks, release of pollutants to the surface, etc
2) Nature, volume and level of pollution	Sections depicting soil layers, aquifers, groundwater levels (min./max.), distribution of pollution in soils, groundwater, soil air, description of the product phase), etc
3) Conditions at the site, the possibility of transformation and distribution of pollutants	Sections depicting soil layers, aquifers, groundwater levels (min./max.), distribution of pollution (soil, groundwater, soil air, description product phases), etc
4) Potential pathways of exposure of recipients to pollutants	Patterns of distribution of pollutants in various environments and areas, etc
5) Potential recipients of the negative effects of pollutants	Plans of settlements, residential zones, etc

The technical capability includes the possibility of processing, cleaning, carrying out construction and protective measures. The organizational opportunity includes the availability of qualified personnel, necessary materials and equipment, obtaining the necessary permits, organizing access to the construction site, etc. [5].

- capital costs, operating costs, operating costs, costs of monitoring ongoing activities at current prices;
- taking into account proposals and comments on the restoration work plan received from public, scientific and other organizations.

The next important step is to develop alternative options for remediation of the contaminated site. With the help of technology selection systems, all available technologies must be investigated in accordance with standardization documents. To do this, all rehabilitation options under consideration must be subjected to a qualitative comparative test, in which the following criteria are used:

- time for reclamation;
- price of reclamation technologies;
- technology productivity;
- negative environmental consequences;

– time and price of monitoring [2].

It is also necessary to take into account at this stage the properties of the soil, geological and pollution conditions, and the planned functional use of the site. To systematize the consideration of the impact of certain criteria and restrictions on the choice of restoration technologies, it is proposed to use the matrix method.

3 Results and Discussions

The procedure for selecting the best available rehabilitation technologies is given in Tables 2, 3 and 4.

Table 2 Selection of rehabilitation technologies, taking into account restrictions on their use [5, 17]

Rehabilitation technologies	Volatile Organic Compounds (VOCs)		
	soil	Underground water	Soil air
Soil steam extraction	●		●
Grounds heating	●		
Contaminated grounds incineration	●		
Bioremediation	●	●	
Excavation and off-site placement	●		
Extraction and processing		●	
Air cleaning in the well		●	
Activated carbon granules		●	
Chemical treatment		●	

Table 3 Selection of rehabilitation technologies, taking into account restrictions on their use [5, 17]

Reclamation methods	Metals		
	Grounds	Underground water	Ground air
stabilization of conditions on the site	●		
Excavation and off-site disposal of waste	●		
Extraction and processing	●		
Chemical deposition		●	
Ion exchange/absorption		●	

Table 4 Selection of rehabilitation technologies, taking into account restrictions on their use [5, 17]

Reclamation methods	Polychlorinated biphenyls /Pesticides		
	grounds	Underground water	grounds air
grounds heating	●		
Burning	●		
Excavation and off-site disposal of waste	●		
Pumping and cleaning		●	
Coal cleaning		●	

The selection of rehabilitation measures and the preparation of an environmental restoration plan is a complex, multi-component process that is influenced by many factors. In order to take into account, including the indirect and indirect influence of various factors, step-by-step matrices are used.

The pollutants in grounds can be in various states: gaseous, liquid, dissolved in water, solid, viscous and etc. Such factors make it difficult to assess the effectiveness of a particular method, rehabilitation technology and the use of general indicators.

The indicators for the purpose of preliminary assessment of rehabilitation technologies planned for use, based on the criteria and indicators given above has been developed.

The productivity indicator of reclamation methods is a value characterizing the possibility of using rehabilitation technologies in terms of unit costs and time for complete rehabilitation.

$$K\text{iep} = 1/\sum_{n}^{i=1}(\text{Qu}ci\ \text{t}pri)kcc$$

where: K iep – indicator of environmental and economic productivity of reclamation methods;

Q uci - level of unit costs when using i-method (Table 5);

t pri - period of complete reclamation using the i-method (Tables 6 and 7);

k *cc* - climatic conditions [5].

Generalization of Russian and foreign experience in implementing methods for environmental restoration of contaminated areas and historical environmental harm control allows us to assess some comparative indicators used to select the best available restoration technologies. To compare the environmental and economic efficiency of using the physical and chemical method “Soil extraction with steam” and the mechanical method “Excavation and disposal of waste off-site” for cleaning up pollution with petroleum products, we determine the unit costs for both methods. From foreign experience, in the first case they amounted to an average of 0.9 thousand rubles. for 1 m^3 and in the second 11 thousand rubles. for 1t of soil. 1 m^3 of soil, depending on its composition, can weigh on average 1300 kg. The processing time for a “standard” plot, in accordance with foreign approaches, with an area of 0.41 hectares, a depth of 3.04 m and a mass of processed soil of 18,200 tons (or 14,000 m^3) is in the first case more than 9 months, and in the second less than 6 months. The *kву*, coefficient, normative, is taken to be 1.6 for the mid latitudes. We obtain Ketr for the first method is 0.16. For the second method, *K etr* was 0.13. Based on this, we can conclude that the "Soil steam extraction" method, in terms of the proposed approaches and site conditions, is preferable for use than "Excavation and off-site waste disposal" [5, 16, 17].

Table 5 Level of unit costs for the use of rehabilitation technologies when processing 1 ton of soil [16]

Unit Cost Accounting (in points)	price
1	Less than 3 thousand rubles
2	3–6 thousand rubles
3	6–9 thousand rubles
4	More than 9–12 thousand rubles

Table 6 Assessment of the duration of ecological restoration when using rehabilitation technologies to treat contaminated soil, previously removed from the surface of the contaminated area [16]

Average estimate duration ecological restoration (in points)	Time
1	Less than 6 months
2	6–9 months
3	9–12 months
4	More than 12 months

When assessing the data obtained during the implementation of comprehensive development of industrial zones and brown-fields, conditions that need to be taken into account when choosing BAT for restoration of contaminated were identified.

These include the lack of methods for engineering and environmental studies of industrial zones and brownfields, development of a recovery plan, recovery methods and monitoring;

- BAT technology registry;
- collections of basic specific indicators of capital investments for the rehabilitation of contaminated industrial and urban areas.

Table 7 Assessment of the duration of ecological restoration when using remediation technologies to treat contaminated soil at the site of contamination [16]

.Average estimate duration ecological restoration (in points)	Time
1	6 months
2	6–12 months
3	1–3 years
4	More than 3 years

To systematize the consideration of the impact of certain criteria on the choice of restoration technologies, the matrix method can be used. In order to take into account, including the indirect and indirect influence of conditioned factors, use step-by-step matrices.

To determine the necessary environmental and economic efficiency of rehabilitation measures when developing and implementing plans for eliminating accumulated environmental damage and ecological restoration, alternative options for restoration measures should be formed. Such plan options may include an assessment of the necessary costs, the possibility of self-healing of the contaminated environment, its restoration to the state before the emission/discharge of pollutants, and others. The current, planned and future functional use of the restoration site must also be taken into account.

4 Conclusions

In order to develop construction and environmental legislation when implementing comprehensive development of industrial zones and brown-fields projects, it is necessary to amend following state standards and information technology reference books on best available technologies BAT:

supplement GOST R 70082-2022 "Resource conservation. Methodological basis for the development of projects of restoration historical environmental damage" with a review of technologies and recommendations for the application of BAT in accordance with information technology reference books on best available technologies ITS NDT 53-2022, provisions on the procedure for selecting technologies, methods for selecting preferred technologies, evaluating alternative options, and on the procedure for estimating the costs of work;

add to GOST R 59057-2020 "Environmental safety. Earth. Reclamation of contaminated areas" provisions on the requirements for the reclamation of disturbed lands for the purposes of implementing comprehensive development of industrial zones and brown-fields;

add to GOST R 54003-2010 "Environmental management. Assessment of past environmental damage accumulated in places where organizations are located. General provisions" provisions on managing the risks of harm to life, health of individuals and property during comprehensive development of industrial zones and brown-fields projects;

add to GOST R 57446-2017 "BAT. Reclamation of contaminated lands. Restoration of biological diversity" provisions on comprehensive development of industrial zones and brown-fields projects;

add to information technology reference books on best available technologies ITS BAT 53-2022 "Elimination of historical damage", when determining criteria for selecting restoration technologies list of the method of constructing a multi-level matrix of requirements in accordance with GOST R 57700.37-2021 [17].

References

1. Town Planning Code of the Russian Federation. Federal Law of December 29, 2004, No. 190-FZ. Access mode: www.consultant.ru, access date 10 Mar 2024
2. Information and technical reference book on the best available technologies ITS NDT 53-2022. Elimination of objects of accumulated environmental harm. Access mode: www.burondt.ru/ NDTDocsFileDownload.php… Date of access 3 July 2024
3. Ministry of Construction. Monitoring the implementation of the mechanism for integrated development of territories by constituent entities of the Russian Federation. Access mode: https://minstroyrf.gov.ru/trades/kompleksnoe-razvitie-territoriy/1-monitoring-realizatsii-mekhanizma-kompleksnogo-razvitiya-territoriy-subektami-rossiyskoy-federatsii/. Access date 20 Feb 2024.
4. Spitsov, D., Yazhlev, I.: Efficiency of regulation of construction and demolition waste management—AIP Conference Proceedings 2791, 030003 (2023). Access mode: https://doi.org/10.1063/5.0143592, access date 28 Feb 2024
5. Yazhlev, I.K.: Environmental Remediation of Contaminated Industrial and Urban Areas. Monograph. Publishing House ASV, Moscow, p. 279. ISBN: 978-5-93093-909-5 (2012)
6. Decree of the Government of the Russian Federation of May 4, 2018 No. 542 Rules for organizing work to eliminate accumulated environmental damage; Access mode: https://docs.cntd.ru/document/557301410?ysclid=lqqqj1w8ew797480582 Access date 20 Dec 2023.
7. General plan of the Kazan urban district. Appendix to the decision of the Kazan City Duma dated February 28, 2020 No. 5-38.
8. Moscow Government Resolution No. 497-PP dated August 1, 2017 On the Housing Renovation Program in the City of Moscow. Access mode: docs.cntd.ru›document/456082244, access date 7 Mar 2024.
9. Redevelopment of industrial zones. Access mode: https://stroi.mos.ru/renovaciya-promzon?from=cl, access date: 5 Mar 2024.
10. Moscow will continue the comprehensive development of territories. Contact mode: Telegram: Contact @mos_sobyanin, date of contact: 16 Feb 2024
11. Objects of the industrial zone of St. Petersburg. Access mode: https://www.spb.kp.ru/daily/27460/4714846/, access date: 24 Feb 2024.
12. Law of St. Petersburg of June 30, 2022 No. 444-59. On amendments to some laws of St. Petersburg on issues of integrated development of territories in St. Petersburg. Access mode: http://publication.pravo.gov.ru/Document/View/7800202207040011?ysclid=lqqpgvd1p8960078236, date of access: 25 Feb 2024

13. The General Plan of Yekaterinburg will adjust the industrial zones. Access mode: https://ural.octagon.media/ekonomika/genplan_ekaterinburga_skorrektiruet_promzony.html?ysclid=lqqpu75af8945600664, access date: 20 Feb 2024.
14. GOST R 57700.37-2021. Computer modulations. Digital twins of products. General provisions. Access mode: docs.cntd.ru›document/1200180928. Access date 3 Oct 2024
15. Drugov, Y.S., Rodin, A.A.: Monitoring of organic compounds in the natural environment. Practical guide. Series: Methods in chemistry. Monograph, -M.; BINOM, Knowledge Laboratory (2009)
16. Grechishcheva, N.Y., Meshcheryakov, S.V., Rybalsky, N.G., Barsov, S., Miertus, A.R.: Handbook. Technologies for Restoration of Soils Contaminated with Oil and Petroleum Products. -M.: REFIA, NIA-Priroda (2001)
17. CP-51 / Soil Cleanup Guidance. Access mode: www.extapps.dec.ny.gov>docs>remediation_hudson_pdf, access date: 10 Mar 2024

Research of Criteria for Improving Low-Rise Housing Construction

G. V. Mikheev(✉)

Kuban State Technological University, Krasnodar, Russia
Mgstyle77@yandex.ru

Abstract. The article examines the factors of the implementation of low-rise housing construction projects in the conditions of the modern economic space of Russia. The aim is to improve and develop theoretical and methodological solutions for the development of new methods and tools for managing construction and development activities in the context of modern digitalization of society. The current state of construction and development activities in Russia is in a turbulent situation, which is due to various factor manifestations and circumstances emerging in the conditions of the modern economic space. Within the framework of the study, instrumental and methodological elements of low-rise housing construction management have been developed, including a matrix of management mechanisms for optimizing costs in low-rise construction and a management model for low-rise housing construction in the conditions of modern economic space. The proposed tools will help ensure the smooth operation of business entities involved in construction and development during periods of high competition. They will also create the necessary conditions for achieving efficiency and effectiveness in business activities and promoting low-rise housing development.

Keywords: Construction · Low-Rise Housing · Construction Technologies · Organization of Construction

1 Introduction

Construction and development activities have their own specific criteria and parametric features characteristic of the modern engineering and digital component of business activities, which are important to take into account, study and consider when forming, improving and developing instrumental and methodological solutions to achieve efficiency indicators and performance coefficients in the implementation of various housing projects.

The evolution of construction management processes has led to the complication of the implementation of individual stages and phases of development projects in the conditions of the modern economic space of Russia. Among the areas of construction and development activities, various factorial circumstances and negative phenomena have emerged, provoked by the variability of the industry market and the turbulence of economic systems as a whole.

S. V. Klyuev et al. (Eds.): ICRAAC 2024, LNCE 627, pp. 217–225, 2025.
https://doi.org/10.1007/978-3-031-82938-3_26

The construction and development industry has faced various complications and factor risks that can affect both large construction companies and medium and small business entities of the construction industry.

Management barriers affected not only large multi-apartment construction projects, but also low-rise housing projects. It should be noted that low-rise housing construction has certain priority properties and characteristics, as it contributes more to the development of prerequisites for providing citizens with their own residential real estate.

2 Methods and Materials

The improvement of existing methodological, theoretical and methodological solutions and the development of new elements of tools for the possibility of optimization, adaptation, modification and reengineering of management processes within the framework of low-rise housing construction projects is the main goal of this study.

Considering the properties and criteria for the development and implementation of low-rise housing construction projects in the conditions of the economic space of Russia, the following can be distinguished:

- formation of prerequisites for solving issues of providing affordable housing to various groups and target audiences of the population, taking into account the regional and spatial characteristics of the territory of residence [1, 2];
- opportunities for the implementation of personal parameters, preferences and properties of the operation of the future project of a low-rise construction facility [3];
- achieving the necessary level of financial balance and availability of funds to enable and ensure the implementation of the project, taking into account the modern environment and factors of the spatial complex, within which the construction of a low-rise building will be implemented [4];
- research, analysis, definition and consideration of managerial, operational-strategic, organizational, environmental, technological, engineering, technical, operational components of the future project, taking into account the parameters of the modern environment [5];
- opportunities for the introduction of modern engineering service systems for high-quality operation of a low-rise construction facility [6];
- formation of prerequisites for the introduction of energy-efficient and resource-saving technologies at all phases, stages and stages of the implementation of a low-rise real estate project and its further operation within the framework of the concept of sustainable development of society and maintaining the criteria of the green economy as a whole [7];
- the possibility of using artificial intelligence technologies in the development and design of a future low-rise housing construction facility and neural networks [8];
- improving the quality of design, development and implementation of construction production processes through specialized information systems and information modeling technologies, programs for organizational, economic and financial control [9, 10];
- using specialized economic systems to improve the efficiency and effectiveness of low-rise housing construction projects.

3 Results and Discussion

The properties and criteria presented above for the development and implementation of project activities within the framework of low-rise construction and the formation of affordable housing in the conditions of the modern economic space of Russia demonstrate to us the multidimensional and complex economic nature of the process management system within the framework of construction and development activities. These circumstances indicate the need to develop and improve theoretical and methodological tools to support the activities of low-rise housing construction and the possibility of forming prerequisites for the development of the industry in the conditions of the modern economic space of Russia.

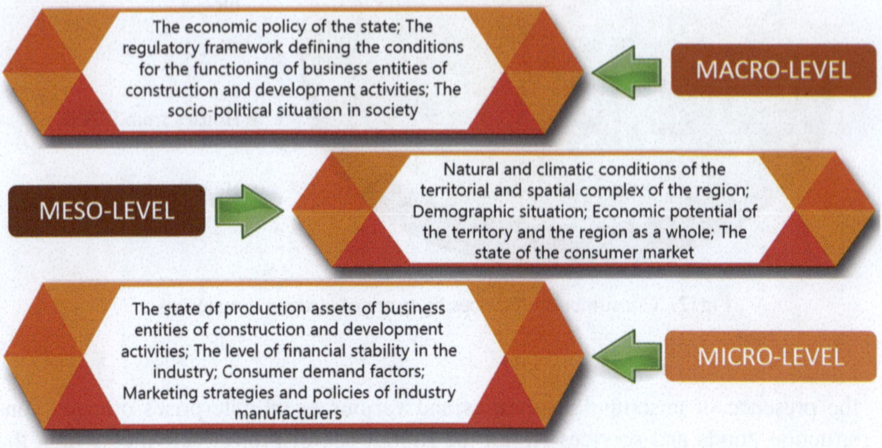

Fig. 1. The impact of various phenomena on construction and development activities within the factor levels of the modern economic space.

The conducted research allows us to form a scheme of the impact of various phenomena on construction and development activities within the framework of factor levels of the modern economic space (see Fig. 1).

The conducted research revealed the following data and indicators see Figs. 2, 3, and 4).

Let's take a closer look at the impact of negative phenomena on low-rise housing construction in the conditions of modern economic space:

- variability in the availability and cost of materials in the wholesale and retail markets of the construction industry;
- difficulties in the supply of analog materials and components;
- variability and high levels of lending rates;
- difficulties in the formation of society's savings in the conditions of the modern financial and economic period;
- the presence of barriers and obstacles on the part of state and municipal institutions, which increases the time required for registration and obtaining permits for

project activities within the framework of low-rise construction and achieving quality indicators of affordable housing;
- risks from financial and economic institutions, turbulence of various indicators and indices, deter the population from participating in the implementation of low-rise housing construction projects in the current consumer real estate market;

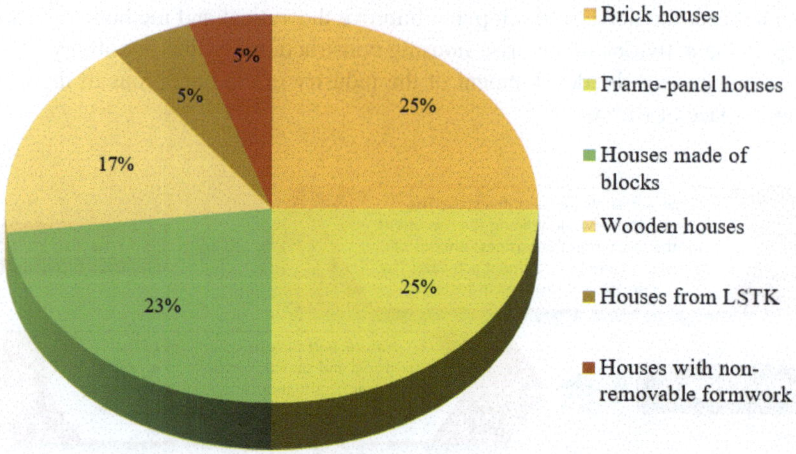

Fig. 2. Consumer preferences by types of structural materials.

- the presence of unscrupulous entities and various agent enterprises offering construction goods and services within the framework of project activities within the framework of low-rise construction and achieving quality indicators of affordable housing;
- the lack of anti-corruption programs for the implementation of high-quality protection and the development of economic and information security measures for business entities of construction and development activities in the conditions of modern economic space;
- high costs of using the attracted equipment, as well as various specialized devices, systems and mechanisms necessary at various phases, stages and stages of construction production to carry out project activities within the framework of low-rise construction and achieve quality indicators of affordable housing;
- the shortage of personnel among specialists and engineers of construction and development activities complicates the development and quality of design estimates and information models of the future low-rise housing construction;
- there is no financial opportunity for savings, as a result, a decrease in the overall level of purchasing power and the implementation of project activities within the framework of low-rise construction and the achievement of quality indicators of affordable housing;

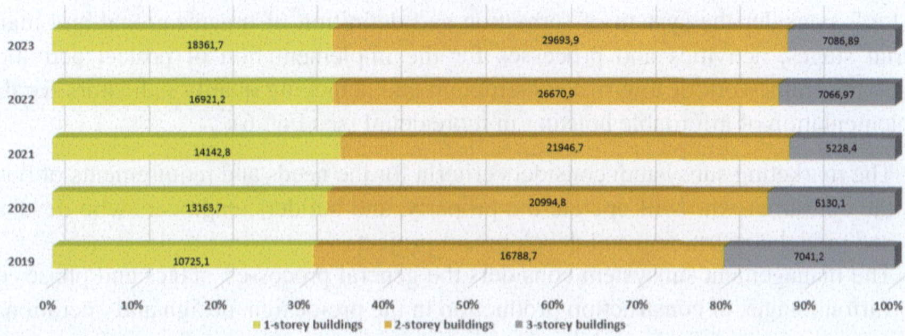

Fig. 3. Commissioning of the total area of low-rise residential buildings by number of storey's in the Russian Federation, thousands of square meters.

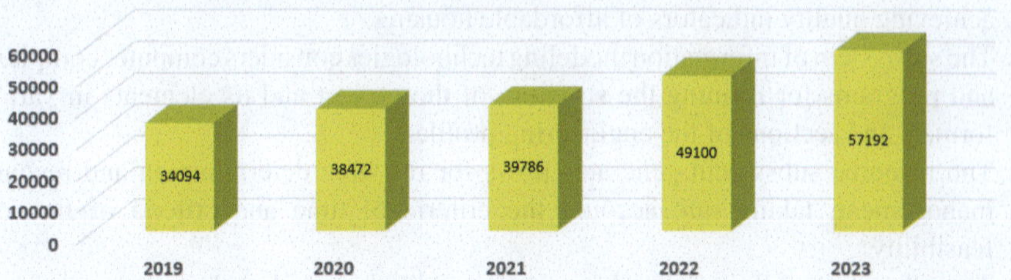

Fig. 4. Commissioning of low-rise housing for 2019–2023, thousand square meters.

- the bonded conditions of mortgage lending models for consumers make it impossible to develop low-rise housing construction in the conditions of the modern digital economy;
- the high level of social tension in connection with military operations in various territories of Russia negatively affects the economic development of territorial and spatial complexes of regions and individual municipalities of the country.

As part of the research, we have developed a matrix of management mechanisms for optimizing costs in low-rise construction, taking into account the phasing and stage-by-stage implementation of processes and their features in the conditions of modern economic space (see Fig. 5).

The matrix is formed from certain components and constituent elements of the implementation of construction and development activities of low-rise housing construction. The stages are divided into design phases, construction stages, commissioning processes, directions and stages of use of information technologies and additional software and computer complexes, controlling, adaptation, modification and reengineering processes, depending on the assessment and analysis of data on the achievement of efficiency and effectiveness indicators for each block.

The matrix also presents optimization mechanisms and features of organizational and managerial decision-making in the implementation of project activities within the framework of low-rise construction and achieving quality indicators of affordable housing in a modern digital environment.

Let's consider the system of formation and definition of organizational and managerial stages, activities and processes for the implementation of project activities within the framework of low-rise construction and achieving quality indicators for the implementation of affordable housing in more detail (see Fig. 6).

1. The marketing subsystem considers criteria for the needs and requirements of both the consumer who will operate the property, and builders, engineers who develop and model construction and development projects.
2. The management subsystem considers the general processes, stages and phases of various stages of construction production in the pre-design, design and operational periods of construction and development activities.
3. The R&D subsystem considers the possibilities of applying new techniques, innovations, developments, systems aimed at forming positive solutions for the implementation of project activities within the framework of low-rise construction and achieving quality indicators of affordable housing.
4. The subsystem of information modeling technologies considers computer complexes and programs for forming the structure of the project and its elements in various formats and sections of the engineering profile.
5. The resource subsystem generates flows for resource determination and resource management, taking into account the criteria of time and criteria of financial feasibility.
6. The subsystem of financial and economic condition considers the processes of asset and cash flow management, taking into account the operational, strategic state, position and phasing of the project.
7. The communications subsystem takes into account the processes of interaction between all participants in the implementation of project activities within the framework of low-rise construction and the achievement of quality indicators of affordable housing in the context of the implementation of processes for the digitalization of the economy.
8. The controlling subsystem considers the criteria for the formation of processes of various directions, taking into account the coordination of production and implementation of work according to operational and strategic plans.
9. The subsystem of organizational and technological criteria ensures the compatibility of making various decisions on the formation of technical, engineering, technological processes and their consistency as part of the methodological, instrumental and methodological structure of management decisions, taking into account the stages and stages of project activities within the framework of low-rise construction and achieving quality indicators of affordable housing.
10. The risk forecasting subsystem considers possible negative circumstances, force majeure manifestations that may affect the effectiveness of a low-rise housing construction project in various periods and stages of construction and development activities.

Figure 3 shows the author's development of a management model for low-rise housing construction in the conditions of modern economic space. The study of the system and elements of the implementation of project activities within the framework of low-rise construction and the achievement of quality indicators of affordable housing allows

Stage	Optimization mechanisms	Specificity
Engineering	- Selection of optimal design solutions - Development of cost estimates - Control over changes in cost estimates	- Analysis of various options for planning solutions - Selection of energy-efficient materials and technologies - Optimization of engineering systems - Preparation of detailed cost estimates - Carrying out an examination of the estimated documentation - Accounting for limited budgets of developers - Use of standard projects and materials - Involvement of family members and friends in the construction
Construction	- Planning and organization of construction - Procurement management - Management of construction and installation works	- Development of a construction schedule - Providing the construction site with the necessary resources - Monitoring the progress of construction - Holding contests among suppliers - Direct purchases of materials from manufacturers - Use of wholesale discounts - Implementation of construction control - Prevention of downtime and defects - Employee motivation
Exploitation	- Implementation of an energy saving system - Regular maintenance	- Installation of energy-saving appliances and lamps - Adjustment of the heating system - Use of alternative energy sources - Control over the technical condition of engineering systems - Early troubleshooting - Use of original spare parts
Information technology	- Application of software packages for project management - Use of computer-aided design (CAD) systems - Use of supply chain management (SCM) systems - Use of energy consumption monitoring systems	- Allows you to plan and monitor the progress of construction - Allows you to create 3D models of objects, optimize their design and reduce material costs - Allows you to optimize the purchase of materials, reduce their cost and delivery time - Allows you to identify sources of energy losses and take measures to eliminate them
Monitoring and evaluation of effectiveness	- Development of a system of performance indicators - Data collection and analysis - Adjustment of management decisions	- Cost of construction - Construction dates - Quality of construction - Operating costs - Regular collection of data on actual costs - Comparison of actual costs with planned ones - Analysis of the causes of deviations from the planned indicators - Taking measures to eliminate the causes of deviations from planned indicators - Improvement of the cost management system

Fig. 5. Matrix of management mechanisms for cost optimization in low-rise construction.

us to formulate the following conclusion: the formation of processes within the management system of low-rise housing construction takes into account various criteria and factor influences, processes are introduced into the management system from profiled subsystems, ranked according to certain properties and directions of construction and development activities in order to achieve the most effective and efficient implementation of a housing construction project in the conditions of modern economic space.

4 Conclusions

The presented management model of low-rise housing construction in the conditions of modern economic space is a tool for the formation of processes within the framework of construction and development activities, for the correct functioning of business entities of the construction industry. This model allows us to take into account the specificity of the processes formed taking into account the circumstances and manifestations at various stages of the implementation of a low-rise housing construction project.

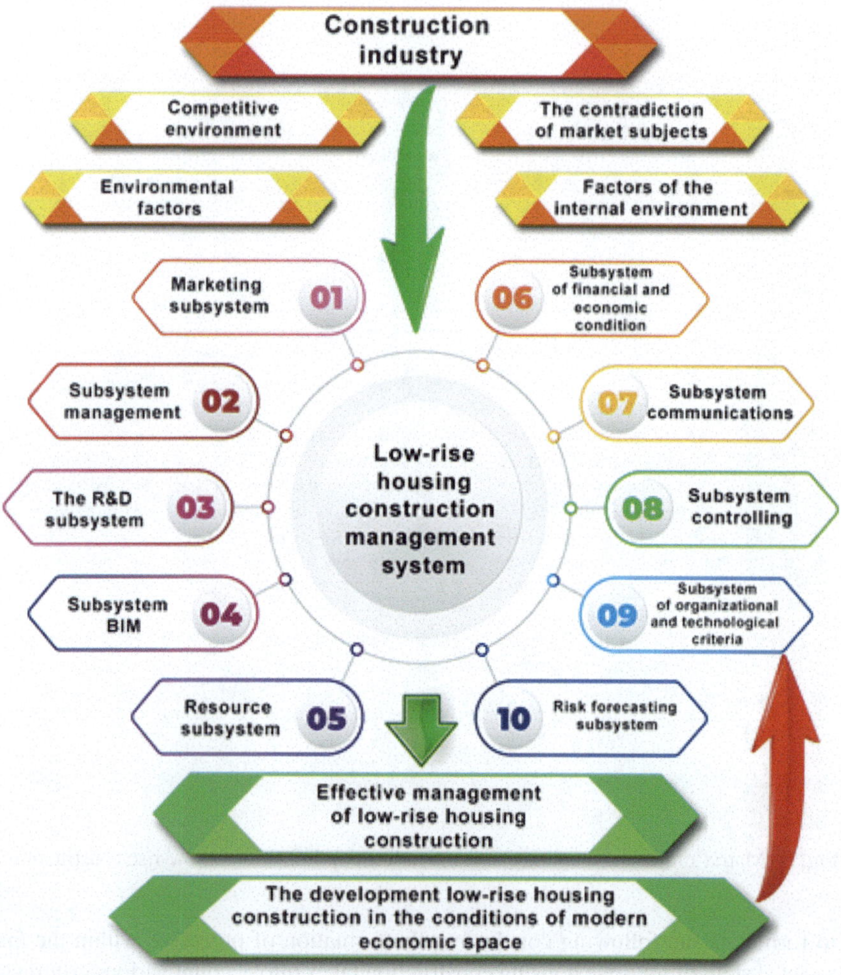

Fig. 6. The management model of low-rise housing construction in the conditions of modern economic space.

The management processes of low-rise housing construction within the framework of the presented model can be adapted taking into account the modern economic space. Adaptation, modification and reengineering of processes occurs when there is a certain need, within which the process and/or management decision should be aimed at a positive and effective result in the future. If the result is not achieved, then the process needs to be rebuilt taking into account the stated criteria and the required indicators.

The proposed instrumental and methodological elements of low-rise housing construction management, including a matrix of management mechanisms for optimizing costs in low-rise construction and a model for managing low-rise housing construction in the conditions of modern economic space, will contribute to the decent functioning of business entities of construction and development activities during periods of competition, and it will also allow to create prerequisites for achieving indicators of efficiency and effectiveness of business activities and the development of low-rise housing construction in general.

Acknowledgement. This scientific research was conducted in the laboratory of the Department of Technology, Organization, and Construction Economics and Property Management at the Institute of Construction and Transport Infrastructure at Kuban State Technical University.

References

1. Gusakova, N., Gusakov, A., Prokhorova, Y., Karakozova, I.: Creation of programs for sustainable administration of low-rise housing construction programs in remote areas with special climatic conditions. J. Law Sustain. Dev. **11**(1), e0264 (2023). https://doi.org/10.37497/sdgs.v11i1.264
2. Bezrukikh, O.A.: A model for managing the development of low-rise housing construction using the method of discursive design. In: IOP Conference Series: Materials Science and Engineering, p. 012088 (2020). https://doi.org/10.1088/1757-899X/880/1/012088
3. Korol, E.: Calculation of energy consumption in the construction of low-rise. E3S Web Conf. **457**, 02054 (2023). https://doi.org/10.1051/e3sconf/202345702054
4. Fediuk, R., Amran, M., Klyuev, S., Klyuev, A.: Increasing the performance of a fiber-reinforced concrete for protective facilities. Fibers **9**(11), 64 (2021)
5. Amran, M., Fediuk, R., Klyuev, S., Qader, D.N.: Sustainable development of basalt fiber-reinforced high-strength eco-friendly concrete with a modified composite binder. Case Stud. Const. Mater. **17**, e01550 (2022)
6. Sekisov, A., Solovyova E., Mikheev G.: Solving the problem of noise reduction as a factor in ensuring safety in construction **2911**, 20047 (2023). https://doi.org/10.1063/5.0163141
7. Klyuev, S., et al.: Technogenic fiber wastes for optimizing concrete. Materials **15**(14), 5058 (2022)
8. Harichandran, A.: Relevance of deep sequence models for recognising automated construction activities: A case study on a low-rise construction system. J. Inf. Technol. Const. **28**, 458–481 (2023). https://doi.org/10.36680/j.itcon.2023.023
9. Rankin, K.H., Arceo, A., Isin, K., Saxe, S.: Embodied GHG of missing middle: Residential building form and strategies for more efficient housing. J. Ind. Ecol. (2024). https://doi.org/10.1111/jiec.13461
10. Sekisov, A.N., Ovchinnikova, S.V., Grebneva, V., Chernyshova, M.: Modern directions of low-rise housing construction development in the world: Economic and technological aspect. In: IOP Conference Series: Materials Science and Engineering, p. 042035 (2020)

The Role of Large Language Models Within Construction Facilities Lifecycle Data Management

Eugeny Chaplygin and Konstantin Losev(✉)

National Research Moscow State University of Civil Engineering (NRU MGSU), Moscow 129337, Russia
LossevKY@mgsu.ru

Abstract. **Introduction**: The modern construction industry faces the challenge of managing vast volumes of fragmented data, which complicates its integration and analysis across all stages of the construction lifecycle. The use of artificial intelligence, particularly large language models (LLM), offers a promising approach to address these issues. The relevance of the study lies in the need to optimize data processing workflows using LLM to improve the efficiency of construction project management. **Materials and Methods:** The study examines three methods for applying LLM: Retrieval-Augmented Generation (RAG) for unstructured data, Text-to-SQL for structured data in relational databases, and Knowledge Map for processing sensitive unstructured data. **Results:** Testing demonstrated that RAG effectively retrieves information from unstructured sources, Text-to-SQL ensures precise access to data in relational databases, and Knowledge Map minimizes errors when handling hierarchical data structures. These methods complement each other, enabling LLM to work with a wide range of data types and improving the accuracy of the responses. **Conclusion:** The application of LLM in construction data management opens new opportunities for automation and improving the precision of data processing. Further integration of LLM with advanced technologies, such as digital twins and the Internet of Things, has the potential to create more efficient systems for managing construction projects.

Keywords: Construction Data Management · Large Language Models (LLM) · Retrieval-Augmented Generation (RAG) · Text-to-SQL · Knowledge Map · Data Integration · Artificial Intelligence (AI) · Construction lifecycle

1 Introduction

The modern construction industry is closely connected with technological progress, and therefore faces challenges related to the management of huge amounts of data that arise at all stages of the life cycle of objects [1]. With the constant growth in the scale and complexity of construction projects, data becomes critically important as it serves as the basis for management decisions and process optimization. These data include project documentation, resource data, operational data, financial reports, schedule and change information, and sales data. Over time, the amount of data increases exponentially, making it more difficult to integrate and process [2].

S. V. Klyuev et al. (Eds.): ICRAAC 2024, LNCE 627, pp. 226–238, 2025.
https://doi.org/10.1007/978-3-031-82938-3_27

Much of the data created in construction projects is stored in various information systems such as Common Data Environment (CDE), Electronic Document Management (EDM) systems, Enterprise Resource Planning (ERP) systems, Customer Relationship Management (CRM) systems, etc. However, this data often remains unconnected and siloed, making it difficult to use and analyze effectively. As a result, many organizations face problems when trying to combine information from different sources, resulting in wasted time and resources, as well as errors in the decision-making process [3].

One of the main problems of this approach is the fragmentation of data and the lack of a single standard for their processing and storage. This results in different project participants working with disparate data, which makes it difficult to coordinate and integrate information. For example, data can be both structured (SQL, XML, Excel) and unstructured (DOC, JPEG, PDF, MP4). This variety of formats complicates data processing and analysis, requiring the use of specialized tools and techniques to handle each type of information.

Thus, integrating data and making it available to all participants in the construction process are key factors in successful construction project management. Disparate information stored in different systems not only hampers operational decision-making, but also reduces the overall efficiency of the project. In the context of complex and multifaceted construction processes, it is necessary to create a unified information environment that integrates all data, from the project concept to its operation [4].

One of the main requirements for such integration is the creation of a unified knowledge base that collects all company data in one place and provides convenient access to them. Information gaps between design, engineering and operational data can lead to costly errors, delays and resource overruns, so it is extremely important to maintain a common structural exchange of necessary information between subsystems and participants [5].

In recent years, the overall innovation development is largely related to machine learning and artificial intelligence (AI) technologies, becoming increasingly important in various industries, including construction [6]. The use of generative models and, in particular, large language models (LLM) is one of the most promising directions in the field of data management. These models are able to process large amounts of information, automate the processes of data search and analysis, providing relevant answers to user queries, faster and more accurately than humans would do, which reduces the time for analysis and decision making [7].

It is expected that in the coming years, the use of AI will contribute to the development of intelligent decision support sub-systems that can not only analyze data, but also predict potential risks and offer optimal solutions based on a vast amount of historical data and current conditions [8–10].

In general, AI plays a key role in developing new methods to automate processes from concept to building utilization. For example, automatic generation of design documents based on customer requirements and regulations, intelligent risk management and prediction of their impact on construction progress in real time are all made possible by AI enhancements [11, 12].

The prospects for further development of AI are also related to its integration into broader life-cycle management ecosystems, including interaction with other advanced technologies such as Digital Twins and the Internet of Things (IoT). The synergy of these technologies can significantly enhance the capabilities of construction project management, creating complex systems that can dynamically adapt to changes and provide users with valuable information in real time [13, 14].

Drawing conclusions from all of the above, it can be predicted that LLMs can significantly change many industries, including the construction industry, significantly increasing its efficiency, adaptability and resilience to the challenges associated with big data management [15].

With the rapid growth of data volumes from various systems and platforms in the construction industry, traditional data management methods are facing limitations. Today, the construction industry uses various specialized information systems and data management solutions, some of the key ones being: project management systems (PLM and BIM), enterprise resource management (ERP) systems, customer relationship management (CRM) systems, and decision support systems (DSS).

ERP systems provide management of core business processes covering budgeting, resource planning and work control. These systems integrate key enterprise data into a single platform, which reduces the probability of errors associated with duplication of information and increases the transparency of processes [16]. However, they focus primarily on structured data and have difficulty working with unstructured information. This limits their ability to interact effectively with other systems, such as PLM and CRM, and reduces their adaptability to real-time data changes.

PLM and BIM systems play a critical role in managing project documentation, coordinating participants, and visualizing design decisions. These systems improve planning accuracy and reduce errors by visualizing and analyzing data in a single model. However, like ERP, PLM and BIM face challenges in integrating data from other sources and platforms, especially unstructured data [17]. The heterogeneous data formats require significant efforts to normalize and combine them into a single information environment.

CRM systems provide customer and partner relationship management, which is essential in the context of large construction projects where maintaining communication and contract management are fundamental. At the same time, integrating CRM with other systems such as ERP and BIM can be difficult due to differences in data formats and the need for manual processing of information.

DSS provide tools to analyze large amounts of data and help you make informed decisions. These systems integrate data from various sources, such as ERP, PLM, CRM and others, to generate analytical reports and forecasts, thereby improving planning accuracy and reducing the likelihood of errors. However, handling large amounts of data, especially unstructured data, remains a significant challenge for these systems, requiring additional tools to process it.

Collectively, all of the management systems described above represent important elements of data management in the construction industry, but their integration and ability to handle unstructured data remain key challenges, emphasizing the need for more flexible and efficient solutions. In such conditions, there are increasing demands on decision makers, who always act on the basis of knowledge and thought models of the relevant subject areas, and must be proficient in developing specialized automated technology tools, including in conditions of uncertainty in decision making [18]. In the conditions of the digital economy, modernization of the construction industry requires a completely new mental approach to the search for optimal and effective means, as well as criteria of life cycle management system functioning.

2 Methods and Materials

Let us consider the application of LLM in construction data management for optimization and automation of information processing and decision-making processes.

LLM (Large Language Model) is a large language model trained on huge amounts of textual data, which is capable of understanding and generating natural language. These models are able to analyze the construction documents, automatically extract the required data and propose solutions based on extensive information [19]. However, it is important to note that LLMs, even trained on large datasets related to the construction industry, do not contain private information of individual companies and rely on publicly available data for their answers. This implies that large language models should be trained on their own data in order to adapt them to different domains [20, 21].

Let us consider the pre-learning method under consideration is RAG (Retrieval-Augmented Generation), which allows the model to refer to an external database (DB) for additional information when generating responses [22]. Unlike purely generative models, RAG makes the model more flexible and able to handle actual data that was not part of the original training set. This is achieved by integrating it with vector databases optimized for working with vector representations of information, which significantly improves the accuracy of responses to user queries.

The use of vector databases is particularly effective when working with unstructured data such as text documents. Before data is placed in a vector database, it undergoes several stages of pre-processing. For example, large documents are divided into text fragments from which unnecessary characters, spaces and artifacts interfering with correct work with text are removed. Each processed fragment is converted into a JSON-object that contains not only text, but also metadata - information about the document, date of its modification, file type and a unique hash key for identification. This data helps to efficiently find the necessary fragments in the database. Thus, for each document in the database, a JSON file is generated that contains JSON objects with its content.

Structured data such as tables present in text documents are converted into HTML and included in JSON objects to preserve their format. Images and other multimedia data containing important information are described by text and also included in JSON structure.

Thus, the RAG method extracts relevant fragments from a vector database based on a user query. The method compares the vector representation of the query with the data in the database, selecting the most relevant fragments. These are then fed to the LLM along with the query, allowing the model to generate accurate and meaningful answers based on data that was not available during its initial training phase, which greatly enhances the capabilities and quality of its performance.

This approach is particularly effective for processing heterogeneous and distributed data stored in different information systems. RAG is able to combine the capabilities of LLM and the wealth of data accumulated in construction companies, presenting them in a single format for analysis, which significantly improves the quality of project management, accelerates decision-making and makes this method one of the most promising for implementation in the industry.

Despite the prospects, successful application of RAG requires careful data preparation. The better JSON-objects are structured and enriched with metadata, the more accurate will be the process of data extraction from vector databases and their subsequent processing by the model. However, there are several serious limitations of the method. For example, in cases where the user's query and the extracted data do not match, the model may generate incorrect responses. This phenomenon is known as hallucinations, where the model comes up with nonexistent information or erroneous conclusions. Due to the possibility of hallucinations, the RAG method is not recommended for use with critical or sensitive data that require extreme accuracy.

Sensitive data such as financial statements, contract data, and project specifications are most often structured data stored in SQL relational databases or Excel spreadsheets. To provide access to this data using LLM, Text-to-SQL technology is used, which converts users' text queries formulated in natural language into valid SQL queries [23, 24]. For this purpose, LLM receives as input not only the user's query, but also the database table schema, as well as examples of row values (if necessary). This allows the model to build accurate and appropriate queries based on the structure of the data contained in the database. Thus, the application of Text-to-SQL makes the work with data more accessible to a wide audience, including managers and specialists who do not have technical skills. Unlike RAG, when using Text-to-SQL, the final answer is not provided by the LLM itself, but by the database after the execution of SQL-query. This significantly increases the accuracy of answers, as the results are retrieved directly from the relational database, which minimizes the likelihood of incorrect answers. Although Text-to-SQL greatly simplifies access to data, there is a possibility that the LLM can misinterpret the user's query and generate an incorrect SQL-query. This could result in incorrect data or errors in the output. However, this risk is not unique to LLMs - even a human can incorrectly generate an SQL query if the task is misinterpreted.

Despite the availability of methods such as Text-to-SQL and RAG, there is still a category of sensitive unstructured data that requires particularly high precision. To work with such data can be used Knowledge Map approach, which creates a structured map that allows LLM to find the necessary text fragments with minimal error probability. Knowledge Map uses pre-prepared JSON objects similar to those used in the RAG method. These objects are document segments containing text and metadata (e.g., modification date, document type, and unique hash). However, for Knowledge Map applications, additional map files are created that include a brief description of each data segment.

The process of creating such maps is as follows: for each JSON file generated for RAG, a corresponding JSON file containing a brief description of each JSON object is generated. Metadata, such as a hash linking it to the original data segment, is also added to this file. Next, a new JSON file is co-created based on these files, which represents the next level map. This file includes brief descriptions of all JSON files in the same directory, as well as unique hashes to identify each of them. The process is repeated at each level of the hierarchy until a master JSON file is created, which can be used to easily find the desired JSON objects in the created hierarchy.

When a user makes a query, LLM uses Knowledge Map to navigate through the data. Along with the query, a "piece of the map", a JSON file is sent to the user, which indicates the route to the desired data. The model starts at the top level of the map and goes down, going from directory to directory, from file to file, until it reaches the desired piece of text. In this way, the LLM does not just generate an answer, but follows a predefined map that points the way to the data it is looking for.

One of the main advantages of Knowledge Map is the increased accuracy of the model's answers. Unlike RAG, where the model relies entirely on JSON objects extracted from a vector database, in this case LLM is guided by a structured data map that leads it to one specific object. This minimizes the possibility of errors or hallucinations (i.e., cases where the model generates incorrect answers). Since the model follows a strictly defined data structure, the entire process can be easily tracked and verified. This makes Knowledge Map an ideal tool for auditing and validating answers, especially when absolute accuracy is required (for example, when dealing with legal documents or important contracts).

Despite its obvious advantages, the Knowledge Map method has its limitations. One of the main difficulties is the complexity and laboriousness of creating a hierarchical map structure during the data preparation phase. For each document, it is necessary to generate short descriptions and build a multi-level hierarchy, which requires considerable effort, especially when dealing with large amounts of data. While some of this process can be automated using LLM, human auditing remains critical to ensure data accuracy and consistency.

In addition, even when using Knowledge Map, there is a risk of error at the stage of query interpretation by the user. If the query is phrased incorrectly or too generalized, the model may take the wrong path in the data hierarchy, leading to the wrong answer. This emphasizes the importance of careful query formulation and regular validation of results to minimize errors in critical data scenarios [25].

The integration of RAG, Text-to-SQL, and Knowledge Map methods can significantly improve LLM's data handling performance, especially in complex and multi-tiered management systems such as construction projects. Each of these methods addresses the unique challenges of processing different types of data: structured and unstructured. Let's consider possible approaches to their integration.

The first approach is based on selecting the best method based on the type of data. In this scenario, the LLM analyzes the incoming request and, based on the information provided about the data of each category, selects one of the three methods most suitable for processing the requested information. This approach is ideal in cases where the data is clearly categorized, each of which is responsible for a specific type of information. The LLM analyzes the request, determines which category the data belongs to, and uses the appropriate method for processing. This approach is most effective when data from different categories do not overlap, minimizing conflicts and increasing processing accuracy.

In the second approach, if data may overlap or be distributed among several methods, an asynchronous approach is used. In this case, the LLM performs parallel data searches in all methods simultaneously. This allows the model to analyze potential sources of information independently. At the final stage, the model retrieves the extracted data from each group and either combines them into a single answer if they are complementary, or selects the best result if the data from different sources overlap or represent alternative answers to the query. This approach is particularly effective in systems where data may be duplicated or distributed across different sources and the most complete and accurate answer is required.

Combining RAG, Text-to-SQL and Knowledge Map allows LLM to work with a wide range of data, including both structured and unstructured sources. This provides more accurate and meaningful answers to user queries. This system can adapt to any type of data and tasks, choosing the best method depending on the structure of the query and the characteristics of the data.

3 Results and Discussion

To test the described RAG, Text-to-SQL and Knowledge Map methods, a series of tests were conducted to demonstrate the advantages of each method in the context of processing different types of data related to construction information management.

The RAG method was tested using unstructured data taken from the one-page website digital.pioneer.ru. The data was automatically collected, split into fragments and converted into JSON objects, which were then placed into a vector database. The model processed queries using RAG to extract accurate answers from the database. The test query was "How can I contact Pioneer Digital?" (see Table 1).

Table 1. Comparison of the base model and the model using the RAG method.

Base model response:	Response using the RAG method:
"You can contact Pioneer Digital AI through a number of different channels. Here are a few ways: Website: Visit the company's official website at www.pioneerdigitalai.com where you will find contact information including email address, phone numbers and office addresses Email: Send an email to info@ pioneerdigitalai.com with your questions or inquiries Social media: The company has profiles on social media such as LinkedIn and Twitter. You can send a private message or leave a comment on one of their posts."	"To contact the company "PIONEER DIGITAL" you can use the following contact information: Address: Moscow, Malaya Pirogovskaya St., 3 Working hours: Monday - Friday: 09:30 - 18:30 Telephone number: 7 (495) 502-9559 E-mail: digital@pioneer.ru You can also use the feedback form on the company's website by clicking on the "SEND" button in the "CONTACT US" section. After sending the request, the company will contact you as soon as possible."

As can be seen from the results, the baseline model, without access to information from the website, generated an incorrect response by fictionalizing the website, email and social profiles. In contrast, the RAG method was able to extract accurate contact information relevant to the query directly from the company's website. This clearly demonstrates the advantage of RAG in dealing with unstructured data extracted from external sources and its ability to generate accurate answers based on relevant information.

The next step tested the Text-to-SQL method, which converts users' text queries into SQL queries to extract data from relational databases. The test data used was a room explication (see Table 2).

Table 2. Explanation of premises Building No. 3.

Room number	Name	Area m^2	Room category
1	Tambour	2.47	-
2	Corridor	5.57	-
3	Living room	17.15	-
4	Kitchen	8.02	-
5	Bathroom	4.93	-

The test query used was "Which rooms in the third house are larger than 8 m2?". The result of the query:

Generated SQL query:

"SELECT [Room Number], [Name], [Area m2], [Room Cat. Room] FROM [Explanation of premises house #3] WHERE [Area m2] > 8;"

SQL query result:

"Room number | Name | Area m^2 | Room Cat.

----

3 | Living room | 17.15 | -
4 | Kitchen | 8.02 | -".

This test demonstrated the model's ability to generate SQL queries that not only request general information, but also apply data filtering by specified conditions, which significantly simplifies access to structured data in relational databases and increases the accuracy and speed of query processing.

To test the Knowledge Map method, we created a hierarchical data structure that simulates real work with documents and their hierarchy. The data structure shown in (see Fig. 1) used four JSON files, each containing two JSON objects with floor, room, and area information. These files were located in different directories. At higher levels of the hierarchy, files containing only floor and room data were created, and at the highest level, a file containing only floor information was created.

The following query was asked for the test: "What is the square footage of the bedroom on the second floor?". The results of the query:

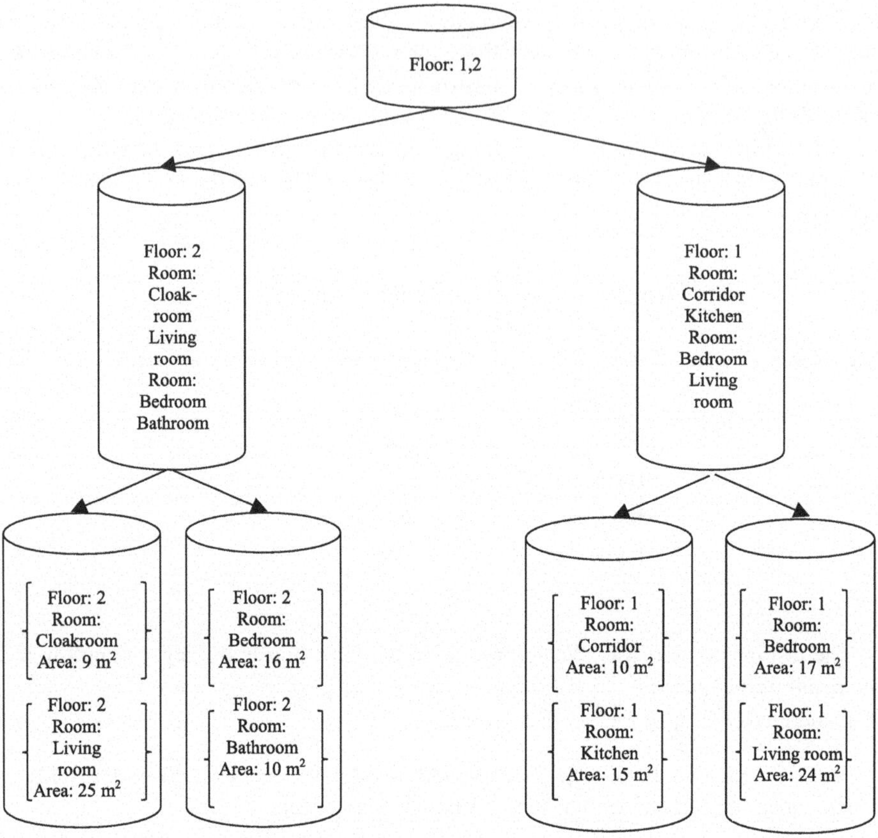

Fig. 1. Structure of Knowledge Map.

- First query result: "Floor: 2;"
- Second query result: "Floor: 2; Room: Bedroom;"
- Result of the third query: "The area of the bedroom on the second floor is 16 m2"

The Knowledge Map method allowed the model to traverse the hierarchical data structure, starting with a common floor file and sequentially finding the desired room and its area. In contrast to the RAG method, Knowledge Map ensured that the model followed a predefined data map, which minimized the probability of errors and provided a traceable path for the model to follow.

The test results show that the use of pre-training methods such as RAG, Text-to-SQL and Knowledge Map can bring new information to the model, significantly improving the quality and accuracy of the model's responses depending on the data type and query context:

- RAG has been shown to be highly effective on distributed and unstructured data such as web pages and documents. This method allows the model to extract relevant information from external sources, significantly improving the accuracy of responses.

- Text-to-SQL proved to be most useful when working with structured data in relational databases, where it is important to accurately retrieve data using SQL-queries, providing fast and correct access to the necessary data.
- Knowledge Map has proven its value when working with unstructured data that requires high precision. It tracks each step of the model through the query process, minimizing error rates and ensuring that the hierarchical structure of the data is followed.

Let us consider the share of data types, as well as the share of structured, unstructured and sensitive unstructured data for DSS systems (Table 3).

Table 3. Share of data types for DSS systems.

Data forms	Incoming data	Percent	Data types
Calculated indicators and metrics	Cost of projects	25%	Structured
	Project Budget - Plan		
	Project Budget - Fact		
	Technical and economic indicators		
	Deadlines for completion of work		
	Percentage of completion		
	Need for resources		
	Number of orders for construction and installation works		
Modeling and analysis results	Risk analysis	15%	
	Scenario modeling		
Financial reports and data	Balance sheets	5%	
	Profit and loss statements		
Reporting documents	Project text reports	20%	Unstructured
	Presentations and summaries		
Multimedia materials	Photographs of objects	10%	
	Video reports		
Strategic plans and forecasts	Business plans	15%	Unstructured sensitive
	Long-term forecasts		
Legal and contractual documents	Contracts and agreements	10%	
	License documents		

4 Conclusions

The use of LLM in data management in the construction industry represents a significant step forward in optimizing the processing and analysis of information at all stages of the construction lifecycle. Application of LLM helps to overcome the barriers associated with data fragmentation and limitations of existing systems such as ERP, PLM/BIM, and CRM, which face difficulties in working with unstructured information and integration of data from different sources. The analysis and testing conducted in this article leads to the following conclusions (Fig. 2):

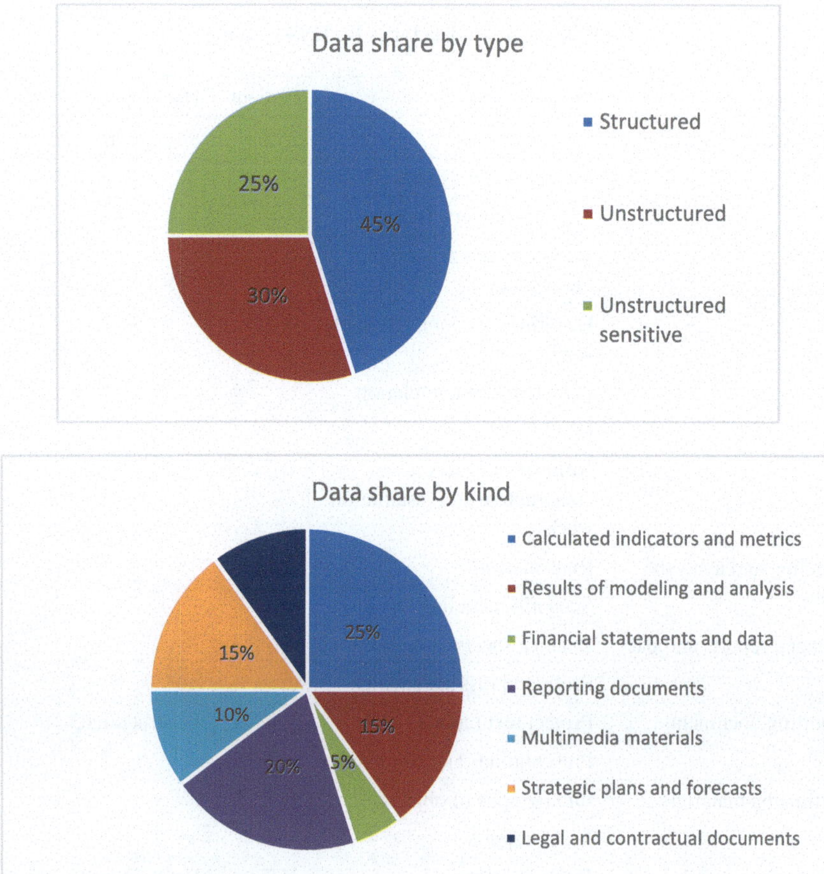

Fig. 2. Share of data by type and by kind in DSS system (clockwise).

1. The evolution of data management in the construction industry has gone from paper-based to computer-based systems and is now moving towards the use of artificial intelligence and large language models (LLM).

2. Today's construction industry is faced with the challenges of managing vast amounts of disparate data at all stages of the facilities lifecycle. Traditional data management methods have limitations in processing and integrating information from various sources.
3. The following methods of applying LLM to work with data are considered:
   - RAG (Retrieval-Augmented Generation) - effective for working with unstructured data;
   - Text-to-SQL – is suitable for structured data in relational databases;
   - Knowledge Map – can be used to work with sensitive unstructured data requiring high accuracy.
4. LLM application represents a promising direction in the field of construction data management. Models are able to process large amounts of information, automate the processes of data search and analysis, providing fast and accurate answers to user queries.
5. Despite the significant advantages, the application of LLM in construction data management comes with certain challenges, such as the need for careful data preparation, risks of incorrect query interpretation and potential "hallucinations" of models.

As a result, the application of LLM in construction lifecycle data management is a promising trend that can revolutionize approaches to information processing and analysis in the construction industry, increasing its efficiency, adaptability and resilience to modern challenges. In the future, LLMs will become an integral part of the construction industry, integrating with advanced technologies such as digital twins and the Internet of Things, leading to comprehensive management systems capable of dynamically adapting to change and providing optimal solutions in real time.

References

1. Lavrinenko, L.I., Smyslova, D.A.: Automation of the construction objects lifecycle management at different stages. In: Proceedings of the VIII International Student Construction Forum, vol. 2, pp. 287–290. Belgorod State Technological University. EDN OSGOML, Belgorod (2023)
2. Glukhov, N.K.: Application of artificial intelligence for lifecycle analysis of a technological project. In: Proceedings of the 13th International Youth Scientific Conference "Youth and the 21st Century", vol. 3, pp. 348–351. ZAO "University Book." EDN USNYMZ, Kursk (2024)
3. Radl, J., Kaiser, J.: Benefits of implementation of Common Data Environment (CDE) into construction projects. IOP Conf. Ser. Mater. Sci. Eng. **471** (2019)
4. Brusselaers, N., et al.: Economic, social and environmental impact assessment for off-site construction logistics: the data availability issue. IOP Conference Series: Earth and Environmental Science (2020)
5. Gariaev, N.A., Gariaeva, V.V.: Big data technology in construction. E3S Web Conf. **97**, 1032 (2019)
6. Losev, K.Y., Losev, Y.G.: On the methodology of automation of the building lifecycle. Bull. Eurasian Sci, EDN NWOLOK **14**(1) (2022)
7. Tolmachev, A.E., Pakhomov, A.A., Gaev, L.V.: The future of artificial intelligence: from machine learning to generative models. In: Proceedings of the All-Russian Scientific and

Practical Conference on Cooperation of Science and Society for Innovation Development, pp. 60–63. Aeterna. EDN CHWACG, Ufa (2024)

8. Krestelev, D.I., Losev, K.Y.: Predictive analytics algorithms for building engineering systems parameters. Constr. Archit. **11**(3), 15 (2023)
9. Brown, M., Williams, R.: Project resource optimization using artificial intelligence: a case study approach. In: Proceedings of the International Conference on Project Management, pp. 78–92 (2023)
10. Chen, L., Patel, S.: Risk management in technological projects: machine learning perspectives. In: Proceedings of the International Conference on Artificial Intelligence in Project Management, pp. 211–226 (2024)
11. Wang, C., Kim, Y.: Project risk forecasting: using artificial intelligence for proactive decision making. Int. J. Project Manage. **30**(1), 45–63 (2022)
12. Garcia, A., Johnson, M.: Collaboration in the digital age: AI-supported tools for project teams. Harv. Bus. Rev. **65**(5), 112–128 (2023)
13. Shananin, V.A., Losev, K.Y.: Creating digital twins in construction using artificial intelligence. Innov. Investment **6**, 537–560 (2023)
14. Ivanova, I.B., Vasileva, A.Y.: "Digital Twin" of a building: differences from BIM technologies and sources of efficiency in housing and utilities. Socio-Economic Manag. Theory Pract. **2**, 43–49 (2021)
15. Generative and Discriminative Models. LearnOpenCV. Retrieved from https://learnopencv.com/generative-and-discriminative-models/
16. Shiryaev, N.: PLM/PDM/ERP: realities and prospects. J. "CAD and Graphics" **12**, 16–20 (2007)
17. Volkov, A.A., Gusakova, E.A., Ovchinnikov, A.N.: Development of programming environments for lifecycle information flows in construction. In: Proceedings of the All-Russian Scientific and Practical Conference on Cyber-Physical Construction Systems, pp. 51–57. National Research Moscow State University of Civil Engineering. EDN MURLFW, Moscow (2019)
18. Losev, Y.G., Losev, K.Y.: Foundations for formalization of automated technologies for lifecycle management of construction objects. Constr. Archit. **10**(4), 86–90 (2022)
19. Liu, Y., Han, T., Ma, S., Zhang, J., Yang, Y., Tian, J.: Summary of chatgpt-related research and perspective towards the future of large language models. Meta-Radiology, 100017 (2023)
20. Kutsenok, L.A., Korablev, Y.A.: Applying the generative pre-trained transformer model to the task of technical document analysis. In: Proceedings of the A.A. Vavilov Youth School on Control Problems in Technical Systems, vol. 1, pp. 8–12, EDN DCQVIJ (2024)
21. Golikov, A.A., Akimov, D.A., Romanovsky, M.S., Trashchenkov, S.V.: Aspects of creating a corporate question-answer system using generative pre-trained language models. Litera **12**, 190–205 (2023)
22. Gao, Y., et al.: Retrieval-Augmented Generation for Large Language Models: A Survey. arXiv (2023). https://doi.org/10.48550/arXiv.2312.10997
23. Qin, B., et al.: A survey on text-to-SQL parsing: concepts, methods, and future directions. arXiv (2022). https://doi.org/10.48550/arXiv.2208.13629
24. Gao, D., et al.: Text-to-SQL empowered by large language models: a benchmark evaluation. arXiv (2023). https://doi.org/10.48550/arXiv.2308.15363
25. Liu, X., et al.: GPT understands, too. AI Open (2023)

Algorithms of Technical Solutions for Dust Reducing in Point-Pattern Housing Development

Svetlana Manzhilevskaya(✉)

Don State Technical University, Rostov-On-Don 344001, Russia
smanzhilevskaya@yandex.ru

Abstract. The article discusses methods of effective dust control in residential areas with point-pattern housing development, aimed at reducing the cost of dust pollution and the need for dust cleaning on the construction site. The main focus is on reducing harmful emissions from construction work and vehicles, which contributes to environmental protection. These measures not only ensure an increase in the environmental safety of the territory, but also contribute to the improvement of construction technologies.

Methods of controlling dust emissions on construction sites in urban areas must include algorithms for controlling dust pollution, as well as technical strategies aimed at minimizing the spread of dust. Optimization of technical and organizational solutions for managing dust emissions during the construction process helps to reduce overall construction costs and opens up new opportunities for the point-pattern housing development.

Keywords: Fine Dust · Point-Pattern Housing Development · Reducing Construction Costs

1 Introduction

Accelerated economic growth and the development of construction have led to the fact that the problem of atmospheric pollution has become very urgent, especially in urban areas, where it has acquired the scale of a global threat. Exposure to fine particles such as PM2.5 and PM10, if inhaled, can cause serious chronic respiratory diseases and even lead to premature death. These circumstances have prompted experts to pay more attention to the quality of atmospheric air in recent years. Among other industries, the construction industry stands out as a significant source of atmospheric pollution in many countries of the world [1].

During the construction process, including tasks such as excavation, landscaping, finishing and mixing of cement, significant amounts of dust are released [2]. This aspect of the construction industry, marked by its high labor intensity, forces workers to be near various sources of dust pollution. The impact of such a working environment not only harms the environment, but also poses threats to the health of people employed at the construction site.

S. V. Klyuev et al. (Eds.): ICRAAC 2024, LNCE 627, pp. 239–247, 2025.
https://doi.org/10.1007/978-3-031-82938-3_28

In light of the harmful effects of construction dust on the health of workers and residents in the immediate vicinity, the issue of environmental protection, as well as human safety, has become the subject of special attention in the construction industry and academia [3]. There is an urgent need to implement protective measures.

2 Methods and Materials

There are three key sources of dust pollution on construction sites that pose a threat to both the health of workers and residents of surrounding areas. Firstly, it is dust that occurs as a result of the operation of construction equipment and machinery. Secondly, the general level of dustiness of the air, characteristic of the area. And finally, the dust released during the execution of various construction works. Most of the construction equipment runs on fuel, which leads to emissions into the atmosphere not only of dust, but also of other pollutants. The introduction of an electric drive into construction equipment and the transition to electricity instead of traditional fuels will not only reduce emissions of harmful substances into the atmosphere, but also reduce operating costs. This will ensure more efficient combustion of fuel; help reduce its consumption and reduce emissions of harmful substances by 3–5 times [4].

Welding machines, compressors, hoists, pumps, as well as bulldozers and excavators can operate efficiently on electric energy, which contributes to improving the environmental situation on construction sites. For the successful implementation of this technology, it is necessary to develop algorithms of technical solutions to reduce air pollution. One of the important aspects is to study the effects of dust on the environment and the health of workers at the construction site. Our goal is to analyze and develop measures to reduce the impact of dust pollution on the environment and human health.

Dust is a serious problem for many production lines, since the use, extraction, transportation or processing of bulk materials is inevitably accompanied by the formation of dust. This process cannot be excluded from production, but it is unacceptable to work in conditions of constant presence of dust in the air. Dust quickly clogs the lungs and organs that filter the air, which increases the risk of allergic reactions and other health problems. To solve this problem, an effective solution is fine spraying of water, which significantly improves air quality and reduces dust content. Small dust particles of 10 microns and less than 2.5 microns are considered especially dangerous. Studies have also shown that particles of 10 microns in size have a significant impact on the formation of smog, which is typical for highly industrial and densely populated cities [5].

These particles are already present in the air, but their level is within the normal range. Approximately 92% of the dust comes from various sources, with about 30% of them related to construction sites. If the concentration of these particles increases by only 5–10 mcg/m^3 in an urban environment, the number of patients with chronic respiratory diseases will increase by 7% [6]. Dust penetrates into all equipment, causing it to wear out quickly. Negative consequences can also manifest themselves in the form of failure of control mechanisms due to ingress of dust particles. The use of the engine in conditions of high dust content leads to a reduction in its service life by 2–3 times. In addition, dust contributes to corrosion and complicates the repair process.

Measurements of dust emissions on construction sites in point-pattern housing development were carried out on construction site in Rostov-on-Don using a Handheld 3016 particle counter. The data obtained showed the average concentrations of particles PM2.5 and PM10 is given in Table 1 [7].

Table 1. The PM2.5 and PM10 concentration in point-pattern housing development.

Types of construction work	Average concentration, mcg/m^3
Preparatory work	2
Earthworks	3.9
Reinforced concrete works	3.5
Reinforcement works	1.5
Work on the installation of the frame	2.2
Stone installation works	3.8
Landscaping works	2.5
Dismantling works	1.1

The analysis of dust pollution made it possible to develop the cost-effective organizational and technical solutions to reduce it on the construction site, taking into account the tightness of urban conditions in point-pattern housing development.

3 Results and Discussions

In the course of the study, algorithms and models were developed to reduce dust pollution with finely dispersed dust particles PM2.5 and PM10 on construction sites and residential areas. Using the mobile technical solutions dustproof screen and water-misting gun [8] an algorithm of operation was developed, shown in Fig. 1, for dust control on the construction site in point-pattern housing development.

The standard model for dust control in the air of point-pattern housing development requires the following actions:

- identify sources of pollution (types of work);
- determine the dispersed composition of dust (size from 0.1 microns to 250 microns);
- determine the concentration by fractions in mcg/m^3;
- compare with the standards of maximum permissible concentration (MPC);
- conduct an assessment of air pollution;
- predict the ecological state;
- equip the site with dust-collecting and dust-proof devices and equipment;
- regulate emissions;
- develop recommendations on dust collection during construction work;
- assess the economic damage resulting from air pollution by dust.

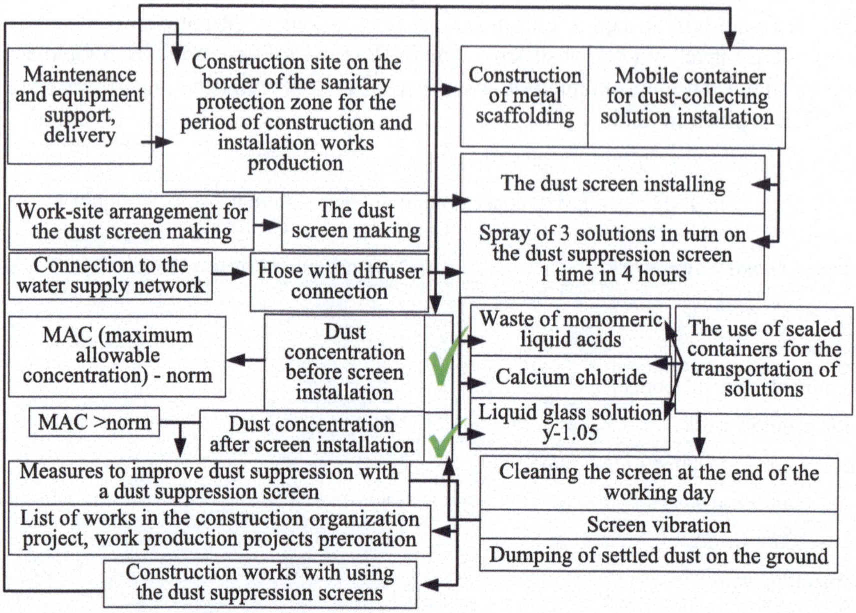

Fig. 1. The algorithm of the dustproof screen operation at the construction site on the border of the sanitary protection zone and the residential area.

The analysis of the effectiveness of the adopted technological solutions is presented in Table 2.

The effectiveness of the application of technical solutions for dust suppression during the construction period is shown in Figs. 2 and 3. Figure 4 shows an approach to solving technical problems related to the organization of construction and, in particular, measures to reduce dust emissions into the air of the point-pattern housing development.

4 Conclusions

This study presents algorithms for applying the most effective dust suppression methods in point-pattern housing development, which significantly reduce labor costs when installing them in a polluted area. Improving the working area conditions not only reduces equipment wear, but also reduces the risk of workers becoming ill, which helps to increase the profits of the construction company. Analysis of data on atmospheric air pollution with fine dust particles PM2.5 and PM10 during construction in the area showed that the situation on the construction site is unfavorable due to air pollution caused by construction work. To reduce this impact, effective dust suppression methods are proposed, which not only significantly reduce installation costs in the pollution zone, but also improve working conditions, prolong the service life of equipment. The reduction of dust emissions with the use of a dust shield reaches 92%, and thanks to the use of a fog gun, the concentration of suspended solids is reduced to 60%. The developed algorithms for dust protection measures on the construction site and technical solutions for

Table 2. Analysis of the effectiveness of patented environmental protection measures.

Name of the activities	**Efficiency factors**	
	Environmental	**Economic**
1. Making a dustproof screen according to the drawing, taking into account the size of the scaffolding and its installation	Total reduction of dust emissions by half	Increase in the daily output of employed workers by about 20%
2. The use of calcium chloride water, liquid glass for the preparation of solutions	Significant reduction of emissions	Additional costs amount to up to 5% of the cost of preparatory work on dust suppression
3. The dustproof screen application for using solutions and then dumping them	Soil pollution is reduced	Reduction of material losses
4. The screen cleaning from the settled dust by hydrogenation	Reducing the dustiness of the workplace air	The radius of the sanitary protection zone is reduced, respectively, the cost of work is reduced by reducing the dust concentration by 20%
5. Installation of the water misting gun at the construction site on the border of the sanitary protection zone, where the MPC is normal		Additional costs amount to 0.1% of the cost of construction work
6. Connection to water	Significant increase in dust suppression	Additional costs due to the cost of water consumed
7. The project of arrangement of a winter garden with the help of a green roof design	Reducing dust pollution on the construction site and residential area	It has been experimentally established that the dust concentration decreases by 25%

(continued)

Table 2. (*continued*)

Name of the activities	Efficiency factors	
	Environmental	Economic
8. The use of sealed containers for the transportation of solutions	Elimination of soil pollution by waste	Reduction of material losses
9. Organization of mechanized refueling of construction equipment and fuel transport and collection of waste oil for regeneration	Prevention of soil and water pollution, elimination of possible fires	Reducing fuel costs
10. Compliance with technology, norms and rules of production and improvement of the quality of work performed	Prevention of environmental pollution by waste	Reducing the cost of construction
11. Containerization of transportation and unloading of fragile piece, bulk materials (sand, brick, etc.)	Minimization of air pollution caused by dust from unloading and waste of destroyed materials	Minimizing resource losses
12. Completion of the construction includes thorough cleaning and landscaping, as well as restoration of vegetation cover	Minimization of soil destruction by air and water flows	Improving the development quality of the territory and reducing the cost
13. Shortening the construction period	Reducing the duration of exposure to the environment	Reduction of the cost of work and maintenance costs of dust collecting equipment

construction projects aimed at reducing dust pollution with PM2.5 and PM10 particles can reduce unforeseen costs from failures of construction equipment and the absence of workers due to illness by 10%, predict the ecological state.

Comprehensive dust emission management measures should be implemented on construction sites in urban areas, which should cover both algorithms for reducing dust pollution and technical methods aimed at limiting the spread of dust. Effective optimization of technical and organizational approaches to controlling dust emissions during construction helps reduce overall construction costs and stimulates the improvement of the point-pattern housing development.

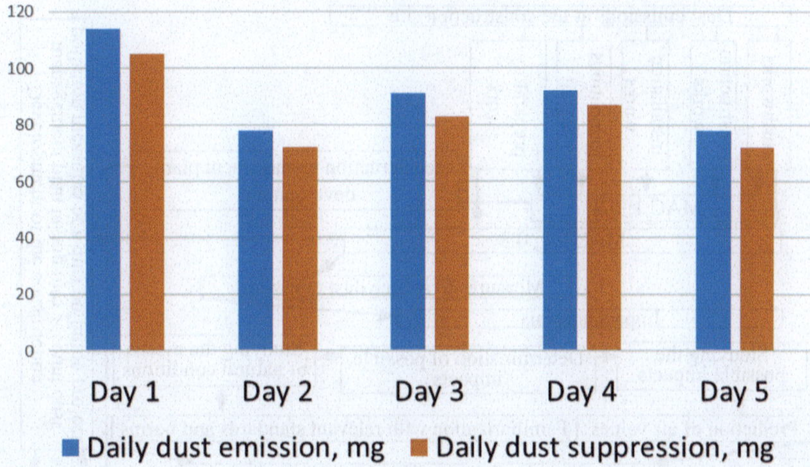

Fig. 2. Effectiveness of dust suppression by a dust proof screen at the construction site.

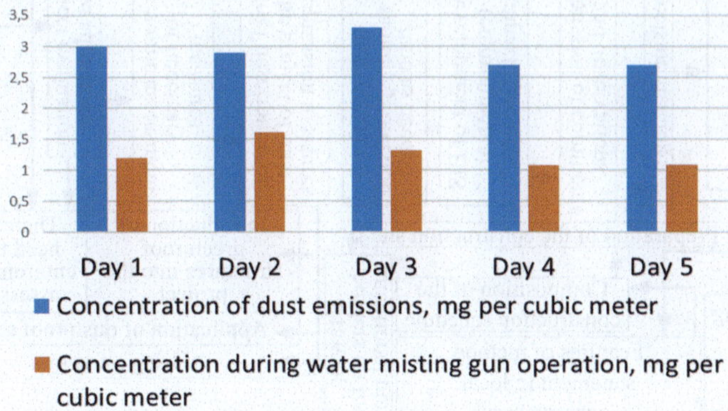

Fig. 3. Effectiveness of dust suppression by a water misting gun at the construction site.

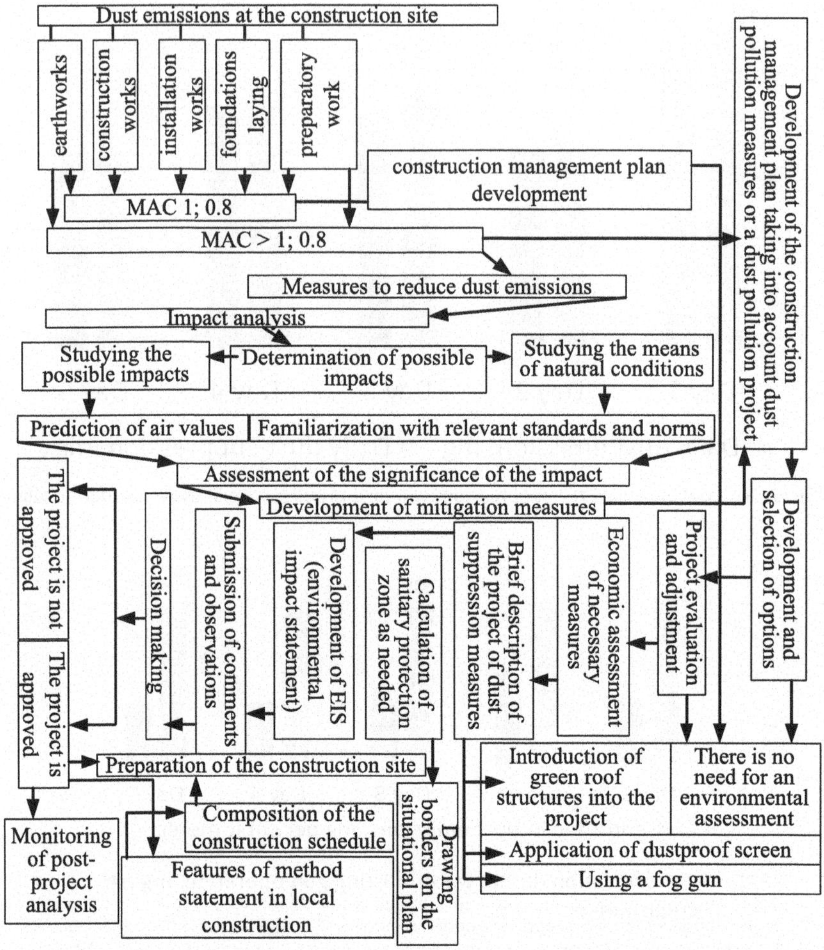

Fig. 4. Algorithm for the technical solutions in the construction management plan.

References

1. Tao, G., Feng, J., Feng, H., Feng, H., Zhang, K.: Reducing construction dust pollution by planning construction site layout. Buildings **12**, 531 (2022)
2. Kumi, L., Jeong, J., Jeong, J., Lee, J.: Empirical analysis of dust health impacts on construction workers considering work types. Buildings **12**, 1137 (2022)
3. Evtushenko, A., Lupinogin, V.: Investigation of air pollution with fine dust during repair and construction work inside premises. IOP Conf. Ser. Earth Environ. Sci. **937**, 022082 (2021)
4. Han, S.-W., Joo, H.-S., Kim, K.-C.: Modification of hybrid receptor model for atmospheric fine particles (PM2.5) in 2020 Daejeon, Korea, Using an ACERWT Model. Atmosphere. 15, 477 (2024)
5. Li, Z., Guo, H.: Time-series monitoring of dust-proof nets covering urban construction waste by multispectral images in Zhengzhou. China. Remote Sens. **14**, 3805 (2022)

6. Yu, B., Lu, X., Fan, X., Fan, P., Zuo, L.: Analyzing environmental risk, source and spatial distribution of potentially toxic elements in dust of residential area in Xi'an urban area. China. Ecotoxicol. Environ. Saf. **208**, 111679 (2021)
7. Manzhilevskaya, S.: Dust pollution in construction sites in point-pattern housing development. Buildings **14**(9), 2991 (2024)
8. Manzhilevskaya, S., Petrenko, L., Azarov, V.: Monitoring methods for fine dust pollution during construction operations. Adv. Intell. Syst. Comput. **1259**, 332–340 (2020)

Connection of Reinforcement with Tuff Concrete Prepared on Portland Cement with and Without Expansion Additives

Beslan Beshtokov[1], Ayes Tsipinov[2](✉), and Inna Zhirikova[2]

[1] Liability Company «SU-22», Nalchik 361520, Russia
[2] Kabardino-Balkarian State University Named After Kh.M. Berbekov, Nalchik 360004, Russia
czi.ayes69@yandex.ru

Abstract. The article discusses the critical role of adhesion between reinforcement and concrete in ensuring the structural integrity of reinforced concrete elements. It emphasizes that without proper adhesion, the appearance of the first crack can lead to increased forces in the stretched zone, further crack opening, reduction in the compressed zone height, and ultimately premature failure of the structure. The text delves into the mechanics of adhesion, describing how the bond between concrete and reinforcement functions to anchor the reinforcement and engage concrete in the load-bearing process. The article also explores the complex stress-strain state in the active bond area, highlighting the importance of maintaining adhesion for structural stability. Furthermore, it examines the limiting states of adhesion failure, both in terms of bearing capacity and operational requirements, and discusses the methods for evaluating adhesion quality through force measurements and stress calculations. The study aims to investigate adhesion of reinforcement in tuff concretes on quartz sands, prepared with different materials, providing insights into the factors influencing adhesion such as surface roughness, friction forces, and concrete characteristics. Overall, the article underscores the significance of adhesion in reinforced concrete structures and the need for careful evaluation and testing to ensure structural safety and longevity.

Keywords: Tuff Concrete · Bearing Capacity · Coupling of Reinforcement · Voltage Reinforcement · Transverse Reinforcement · Concrete

1 Introduction

The adhesion of reinforcement to concrete plays an important role in ensuring the bearing capacity of reinforced concrete structures. In the absence of adhesion, after the appearance of the first crack, the forces in the stretched zone increase along its entire length, which leads to an even greater crack opening, a reduction in the height of the compressed zone and, as a result, an earlier destruction of the structure.

The coupling of reinforcement with concrete is understood as the bond that develops on the contact surfaces between these two materials, ensuring their joint work. In a

S. V. Klyuev et al. (Eds.): ICRAAC 2024, LNCE 627, pp. 248–255, 2025.
https://doi.org/10.1007/978-3-031-82938-3_29

reinforced concrete element, the coupling performs two functions: anchoring the reinforcement (along the length of the seal, in the joints in the overlap) and involving concrete in the work (when the longitudinal stresses in the reinforcement change) [1–3].

The force from the reinforcement to the concrete is delivered through the tangential coupling stresses t_{cm}, in the lac section (the active coupling section or the zone of active redistribution of forces), the absolute values of which, after reaching their maximum in the lac section, begin to decrease.

The normal stress in the armature t s and the displacement of the rod g, as they move away from the loaded end, decrease and at a distance l k from the end are zero.

The lac section is characterized by a rather complex stress-strain state: the cross-sections are deplaned, the reinforcement is displaced relative to the concrete, plastic deformations of concrete develop, cracks occur inside the sample, etc. A violation of adhesion in the active area can occur in the event of one of two limiting states: (SP 63.13330.2018) in terms of bearing capacity and suitability for normal operation.

2 Methods and Materials

Consider the limiting states when pulling or pushing reinforcement from a concrete prism.

In the absence of transverse reinforcement and sufficient strength of concrete, significant annular tensile stresses arise from the ultimate load when pulling or pushing through concrete, which cause it to rupture along the axis of the reinforcement. In this case, the rod begins to slip due to loosening of adhesion and splitting of concrete into several blocks and their sliding. Such destruction is possible without a complete cut of all concrete consoles.

The first limit state is characterized by the destruction of the sample or loss of bearing capacity. It occurs when the pulling or pushing force reaches its maximum value (P_{max}), after which, without increasing the load, the mutual displacement of reinforcement and concrete begins to increase continuously. In the case of insufficient strength of concrete (on the cut) or in the presence of transverse reinforcement in sufficient quantity, the destruction of adhesion occurs as a result of the cut of all concrete consoles filling the hollows of the profile. Here, the quality of anchoring is usually assessed by the magnitude of the force corresponding to the beginning of the shift of the unloaded end of the reinforcement or by the magnitude of the force or the conditional tension of the coupling, at which a complete violation of the coupling occurs.

In addition, the onset of the first limit state is also possible due to fatigue failure as a result of shearing of all concrete consoles filling the profile depressions. In this case, adhesion can be assessed by endurance tests similar to concrete tests according to GOST 24545–2021.

The second limit state is characterized by operational requirements that are determined by compliance, deformability and crack resistance. Compliance and deformability are determined by the fact that certain geometric values characterizing the deformed state of reinforcement and concrete in the active bond area reach unacceptable, for some reason, limit values. They are usually estimated by the magnitude of the relative displacement of the loaded and unloaded ends of the reinforcement.

Crack resistance is characterized by the magnitude of the force corresponding to the moment of occurrence of longitudinal or transverse cracks in the active bond area or their permissible opening width. However, in the absence of adequate transverse reinforcement, after the occurrence of cracks, the sample usually quickly passes to the first limit state.

The adhesion of smooth and ribbed reinforcement to concrete is caused by the action of physical and chemical factors:

1. By engaging the micro-roughness of the reinforcement surface and the pro file protrusions with the concrete and the corresponding resistance of concrete consoles to crushing, shearing and the action of the main tensile stresses.
2. By bonding (adhesion) of reinforcement to concrete during hardening of cement paste.
3. Friction forces along the contact surface of the reinforcement and concrete, caused by radial pressure from concrete shrinkage.
4. Friction caused by transverse compression of concrete by an external load (support pressure, pressing of reinforcement to concrete when it bends in bending elements).

These factors are mainly determined by the strength capabilities of concrete, its composition, reinforcement profiling parameters, type and quantity of binder, size and type of filler, concrete shrinkage, etc.

In connection with the above, the task was set to study the adhesion of reinforcement with tuff concretes on quartz sands, prepared both on ordinary Portland cement and with expanding additives. Guided by the recommendations, the criteria for adhesion of reinforcement to concrete were taken as the value of the force N and the conditional adhesion stress τcm, at which a complete failure of adhesion occurred when pushing through reinforcement from a concrete prism measuring 10 * 10 * 20 cm. Concrete prisms for testing were made in metal forms with one centrally located reinforcement bar of periodic profile class A-400 pre-cleaned from dirt and oil.

In the manufacture of samples, the reinforcing bar was positioned horizontally, perpendicular to the direction of concreting.

The length of the sealing (anchoring), taking into account the piercing of concrete with reinforcement at the unloaded end, was 19 cm.

The studied compositions of tuff concrete are shown in Table 1.

It should be noted that the results of the punching test are 2–8% higher compared to pulling [2], and the conditions for transferring load from reinforcement to concrete in this case are closer to the condition for transferring forces from reinforcement to concrete under seismic loads, in prestressed structures, columns, monolithic cores, pile foundations, etc.

The experimentally obtained results are shown in Table 2.

Table 1. Compositions of tuff concrete.

№ compositions	Material consumption per 1 m^3 of tuff concrete							Concentration of coarse filler K.K.Z.,	Mobility of tuff-concrete mixtures, cm	Density of freshly laid concrete kg/m^3
	PC,kg/m^3	PC + EA, kg/m^3	R, kg/m^3	Q.s, Kg/m^3	T.S. kg/m^3	*Wate*, kg/m^3	Mel, kg/m^3			
1	–	350	669	826	–	238	2.8	0.35	5.5	2085
2	–	350	765	707	–	236	2.8	0.45	6	2060
3	–	350	860	589	–	236	2.8	0.55	6.5	2040
4	–	430	669	759	–	239	3.44	0.35	6	2100
5	–	430	765	640	–	237	3.44	0.45	6	2075
6	–	430	860	522	–	237	3.44	0.55	6.5	2050
7	–	510	669	692	–	241	4.08	0.35	6	2115
8	–	510	765	573	–	240	4.08	0.45	6.5	2090
9	–	510	860	455	–	240	4.08	0.55	6.5	2070
10	350	–	765	707	–	236	2.8	0.45	6.5	2060
11	430	–	765	640	–	237	3.44	0.45	6.5	2075
12	510	–	765	573	–	240	4.08	0.45	7	2090
13	–	430	669	–	575	310	3.44	0.35	5	1985
14	430	–	669	–	575	310	3.44	0.35	5.5	1985
15	–	430	765	–	519	295	3.44	0.45	5	2010
16	430	–	765	–	519	295	3.44	0.45	6	2010

3 Results and Discussion

According to the results of the experiment, it was found that the adhesion of reinforcement to concrete is mainly determined by the compressive strength of concrete.

The adhesion strength of reinforcement to concrete increases with increasing concrete strength. In the studied compositions, the effect of the concentration of a coarse (tuff) filler or an expansion additive was not detected (Fig. 1).

The obtained results of the adhesion strength of reinforcement to concrete on porous aggregates of Kabardino-Balkaria prepared on quartz sands are described by the dependence:

$$\tau_{\mathrm{cm}} = 0.255 * \mathrm{R} + 2.5 \quad (1)$$

where: R is the compressive strength of concrete, MPa

τ_{cm} is the average adhesion strength of reinforcement to tuff concrete, MPa.

The measure of the spread of the experimental values relative to the dependence was $\varepsilon = \pm 0.036$.

From the test results shown in Figs. 1 and 2, from the Table 2 it can be seen that with the same strength of concrete, there is a slight increase in the adhesion strength of samples stored for a long time (180 days). Compared to the samples stored for 28 days. This phenomenon is more evident with lower concrete strengths.

It should be noted that the loads pushing through the reinforcement (A 400 d = 10 mm) from a concrete prism, with a concrete strength above 20 MPa, already lie beyond the yield strength of reinforcing steel.

The results obtained were compared with the results obtained by other authors by pulling rebar out of a concrete prism, who conducted studies on heavy concrete and expanded clay concrete. Thus, dependence of the genus was proposed

$$\lambda_{ak} = 0.75\frac{\tau_{cm}}{R} \quad (2)$$

For the anchoring of 19 cm, this dependence was transformed and compared with the results obtained (Fig. 1).

Also provides data on the adhesion of reinforcement to expanded clay concrete, the value of which is also given in comparable forms [4–7]. Figure 1 shows that the adhesion strength of the reinforcement to tuff concrete is 2.5 MPa higher than with equal-strength heavy concrete.

This is explained by the fact that the cavities of the reinforcement profile are filled not with concrete, but with its mortar component. The amount of adhesion of periodic profile reinforcement to concrete is largely determined by the shear strength of the mortar component. Since the compressive strength of light and heavy concretes is equal to the strength of the mortar component of light concrete by 30–50% higher than the strength of heavy concrete, the tensile strength will also be higher.

Therefore, as follows from the Mersch formula, the shear strength in this case is higher by 30–45%, which is in good agreement with the magnitude of the increase in adhesion strength by 35%. For an indirect assessment of the endurance and crack resistance of the reinforcement coupling with tuff concrete, the value of conditional

Table 2. The experimentally obtained results.

№ compositions	Compressive strength, R_c, MPa		Adhesion strength R_{ad}, MPa	Adhesion strength, R_{ad}, MPa	L_{cond}, cm	L_{cond}, cm
	28d	180 d	28d	180 d	28	180
1	28	33	11	11.6	7	2
2	30	35.5	9	11.7	6.5	2
3	34	38	10.85	13.2	7	1.5
4	41	47	13.7	14.9	5.5	–
5	42	48	13	–	5	–
6	42	47	13	13.45	5	0.8
7	44.8	49	13.4	14.67	3.5	1.4
8	45.3	51	14.5	15.43	3	1.3
9	45.6	50	13.7	13.45	3	1.3
10	23	28	8.8	11.9	10	2.5
11	35.5	42	12.15	12.7	5.5	1.8
12	40.6	46	13.3	13.45	2.5	1.7
13	28	36.5	10	–	–	–
14	24	33	11.15	–	–	–
15	30	38	9.6	–	–	–
16	25	34	7.34	–	–	–

coupling Lsl is proposed, numerically equal to the length of the visually visible section of the consoles on concrete samples after punching tests. In addition, this difference is explained by the form of sample testing: in the experiments of these authors, the samples were tested by holding reinforcement from a concrete prism, when in this work experimental results were obtained by pushing reinforcement from a concrete prism.

In [8–10] it was noted that the adhesion strength of reinforcement to tuff concrete is not lower than the adhesion strength of reinforcement to concrete on dense aggregates. However, the adhesion strength of reinforcement with ash concrete and shell concrete on porous sands is 10–15% lower than with heavy concrete [11–14].

4 Conclusions

1. The adhesion of reinforcement to concrete is largely determined by the strength of the concrete itself. The type of binder (the presence of an expanding additive) and the concentration of coarse aggregate do not significantly affect the adhesion of reinforcement to concrete.
2. The dependence between the adhesion of reinforcement during punching and the strength of concrete has been revealed.

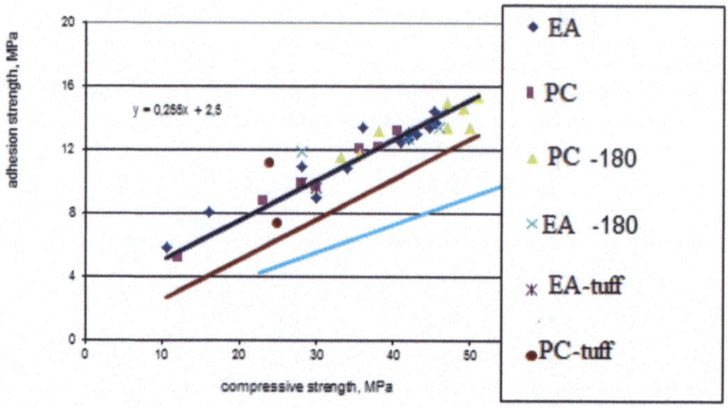

Fig. 1. Dependence of the adhesion strength of reinforcement with concrete on the strength of tuff concrete.

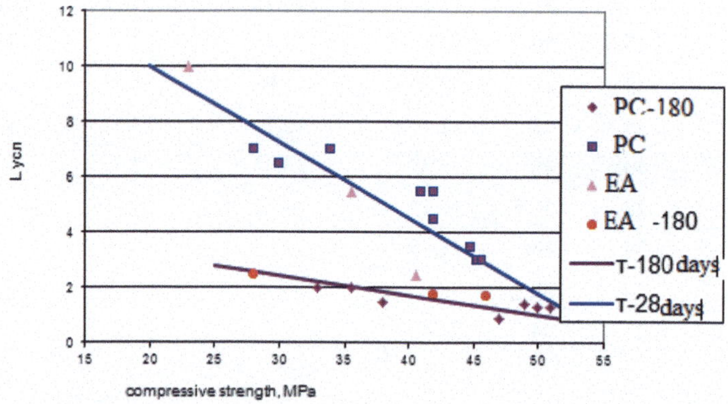

Fig. 2. Dependence of L_c on the cubic strength of tuff concrete 28 and 180 days old.

3. It is possible to evaluate (indirectly) the endurance and crack resistance of the coupling by (the value of the conditional coupling) l_{sl}, determined by the length of the visually visible section of the consoles on concrete samples during penetration tests.
4. In tuff concrete of the studied compositions, the strength decreases with an increase in the strength of concrete.
5. In tuff concrete with RD, the adhesion endurance and crack resistance turned out to be higher compared to tuff concrete on conventional Portland cement.
6. The dependence describing the adhesion strength of reinforcement with tuff concrete hardened in normal humidity conditions is also valid for samples subjected to early freezing.
7. In the case of an equal-strength comparison.

References

1. Hwan, O.B., Kim, S.H.: Realistic models for local bond stress-slip of reinforced concrete under repeated loading. J. Struct. Eng. **133**, 216–224 (2007)
2. Sang, G.K., Abuodha, S.O., Mwero, J.N.: Investigating the potential use of tuff aggregates to produce lightweight concrete. Int. J. Sci. Res. Publ. **10**(9), 458–478 (2020)
3. Hezhev, T.A., Zhurtov, A.V., Tsipinov, A.S., Klyuev, S.V.: Fire resistant fibre reinforced vermiculite concrete with volcanic application. Mag. Civ. Eng. **80**(4), 181–194 (2018)
4. Walid, F.E., Safwat, A., Ahmed, H.E.S., Abd Al-Kader, A.A.: Concrete Behaviour with Volcanic Tuff Inclusion. Civ. Eng. Arch. **9**(5), 1434–1441 (2021)
5. Chung, S.Y., Elrahman, M.A., Stephan, D., Kamm, P.H.: The influence of different concrete additions on the properties of lightweight concrete evaluated using experimental and numerical approaches. Constr. Build. Mater. **189**, 314–322 (2018)
6. Jehad, A.Z., Kamel, K.A.: Effect of volcanic tuff on the characteristics of cement mortar. Al-Huson University College, Al-Balqa' Applied University 279–284 (2014)
7. Bogas, J.A., Cunho, D.: Non-structural lightweight concrete with volcanic scoria aggregates for lightweight fill in building's floors. Constr. Build. Mater. **135**, 151–163 (2017)
8. Alkheder, S., Obaidat, Y.T., Taamneh, M.: Effect of olive waste (Husk) on the behavior of cement paste. Case Stud. Constr. Mater. **5**, 19–25 (2016)
9. Uçar, B., Öteyaka, A., Uz, B., Yılmaz, V.: Properties of zeolitic tuff (clinoptilolite) blended Portland cement. Build. Environ. **42**(11), 3808–3815 (2007)
10. Uzal, B., Turanli, L.: Studies on blended cement containing a high volume of natural pozzolans. Cem. Concr. Res. **33**(11), 1777–1781 (2003)
11. Khezhev, T.A., Pukharenko, Y., Khezhev, K., Klyuev, S.V.: Fiber gypsum concrete composites with using volcanic tuff sawing waste. ARPN J. Eng. Applied Sci. **13**(8), 2935–2946 (2018)
12. Klyuev, S.V., Khezhev, T.A., Pukharenko, Y.V., Klyuev, A.V.: Fiber concrete on the basis of composite binder and technogenic raw materials. Mater. Sci. Forum **931**, 603–607 (2018)
13. Klyuev, S.V., Khezhev, T.A., Pukharenko, Y.V., Klyuev, A.V.: Experimental study of fiber-reinforced concrete structures. Mater. Sci. Forum 945 MSF, 115–119 (2018)
14. Klyuev, S.V., Klyuev, A.V., Khezhev, T.A., Pucharenko, Y.: Technogenic sands as effective filler for fine-grained fibre concrete. J. Phys. Conf. Ser. **1118**, 012020 (2018)

Lightweight Composite Gypsum Materials

K. S. Petropavlovskii[1], T. B. Novichenkova[1], V. B. Petropavlovskaya[1(✉)], M G Sulman[1], and D Sunjidmaa[2]

[1] Tver State Technical University, Emb. Af. Nikitina, 22, Tver 170026, Russia
kspetropavlovsky@gmail.com

[2] Mongolian University of Science and Technology, Building T. No. 104, Ulaanbaatar, Mongolia

Abstract. The paper studies gypsum systems with fillers based on glass or vitrified microspheres. Such fillers in building mixtures allow to reduce density and improve thermal performance. The problem can be solved by combining a high-strength matrix base and a maximally lightweight filler in the composite. Such fillers can be vitrified spheres with micro-scale particle sizes, hollow microspheres, and aluminosilicate microfillers made from natural or man-made raw materials. The aim of the project was to study the possibility of using such fillers in building mixtures with reduced density and improved performance characteristics. Also important aspects in the synthesis of such mixtures are traditional physical and mechanical properties - compressive strength and bending strength, as well as a number of performance quality indicators. During these studies, an analysis of the test results of gypsum composites was conducted. Their synthesis on the basis of a reinforced gypsum matrix allows obtaining a material with unique properties, low density and sufficiently high strength. The use of a traditional initial gypsum binder increases the economic efficiency of the resulting material.

Keywords: Ash · Microspheres · Gypsum Binder · Reinforced Matrix · Filler · Reinforcement · Composite

1 Introduction

Improving the thermal performance of dry building mixes requires the creation of composite materials with high performance properties. To solve this problem, the direction of synthesis of modern composites with targeted formation of properties based on the synergistic effect is developing in construction materials science [1]. In materials science, a composite is a material that is made up of two or more materials with different characteristics and gives a light-weighted material with high strength and stiffness, resistant to corrosion that could be caused by chemical or environmental factors, easy to form, and highly resistant to creep [2]. A composite as a system includes two main elements - a continuous matrix and a reinforcing element [2, 3]. The reinforcing element is distributed in the volume of the matrix and primarily determines the deformation properties of the resulting composite. Whereas the matrix in this system plays the role of a binder,

S. V. Klyuev et al. (Eds.): ICRAAC 2024, LNCE 627, pp. 256–261, 2025.
https://doi.org/10.1007/978-3-031-82938-3_30

and also restrains and forwards stress that is applied to the composite. It consists of a reinforce material as its main component and matrix as its complement [2]. Composites categorized by its matrix are divided into Polymer Matrix Composite (PMC), used mostly in home appliances, Metal Matrix Composite (MMC), used mostly in automotive and airplane components and Ceramic Matrix Composite (CMC), used mostly on spaceships and rockets components as heat protection [3]. In building composites, mineral or polymer matrices are most often used as a matrix material. Mineral binders (cement, gypsum) have high performance properties, fire and environmental safety. To improve their physical and mechanical properties, fibers, microspheres, whiskers, etc. are used as reinforcing substances in composites based on them [2, 4, 5]. It is known that the use of hollow glass microspheres (HGM) helps to reduce the density and improve the thermal insulation properties of composites [2], including those based on gypsum [2]. Coal aluminosilicate or high-calcium microspheres from thermal power plant ash can be used due to their exceptional properties, such as low bulk density, high heat resistance, high processability and high strength [6–8].

The combination of fibrous and sphere-like microdispersed fillers can improve [9, 10] the performance properties of composites based on traditional mineral binders and concretes based on them [10].

2 Methods and Materials

During the experimental studies, the main initial component was gypsum binder β-modification, grade G-5 (Decor-1, Peshelansky gypsum plant, Bebyaevo settlement, Nizhny Novgorod region). In order to obtain a lightweight composite with improved performance properties, aluminosilicate and hollow glass microspheres and aluminosilicate granules were used as microfillers. Aluminosilicate microspheres are a product of processing ash and slag mixture of the Kashira fuel station in the Moscow region. Technical characteristics of aluminosilicate microsphere powders are given in Table 1.

Table 1. Aluminosilicate ash microfiller.

№ p/p	Name of indicators	Units of measurement	Meanings
1	Bulk density	kg/m^3	995 - 1005
2	True density	%	From 60
3	Humidity	%	Up to 0.3
4	Specific surface area	$м^2/kg$	203,5

Glass hollow microspheres with a diameter of 0.1 - 50 μm with a vacuum inside. The main indicators of the glass microfiller are given in Table 2.

All quality indicators of the starting substances and materials, as well as the obtained gypsum composites with microfillers, were determined using standard methods in accordance with the relevant regulatory documents of the GOST level.

Table 2. Glass microfiller.

№ p/p	Name of indicators	Units of measurement	Meanings
1	Bulk density	kg/m^3	1700 – 1750
2	Volume filling coefficient	%	From 60
3	Humidity	%	Up to 0.3
4	Buoyancy	%	75

The true density of the powders was determined using a Le Chatelier flask (Fig. 1a). The average particle size in the composition of the gypsum binder and filler powders was determined by the air permeability method using the PSH-11M device (Fig. 1b). The bulk density of ash was determined in accordance with the methods given in GOST 8735 and GOST 8269.0. The granulometric composition of the powders was evaluated using a laser particle analyzer «Analysette 22» made in Germany.

When studying the physical and mechanical characteristics of lightweight gypsum composites, depending on the microfillers, their content varied in an amount from 0 to 20%. Strength indicators after 3 days of hardening were calculated based on the destructive load indicators on a hydraulic press for gypsum composite cube samples. Research of composites based on a gypsum binder matrix and reinforcing microfillers were carried out on samples measuring 20x20x20 mm, according to the methods of GOST 23789; GOST 8462; GOST 7025.

3 Results and Discussions

At the first stage of comparative experiments, the effect of aluminosilicate microfiller from the ash of the Kashirskaya State District Power Plant was investigated. The results of the studies of the compressive strength and average density are shown in Figs. 2 and 3.

The obtained experimental data on the study of the strength and density of the gypsum composite with the addition of glass microfillers were obtained during the second stage of the comparative analysis and are shown in this paper in Fig. 4.

During this study, it was found that the introduction of the aluminosilicate component in the studied range of variation does not lead to a drop, but to some increase in the density and compressive strength of the gypsum composite (Figs. 2 and 3). Thus, the introduction of 10% aluminosilicate microfiller allows increasing the strength to 11%, with a density change of less than 5%.

An increase in the percentage of microfiller by more than 10% has a negative effect on strength. However, the decrease in strength is not significant - does not exceed the standard deviation. The increase in density is less than 1%. However, a different dependence was established for a gypsum composite with glass microfiller (Fig. 3). The average density decreases monotonically over the entire studied range of filler content change, reaching a minimum value of 998 kg/m^3 with a microfiller content of 20%. However, the ultimate strength repeats the nature of the density change. Only the decrease in strength is more significant. It is more than 50% in the studied range.

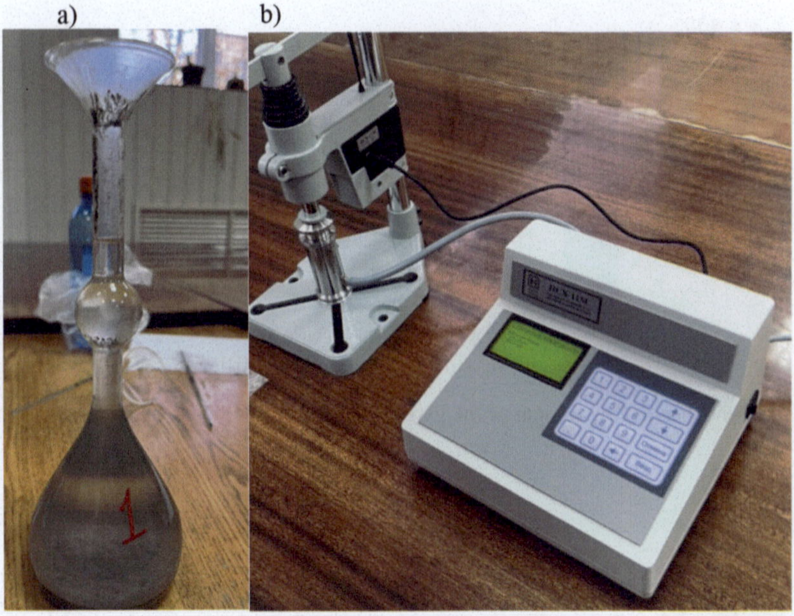

Fig. 1. Testing of aluminosilicate microsphere powders: a) determination of true density; b) determination of specific surface area and average particle size in the powder composition using the PSH-11M device.

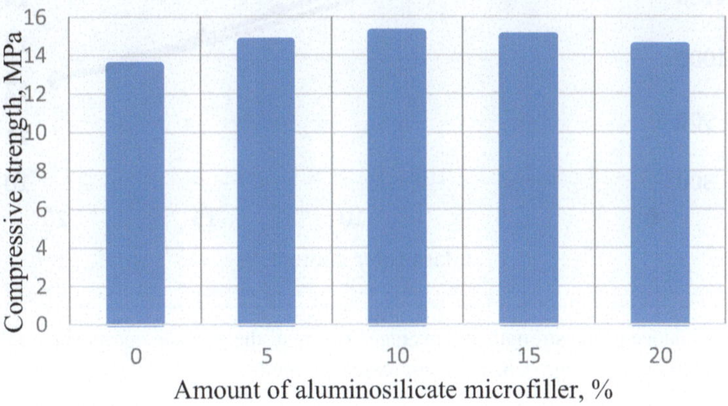

Fig. 2. Dependence of the strength of the gypsum composite on the content of aluminosilicate microfiller.

Comparative analysis showed that the glass-reinforced composite has a density of about 1000 kg/m^3 with a strength of 14.3 MPa, while the gypsum composite with aluminosilicate filler with the same percentage content has a density of 1270 kg/m^3 with a strength of 15.2 MPa. Consequently, the specific strength of the glass composite will be

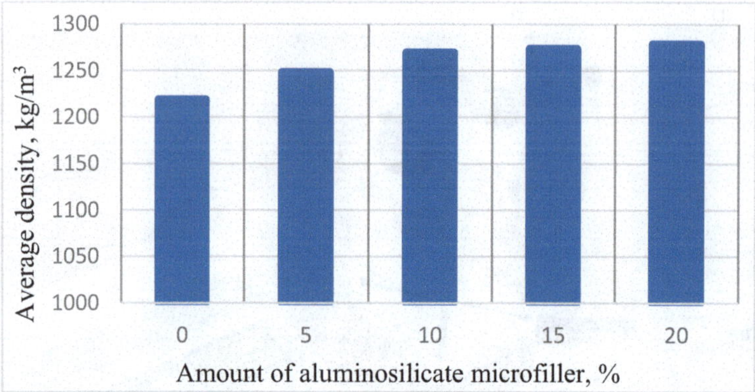

Fig. 3. Dependence of the strength of the gypsum composite on the content of aluminosilicate microfiller.

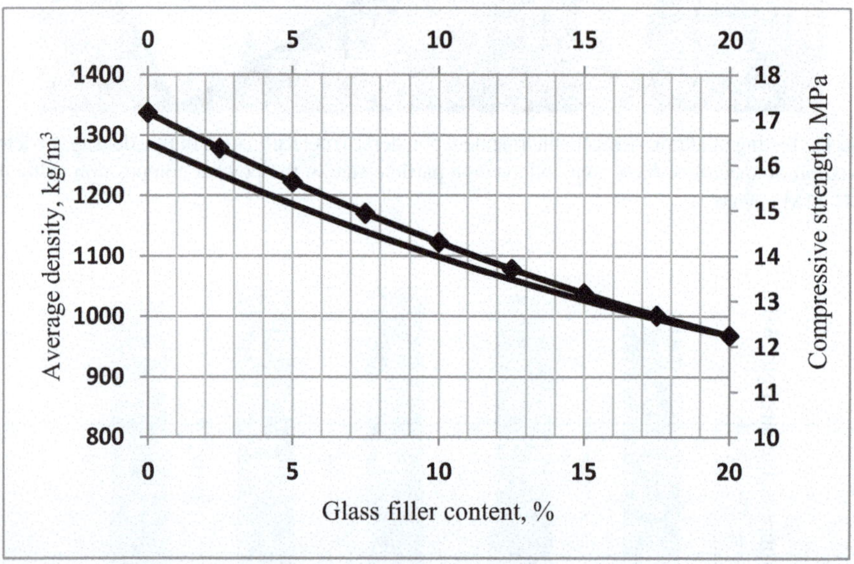

Fig. 4. Dependence of the strength and average density of the gypsum composite on the content of glass microfillers (Average density, Compressive strength).

11% higher than the specific strength of the gypsum composite with the addition of ash microfiller.

4 Conclusions

Thus, the conducted studies confirmed the possibility of obtaining lightweight gypsum composites for the production of building mixtures based on glass microfiller. The material is characterized by sufficient strength characteristics even in the case of using the maximum volume of glass microfiller in the studied range - 20%. In addition, other quality indicators of the glass-reinforced material, for example, insulating properties, are also improved. The use of aluminosilicate microspheres in combination with a gypsum matrix allows improving performance indicators and reducing water absorption of the reinforced gypsum composite. However, this does not allow obtaining lightweight composites without the introduction of additional lightweight fillers. In general, aluminosilicate microspheres are no less valuable components of building composites based on gypsum binder due to their greater technological effectiveness and ease of processing. Therefore, the research will be continued. The resulting lightweight gypsum materials open a new direction in the production of materials for the formation of an energy-efficient, comfortable and safe human environment.

Acknowledgement. The research was supported by Russian Science Foundation (project No. 24-49-03004).

References

1. Klyuev, A.V. et al.: Experimental studies of the processes of structure formation of composite mixtures with technogenic mechanoactivated silica component. Construction Materials and Products **6**(2), 5–18 (2023). https://doi.org/10.58224/2618-7183-2023-6-2-5-18
2. Sutikno, Berata, W., Wijanarko, W.: Hollow glass microsphere-epoxy composite material for helmet application to reduce impact energy due to collision. In: AIP Conference Proceedings, **1855**(1), 030013 (2017)
3. Sweity, Y., Petropavlovskaya, V., Novichenkova, T., Petropavlovskii, K.: Influence of bed ash on the rheology and properties of gypsum building mixtures. In: XII International Scientific and Practical Forum - «Environmentally sustainable cities and settlements: problems and solutions» (ESCP-2023), p. 03012. E3S Web of Conference (2023)
4. Klyuev, A.V., Kashapov, N.F., Klyuev, S.V., Lesovik, R.V., Ageeva, M.S., Fomina, E.V., Ayubov, N.A.: Development of alkali-activated binders based on technogenic fibrous materials. Construction Materials and Products **6**(1), 60–73 (2023). https://doi.org/10.58224/2618-7183-2023-6-1-60-73
5. Tkach, E., Bichaev, M.: Properties of epoxy composites with halloysite nanotubes subjected to tensile testing. In: Akimov, P., Vatin, N. (eds.) Proceedings of FORM 2021, LNCS, vol. 170, pp. 73–83. Springer, Cham (2022)
6. Petropavlovskaya, V., Zavadko, M., Novichenkova, T., Petropavlovskii, K., Sulman, M.: The use of aluminosilicate ash microspheres from waste ash and slag mixtures in gypsum-lime compositions. Materials **16**(12), 4213 (2023)
7. Fediuk, R., et al.: Nanomodified basalt fiber cement composite with bottom ash. Journal of Composites Science **7**(3), 96 (2023)
8. Miryuk, O., Fediuk, R., Amran, M.: Porous fly ash / aluminosilicate microspheres-based composites containing lightweight granules using liquid glass as binder. Polymers **14**(17), 3461 (2022)
9. Tkach, E.V., Temirkanov, R.I., Tkach, S.A.: Comprehensive study of modified concrete based on activated microsilicon together with micro-reinforcing fiber to improve performance. Bulletin of the Tomsk Polytechnic University. Geo Assets Engineering. **332**(5) 215–226 (2021)
10. Alhokabi, A., Hasan, M., Amran, M., Fediuk, R., Vatin, N., Alshaeer, H.: The effect of pofa-gypsum binary mixture replacement on the performance of mechanical and microstructural. Properties enhancements of clays. Materials **15**(4), 1532 (2022)

Influence of the Type of Structural Scheme of a Multi-Stored Residential Building on the Value of Maximum Floor Acceleration

S. V. Klyuev(✉), V.A. Beliy, and T.V. Nazmeeva

Belgorod State Technological University Named After V.G. Shukhov, Belgorod, Russia
Klyuyev@yandex.ru

Abstract. The paper deals with the influence of the choice of material for the structural scheme of a multi-stored residential frame building on the behavior of the building when resisting the action of wind pulsation load. In the course of the study, a comparative analysis of floor accelerations for two variants of the structural design of the building frame was performed and the influence of the structural design of the building on the value of the maximum acceleration and resistance to pulsation wind load was evaluated. Residential multi-stored building with steel frame and residential multi-stored building with monolithic reinforced concrete frame were considered. Nowadays, the construction of steel-framed residential buildings is actively developing in the world and it is important to evaluate the comfort of people's stay in such a building. In the course of the study, calculations of the structural system of the building are carried out, maximum floor accelerations are determined, conclusions are drawn about the influence of the material on the behavior of the frame and recommendations on the application of compensating measures are given. The steel-framed building is a frame-rod system and is a low-frequency building; the value of the first natural vibration frequency f_1 does not exceed 0.2 Hz. To ensure the economic efficiency of construction has a reduced metal intensity and as a consequence, reduced mass, which reduces the resistance of the frame to the action of the pulsation component of the wind load and increases the value of the maximum acceleration. The article analyzes the reasons for reducing the resistance of the frame to the action of the pulsation component of the wind load and increases the value of the maximum acceleration for the building on the steel frame, which is associated with the structural features of the building and metal intensity, and proposed compensatory measures to ensure compliance with the requirement to ensure dynamic comfort.

Keywords: Steel frame · Pulsating wind load · Logarithmic decrement · Acceleration of floors · Wind · Comfort of living

1 Introduction

Steel construction is a fast and high-tech process. The advantages of steel construction are in the reduction of construction time due to the transfer of most of the manufacturing processes of structures from the construction site to the factory. Currently, steel

S. V. Klyuev et al. (Eds.): ICRAAC 2024, LNCE 627, pp. 262–272, 2025.
https://doi.org/10.1007/978-3-031-82938-3_31

structures are widely used in the construction of high-rise and rapidly erected buildings, in the construction of outrigger floors to minimize the movement of the building as a whole [1–3]. The advantages of steel construction structures can also be in demand in the construction of multi-stored residential buildings of mass construction, where now mainly monolithic reinforced concrete is used. But steel construction is quite expensive. For the introduction of steel in mass construction, a steel-framed building must be cost effective [4–6], one possible way to increase cost effectiveness is to reduce metal intensity, which in turn can reduce the load-bearing capacity characteristics of the building compared to a heavy monolithic reinforced concrete frame.

In the course of the study, a comparative analysis of floor accelerations for two variants of the structural solution of the building frame was performed and the influence of the structural solution of the building on the value of the maximum acceleration and resistance to pulsation wind load was evaluated. A residential multi-stored building with a steel frame and a residential multi-stored building with a monolithic reinforced concrete frame were considered [7–9].

2 Methods and Materials

The comfort of residents in an apartment building is usually assessed by the maximum floor acceleration [10, 11] caused by the building oscillation from the action of the pulsation component of the wind load, which depends on the frequency of the building itself. The structural scheme of the building affects its own frequency and in order to ensure the requirement of occupant comfort, additional structural measures may be required, which will affect the cost and material intensity of the building.

A 17-storey tower residential building of mass construction with a basement and non-residential public premises (store) on the ground floor was selected for analysis; there is no underground parking; the basement is used for technical purposes. Building dimensions in axes: 21.04 x 33.3 m. The maximum height of the building up to the top of the parapet is 54.3 m. The height of the typical floor is 3.0 m, the height of the ground floor is 4.2 m, and the height of the basement is 3.3 m. Two types of structural scheme with monolithic reinforced concrete and steel frame were developed for such a building.

The structural system of a multi-stored building with a monolithic reinforced concrete frame is a frame with rigid nodes in all directions. The vertical load-bearing elements are walls and pylons. The thickness of walls is not more than 200 mm. This system is characteristic for residential buildings made of monolithic reinforced concrete, as it allows for optimal reinforcement and concrete consumption for a residential building. The technical solution of the building is simple in plan, with equal height typical floors and coaxial arrangement of walls in height. The stair and elevator unit is located in the center of the building.

The overall stability of the monolithic reinforced concrete building and horizontal loads is ensured by the joint work of vertical load-bearing structures of the frame, rigidly clamped in the foundation and horizontal disks of slabs, rigidly connected with vertical structures.

Foundation - monolithic reinforced concrete slab. Walls and elevator shaft - monolithic reinforced concrete. Floor slabs - beamless in the form of a smooth slab, 180 mm thick. Structures of the whole building are made of concrete with characteristics R_b=18.5 MPa, R_{bt}=1.55 MPa. Exterior walls are self-supporting made of aerated concrete blocks 200 mm thick and clad with a system of curtain-type ventilated facade made of porcelain stoneware.

The structural system of a multi-stored steel-framed building is a frame and ligament frame consisting of columns, beams, links and slab disks. The spatial rigidity and stability of the structural scheme of the building is ensured by the joint operation of the frame and links by combining them with rigid slab disks. To create the slab disk, a composite solution was used with the pouring of monolithic reinforced concrete over corrugated sheet as permanent formwork and resting the corrugated sheet on the beams. Based on the results of work with the structural scheme of the building, the lowest possible metal intensity was obtained.

Columns - welded I-beams made of high-strength steel with yield strength not less than 440 MPa. Sheet thickness from 6 to 30 mm. Beams - welded I-beams of different heights, the lower flange of the beam is 80 mm wider than the upper flange for laying corrugated sheet. Ties - bent-welded square profiles. The ties are integrated into the layout. The material of beams and ties is steel with yield strength not less than 350 MPa. Metal intensity for the total area of the building was 42 kg/m^2. The frame elements are bolted together, Fig. 1, with minimal use of factory and assembly welding to ensure all-season installation and high construction speed.

The structural solution of the frame does not affect the layout, both building solutions have the same layout, engineering and finishing, the buildings differ only in slab thickness. In the case of a composite slab solution in a steel-framed building, it is quite difficult to ensure a slab thickness of 180–200 mm, as in the monolithic version. The distance between the columns is quite large; the floor beams have a certain height to ensure the required deflections.

In the course of the study, a calculation scheme was developed in SCAD and ETABS programs for each structural variant of the building, Fig. 2, and the calculation was performed. The calculation was performed for all types of loads acting on the building, including the pulsation component of the wind load.

Wind is a flow of air moving predominantly in a horizontal direction in a chaotic and turbulent manner. The occurrence of turbulence is influenced by the roughness of the earth's surface as the wind moves. Wind motion is associated with chaotic changes in wind speed. Random number theory is used to describe the processes of wind motion.

The ripple component of wind speed, V_g, is a centered random variable and is defined as

$$V_g = V - V_M,$$

there V – wind speed, m/s; V_m – average wind speed, m/s.

The ripple component of the wind load, w_m, is defined as.

$w_g = 0,5\frac{V_g}{V_m}w_m$, there

$$w_m = 0,5 \cdot C \cdot \rho \cdot V_m^2,$$

w_m – average wind load component, kPa; ρ - air density, kg/m^3, C – aerodynamic coefficient.

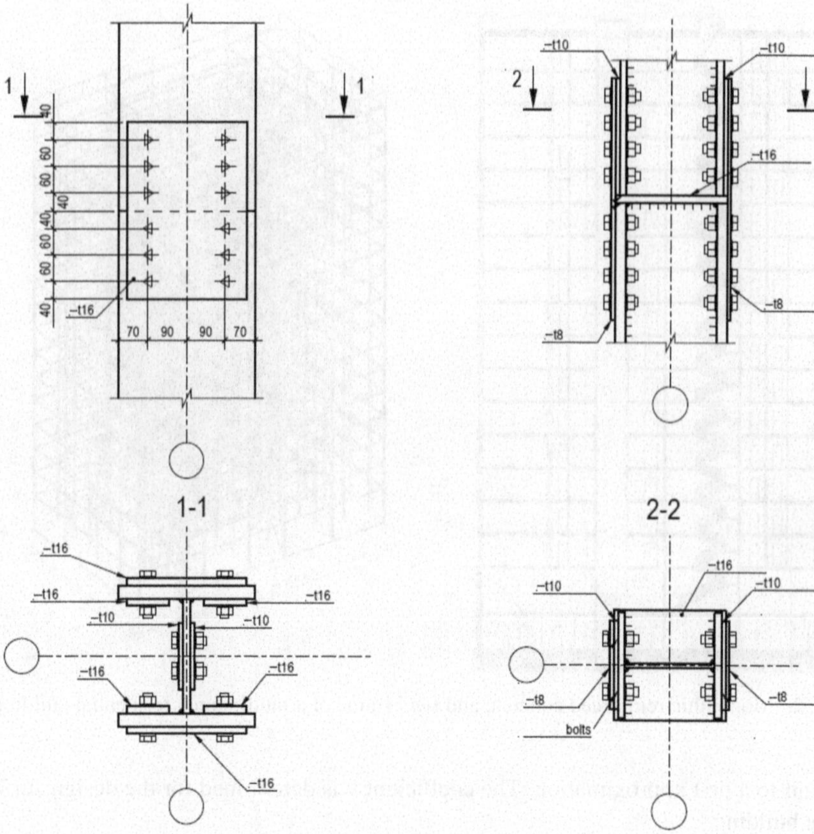

Fig. 1. Joints of steel columns to each other in height.

Depending on the type of building in terms of frequency - low-frequency or high-frequency building [6], which is determined by the first frequency of natural vibrations f_1, Hz, we have two design cases of calculation for ripple load in the regulatory documentation, taking into account the dynamic effect caused by the ripple load and without taking into account this dynamic effect.

1st calculation case $f_1 > f_{\lim}$, then the dynamic coefficient of the ripple load is $\xi = 1$ and for the ripple component of the wind load w_g the static calculation is carried out according to the formula:

$$w_g = w_m \varsigma(z_e)\nu,$$

there $f_{\lim}$ - limit value of natural oscillation frequency, Hz; ζ(ze) – wind pressure ripple coefficient for equivalent height ze, depending on the type of terrain; ν - spatial correlation coefficient of wind pressure pulsations.

The meaning of the coefficient ν is that it takes into account the random variation of wind speed not only in time, but also in space. On a surface interacting with wind flow, the ripple pressures are different at different points, the coefficient ν takes this effect into

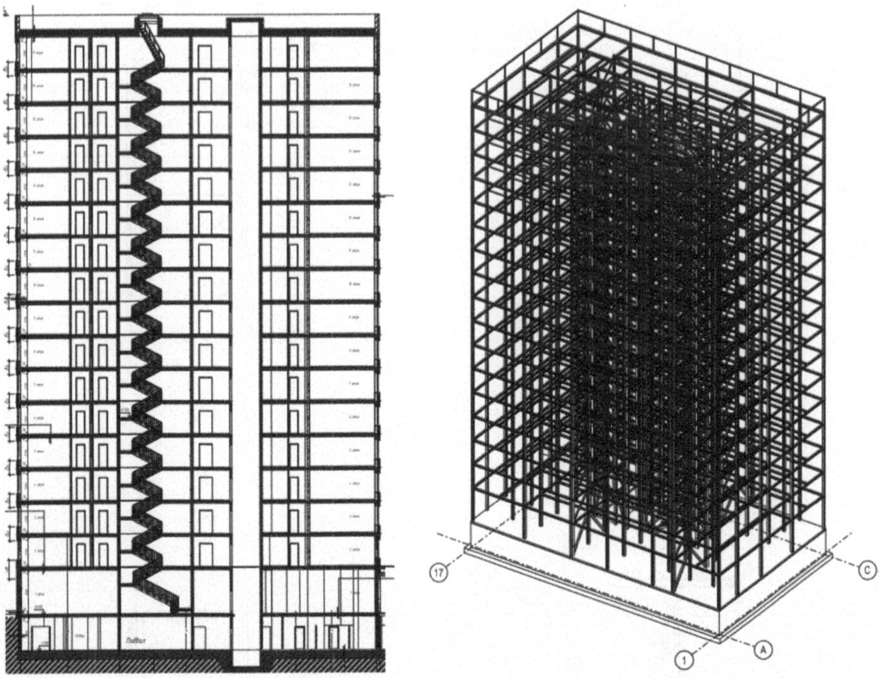

Fig. 2. Monolithic reinforced concrete and steel frame of a multi-storey residential building.

account to a first approximation. The coefficient ν is determined for the design surface of the building.

The limiting value of the natural oscillation frequency $f$lim is determined from the dimensionless period Tg,lim.

$$f_{lim} = \frac{\sqrt{w_0 k(z)\gamma_f}}{940T_{g,lim}},$$

there $T_{\text{g,lim}}$ is determined depending on the value of the total logarithmic decrement of oscillations δ.

The value of the limiting frequency depends on the wind region. Since the spectral densities of wind speed are different for different wind regions. The cutoff frequencies are also different, which in this case are called the limiting frequencies. It is assumed that the spectral density beyond the cutoff frequency is negligibly small.

The 2nd calculation case, when $f_1 < f_{\text{lim}} < f_2$, then $\xi > 1$ and for the ripple component of the wind load w_g is calculated taking into account the dynamism according to the formula:

$$w_g = w_m \xi \varsigma(z_e)\nu$$

there ξ – the coefficient of dynamism, determined depending on the total logarithmic decrement of oscillations δ and dimensionless period $T_{\text{g,1}}$, which is determined for the first natural frequency f_1, Hz;

$$T_{g,1} = \frac{\sqrt{w_0 k(z)\gamma_f}}{940 f_1}$$

there w_0 - standard value of wind pressure, kPa, $k(z_{эk})$- coefficient, taking into account wind pressure variation for height $Z_{эk}$, γ_f – load reliability coefficient

For low-frequency buildings, the ripple load is calculated using random function theory, and the dynamic effects of the ripple load are identified. For high-frequency buildings, the ripple load is also taken into account by static calculation.

3 Results and Discussions

To find the maximum floor acceleration for the two building variants, we do the following:

1. We determine consecutively for reinforced concrete and steel frame the values of floor displacements from the action of only the ripple component of wind, Fig. 3.

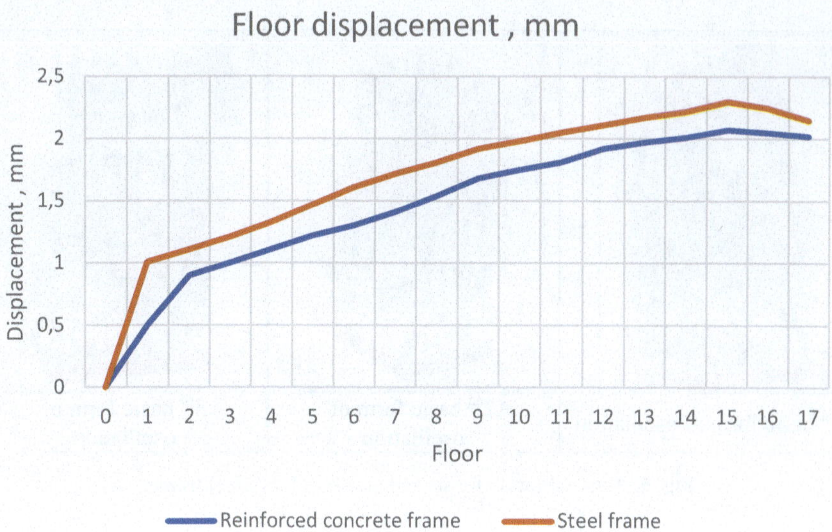

Fig. 3. Graph of floor displacement from the action of only the ripple component of wind load.

2. Next, we perform modal analysis and determine the main vibration shapes for the monolithic and steel versions, Figs. 4 and 5. In the reinforced concrete version, the torsional form was the third, and in the steel version the first, which is not a violation, but characterizes the distribution of weight and stiffness of the building. For each of the forms we determine the period of vibration and, respectively, the first frequency of natural vibrations f_1, Hz, Table 1.

As it is shown in Table 1, monolithic reinforced concrete frame is high-frequency and steel frame is low-frequency.

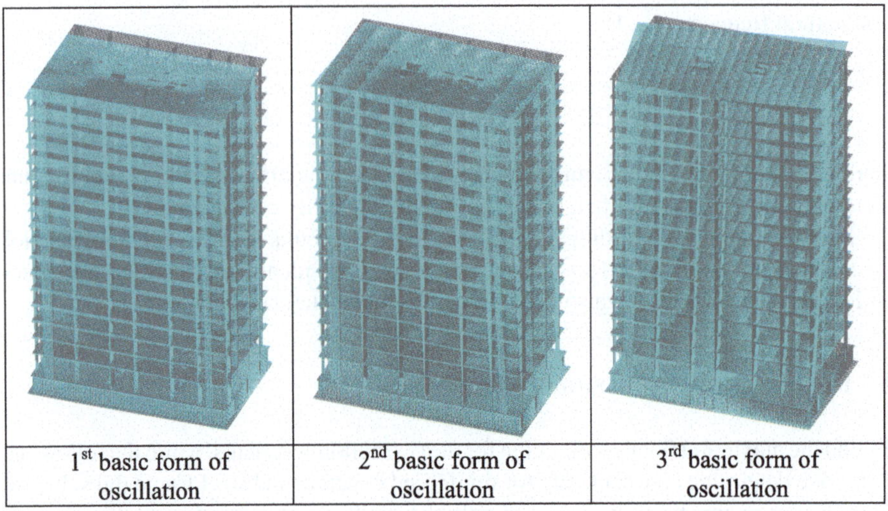

Fig. 4. First vibration forms and periods of monolithic reinforced concrete frame.

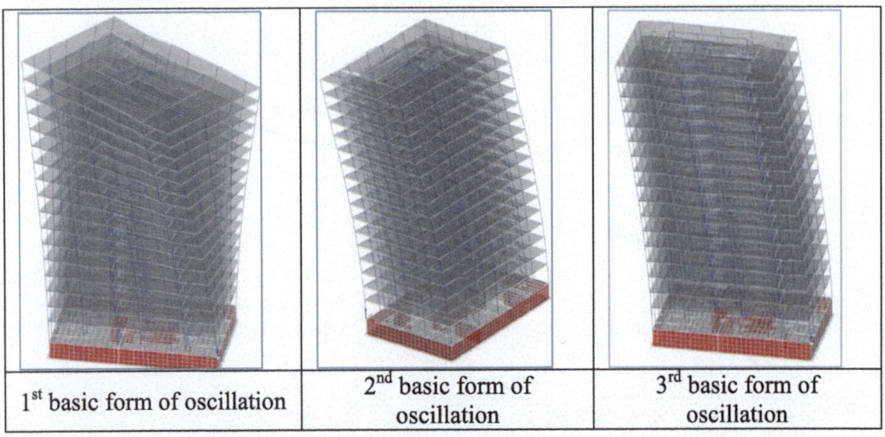

Fig. 5. First vibration forms and periods of the steel frame.

Table 1. Comparison of wind parameters for two frame variants.

Type of frame	T, s	f_1, Hz	$f_{\lim}$, Hz	w_0, kPa
Monolithic reinforced concrete	0.915	1.092	0.801	0.23
Steel	6.1	0.164	0.801	0.23

For the monolithic frame there is the 1st calculation case f_1=1.092>$f_{\lim}$=0.801 and the calculation is carried out without taking into account the dynamic effects of the ripple load, and for the steel frame of the building there is the 2nd calculation case when $f_1=0.164<f_{\lim}=0.801<f_2=0.824$, then $\xi>1$.

3. Then we find the maximum acceleration $a_{\max,i}$ for each oscillation form by the formula

$A_{\max,i}=A_i*\omega_i^2$,

there A_i – amplitude of the i-th oscillation form, m; ω_i – the circular frequency of the i-th form of oscillations, rad/s.

4. Then we determine the maximum acceleration of the characteristic points as the sum of the maximum accelerations for all the main vibration forms.

The maximum accelerations of each variant of the building frame are determined programmatically and summarized in a general graph, Fig. 6.

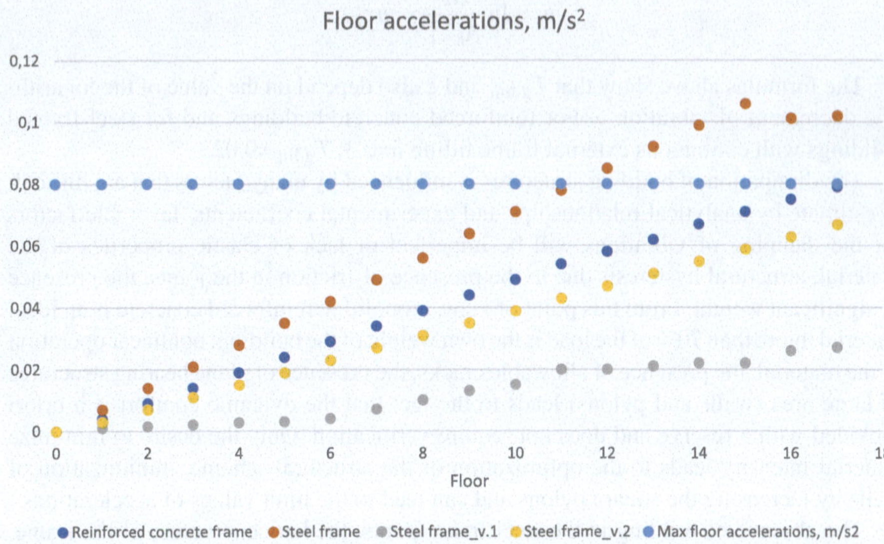

Fig. 6. Maximum floor accelerations for monolithic reinforced concrete and steel building frame variants.

The comfort level of residents staying in apartment buildings under the action of wind load (dynamic comfort) is estimated by the equation.

$$\alpha_{\mathbf{max}} < \alpha_{\mathbf{c,max}}$$

there $\alpha_{\max}$ - maximum acceleration of the building floor m/s^2, $\alpha_{c,\max}$=the limit value of maximum floor acceleration at which a person does not experience physical discomfort and dizziness, $\alpha_{c,\max}$ =.08 m/s^2, the value of $\alpha_{\max}$ is determined from the design value of the wind load $w_c = 0,7w_g$.

The acceleration graph shows that the acceleration of the upper floors exceeds the permissible limits. To ensure the required comfort we will increase the stiffness of the building by increasing the cross-sections and the number of ties. For this variant on the graph the line Steel frame_v.1 is given, maximizing the stiffness, it was possible to achieve a minimum of accelerations, but the metal intensity of the steel frame was 85.7 kg/m^2, which is 2 times higher than the original, so it is necessary to look for other options to ensure comfort.

To select options for providing the required resident comfort in a steel building, a closer look at what influences system vibrations and how to control them is necessary.

As we can see that a steel building is light and has a large enough period of vibration that becomes perceptible to humans. And in this case it becomes important that the vibrations in the system should be damped as quickly as possible. The ability of a system to damp oscillations or dissipation is characterized by the damping decrement or logarithmic decrement of oscillations δ. The logarithmic decrement of oscillations δ is the logarithm of the ratio of two successive deviations

$$\delta = \ln \frac{A_k}{A_{k+1}} = n\tau_1$$

The formulas above show that $T_{g,\mathrm{lim}}$ and ξ also depend on the value of the logarithmic decrement of vibration δ. For reinforced concrete buildings and for steel-framed buildings with continuous external frame filling δ=0.3; $T_{g,\mathrm{lim}}$=0.023.

The dissipation of building structures is influenced by many factors that are difficult to estimate by analytical relationships and experimental coefficients, favorable factors for the damping of vibrations will be imperfect or lack of elastic properties of the material, structural hysteresis due to the presence of friction in the joints, the presence of significant weight. From this point of view, monolithic reinforced concrete is an ideal material, more than 70% of the load is the own weight of the building, nonlinear operation of the material, the presence of allowable cracks, the presence of plane bearing structures of large area (walls and pylons) leads to the fact that the dynamic comfort is a priori provided with a reserve and does not require verification. Only the desire to minimize material intensity leads to the optimization of the structural scheme, minimization of walls by increasing the square pylons and can lead to the limit values of accelerations.

The absence of welding in the steel frame joints, the lattice structure of the frame, the absence of vertical wall elements, the elastic stage operation, and the low mass are antagonists of energy dissipation in the system. Thus, to ensure the fulfillment of the comfort requirement it is necessary to apply compensating structural measures.

For any model there is a critical value of the parameter α, exceeding which leads to the disappearance of oscillatory processes in the model and to the appearance of damped aperiodic processes. But from a certain point, increasing the stiffness of the system by introducing additional stiffness diaphragms has no effect, and to ensure the parameter α, it is necessary to introduce additional sources, such as vibration dampers, into the system [12–15].

When calculating building structures for wind loads, we are talking about the total logarithmic decrement of vibrations, which is the sum of the logarithmic decrement of structural damping δ_s and the aerodynamic logarithmic decrement δ_a. In case of using different damping devices, for example, vibration dampers of different types, it is necessary to additionally take into account the corresponding logarithmic decrement δ_d, obtained on the basis of research and dimensionless period $T_{g,1}$. When applying an impact vibration damper, the total logarithmic decrement of vibrations will be equal to.

Thus, it is necessary to design and introduce into the system such an impact vibration damper that will provide the required resident comfort, Fig. 6.

Due to the fact that wind pressure varies according to the wind region, it is obvious that for buildings in different wind regions, appropriate vibration dampers should be used.

4 Conclusions

Steel-framed buildings are an actual solution for mass residential development, when it is required to ensure the construction of the building in a short period of time, as steel construction is all-season and practically does not depend on climatic factors, except for wind load.

All-season construction is ensured by the use of bolted joints to connect the elements of the building frame and minimization of welding at installation, ensuring the minimum required metal intensity makes the building quite light and cost-effective. But these factors worsen the building's ability to dissipate energy and reduce the resistance to wind ripple load in terms of oscillations and the values of floor accelerations increase by 20% compared to reinforced concrete frame.

In the process of designing the structural scheme of a steel building, it is necessary to find a balance between ensuring metal intensity and resistance to ripple wind load, in case the installation of diaphragms is impossible for structural reasons, the use of impact vibration dampers is recommended.

As a result, we can conclude that the material of the structural system of the building has an impact on the acceleration of floors and the comfort of people staying in the building, but through the application of compensating measures this difference is leveled out and comfort is ensured. The choice of material for the building is dictated only by economic indicators.

References

1. Zelentsov, L., Mailyan, L.: Creation of intelligent management systems in construction. MATEC Web of Conferences **106**, 08051 (2017)
2. Mailyan, D., Mailyan, L.: Ecologically safe and techno economically efficient reinforced concrete constructions of equal resistance. MATEC Web of Conferences **73**, 04020 (2016)
3. Zelentsov, L., Mailyan, L., Triputa, I.: Adaptive organizational and technological modelsin the systems of construction management. MATEC Web of Conferences **106**, 08052 (2017)
4. Fediuk, R., Amran, M., Klyuev, S., Klyuev, A.: Increasing the performance of a fiber-reinforced concrete for protective facilities. Fibers **9**(11), 64 (2021)
5. Klyuev, S., et al.: Fresh and mechanical properties of low-cement mortars for 3D printing. Constr. Build. Mater. **338**, 127644 (2022). https://doi.org/10.1016/j.conbuildmat.2022.127644
6. Klyuev, S., et al.: Phase formation of mortar using technogenic fibrous materials. Case Stud. Constr. Mater. **16**, e01099 (2022)
7. Klyuev, S., et al.: Technogenic fiber wastes for optimizing concrete. Materials **15**(14), 5058 (2022)

8. Amran, M., Fediuk, R., Klyuev, S., Qader, D.N.: Sustainable development of basalt fiber-reinforced high-strength eco-friendly concrete with a modified composite binder. Case Stud. Constr. Mater. **17**, e01550 (2022)
9. Hao, J.-P., Sun, X.-L., Xue, Q., Fan, C.-L.: Research and applications of prefabricated steel structure building systems January (2017) https://doi.org/10.6052/j.issn.1000-4750.2016.08.ST14
10. Rybakov, V., Nazmeeva, T., Zhang, Y., Rayimova, I.: Seismic performance of the buckling-restrained brace outrigger. AIP Conferences Proceeding **2612**, 040001 (2023)
11. Mohsen, Soori, Fooad, Karimi Ghaleh, Jough: Artificial Intelligent in Optimization of steel moment frame structures. A Rev. Int. J. Struct. Eng. **18**(3), 141–158 (2024)
12. Athanasiou, A., Tirca, L., Stathopoulos, T.: Performance-based wind and earthquake design framework for tall steel buildings with ductile detailing. J. Wind. Eng. Ind. Aerodyn. **240** (2023). https://doi.org/10.1016/j.jweia.2023.105492
13. Xiaofang, Guo: Simulation of dynamic characteristics of buildings under wind loads and innovative design methods for wind-resistant structures. Appl. Math. Nonlinear Sci. **9**(1) (2024). https://doi.org/10.2478/amns-2024-2787
14. Velikanov, P.G., Artyukhin, Y.: Research on the dynamics of multi-storey buildings. Geosystems Transit. Zones **7**(3), 304–315 (2023). https://doi.org/10.30730/gtrz.2023.7.3.304-315
15. Youssef, R. Hamaty, Alexander, R. Tusnin: Operation of braced steel frameworks with regard to assembly and fabrication errors. Constr. Sci. Educ. **13**(1), 22–40 (2023). https://doi.org/10.22227/2305-5502.2023.1.2

Analysis of Investment and Construction Projects on the Basis of Markov Decision Processes

L. D. Mailyan(✉)

Don State Technical University, Rostov-On-Don 344001, Russia
19liechka84@mail.ru

Abstract. The approaches discussed in this paper undoubtedly make it possible to solve promptly the whole set of problems that objectively exist in the implementation of ICP, taking into account the factors of internal and external uncertainty. In fact, by implementing a machine learning model with reinforcement learning based on Markov decision process, we get a framework of decision support system in which the model itself can quickly analyze the progress of the project, build very flexible and accurate strategies with minimal errors, and such a model can include any internal contours: regression, classification, clustering, deep learning neural network models. As a result, the parameter of organizational and technological reliability of ICP significantly increase, which was required to obtain.

Keywords: Agent · Reward · Problems · Tasks · Knowledge · Models · System · State · Parameters · Transitions · Work · Control

1 Introduction

Nowadays, intelligent systems that can analyze and forecast various time and resource parameters of both individual tasks and the entire complex are increasingly used to manage construction projects. This is an important aspect in the work of organizations engaged in construction. This process is an integral part of the forecasting process. This process is an integral part of the forecasting process. It includes the formation of a system of both quantitative and qualitative indicators to assess the performance of investment projects (IP), which make it possible to evaluate the effectiveness of their activities, as well as to determine development trends in the future. The importance of forecasting the parameters of investment and construction projects (ICP) is increasing due to the fact that the processes of erecting buildings and structures are constantly becoming more complex and competition in the contracting market is growing. The key factor affecting the competitiveness of a construction organization is the ability of its managers to apply modern digital technologies that combine forecasting and development models in the process of management decision-making. In the process of selecting models for forecasting, it is necessary to take into account the peculiarities of construction production, including many factors that can be both external and internal. Factors affecting the

S. V. Klyuev et al. (Eds.): ICRAAC 2024, LNCE 627, pp. 273–281, 2025.
https://doi.org/10.1007/978-3-031-82938-3_32

internal capabilities of a construction organization include: the level of technical equipment, technologies and software used, ensuring the performance of work using modern technologies and information technologies; financial stability; competence of personnel and their ability to learn new competitive technologies. These factors influence the fact that certain technical and economic risks may be identified in the implementation of ICP [1]. They also influence the choice of forecasting models, planning systems and management decision-making. Unfortunately, in the absence of information about the influence of certain factors in the development of the forecast, it is possible to obtain incorrect results [2]. This can lead to the fact that the actual performance of individual projects and activities of the construction organization may significantly differ from the forecast values.

2 Methods and Materials

Long-term nature of construction products creation and, as a consequence, a large level of work-in-progress dictates the need to develop and use models of construction production organization, both at short-term planning intervals (week, month) and at longer intervals (quarter, year). In this case, there is a need for models that allow redirecting the types of work depending on internal and external uncertainties, in particular, allowing at discrete points in time to choose not only individual types of work, but also to determine short-term and long-term strategies.

Taking into account the multi-criteria nature of the construction management system, there is a need to develop a methodology for evaluating the results of the performance of certain ICP tasks, both at the design and construction stages, based on the application of modern digital technologies.

There is a huge number of models and algorithms for ICP management at any stages of construction production [3], but errors associated with the uncertainty of such factors as the impact of weather, fulfillment of obligations by subcontractors, uniformity of deliveries of equipment and raw materials to the construction site and many others lead to disruption of the schedule and construction and create a threat to the parameters of organizational and technological reliability as a whole at all stages of the life cycle.

Existing methods for managing ICP complexes have the following shortcomings:

- all ICP parameters described are random functions with distribution laws that change during operation, and the models themselves are based on point estimates, which is not always fair, and as a result, decisions are lagging and do not lead to the required results;
- construction project schedule has an extremely inflexible management character, which leads to untimely and incorrect actions (often just a reaction to unfavorable external and internal factors), when it is difficult and expensive to correct anything;
- short-term losses do not always lead to losses in general, so working to correct constantly the results obtained in the long term may have a negative impact on the whole course of work, etc.

Thus, in order to increase organizational and technological reliability in the performance of ICP, flexible technologies are needed that enable rapid changes in the processes of work implementation analysis and, on their basis, build new flexible strategies in the short and long term. At the same time, any construction project has the property of discreteness in terms of the stages of implementation and evaluation of results. Consequently, a mechanism is needed to intelligent decision support in ICP management using machine learning models.

The most promising from this point of view are reinforcement learning (RL) models, which fully allow very flexible strategy building for decision makers (DM) when managing complex systems, which certainly includes construction production. In such models, there are no targets (responses) in advance, and in the course of training it is necessary to build strategies of an agent so that the sum of its gains is maximized [4].

At the same time, the random nature of all variables included in the training model is taken into account, and depending on this, it is possible to flexibly and quickly change strategies in the short term without changing the main thing - the strategic goal for us is the unchanging value of organizational and technological reliability of the ICP. Such requirements are fully met by Markov making processes [5], when the following four parameters are created (*S, A, R, P*):

S – the set of possible states of the agent (stages of ICP schedule realization);

A – the set of available actions of the agent (all possible actions when realizing ICP);

$R : S \times A \to \mathbb{R}$ – reward function for transition from state S to S' when a certain action *a* is chosen.

$P : S \times A \to \pi(S)$ – transition function between states and is a set of probability distributions over *S* (probabilities of subsequent transitions do not depend on previous transitions in any way), i.e.:

$$P_k\big(S_{t+1} = s' | S_t, a_t, r_t, S_{t-1}, a_{t-1}, r_{t-1}, \ldots, S_0, a_0\big) = P_k\big(S_{t+1} = s' | S_t, a_t\big) \quad (1)$$

Thus, the probability of an agent's transition to state s' at step t depends only on the current state S_t and the selected action a_t, and the described set of MMP rules is ideal for analyzing the performance of ICP execution and selecting the necessary actions to adjust it if necessary.

3 Results and Discussions

The selection of ICP management strategies was based on the **Construction_Data_PM_Forms_All_Projects.csv** dataset [6], which is a dataset composed of various projects with assessments of the quality of work and safety measures at the construction site. A fragment of the dataset is shown in Fig. 1.

The dataset contains information about the following predictors: project Id, status, location, name, creation date, type, status change, tasks in progress, total tasks. A total of 12,424 projects were analyzed.

Since the number of steps in the construction project is known in advance, it falls under a finite Markov decision process, and the task aimed at delivering the finished project will represent the terminal vertex of the MDP graph. In this case, for training purposes, we will estimate the expected benefit after step t:

$$G_t = r_{t+1} + r_{t+2} + \ldots + r_T \quad (2)$$

	Ref	Status	Location	Description	Created	Target	Type	To Package	Status Changed	Association	OverDue	Images	Comments
0	T1.23963030	Open	JPC Project Management>EHS Management>01 Inspe...	task raised in incorrect location of this form...	14/09/2020	NaN	Safety Notice (Amber) - General Issue	Main Contractor	14/09/2020	FormAnswer	False	NaN	NaN
1	T116412.200	Closed	QC & BC(A)R>ITP 02 Architectural & M&E Service...	Metsec	14/09/2020	NaN	JPC - Progress Photo	Ceilings & Partitions	14/09/2020	NaN	False	True	False
2	T141663.27	EHS Good Observation	JPC Project Management>EHS Management>01 Inspe...	Good clear exclusion zones and access through ...	14/09/2020	NaN	Safety Notice (Green) - Good Observation	Main Contractor	14/09/2020	FormAnswer	False	True	False
3	T116412.199	Closed	QC & BC(A)R>ITP 02 Architectural & M&E Service...	RC walls	14/09/2020	NaN	JPC - Progress Photo	Precast Concrete	14/09/2020	NaN	False	True	False

Fig. 1. Fragment of the analyzed dataset Construction_Data_PM_Forms_All_Projects.csv.

there: r_{t+i}– set of DM's rewards when choosing action a_k for the current state and r_T – reward for choosing the action that brings the system to a terminal state.

Since the interpretation of the expected benefit results is rather complex, we will use the present expected benefit to choose an ICP execution strategy.

$$G_t = r_{t+1} + \gamma r_{t+2} + \ldots + \gamma^{T-t-1} r_T \tag{3}$$

there: $\gamma \in [0, 1]$, represents the discount factor. This parameter can be used to control the choice of ICP implementation strategies: if γ is close to zero, then it is favorable for the DM to get the maximum benefit at the current step (project task); otherwise, the DM seeks to implement the entire project within a specified time and with a specified quality (long-term strategy).

The next task is to determine the value of states and actions based on the present benefit under the chosen construction project strategy, depending on the random factors discussed earlier.

Then, the value function of states S (realizable project tasks) when using the π strategy:

$$\upsilon_\pi(s) = M_\pi\left(\sum_{k=0}^{T} \gamma r_{t+k+1} | s_t = s\right) \tag{4}$$

there: $s \in S$, i.e. the state of the project schedule can only be described by the available set of tasks.

Thus, by changing the state, the tasks change and thus the value of the state in the short term and for the project as a whole changes. Now let us consider the action value function a for task S using strategy π:

$$q_\pi(s, a) = M_\pi\left(\sum_{k=0}^{T} \gamma r_{t+k+1} | s_t = s, a_t = a\right) \tag{5}$$

Expressions 4 and 5 represent the mathematical expectation of the sum of DM's rewards for selected actions in the implementation of construction project tasks, taking into account the discount factor and factors of internal and external uncertainty [7, 8].

Since we generally know the main parameters of the construction project, we can apply recurrence expressions for calculating value functions [9]. Then, provided that we know the probabilities of all transitions from task s to task s' when performing work a:

$$P^a_{ss'} = P(s_{t+1} = s' | s_t = s, a_t = a) \tag{6}$$

All rewards to the DM for choosing a particular job have been identified:

$$R^a_{ss'} = M_\pi (r_{t+1} | s_t = s, a_t = a, s_{t+1} = s') \tag{7}$$

Now we can write the Bellman equations for estimating the value functions of states and actions when using the chosen strategy:

$$\upsilon_\pi (s) = \sum_{a \in A(s_t)}^{T} \pi (a|s) \sum_{s' \in S} P^a_{ss'} (R^a_{ss'} + \gamma \upsilon_\pi (s')) \tag{8}$$

$$q_\pi (s, a) = \sum_{s' \in S} P^a_{ss'} (R^a_{ss'} + \gamma \upsilon_\pi (s')) \tag{9}$$

In the case of estimation of the value function of task fulfillment in a construction project, the Bellman equation expresses the dependence of the value of some task on the values of subsequent tasks [10, 11]. In the case of work value functions - dependences of the value of a pair (task, work) on the values of subsequent such pairs.

To assess the value of states and actions, we initially construct the MDP graph for the execution of the construction project. A fragment of the graph is shown in Fig. 2.

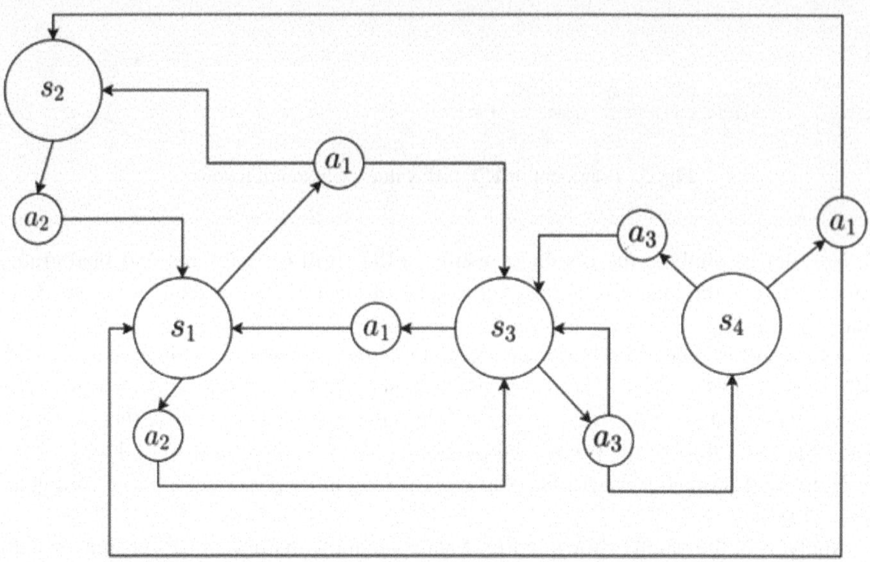

Fig. 2. Fragment of the MDP graph for ICP implementation.

For all states we know the probability distributions of choosing available actions $\pi(a|s)$, the probability distributions of transitions to the next states $P^a_{ss'}$ and the mathematical expectations $R^a_{ss'}$ of the rewards received in this case. We also determine the discount factor γ.

Now our task has been reduced to determining the values of all states under a particular chosen strategy, which will be determined by a function of internal and external uncertainty factors, which can be obtained, for example, by a multiple regression equation.

Next, using the Gym framework [7] we build a notebook in the Google Colab environment to solve our problem. As a strategy we used ε - greedy strategy with parameter epsilon = 0.1, discount factor gamma = 0.8, parameter random_seed = 3, number of game wins - 10,000.

The results obtained for the value assessment part of the ICP tasks are summarized in Fig. 3.

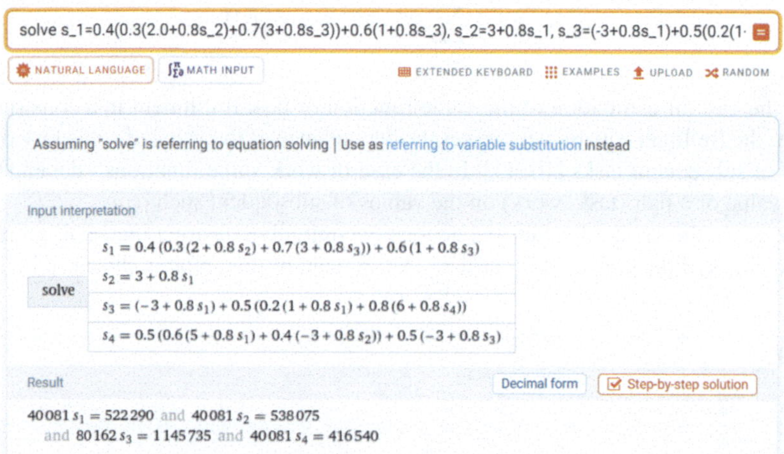

Fig. 3. Fragment of ICP task value assessment results.

Now let us analyze the obtained results in the form of diagrams and histograms. Figure 4 shows the analysis of task value estimation in the projects of the considered dataset.

As a result of analyzing Fig. 4 we see that the highest value was given to tasks 1328 - 30.2% and 1330 - 29.7%. The rest of the tasks received significantly lower value. This means that with an ε - greedy strategy, the DM will select predominantly those jobs that match the task range of the prioritized plan and only then the others in rank order.

Next, let us consider the results obtained by the quality parameters of the completed works (Fig. 5).

Analysis of the results shown in Fig. 5 shows that the histograms differ sharply from each other and along with the leaders unsatisfactory data were obtained, which gives us a reason for further refinements and adjustments.

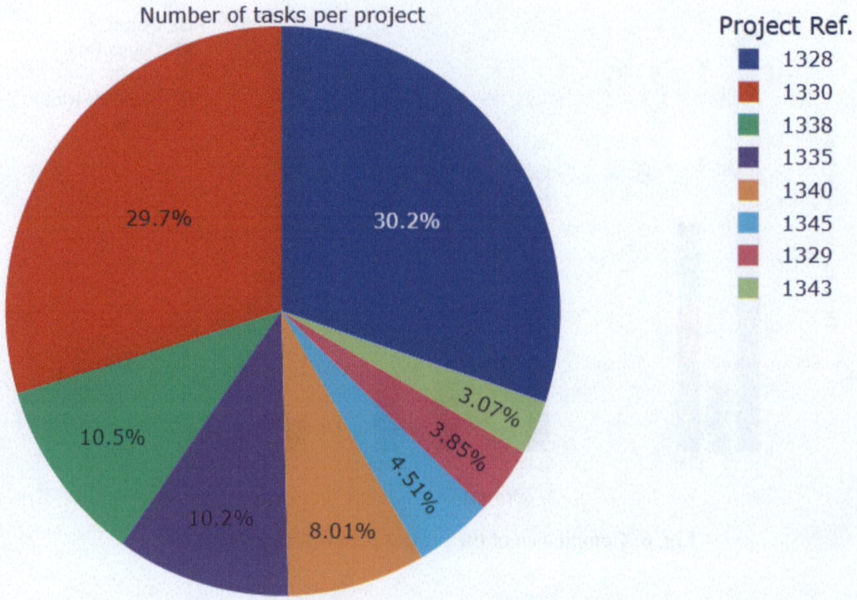

Fig. 4. Value assessment analysis of ICP tasks.

Fig. 5. Results obtained by quality parameters of the performed works.

Finally, we can assess the performance of organizational and technological reliability parameters of the considered ICPs (Fig. 6).

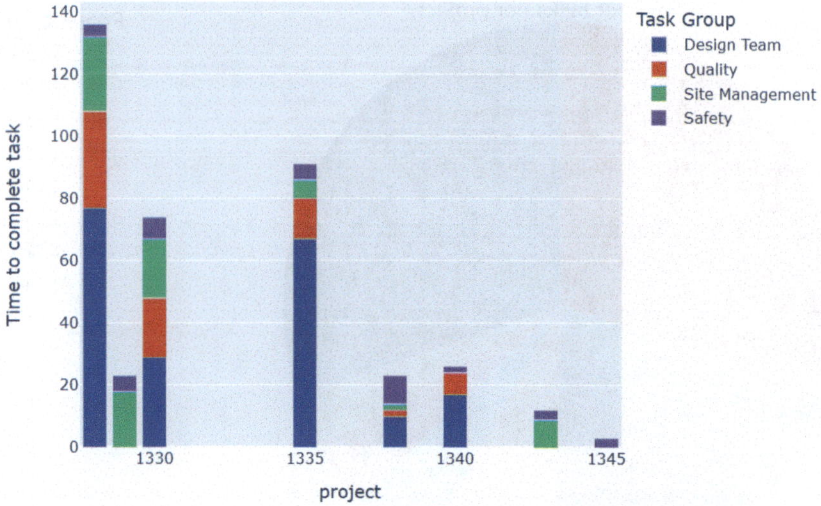

Fig. 6 Compliance of the project parameters of the OTR

Approximately 30% of the tasks fully meet the requirements of the OTR, further - a significant scatter, which suggests that the strategy we have chosen for this project is far from perfect and it is necessary to change the parameters of the transition probability functions from one task of the schedule to another.

4 Conclusions

The approaches discussed in this paper undoubtedly make it possible to solve promptly the whole set of problems that objectively exist in the implementation of ICP, taking into account the factors of internal and external uncertainty.

As a result of analyzing the value assessment of ICP tasks, we see that tasks 1328 - 30.2% and 1330 - 29.7% received the highest value. The other tasks received many times less value. This means that in the ε-greedy strategy, the DM will select mainly those tasks that correspond to the range of tasks in the prioritized plan and only then the others in the ranking order.

Analysis of the results obtained by the quality parameters of the completed works shows that the histograms differ sharply from each other and unsatisfactory data were obtained along with the leaders, which gives us a reason for further refinements and adjustments.

Approximately one third of the tasks fully meet the requirements of the OTR, and then there is a significant scatter, which suggests that the strategy we have chosen for this project is far from perfect and requires changing the parameters of the transition probability functions from one task of the schedule to another.

In fact, by implementing a machine learning model with reinforcement based on the Markov decision process, we get a framework for a decision support system in which the model itself can quickly analyze the progress of the project, build very flexible and accurate strategies with minimal errors, and such a model can include any internal contours: regression, classification, clustering, deep learning neural network models.

As a result, the parameter of organizational and technological reliability of ICP significantly increase, which was required to obtain.

References

1. Belousov, V.E., Konchakov, S.A.: Algorithm for operational determination of object states in multilevel technical systems. Econ. Manag. Control. Syst. Publishing house "Scientific Book" **32**(17), 227–232 (2015)
2. Zelentsov, L., Mailyan, L.: Creation of intelligent management systems in construction. MATEC Web Conf. **106**, 08051 (2017)
3. Zelentsov, L.B., Mailyan, L.D., Pirko, D.V.: Methodology of making organizational and technological decisions at the stage of operational management of construction operations based on the forecasting system. IOP Conf. Ser. Mater. Sci. Eng. **2131**(2), 022114 (2021)
4. Zelentsov, L., Mailyan, L., Triputa, I.: Adaptive organizational and technological modelsin the systems of construction management. MATEC Web Conf. **106**, 08052 (2017)
5. Klyuyev, S.V., Klyuyev, A.V., Sopin, D.M., Netrebenko, A.V., Kazlitin, S.A.: Heavy loaded floors based on fine-grained fiber concrete. Mag. Civil Eng. **38**(3), 7–14 (2013). https://doi.org/10.5862/MCE.38.1
6. Mailyan, D., Mailyan, L.: Ecologically safe and techno economically efficient reinforced concrete constructions of equal resistance. MATEC Web Conf. **73**, 04020 (2016)
7. Mailyan, D.R., Mailyan, L.D.: Conditions of maximum economic efficiency and environmental safety of reinforced concrete columns with precompressed reinforcement. Mater. Sci. Forum 931 MSF, 269–274 (2018)
8. Klyuyev, S.V., Guryanov, Y.V.: External reinforcing of fiber concrete constructions by carbon fiber tapes. Mag. Civ. Eng. **36**(1), 21–26 (2013). https://doi.org/10.5862/MCE.36.3
9. https://www.kaggle.com/code/kscoder77/construction-project-data-analysis. Construction Project - data analysis from field apps related to construction and PM works
10. https://www.gymlibrary.dev. Gym is a standard API for reinforcement learning, and a diverse collection of reference environments
11. Amran, M., Fediuk, R., Klyuev, S., Qader, D.N.: Sustainable development of basalt fiber-reinforced high-strength eco-friendly concrete with a modified composite binder. Case Stud. Constr. Mater. **17**, e01550 (2022)

Factors Affecting Labor Productivity in Overhaul Operations in the Arctic Zone

Azariy Lapidus(✉), Tembot Bidov, Rustam Fatullaev, and Anastasia Borovkova

National Research Moscow State University of Civil Engineering, 129337 Moscow, Russia
BidovTH@mgsu.ru

Abstract. The purpose of this study was to identify and describe the factors affecting labor productivity in major repairs in the Arctic zone. The study considers a set of interrelated factors including both objective (climatic conditions, complex logistics, underdeveloped infrastructure, quality of materials and equipment) and subjective (personnel qualifications, working conditions, coordination and planning). In the course of the study, the expert method with factor coding was applied to identify the most significant factor. Consistency of expert opinions was calculated, which is close to one and significance of expert opinion, which showed that the opinions of experts are consistent and significant. The most significant factors that will require minimal costs in implementation and significantly increase productivity in the shortest possible time are: coordination and planning of work in the organization, qualification of personnel. The results of the study indicate the introduction of lean production technology using the following methods: optimization of technologies and methods of work production, identification of time losses during overhaul, determination of their sources, introduction of parallelism of work and connection of separate technological operations.

Keywords: Overhaul · Labor Productivity · Labor Productivity Factors · Arctic Zone · Lean Manufacturing

1 Introduction

The Arctic zone is a region with extreme climatic conditions that significantly affect all aspects of industrial and construction activities. Major repairs in this region present a number of challenges, including low temperatures, polar night, inaccessible areas, and poor infrastructure. In such an environment, improving labor productivity becomes critical for successful project delivery, shortening project timelines, and reducing costs [1–6]. The relevance of this topic is due to the extreme climatic conditions of the region, which significantly complicate construction and repair work. It is important to note that downtime in construction and installation work caused by unfavorable weather conditions leads to significant financial losses and often affects the safety of operation of facilities. Moreover, the difficulties of delivery of materials and equipment to the sites and the shortage of highly qualified personnel require the development of integrated solutions

S. V. Klyuev et al. (Eds.): ICRAAC 2024, LNCE 627, pp. 282–291, 2025.
https://doi.org/10.1007/978-3-031-82938-3_33

that will allow effective management of construction processes in such conditions [7–13]. The scientific research is aimed at identifying the factors affecting labor productivity during major repairs in the Arctic zone. The introduction and practical application of the proposed solutions will contribute to a more effective development of the Arctic zone, improve safety and ensure sustainable development of a strategically important region [14–16].

2 Methods and Materials

Labor productivity is an indicator that characterizes the effectiveness of labor (Table 1).

Labor productivity can be measured by the amount of output produced by an employee per unit of time. In the construction industry, this indicator is measured by labor intensity and output [17–19].

On this basis, according to the authors, it is necessary to create a methodology based on the study of factors affecting labor productivity in capital repairs in the Arctic zone, optimization of processes and methods of work production by "smoothing" time and technological downtime [20].

To achieve the set goal and to consider the possibility of implementing lean production methods, the following tasks were set:

- To form and describe the factors affecting labor productivity in Arctic conditions.
- To conduct an expert survey and identify the most significant factors affecting labor productivity in the overhaul in the Arctic zone.

At this stage, the factors affecting labor productivity in overhaul in the Arctic are considered and described (Table 1).

The expert method was used to select the factors affecting labor productivity during capital repairs in the Arctic zone and to determine their weighting.

3 Results and Discussion

The coding of factors affecting labor productivity for overhaul in the Arctic Zone is presented in Table 2.

At the first stage a survey of experts was conducted, the number of which should be at least 6, a survey questionnaire is prepared [21–27]. As experts were selected specialists from leading organizations whose activities are related to capital repairs [27–31]. The questionnaire asked the following question: Which factors would require minimal cost for the organization to implement and/or optimize, but would significantly increase productivity in the shortest possible time. Experts are asked to rank the predetermined factors presented in Table 2.

In the questionnaire the most significant indicator is indicated $R = 1$, the least significant $R = 8$. Experts conducted ranking - placement of factors by rank (category of significance) determined by the expert. The contribution of each factor is evaluated by the value of the rank, taking into account their expected impact on labor productivity during overhaul.

Table 1. Factors affecting labor productivity in the Arctic conditions.

№	Name of factor	Description
1	Climatic conditions	Low temperatures, strong winds and snowfalls reduce the efficiency of employees, increase the risk of injury, require additional costs for heat and wind protection of places of work and recreation, use of equipment suitable for such conditions
2	Climatic conditions	In the Arctic region there are long periods of polar night and polar day, which affects the psychological state of workers, the quality of their work and their productivity. Therefore, ensuring the necessary working conditions and safety of workers in difficult climatic conditions requires additional costs and control of the work performed. In addition, special requirements are imposed on clothing and personal protective equipment
3	Logistics	Difficult accessibility of work sites, high transportation costs, underdeveloped or absent transportation infrastructure, lack of access roads and long delivery times for cargo and equipment increase the cost of projects and require special construction and planning of logistics chains
4	Infrastructure	Absence or lack of necessary infrastructure (roads, housing, transport) hampers access to the site of works, delivery of materials, organization of accommodation for employees and their everyday life
5	Staff qualifications	Working in the harsh conditions of the Arctic requires high competence, knowledge and skills from specialists. Employees must be prepared and trained to work in the specific conditions of the north, otherwise the efficiency of their labor critically decreases and the project becomes unprofitable
6	Equipment and technology	Low temperatures require the use of special equipment and technologies designed to work in extreme conditions, and the introduction of new technologies using AI

(continued)

Table 1. (*continued*)

№	Name of factor	Description
7	Quality of materials and products	Materials and products with special properties for use at sub-zero temperatures, as well as high quality control requirements for their application
8	Coordination and planning	When organizing the work process, clear planning and coordination of works, as the work is limited by daylight hours, or proper cladding and organization of lighting of territories and places of work

Table 2. Coding of factors affecting labor productivity.

N	Factors affecting labor productivity	Factor code
1	Staff qualifications	X1
2	Climatic conditions	X2
3	Equipment and technology	X3
4	Logistics	X4
5	Infrastructure	X5
6	Working conditions	X6
7	Quality of materials and products	X7
8	Coordination and planning	X8

The sum of ranks of each expert horizontally should be constant

$$\sum R_{ji} = 0.5n(n+1) = 36 \tag{1}$$

where: n - number of factors

The ranks obtained for all questionnaires are entered into a summary table - ranking matrix (Table 3).

At the second stage, the expert opinions are processed to determine their consistency and significance.

The consistency of expert opinion was determined using the concordance coefficient W:

$$W = \frac{\sum (S_i - \overline{S})^2}{\frac{1}{12}m^2(n^3 - n) - m\sum_{i=0}^{n} T_j} = \frac{1540}{\frac{1}{12}6^2(8^3 - 8) - 6 * 2} = 0.98 \tag{2}$$

where S is the sum of squares of deviations;

T_j - similarity index;

m - number of interviewees;

Table 3. Matrix of ranks for factors affecting labor productivity during major repairs.

Expert	X1	X2	X3	X4	X5	X6	X7	X8	$\sum R_{ji}$	t_j
1	2	3	4	5	6	7	8	9	10	
1	1.5	8	3	4	6	7	5	1.5	36	0.5
2	2	8	3.5	3.5	7	6	5	1	36	0.5
3	2	8	3	4	7	5	6	1	36	0
4	1.5	8	3	4	6.5	6.5	5	1.5	36	0.5
5	1	7.5	3	4	7.5	6	5	2	36	0.5
6	2	8	3	4	6	7	5	1	36	0
S_i	10	47.5	18.5	23.5	40	37.5	31	8	216	2
$S_i - \overline{S}$	-17	20.5	-8.5	-3.5	13	10.5	4	19	-	
$(S_i - \overline{S})2$	289	420	72.3	12.3	169	110	16	361	1540	
m*n-S_i	38	0.5	29.5	24.5	8	10.5	17	40	168	
J_i	0.226	0.003	0.176	0.146	0.047	0.062	0.101	0.238	1	

n - number of factors.

Thus, the coefficient of concordance is equal to W = 0.98, close to 1, hence, the experts' estimates have high consistency.

It is possible to use the concordance coefficient after assessing its significance.

The significance of the concordance coefficient W was determined using χ^2 distribution (Pearson distribution) with the number of degrees of freedom f = k - 1 (Fig. 1):

$$\chi_R^2 = W * m(n-1) = 0.98 * 6 * (8-1) = 41.2 \tag{3}$$

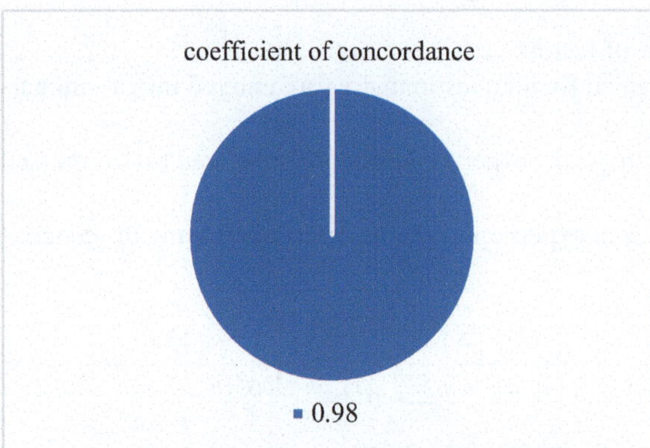

Fig. 1. Diagram coefficient of concordance.

Since χ_R^2=41.2 > χ_T^2 = 14.1, therefore the concordance coefficient W is significant with probability $P_d = 0.95$.

The weighting coefficient of each indicator is determined by the formula

$$J_i = \frac{m * n - S_i}{0.5mn(n-1)} \quad (4)$$

Figure 2 shows a diagram of the distribution of weighting coefficients for factors affecting labor productivity in capital repairs.

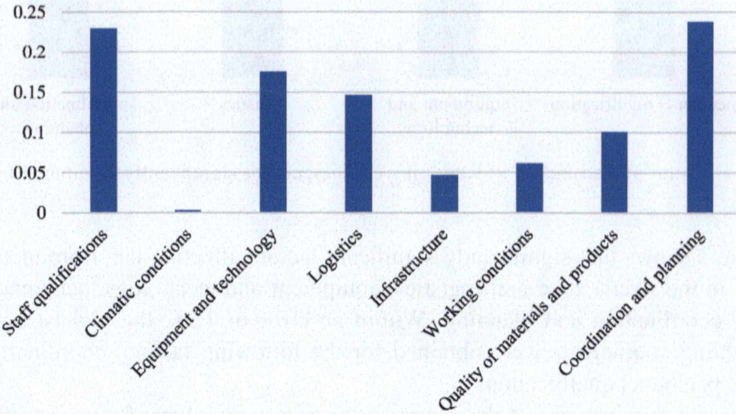

Fig. 2. Diagram of distribution of weighting coefficients for factors affecting labor productivity.

To determine the significantly significant factors affecting labor productivity in overhaul, it is necessary that the inequality is fulfilled:

$$J_i > \frac{1}{n} = 0.125 \quad (5)$$

where: J_i- weighting coefficients of factors affecting labor productivity;

n - number of factors.

Consequently, the most significant factors (Table 1) for which inequality (5) is fulfilled are:

X1 - personnel qualification;

X3 - equipment and technology;

X4 - logistics;

X8 - coordination and planning;

The weighting coefficients of essentially significant factors are determined by the formula:

$$J_{oi} = \frac{J_i}{\sum J_i} \quad (6)$$

The weighting coefficients of the most significant factors affecting labor productivity in capital repair are presented in Fig. 3.

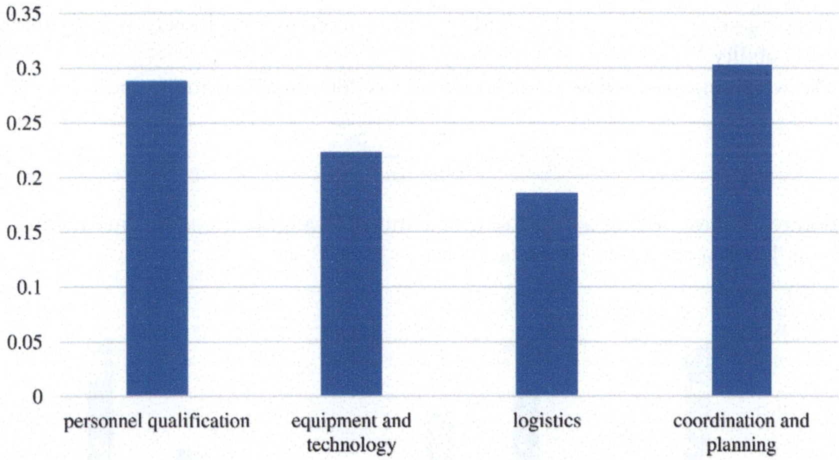

Fig. 3. Diagram of distribution of weighting coefficients for significantly significant factors.

Figure 2 shows that significantly significant factors affecting labor productivity in overhaul in the Arctic zone are: logistics, equipment and technology, personnel qualification, coordination and planning. Within an error of 10%, the highest values of the weighting coefficients were obtained for the following factors: coordination and planning, personnel qualification.

According to the results of the expert survey it was found that for experts the most important indicator (having the highest value of the weighting coefficient) is coordination and planning, qualification of personnel.

4 Conclusions

The study of labor productivity in capital repair of apartment buildings located in the Arctic zone is an important step to improve the efficiency and quality of work performance. As a result of the work, the factors affecting labor productivity in capital repairs in the Arctic zone were identified and described: climatic conditions, complex logistics, undeveloped infrastructure, quality of materials and equipment, staff qualifications, working conditions, coordination and planning. An expert survey was conducted and it was revealed that the most significant factors that will require minimal costs in implementation and significantly increase labor productivity in the shortest possible time are: coordination and planning of work in the organization, qualification of personnel. The results of the study indicate the need to implement the methodology of lean production: optimization of processes with reducing the loss of time for technological operations, introduction of parallelism of work and connection of separate technological operations.

References

1. Lapidus, A.A.: A method of increasing labour productivity in construction. Vestnik MGSU **9**(8), 1365–1372 (2024)
2. Trofimova, N.V., Mamleeva, E.R., Sazykina, M.Yu., Shaikhutdinova, G.F.: Labor productivity in the regions of the Russian Federation: essence, factors and reserves of growth. Bull. USPTU Sci. Educ. Econ. Ser. Econ. **2**(40), 111–121 (2022). https://doi.org/10.17122/2541-8904-2022-2-40-111-121
3. Fatullaev, R.S., Ledovskikh, L.I., Borovkova, A.E., Sedov, D.S.: Comparative analysis of the standard labor intensity of capital repairs in Russia and abroad. Constr. Prod. **4**, 169–175 (2023). https://doi.org/10.54950/26585340_2023_4_169
4. Mishchenko, V.Ya., Gorbaneva, E.P., Sevryukova, K.S.: Foreign and Russian experience in carrying out capital repairs of housing stock taking into account energy efficient measures. Sci. J. Constr. Archit. **1**(67), 62–73 (2020). https://doi.org/10.25987/VSTU.2020.57.1.006
5. Korol, O.A.: Development of the methodology of calendar planning in the system of organizational and technological preparation of capital repair of multi-apartment residential houses. J. Phys.: Conf. Ser. 012086 (2020). https://doi.org/10.1088/1742-6596/1425/1/012086
6. Krasnovsky, B.M.: Organizational and technological potential of solutions for a construction project - a tool for increasing the efficiency of construction organization. Constr. Prod. **2**, 140–143 (2020). https://doi.org/10.54950/26585340_2020_2_140
7. Skibniewski, M.J., Zavadskas, E.K.: Technology development in construction: a continuum from distant past into the future. J. Civ. Eng. Manag. **19**(1), 136–147 (2013). https://doi.org/10.3846/13923730.2012.756060
8. Tamošaitienė, J., Gaudutis, E.: Complex assessment of structural systems used for high-rise buildings. J. Civ. Eng. Manag. **19**(2), 305–317 (2013). https://doi.org/10.3846/13923730.2013.772071
9. Petrosyan, R.S.: Formation of an organizational and technological mechanism for increasing the manufacturability of work during major repairs of buildings. Constr. Sci. Educ. **13**(1), 84–97 (2023). https://doi.org/10.22227/2305-5502.2023.1.6
10. Zavadskas, E.K., Turskis, Z., Šliogerienė, J., Vilutienė, T.: An integrated assessment of the municipal buildings' use including sustainability criteria. Sustain. Cities Soc. **67**(1), 102708 (2021). https://doi.org/10.1016/j.scs.2021.102708
11. Yevseiev, S., Rzayev, Kh., Korol, O., Imanova, Z.: Development of mceliece modified asymmetric crypto-code system on elliptic truncated codes. East.-Eur. J. Enterpr. Technol. **4–9**(82), 18–26 (2016). https://doi.org/10.15587/1729-4061.2016.75250
12. Lapidus, A.A., Fatullaev, R.S., Bidov, T.Kh., Nikolenko, D.M.: Analysis of the cost of work performed on major repairs of common property in multi-apartment residential buildings in the constituent entities of the Russian Federation. Constr. Prod. **2**, 3–7 (2023). https://doi.org/10.54950/26585340202323
13. Kagazezhev, A.Yu., Fatullaev, R.S., Khubav, A.O., Shesterikova, Ya.V.: Analysis of the main problems of planning capital repair programs. Perspect. Sci. **12**(159), 81–86 (2022). https://moofrnk.com/perspektivy-nauki/arhiv/vyipuski-za-2022-god/
14. Fatullaev, R.S.: Consumer quality of an apartment building as a parameter influencing the composition of organizational and technological solutions when carrying out major repairs. Sci. Bus. Ways Develop. **2**(92), 149–155 (2019). http://globaljournals.ru/nauka-i-biznes/arhiv/vyipuski-za-2019-god/
15. King, O.A.: Conceptual basis for the formation of a regulatory framework for major repairs of common property of multi-apartment residential buildings. BLS: Bull. Constr. Technol. **11**(1011), 20–21 (2018). http://bstmag.ru/archive

16. Mishchenko, V.I., Shishkanova, M.V.: Major repairs of apartment buildings: regional aspect. Young Sci. **29**(133), 264–266 (2016). https://moluch.ru/archive/133/
17. Selezneva, A.K., Tolstykh, Yu.O., Uchinina, T.V.: Main types of work on major repairs of multi-apartment residential buildings. Mod. Probl. Sci. Educ. **5**, 438 (2014). https://science-education.ru/ru/issue/view?id=119
18. Abramyan, S.G., Klyuev, S.V., Polyakov, V.G., Sabitova, T.A., Akopyan, G.O., Guseynov, K.M.: Specifics of information model development for functional conversion of offshore oil platforms. Constr. Mater. Prod. **6**(4), 42–57 (2023). https://doi.org/10.58224/2618-7183-2023-6-4-42-57
19. Oleynik, A.K., Filippova, E.V.: Application of modern technologies in the reconstruction of roofs as part of a major overhaul. Effective building structures: theory and practice - article in the collection of conference proceedings 81–88 (2021). https://elibrary.ru/item.asp?id=46350887
20. Kiselitsa, E.P., Shilova, N.N., Shelomentsev, A.G.: Improving the labour productivity of construction organizations in an innovative economy. Russian J. Innov. Econ. **11**(1), 225–238 (2021). https://doi.org/10.18334/vinec.11.1.111807
21. Lapidus, A., Bidov, T., Khubaev, A.: The study of the calibration dependences used when testing the concrete strength by nondestructive methods. MATEC Web of Conferences **117**, 00094 (2017). https://doi.org/10.1051/matecconf/20171170009
22. Lapidus, A.A., Kuzmina, T.K., Bolshakova, P.V.: Improving the procedure for interaction between participants in the construction control system in order to increase labor productivity. Constr. Mater. Prod. **7**(4), 1 (2024). https://doi.org/10.58224/2618-7183-2024-7-4-1
23. Khubaev, A., Bidov, T., Rybakova, A.: Analysis of physical and mechanical properties of vacuum treated claydite-concrete. MATEC Web of Conferences **196**, 04071 (2018). https://doi.org/10.1051/matecconf/201819604071
24. Khubaev, A., Bidov, T., Bzhienikov, A., Nesterova, V.: Modern Russian high-tech construction materials and their application in domestic construction industry (on example of metal-ceramic panels Hardwall). IOP Conf. Ser. Mater. Sci. Eng. **365**, 032005 (2018). https://doi.org/10.1088/1757-899X/365/3/032005
25. Fedosov, S.V., Lapidus, A.A., Narmaniya, B.E., Ayzatullin, M.M.: Cauchy problem for modeling of unsteady mass transfer processes in an unbounded plate by the integral Laplace transform method. Constr. Mater. Prod. **7**(5), 4 (2024). https://doi.org/10.58224/2618-7183-2024-7-5-4
26. Lapidus, A.A., Kuzmina, T.K., Bolshakova, P.V.: Developing administrative and process-related solutions for the benefit of residential construction projects. Constr. Mater. Prod. **7**(6), 2 (2024). https://doi.org/10.58224/2618-7183-2024-7-6-2
27. Lapidus, A. et al.: Development software for the non-destructive control of monolithic structures in housing construction. E3S Web Conf. **258**, 09003 (2021). https://doi.org/10.1051/e3sconf/202125809003
28. Khubaev, A.O.: Determination of the efficiency of the production process of winter concreting based on field studies. E3S Web Conf. **389**, 06012 (2023) https://doi.org/10.1051/e3sconf/202338906012
29. Zholtikov, A., Zinovkin, A., Khubaev, A.: Improving the efficiency of applying the method of slab and floor lifting. E3S Web Conf. **389**, 01067 (2023). https://doi.org/10.1051/e3sconf/202338901067

30. Bidov, T.H., Khubaev, A.O., Shabanova, A.A.: Organizational and technical modeling of the complex system of concrete works production in winter period during the construction of residential buildings. Constr. Geotech. **12**(2), 15–25 (2021). https://doi.org/10.15593/2224-9826/2021.2.02
31. Bidov, T.Kh., Khubaev, A.O., Pomytko, E.A., Kotelnikova, A.D.: Methodology of evaluating of factors affecting the efficiency of construction project implementation by general contracting organization in Fit-out direction. Vestnik MGSU **19**(9), 1562–1569 (2024). https://doi.org/10.22227/1997-0935.2024.9.1562-156

Stability of a Compressed Vertical Rod Under the Influence of Bearing Force and Distributed Load

Lyalusya Baragunova, Maryana Shogenova(✉), Oleg Shogenov, and Lyana Kanukoeva

Kabardino-Balkarian State University Named After H. M. Berbekov, Nalchik 360004, Russia
shogenova_mar@mail.ru

Abstract. The article is devoted to a straight rod of variable cross-section. A rod of variable cross-section is loaded with a concentrated force and a distributed load. The boundary between the areas of stability and instability is determined. The study of the stability of deformable systems began a long time ago. Determining the critical loads for compression bars is fraught with significant difficulties. Solving eigenvalue problems is greatly simplified when using numerical methods and modern computer software. A vertically standing column of variable cross-section is considered. The bending of a beam is described by classical theories using Bernoulli's hypothesis. The curved axis of the beam is described using a linear ordinary differential equation. As a result, a simplified solution to the problem was achieved in non-traditional cases when studies of a rod of variable cross-section and load are combined. Two special examples are considered. The boundaries of the zones of stability and instability are determined. The results presented in the figures have a high degree of accuracy.

Keywords: Variable Section Rod · Focused Power · Distributed Load · M-dimensional Euclidean Space · Boundary · Areas of Stability and Instability · Finite Difference Method · Critical Loads · Eigenvalues · Bernoulli's Hypothesis · Euler's problem · Integro-Differential Equation

1 Introduction

A rectilinear rod of variable cross-section, loaded with a distributed load and a concentrated force, is considered in the m-dimensional Euclidean parameter space Em, where the boundary of the regions of stability and instability is determined using the finite difference method.

The authors consider a vertical column used for the construction of multi-story buildings, water towers, television towers, etc. as such a rod. A column is one of the simplest and oldest types of supports for other building structures, transmitting compressive loads and ensuring the stability of the structure.

Depending on the specific requirements of a building project, columns vary in shape and are mainly of two types: constant section columns and variable section columns.

S. V. Klyuev et al. (Eds.): ICRAAC 2024, LNCE 627, pp. 292–300, 2025.
https://doi.org/10.1007/978-3-031-82938-3_34

It is known that determining the critical loads for compressed columns of variable cross-section is fraught with difficulties when they carry an unevenly distributed load or a combination of different loads, etc.

In such cases, solving eigenvalue problems is simplified by using numerical methods. Modern computer software helps solve a wide range of problems and reduce the time for performing routine operations and processing results.

2 Methods and Materials

Let's consider a vertically standing column of variable cross-section in Fig. 1, which is under the action of a variable distributed load and carries a concentrated load P (water tower, observation tower, television tower, etc.). The functions of changing the variable stiffness EJ(x) and the distributed load p(x) will be considered given.

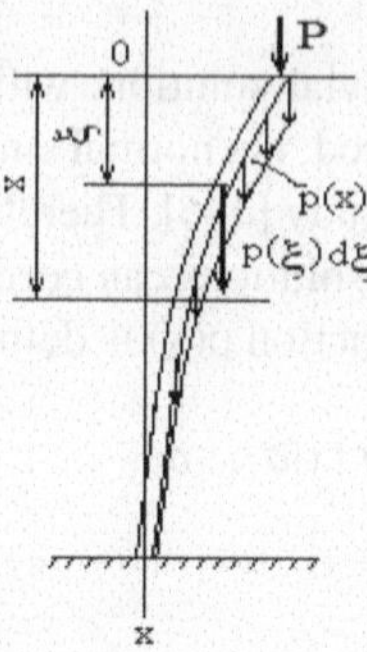

Fig. 1. Calculation scheme. Variable section column.

Let us assume that the bending of the beam is described by classical theory using the Bernoulli hypothesis, and the critical forces [1–3] are determined from the Euler problem with appropriate assumptions. Then the curved axis of the beam is described using a linear ordinary differential equation and in traditional notation has the form

$$EJ(x)v''(x) = M(x) \tag{1}$$

The bending moment on the right side is determined from the equilibrium condition of the upper part, cut off by coordinate x

$$M(x) = P[v(0) - v(x)] + \int_0^x p(\xi)[v(\xi) - v(x)]d\xi \tag{2}$$

When taking (2) into account, Eq. (1) becomes integro-differential

$$B(x)v''(x) = P[v(0) - v(x)] + \int_0^x p(\xi)[v(\xi) - v(x)]d\xi \tag{3}$$

where the designation for variable bending stiffness is introduced

$$\mathrm{B(x)} = \mathrm{EJ(x)}$$

To ensure that various boundary conditions that may occur in such a problem are taken into account, it is advisable to move on to a higher order differential equation. This result is achieved after twice differentiating Eq. (3) with respect to the variable x, as a result of which it takes the form

$$[B(x)v''(x)]'' = -N(x)v''(x) - p(x)v'(x) \tag{4}$$

in which the longitudinal compressive force is indicated

$$N(x) = P + \int_0^x p(\xi)d\xi$$

Equation (4) has an obvious trivial solution, which corresponds to the rectilinear equilibrium vertical position of the rod, i.e. normal simple compression. Such a solution is not the subject of interest of this study [4–6]. Therefore, the further task will be to find such loads P and p(x) that non-zero solutions can correspond to, i.e. curved equilibrium positions. Let us assume that the function p(x) is defined through m-1 parameters

$$p(x) = f(x|a_1,\ a_2,\ \ldots,\ a_{m-1})$$

3 Results and Discussions

Together with the concentrated force parameter am = F, they form an m-dimensional Euclidean space Em, which is divided into two subspaces corresponding to stable or unstable rectilinear equilibrium positions. The set of points belonging to the boundary line of these subspaces will be determined by the eigenvalues of Eq. (4).

Separation of regions of stability (instability) by analytical methods is possible only in the simplest problems, where B(x) = const, p(x) = const, m = 1, 2, and they are well studied. Solving more complex problems in this way involves significant mathematical difficulties or in many cases is impossible. Therefore, we will further use the finite difference method [7–9] and divide the length of the rod into n identical segments with a step h = 1/n, with nodal point numbers i = 1, 2, …, n-1, n, n + 1.

Let's introduce a grid function $y_i \approx v(x_i),\ x_i = (i-1)h$. Then the derivatives in Eq. (4) can be approximately represented in the form of finite difference relations

$$v' = (y_{i+1} - y_{i-1})/2h, \quad v'' = (y_{i+1} - 2y_i + y_{i+1})/h^2$$

due to which it will take the form

$$c_{i,i-2}y_{i-2} + c_{i,i-1}y_{i-1} + c_{i,i}y_i + c_{i,i+1}y_{i+1} + c_{i,i+2}y_{i+2} = 0,\ i = 3,\ 4,\ \ldots,\ n \tag{5}$$

In this case, the procedure of replacing the second derivative with a finite difference relation is applied to the left side of (4) twice. The coefficients of the equation have meanings

$$c_{i,i-2} = B_{i-1},\ c_{i,i-1} = -2(B_{i-1} + B_i) + N_i h^2 - p_i h^3/2,$$
$$c_{i,i} = B_{i-1} + 4B_i + B_{i+1} - 2N_i h^2,\ c_{i,i+1} = -2(B_i + B_{i+1}) + N_i h^2 + p_i h^3/2,$$
$$c_{i,i+2} = B_{i+1}, \quad i = 3,\ 4,\ \ldots,\ n. \tag{6}$$

The system of Eqs. (5) is underdetermined, its coefficient matrix is rectangular in size $n \times (n+4)$, and therefore contains unnecessary unknowns: To eliminate them, we obtain additional equations using boundary conditions

$$y_{-1},\quad y_0,\quad y_{n+1},\quad y_{n+2}.$$

The upper end (x = 0) is free and a force P is applied to it.

1) The bending moment is zero. That's why

$$M(0) = 0 \quad \Rightarrow \quad B(0)v''(0) = 0 \quad \Rightarrow \quad v''(0) = \frac{y_0 - 2y_1 + y_2}{h^2} = 0 \quad \Rightarrow$$
$$\Rightarrow \quad y_0 - 2y_1 + y_2 = 0 \quad \Rightarrow \quad y_0 = 2y_1 - y_2 \tag{7}$$

2) Shear force is determined using the deflection function

$$Q(0) = [B(0)v''(0)]' \tag{8}$$

and from Fig. 2 taking into account the small angle of rotation of the end section

$$Q(0) = -Pv'(0) \tag{9}$$

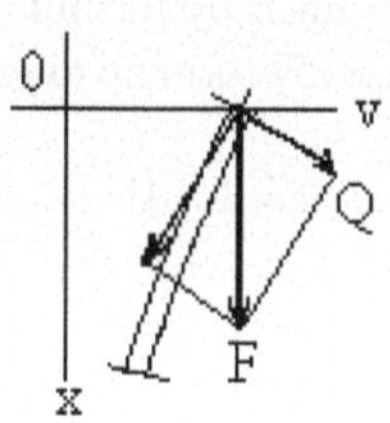

Fig. 2. End section rotation angle.

Equating the right-hand sides of (8) and (9), we obtain

$$[B(0)v''(0)]' = -Pv'(0)$$

Finite-difference approximation of derivatives after elimination using (7) gives the equation

$$c_{11}y_1 + c_{12}y_2 + c_{13}y_3 + c_{14}y_4 = 0 \tag{10}$$

where

$$c_{11} = 4B_2 - 3Ph^2, \quad c_{12} = -8B_2 - B_3 + 4Ph^2, \quad c_{13} = 4B_2 + 2B_3 - Ph^2, \quad c_{14} = -B_3.$$

The lower end (x = l) is clamped, so the displacement and rotation angle are zero.

$$3) \quad v(l) = 0 \quad \Rightarrow \quad y_{n+1} = 0 \tag{11}$$

$$4) \quad v'(l) = 0 \quad \Rightarrow \quad (y_{n+2} - y_n)/2h = 0 \quad \Rightarrow \quad y_{n+2} = y_n \tag{12}$$

For i = 2, Eq. (11) contains the deflection y0 located outside the rod. We replace it using (12) and get

$$c_{21}y_1 + c_{22}y_2 + c_{23}y_3 + c_{24}y_4 = 0 \tag{13}$$

$$c_{21} = -2B_2 + N_2h^2 - p_2h^3/2, \quad c_{22} = 4B_2 + B_3 - 2N_2h^2$$
$$c_{23} = -2(B_2 + B_3) + N_2h^2 + p_2h^3/2, \qquad c_{24} = B_3$$

In the last two equations of system (15), we take into account (1), (2) and exclude y_{n+1} and y_{n+2}. Then they will look like

$$c_{n-1,n-3}y_{n-3} + c_{n-1,n-2}y_{n-2} + c_{n-1,n-1}y_{n-1} + c_{n-1,n}y_n = 0 \tag{14}$$

$$c_{n,n-2}y_{n-2} + c_{n,n-1}y_{n-1} + c_{n,n}y_n = 0 \tag{15}$$

and

$$c_{n,n} = B_{n-1} + 4B_n + 2B_{n+1} - 2N_nh^2$$

the remaining coefficients are determined by formulas (6).

Equations (11), (12), (13), (14), (15) form an algebraic system

$$Cy = 0 \tag{16}$$

where y = {$y_1, y_2, \ldots, y_n$} is a vector whose components are the deflections of the rod, a C-square matrix of size $n \times n$

$$C = \left(\begin{array}{ccccccc} c_{11}\ c_{12}\ c_{13}\ c_{14} & \vdots & & & & & \\ c_{21}\ c_{22}\ c_{23}\ c_{24}\ c_{25} & \vdots & & & & & \\ c_{31}\ c_{32}\ c_{33}\ c_{34}\ c_{35} & \vdots & & & & & \\ c_{42}\ c_{43}\ c_{44}\ c_{45}\ c_{46} & \vdots & & & & & \\ \cdots\cdots\cdots\cdots & \cdots & \cdots & \cdots & \cdots & \cdots & \cdots \\ & \vdots & c_{n-2,n-4} & c_{n-2,n-3} & c_{n-2,n-2} & c_{n-2,n-1} & c_{n-2,n} \\ & \vdots & & c_{n-1,n-3} & c_{n-1,n-2} & c_{n-1,n-1} & c_{n-1,n} \\ & \vdots & & & c_{n,n-2} & c_{n,n-1} & c_{n,n} \end{array} \right)$$

The zero elements of the matrix are not written out here.

Critical parameter values, i.e. the coordinates of the boundary of the stability subspace are determined from the equation

$$\det(C) = 0$$

which is a condition for the existence of a nontrivial solution to Eq. (16).

The algorithm for determining the boundary of the stability region will be as follows. In parameter space ($a_1, a_2, \ldots, a_m$) a Cartesian coordinate system with axes of the same name is introduced $0a_1a_2\ldots a_m$ [10–12]. Origin (0, 0, ..., 0) obviously belongs to the field of stability. A ray is drawn from it, described by equations in parametric form

$$a_1 = l_1\lambda, \quad a_2 = l_2\lambda, \quad \ldots\ldots, \quad a_m = l_m\lambda \tag{17}$$

where l_k - direct direction cosines, λ - changeable parameter, and

$$\sum_{k=1}^{m} l_k^2 = 1$$

Now moving along this beam, i.e. giving λ increasing values, check the sign of the determinant det (C). The change in sign corresponds to crossing the boundary of the stability region, i.e., the transition to the instability region. Several trial calculations refine the boundary point $\lambda = \lambda_{kp}$. Then the corresponding parameter values can be found using the formulas (27). The set of points obtained in this way defines a certain hypersurface, which is the boundary of the subspaces of stability and instability, the union of which forms the parameter space under consideration

$$E_y \cup E_H = E_m$$

Another algorithm could be as follows. Building a dependency graph λ – det (C) on the computer monitor screen, where you can visually determine with a high degree of accuracy the coordinate where the graph intersects the x-axis. This is the critical value of the parameter λ_{kp}. As in the previous case, the critical values of the physical parameters can be found according to (17).

In order to verify the reliability of the results obtained by the proposed algorithms and computer programs in the MatLab software environment, test tests were carried out using two examples in which the exact values of the critical loads are known. In both cases, a vertical rod of constant cross-section with a lower clamped and upper free end was considered.

In the first example, a concentrated force F is applied to the upper end of the rod. It is well known that its critical value

$$P_{kp} = 0.25\frac{\pi^2 EJ}{l^2}$$

In the second example, a uniformly distributed axial load p acts on the rod. Here is the critical value

$$p_{kp} = 7.83\frac{EJ}{l^3}$$

In both cases, numerical values of the coefficients were obtained that were practically no different from those given above [13, 14]. This allows us to conclude that the finite difference method gives eigenvalues of a high degree of accuracy while significantly simplifying the procedure for solving the problem. Let's move on to a more general example. Let's consider a column with initial data

$$l = 20\text{m}, E = 40\text{GPa}, \text{J(x)} = 0.06 + 0.001x^2, \text{m} = 2, p(x) = p = a_1, P = a_2$$

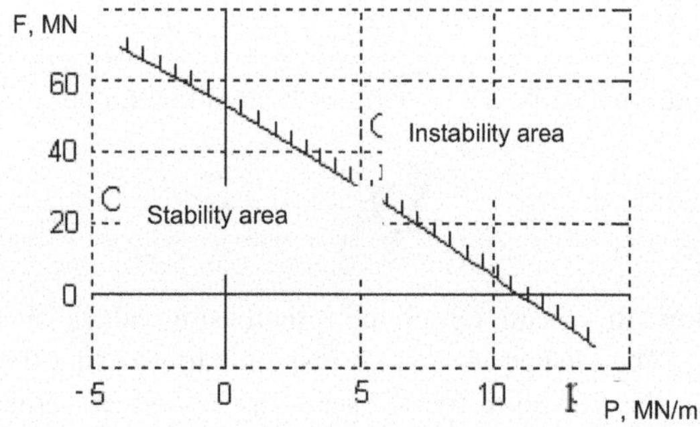

Fig. 3. Counting results.

The parameter space represents a plane with a system of coordinate axes 0a1a2. Calculations carried out using the algorithm described above gave the results presented in Fig. 3. As a result, the subspaces of stability and instability turned out to be open, which was to be expected [15, 16]. The reason is that for simultaneously negative values P и p(x) loss of stability does not occur, due to which the third quarter of the plane entirely belongs to the stability region.

4 Conclusions

The main points of the theory of calculating the stability of rods are outlined, and a simple algorithm for solving problems is given. Determining the critical loads for compressed rods was fraught with significant difficulties, since a column of variable cross-section was considered, loaded with an unevenly distributed load, a concentrated force.

The use of the finite difference method and modern computer programs has significantly simplified the solution of problems on the stability of deformable systems; a relatively simple and universal method has been found for determining the critical forces, based on the capabilities of computer technology. The finite difference method can be used to solve complex problems without resorting to analytical methods.

The article discusses specific examples with source data. As a result, the boundaries of the zones of stability and instability were determined. A figure is presented in which you can see results that have a high degree of accuracy and confirm the reliability of the theoretical and computer calculations used.

References

1. Baragunova, L.A.: Determination of the critical force of a compressed rod by the finite difference method. Bulletin of Kabardino-Balkarian State University, Nalchik (2000)
2. Maslennikov, A.M.: Dynamics and stability of structures. Textbook and practice for higher education institutions. Publishing House Yurait, Moscow (2016)
3. Zolotov, A.B., Akimov, P.A., Sidorov, V.N.: Mathematical methods in structural mechanics (with the basics of the theory of generalized functions). ASV Publishing House, Moscow (2008)
4. Akimov, P.A., Belostotsky, T.B.: And others: Informatics in construction (with the basics of mathematical and computer modeling). KNORUS, Moscow (2017)
5. Samarsky, A.A., Mikhailov, A.P.: Mathematical modeling. Ideas. Methods. Examples. FIZMATLIT, Moscow (2005)
6. Formalev, V.F., Reviznikov, D.L.: Numerical methods. Fizmatlit, Moscow (2006)
7. Akulenko, L.D.: High-frequency natural oscillations of mechanical systems. Prikl. mat. i fur. Moscow **64**(5), 817–832 (2000)
8. Verzhbitsky, V.M.: Fundamentals of numerical methods. High School, Moscow (2002)
9. Ilyin, V.P., Karpov, V.V., Maslennikov, A.M.: Numerical methods of solving problems of structural mechanics. Publishing House ASV, Moscow, Saint-Petersburg State University of Civil Engineering and Architecture, Saint-Petersburg (2005)
10. Baragunova, L.A., Shogenova, M.M.: Stability of multi-span rods. Bulletin of the Dagestan State Technical University. Technical Sciences 49(3) (2022)
11. Lafisheva, M.M., Baragunova, L.A., Tseeva, F.M., Khamukova, L.A: On vibration of multi-span continuous beam in view of rotational inertia. In: Alikhanov, A., Tchernykh, A., Babenko, M., Samoylenko, I. (eds.) Current Problems of Applied Mathematics and Computer Systems. CPAMCS 2023, LNNS, vol. 1044, pp. 82–90. Springer, Cham. (2024)
12. Kulterbaev, K.P., Baragunova, L.A., Shogenova, M.M., Senov, K.M.: About a high-precision graphoanalytical method of determination of critical forces of an oblate rod. In: Proceedings 2018 IEEE International Conference "Quality Management, Transport and Information Security, Information Technologies" (IT&QM&IS), pp. 794–796. St. Petersburg. Russia (2018)

13. Kulterbaev, K., Baragunova, L.A., Lafisheva, M.M.: Calculation of a beam on elastic base analytical and numerical methods. Struct. Mech. Struct. **4**(35), 16–23 (2022)
14. Kulterbaev, K., Lafisheva, M., Baragunova, L.: Solving the Euler problem for a flexible support rod base on the finite difference method. In: Tchernykh, A., Alikhanov, A., Babenko, M., Samoylenko, I. (eds) Mathematics and its applications in new computer systems. MANCS 2021, LNNS, vol. 424, pp. 143–151. Springer, Cham (2022)
15. Kulterbaev, K.P., Baragunova, L.A., Shogenova, MM: Random oscillations of the vertical continual and discrete rod under seismic influences. Mater. Sci. Forum **931**, 66–71 (2018)
16. Shogenova, M.M., Shogenov, O.M., Baragunova, L.A.: Solving the Euler problem on the stability of a rod with non-classical boundary conditions. Engineering Herald of the Don 11 (2021)

Enhancement of Building Structures on the Basis of Studying the Structure of Plant Objects

Irina Mayatskaya[1(✉)], Svetlana Yazyeva[1], Song Xuanzhen[2], Khairulla Muselemov[3], and Irina Protopopova[1]

[1] Don State Technical University, Rostov-On-Don 344022, Russia
irina.mayatskaya@mail.ru
[2] Shandong Jiaotong University, Changqing, Jinan 250357, Shandong, China
[3] Dagestan State Technical University, Republic of Dagestan 367026, Russia

Abstract. The study of the regularities of the structure of plant objects makes it possible to create new in shape and structure, strong and reliable building structures. This makes it possible to improve building structures based on the principles that are outlined in this study. The article considers such a science as plant architectonics, the founders of which were S. Schwendener and V.F. Razdorsky. The article discusses the theory of the application of structural and mechanical principles in the study of plant structure. Studying the patterns of plant structure allows us to see ways to increase the strength of the structure. Plant objects are able to withstand significant external influences on them, since the structure of these objects can be maximally adapted to various mechanical actions. The form and structure of some plant objects, for example, bamboo, are considered. Bamboo architecture is very interesting in its shape and makes it possible to use natural materials. The use of natural materials in architecture and design is very relevant. The use of bamboo in architectural structures allows you to create delicate, durable and unique shaped objects. Examples of plant objects with a frame structure are given. The leaves of the plants form an optimal strong lattice of the main veins. The types of venation of leaf blades are given. The search for optimal design options with natural structures in construction is especially relevant and gives rise to such a direction in science as bionics. It is necessary to continue to develop methods of bionic research in the analysis of plant architectonics. And the introduction of mathematical modeling methods in the design creates conditions for the construction of more durable and reliable building structures.

Keywords: Plant Object · Mathematical Modeling · Bionics · Form · Structure

1 Introduction

The creation of strong and reliable building structures has always remained the most important task in the construction of a wide variety of structures. The improvement of these objects is also possible with the study of the regularities of the structure of natural

S. V. Klyuev et al. (Eds.): ICRAAC 2024, LNCE 627, pp. 301–312, 2025.
https://doi.org/10.1007/978-3-031-82938-3_35

objects, for example, plants [1]. Plant reinforcement is a spatial structure that forms a frame, which affects the strength of a plant object. As you know, many plants are able to withstand a sufficiently large static and dynamic load.

The founders of the theory of the application of structural and mechanical principles in the study of the structure of plants were S. Shvendener and V.F. Razdorsky. By studying the regularities of the structure of plants, one can see ways to increase the strength of the structure, since the structure of these objects can be most adapted to various mechanical influences [2].

When studying the regularities of the structure of plants, the following principles of construction can be distinguished:

1. The principle of joint operation of structural elements, providing high reliability of the structure;
2. The principle of reinforcement, which characterizes the number of fibers, the location and shape of the mechanical tissue;
3. The principle of corrugation, which is widely used in technology to improve the strength properties;
4. Principle of structural building.

Natural plant structures, such as the stems of herbaceous plants or tree trunks, have evolved over a long period of time and have an optimal structure. This makes it possible to use the structural features of plant objects when creating new structures. At the same time, a more detailed study of the structure of plant objects, taking into account their composite structure, and conducting experiments to study the mechanical properties are necessary.

The goal of optimal design is to achieve the greatest efficiency of the project, and in the first place, to achieve the strength and reliability of the structure. The structure of plant objects is very diverse and nature itself chooses the optimal solution. The purpose of this study is to study the regularities of the structure of plant objects and ways to improve building structures based on the data obtained.

Research methods include the following sequences of studying the problem: studying a plant object, for example, the structure of a bamboo stem; selection of essential elements in the structure of a plant object; analysis of those features of the structure that can be used in the building structure and then the choice of the optimal solution with the modified structural elements that were obtained during the study of the first stage. In the process of finding the optimal solution, it is necessary to actively use software systems. This will optimize the design process and get the best solution to the task.

2 Methods and Materials

The principle of mathematical modeling, based on the study of the laws of the structure of biological, and in particular, plant materials, underlies the design of building structures. These principles make it possible to create durable and resistant to external influences building structures. Nature creates optimal designs. The study of natural forms enriches the possibilities of creating interesting solutions.

Bionic research on the study of plant objects is an interesting area. Bionics is the use of the principles and methods of organizing natural objects in the design and improvement of building structures. Building bionics, one of the areas of bionics, studies the structure of natural objects, as well as plants, which makes it possible to find amazing forms [3–7].

In bionic research, questions of the regularities of the structure of plants can be considered, which allow them to adapt to the mechanical influences of the external environment. It was these questions that were considered by Razdorsky V.F. in his book "Architectonics of Plants" (1955). Timiryazev K.A. spoke about the possibility of studying the structure of plants: "It was on the stems that we learned a number of amazing facts proving that they were built according to all the rules of the art of building." Architectonics of plants is a science that studies the regularities of the spatial arrangement of the mechanical and reinforcing tissues of a plant and their effect on effective resistance to the mechanical loads of the environment. At the same time, Razdorsky V.F. considered the strength of the plant in the dynamics of its development. In his book, he described in most detail the doctrine of the structural mechanics of a plant, gave a historical overview of the origin of the science of plant architectonics and the development of the doctrine of the mechanical principle in the structure of the frame of the plant itself. He also outlined the methods and results of experimental studies on the strength, rigidity and stability of plant objects.

For the strength of a plant object in the form of a stem, the arrangement of fibers in the longitudinal and radial directions is very important. The strength of the fibers varies from plant to plant and is transmitted to the plant itself. Fibers of woody varieties have a higher strength compared to the strength of leafy plants. Objects are able to withstand large wind loads and be flexible. So palms and bamboos are a strong structure, and are also flexible structures, which allows them not to break under the action of wind load.

3 Results and Discussion

S. Schwendener's theory of structural and mechanical principles in the design of organs of plant objects presents a classification of structural elements that can increase its strength with various types of resistance. In the constructions of plant objects there are elements that are strong in bending; plant elements that function as columns (trees, plant stems); tensile elements of plants (lianas). There are elements that can experience many different types of complex resistance, such as resist stretching and shearing.

In the process of evolution, plants improve their structure. If an engineer builds structures, then a plants builds itself according to the same principles, only more perfect [8, 9]. The construction of the Ostankino television tower used the principles of the structure of the stem of a herbaceous plant, for example, wheat from the cereal family (Fig. 1). The structure uses a prestressed system of 150 steel ropes similar to the vertical fibers of wheat stalks. And externally, this structure resembles a stem (Fig. 2).

Plants can be adapted to the bending action of the wind, they can have elastic flexibility. In this regard, we can note bamboo and palm (Figs. 3, 4).

S. Schwendener and his colleagues believed that for the strength of plants in bending, it is necessary to ensure bending rigidity with a rational distribution of material. The section of the stem model according to the theory of S. Schwendener is shown in Fig. 5. This design is not accurate for describing a plant model, but the idea of using this scheme for building structures is quite acceptable. Other intermediate layer options can be offered.

Fig. 1. Stalks of wheat.

Fig. 2. Ostankino television tower, Moscow.

Razdorsky V.F. considered plants as part of structures, building a structural-mechanical model of a plant. We are solving the reverse problem. By studying the structural features of plants, such as bamboo or palm trees, we choose the best solutions that can increase the strength of the building structure. It is necessary to consider a specific model for a certain type of plants.

The structural features of cereals and other crops are discussed in this section. Nature is so varied that it is not limited to just one model of a plant object. On Fig. 6 diagrams of the cross section of plants of the sedge family and cereal crops are presented. It is characterized by a hard skin of the stem; inside the shell are longitudinal cavities and sclerenchymal shells of vascular bundles, which are the reinforcing structure of the plant; and also the central channel with a large radius. Palm trees are characterized by

Fig. 3. Bamboo thickets.

Fig. 4. Palm tree on the seashore.

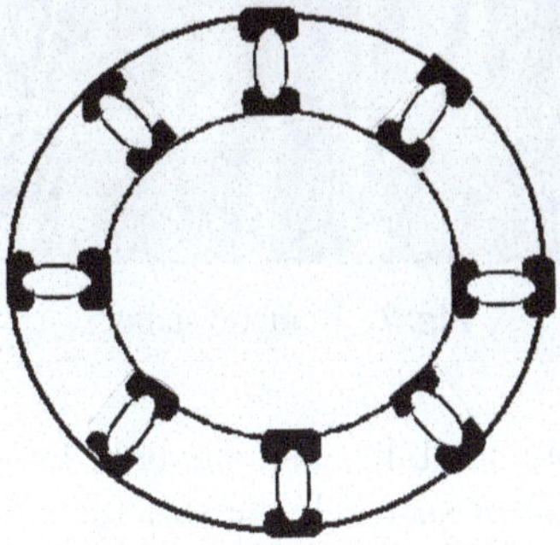

Fig. 5. Stem section model (model of Schwendener).

sclerenchymal cords, which are thin, flexible, elastic threads. They make up a three-dimensional grid, which are directed to the longitudinal axis of the trunk at an angle close to 45 degrees. This construction is very rational and optimal.

Bamboo is the tallest tree-like cereal. The structure of the bamboo and its branches for most of its length is of a hollow-tubular design with diaphragms in the nodes, a central empty cavity, all material is distributed almost evenly along the cross-sectional

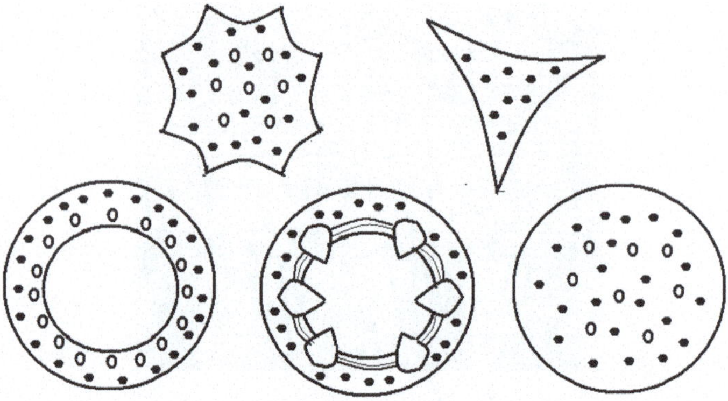

Fig. 6. Models of the cross sections of sedge family and cereal crops.

ring. If young bamboo plants have a less durable structure and the location of the nodes is shown in Fig. 7, then a mature plant has a denser structure and a more stronger structure (Fig. 8). Nodal connections affect the strength of this plant. These elements increase the strength and stability of this object.

Fig. 7. Bamboo stalks.

Bamboo has considerable durability and strength. Its strength is 2.5 times higher than that of oak. An example of such reliable structures is the Anlan bamboo bridge thrown over the Minjiany River (built in the 3rd century). Bamboo cables are used as load-bearing structures in the hanging bridge. It is bamboo nodes that are dampers, which plays an important role in the redistribution of internal force factors under the action of an external load.

There are bamboo bridges in other cities in China and in other countries (Fig. 9). Amazing in form, and at the same time, strong and stable structures exist in the modern world [10]. The bamboo canopy above the water in Huludun (Fengyuan) Park in Taiwan is a curvilinear openwork surface (Zuo Studio, 2019).

Café "Nocenco" in Vinh, Vietnam is an amazing example of bamboo architecture (Vo Trong Ngnia Architecture, 2018). Curvilinear conical surfaces in the form of inverted

Fig. 8. Nodal connection with the partition of a mature bamboo plant.

Fig. 9. Bamboo bridges.

Fig. 10. Bamboo structures in café "Nocenco", Vinh city, Vietnam.

structures as columns are also used here in the internal space of the structure, acting as load-bearing structures. And also beams with curved elements also make the structure more durable and reliable (Fig. 10).

The fantastic shape of the bamboo openwork house "Sharma Springs" (IBUKU architectural bureau, 2012) in Indonesia simply surprises with its lightness and originality.

One of the most important architectural elements is the column. Plant stems and tree trunks are also such a construction [11]. Architects create it with the typical structure for reinforced concrete columns. But in nature there are structures similar to columns, but they have an optimal and rational shape and structure. We need to learn from nature to create such structures.

The improvement of building structures based on the study of the shape and structure of plants is a very interesting area of research. Natural objects have an optimal design. Nature and vegetable world took care of this [8, 9, 12]. According to the theory of V.F. Razdorsky plants are a living organism, according to the laws of structural mechanics, reacting to the action of forces applied to it at the moment, and according to the biological laws of this plant object, which leads to a change in the structure as a whole and its parts. This study is based on the concept of the structure of the plant object itself without taking into account its biological development.

The stems are characterized by conical or cylindrical composite shells with internal structural architecture. This structure can be in the form of filaments along the stem, in the form of a shell with complex geometry. The optimal variant of the inner shell in a composite structure would be a honeycomb or mesh structure. The spiral organization of the architecture of the rod is also interesting, which is typical for some types of palms (Figs. 11, 12).

The armature of a plant may consist of such elements as a series of rings connected to each other by threads or fine mesh fabric, as well as single and double spirals. Plants are also characterized by quite unusual structures: nets that include filaments with different diameters and cross sections, for example, in the form of n-gons, or a set of thin elongated leaf blades.

Examples of plant objects were given above, the use of their characteristic structures in building structures can increase the strength and reliability of structures. Frame structures are observed in many plants (Fig. 13). So the leaf of dicotyledonous plants or trees has a frame structure.

The veins of the sheet form an optimal lattice that gives strength to the surface of the sheet (Fig. 14), similar to that which in modern skyscrapers is provided by the steel lattice structure of the reinforcement for relatively thin external walls and internal partitions.

Plant leaf-stem materials are objects of complex structure with various types of branching (Table 1).

There are two classes of venation: marginal and arcuate. With marginal venation, the main veins reach the ends of the leaves. With arcuate venation, the main veins run almost to the edges of the leaf. But they turn around before reaching the edge. Table 2 shows the main types of venation.

Plant objects are characterized by a stressed state, even when it is not exposed to external influences. Razdorsky V.F. carried out experimental studies to determine the mechanical properties of plants, without depriving them of their natural habitat. But we need to explore the geometric structures of a plant object, such as a palm tree or a bamboo. And to study the possibility of using these features to improve building structures. In this

Fig. 11. Palm trees.

Fig. 12. The surface structure of the palm tree trunk.

case, the method of analogies is applied. We transfer the features of the structure of a plant object to a structure, for example, a composite one. But this must be done at a new level of analysis, not simply by copying the object. The advantage of this method is the elucidation of those structural features of the plant, which can significantly increase the strength properties of the structure under consideration. But this is only the first stage, then you need to find out whether there are optimal solutions to this problem.

Using the knowledge of researchers of bionics in relation to the development of shaping and the optimal structure of plant objects, when designing, it is possible to improve building structures [13–18].

Fig. 13. Maple leaves.

Fig. 14. Surface of leaf.

Table 1. The main types of venation.

Simply venation	Mesh venation	Dichotomous venation
	feathery	
	radial	
	fingered	
	radiant	
	arcuate	
	parallel	

4 Conclusions

The methods of bionic research in the analysis of architectonics of plants, as well as modern software systems, make it possible to find strong and reliable structures of interest in form and structure [19, 20]. It is the development of bionics that allows you to create unique designs not only from traditional materials, but also from natural ones, for example, from bamboo.

Table 2. The structure of plant leaf-stemmed objects.

Sympodial	Dichotomous	Monopodial	
		Symmetrical	Asymmetrical
	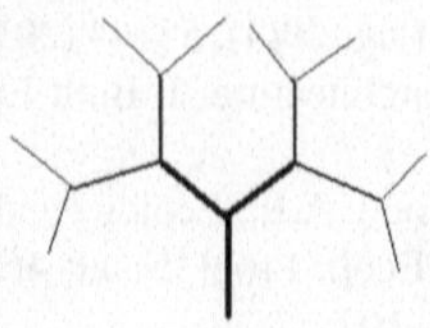		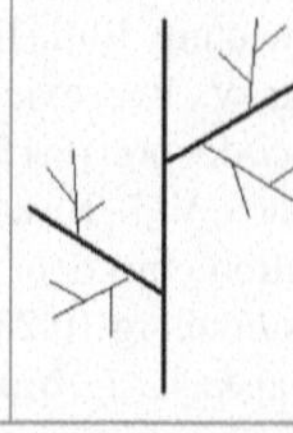

V.F. Razdorsky and his colleagues paid great attention to determining the mechanical properties of both the plants themselves and their parts. During the tests, they determined the modules of elasticity, the limits of strength and elasticity, relative deformation of of the sclerenchyma. It is necessary to continue to conduct research to determine the mechanical characteristics of natural materials.

I would like the designers to use the laws of plant objects when creating structures that are interesting in shape and structure. By analyzing the regularities and properties of plant objects, you can create amazing designs. The shaping has been tested by time and the structure is always rational for a particular plant object. We need to take a closer look at the natural world and make our building structures not only durable, but also interesting in shape.

References

1. Mayatskaya, I.A., Yazyeva, S.B., Yazyev, B.M.: Optimality of structures from the point of view of architectural bionics. Constr. Technog. Saf. **9**(61), 7–11 (2017)
2. Mayatskaya, I.A.: About the possibility of improvement of building constructions taking into account bionic principles. Bull. Volgogr. State Univ. Arch. Civ. Engineering. Ser.: Civ. Engi-Neering Arch. **45**(64), 27–36 (2016)
3. Lebedev, Yu.S., Rabinovich, V.I., Polozhai, E.D.: Architectural Bionic. Publishing Stroiizdat (1990)
4. Lebedev, Yu.S.: Architecture and Bionic. Publishing Stroiizdat (1977)
5. Zherarden, L.: Bionic. Publishing Mir (1971)
6. Villesenor, D.: Architecture and Nature. Publishing Rizzoli (2015)
7. Mayatskaya, I.A., Yazyev, B.M., Yazyeva, S.B., Kulinich, P.B.: Building construction: architecture and nature. MATEC Web Conf. **106**, 01031 (2017). https://doi.org/10.1051/mateconf/201710601031
8. Mayatskaya, I., Eremin, V.: Bionics and the choice of rational structural form. E3S Web Conf. **110**, 01042 (2019). https://doi.org/10.1051/e3sconf/201911001042
9. Mayatskaya, I.A., Yazyeva, S.B., Lapina, A.P., Davydova, V.V.: Architectural bionics and the search for optimal solutions in the unique structures' design. IOP Conf. Ser.: Mater. Sci. Eng. **913**, 022070 (2020). https://doi.org/10.1088/1757-899X/913/2/022070
10. Solanilla Medina, Y.M.: Architectural design of bamboo as an ecological alternative material of the XXI century. Arch. Mod. Inf. Technol. **1**(42), 201–211 (2018)

11. Mayatskaya, I., Yazyev, B., Kuznetsov, V., Tetenkov, N., Klyuev, S., Nabiullina, K.: Features of the development of architectural bionics in the modern world. Lect. Notes Civ. Eng. **436**, 285–293 (2024)
12. Balandin, R.K.: Ecology: Man and Nature. Publishing OLMA-PRESS (2001)
13. Vityuk, E.Y., Humorina, J.E.: Natural technologies as a new principle of shaping in architecture. Bull. Tomsk. State Univ. Arch. Civ. Eng. **20**(4), 55–64 (2018)
14. Sugar, V., Leczovics, P., Horkai, A.: Bionics in architecture. J. Built Environ. **5**(1) (2017). https://doi.org/10.1515/jbe-2017-0003
15. Temnov, V.G., Rusanov, G.E., Bolotin, S.A., Gelfond, A.L.: Ecology and architectonics construction objects of urban environmental. Water Ecol.: Probl. Solut. **4**(72), 95–102 (2017). https://doi.org/10.23968/2305-3488.2017.22.4.95-102
16. Mayatskaya, I., Yazyev, B., Kuznetsov, V., Rusakova, E., Klyuev, S., Sabitov, L.: Use of modern composite materials in construction and repair of structures. Lect. Notes Civ. Eng. **436**, 346–355 (2024)
17. Vishnevskaya, E.V., Tolynbekova, G.A., Nurkusheva, L.T.: Global and regional aspects of sustainable design in the frame of innovative technologies. IOP Conf. Ser.: Mater. Sci. Eng. **698**(3), 033009 (2019). https://doi.org/10.1088/1757-899X/698/3/033009
18. Mayatskaya, I., Yazyeva, S., Gatiev, M., Kuznetsov, V., Klyuev, S., Sabitov, L.: Application of fractal methods in the design of modern structures. Lect. Notes Civ. Eng. **436**, 414–422 (2024)
19. Dymchenko, M.E., Dakoro, M.F., Dadiyan, D.G.: The problem of form in modern architecture. E3S Web Conf. **281**, 02026 (2021). https://doi.org/10.1051/e3sconf/202128102026
20. Mayatskaya, I.A., Yazyev, B.M., Demchenko, D.B., Yazyeva, S.B.: Creation of a comfortable environment in urban and rural settlements based on bionic principles. IOP Conf. Ser.: Earth Environ. Sci. **937**, 042026 (2021). https://doi.org/10.1088/1755-1315/937/4/042026

Numerical-Analytical Method in Solving the Problem of Creep of a Shallow Shell

Svetlana Yazyeva[1], Anton Popov[2], Vladimir Erofeev[3], Batyr Yazyev[1](✉), and Stepan Litvinov[1]

[1] Don State Technical University, Rostov-On-Don 344003, Russia
ps62@yandex.ru
[2] Kazan Federal University, Kazan 420008, Russia
[3] National Research Moscow State University of Civil Engineering, 129337 Moscow, Russia

Abstract.
Objective The paper presents general equations of the moment theory of thin shallow shells with a relatively small rise above the plane of their projection taking into account creep deformation. The problem of the stress-strain state of the shell is considered, with boundary conditions, in particular, the calculation assumed that the shell is connected to diaphragms at the edges, absolutely rigid in their plane and flexible from it. Method. To determine creep deformations, a linear approximation of the first derivative with respect to time (Runge-Kutta method) was used. To verify the solution to the problem, a calculation of a shell made of secondary PVC was performed using the grid method [1]. Result. A program was developed for calculation in the MATLAB package with the ability to vary the initial data and output a graph of the dependence of displacements and stresses on time. As a result of the calculation, it was found that stresses and internal forces in an orthotropic shell of the same shape as for an isotropic one, a redistribution of stresses occurs: normal stresses increase, and tangential stresses decrease. Conclusion. The proposed approach can be applied to the analysis of the stress-strain state and bearing capacity of a reinforced concrete shell as well. There are no restrictions on boundary conditions and the type of loading, and the beam material can be not only polymers and composites for construction purposes, but also concrete.

Keywords: Concrete Creep · Inelastic Coefficients · Relative Deformation · Maxwell-Gurevich Constraint Equation · Bending Moments · Variable Stiffness · Double Trigonometric Series

1 Introduction

Flat procedures are widely used for devices of longer span floors, therefore their preservation is of special interest. The stress state of flat processing is a transition from the unfavorable purely moment stress state of the plate to the favorable momentless state of support [2–4]. Thus, flat structures increase the intermediate position between bending

S. V. Klyuev et al. (Eds.): ICRAAC 2024, LNCE 627, pp. 313–324, 2025.
https://doi.org/10.1007/978-3-031-82938-3_36

slabs and momentless shells, combining the advantage of the former in terms of material distribution over the area transfer with the advantage of the shell in terms of stress distribution over the thickness.

This study considers flat rectangular phenomena in the world of outlines, which are most common in construction practice as a floor device of instantaneous action [5–7].

Even in those cases where a strict fundamental solution is possible, numerical calculation and obtaining the final result are mostly associated with very difficulties.

2 Methods and Materials

2.1 Problem Statement

Let us consider a rectangular in plan infinitely small element of a shallow shell with dimensions $dx \times dy$, on the lateral faces of which forces act in the middle surface (Fig. 1, a), as well as moments and transverse forces (Fig. 1, b). An external transverse load q is applied normally to the middle surface (Fig. 1). The equation of the surface of which has the form:

$$z = f(x, y)$$

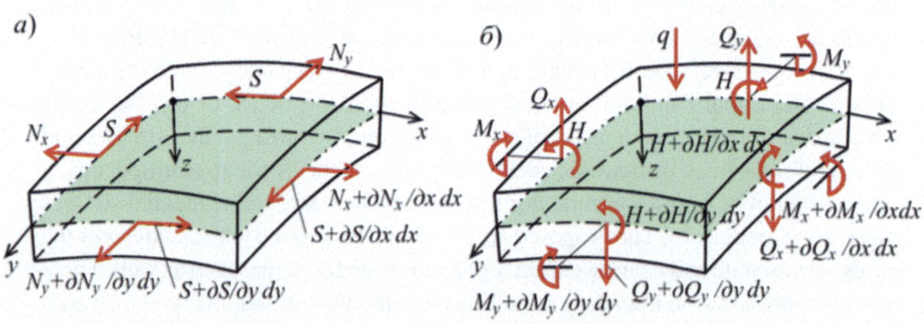

Fig. 1. a) Middle surface of a shallow shell. b) Efforts in a shallow shell element

In the presence of creep, the geometric and static equations of the shallow shell theory remain unchanged. Changes occur only in the physical equations.

Derivation of the resolving equations.

The equilibrium equation has the form:

$$\frac{\partial N_x}{\partial x} + \frac{\partial S}{\partial y} = 0; \frac{\partial S}{\partial x} + \frac{\partial N_y}{\partial y} = 0;$$

$$\frac{\partial Q_x}{\partial x} + \frac{\partial Q_y}{\partial y} - k_x N_x - k_y N_y + q = 0, \tag{1}$$

$$\frac{\partial M_x}{\partial x} + \frac{\partial H}{\partial y} = Q_x; \frac{\partial M_y}{\partial y} + \frac{\partial H}{\partial x} = Q_y;$$

where N_x, N_y are longitudinal forces; S is the shear force; M_x, M_y are bending moments; H is the torque; Q_x, Q_y are transverse forces; q is the surface load.

Let us exclude the transverse forces Q_x, Q_y from these equations by differentiating the last line of Eq. (1) with respect to x, y and substituting into the 2nd line of Eq. (1), we obtain the generalized equilibrium equation

$$\frac{\partial^2 M_x}{\partial x^2}+2\frac{\partial^2 H}{\partial x \partial y}+\frac{\partial^2 M_y}{\partial y^2}-k_x N_x-k_y N_y+q=0 \tag{2}$$

Geometric relationships according to [6, 7], total deformations are the sum of the components of deformations of the middle surface and deformations associated with the change in curvature:

$$\begin{Bmatrix}\varepsilon_x\\ \varepsilon_y\\ \gamma\end{Bmatrix}=\begin{Bmatrix}\varepsilon_x^0\\ \varepsilon_y^0\\ \gamma^0\end{Bmatrix}+z\begin{Bmatrix}\chi_x\\ \chi_y\\ \chi_{xy}\end{Bmatrix} \tag{3}$$

where u, v, w are the displacements of the middle surface in accordance with the coordinates x, y, z.

In (3) the first term is the deformation of the middle surface, the second term is the change in the curvature of the middle surface. Let us introduce the following notations:

$$\begin{Bmatrix}\varepsilon_x^0\\ \varepsilon_y^0\\ \gamma^0\end{Bmatrix}=\begin{bmatrix}\frac{\partial}{\partial x} & 0 & 0\\ 0 & \frac{\partial}{\partial y} & 0\\ \frac{\partial}{\partial y} & \frac{\partial}{\partial x} & 0\end{bmatrix}\begin{Bmatrix}u\\ v\\ w\end{Bmatrix}+\begin{Bmatrix}k_x\\ k_y\\ 0\end{Bmatrix}w;\quad \begin{Bmatrix}\chi_x\\ \chi_y\\ \chi_{xy}\end{Bmatrix}=-\begin{bmatrix}\frac{\partial^2}{\partial x^2}\\ \frac{\partial^2}{\partial y^2}\\ \frac{\partial^2}{\partial x \partial y}\end{bmatrix}w \tag{4}$$

where u, v, w are the displacements of the middle surface in accordance with the coordinates x,y,z.

The equation for compatibility of deformations can be obtained from (4) by differentiating ε_x^0 and ε_y^0 with respect to x and y respectively and summing them up we have:

$$\frac{\partial^2 \varepsilon_x^0}{\partial y^2}+\frac{\partial^2 \varepsilon_y^0}{\partial x^2}=\frac{\partial^2 \gamma^0}{\partial x \partial y}+\frac{\partial^2}{\partial y^2}(k_x w)+\frac{\partial^2}{\partial x^2}(k_y w) \tag{5}$$

If $k_{xy}=-\frac{\partial^2 z}{\partial x \partial y}=0$, from which it follows that k_x and k_y are the principal curvatures, the deformation compatibility Eq. (5) takes the form:

$$\frac{\partial^2 \varepsilon_x^0}{\partial y^2}+\frac{\partial^2 \varepsilon_y^0}{\partial x^2}=\frac{\partial^2 \gamma^0}{\partial x \partial y}+k_x\frac{\partial^2 w}{\partial y^2}+k_y\frac{\partial^2 w}{\partial x^2} \tag{6}$$

Physical relationships taking into account creep deformations based on the first hypothesis of the theory of calculating shells, as well as plates, the dependence of stresses on deformations is taken as:

$$\begin{Bmatrix}\sigma_x\\ \sigma_y\\ \tau_{xy}\end{Bmatrix}=\frac{E}{1-\nu^2}\begin{bmatrix}1 & \nu & 0\\ \nu & 1 & 0\\ 0 & 0 & \frac{(1-\nu)}{2}\end{bmatrix}\left(\begin{Bmatrix}\varepsilon_x^0\\ \varepsilon_y^0\\ \gamma^0\end{Bmatrix}+z\begin{Bmatrix}\chi_x\\ \chi_y\\ \chi_{xy}\end{Bmatrix}-\begin{Bmatrix}\varepsilon_x^{cr}\\ \varepsilon_y^{cr}\\ \gamma_{xy}^{cr}\end{Bmatrix}\right) \tag{7}$$

We will represent longitudinal forces as:

$$\begin{Bmatrix} N_x \\ N_y \\ S \end{Bmatrix} = \begin{Bmatrix} \int_{-h/2}^{h/2} \sigma_x dz \\ \int_{-h/2}^{h/2} \sigma_y dz \\ \int_{-h/2}^{h/2} \tau_{xy} dz \end{Bmatrix} = \frac{Eh}{1-\nu^2} \begin{bmatrix} 1 & \nu & 0 \\ \nu & 1 & 0 \\ 0 & 0 & \frac{(1-\nu)}{2} \end{bmatrix} \begin{Bmatrix} \varepsilon_x^0 \\ \varepsilon_y^0 \\ \gamma^0 \end{Bmatrix} - \begin{Bmatrix} N_1^{cr} \\ N_2^{cr} \\ S^{cr} \end{Bmatrix} \tag{8}$$

where

$$\begin{Bmatrix} N_x^{cr} \\ N_y^{cr} \\ S^{cr} \end{Bmatrix} = \frac{E}{1-\nu^2} \begin{Bmatrix} \int_{-\frac{h}{2}}^{\frac{h}{2}} \begin{bmatrix} 1 & \nu & 0 \end{bmatrix} \begin{Bmatrix} \varepsilon_x^{cr} \\ \varepsilon_y^{cr} \\ \gamma_{xy}^{cr} \end{Bmatrix} dz \\ \int_{-\frac{h}{2}}^{\frac{h}{2}} \begin{bmatrix} \nu & 1 & 0 \end{bmatrix} \begin{Bmatrix} \varepsilon_x^{cr} \\ \varepsilon_y^{cr} \\ \gamma_{xy}^{cr} \end{Bmatrix} dz \\ \int_{-\frac{h}{2}}^{\frac{h}{2}} \begin{bmatrix} 0 & 0 & \frac{(1-\nu)}{2} \end{bmatrix} \begin{Bmatrix} \varepsilon_x^{cr} \\ \varepsilon_y^{cr} \\ \gamma_{xy}^{cr} \end{Bmatrix} dz \end{Bmatrix} \tag{9}$$

The relationships for bending and torsional moments are defined as

$$\begin{Bmatrix} M_x \\ M_y \\ H \end{Bmatrix} = \begin{Bmatrix} \int_{-h/2}^{h/2} \sigma_x z dz \\ \int_{-h/2}^{h/2} \sigma_y z dz \\ \int_{-h/2}^{h/2} \tau_{xy} z dz \end{Bmatrix} = -D \begin{bmatrix} 1 & \nu & 0 \\ \nu & 1 & 0 \\ 0 & 0 & (1-\nu) \end{bmatrix} \begin{Bmatrix} \chi_x \\ \chi_y \\ \chi_{xy} \end{Bmatrix} - \begin{Bmatrix} M_x^{cr} \\ M_y^{cr} \\ H^{cr} \end{Bmatrix} \tag{10}$$

where

$$\begin{Bmatrix} M_x^{cr} \\ M_y^{cr} \\ H^{cr} \end{Bmatrix} = \frac{E}{1-\nu^2} \begin{Bmatrix} \int_{-\frac{h}{2}}^{\frac{h}{2}} \begin{bmatrix} 1 & \nu & 0 \end{bmatrix} \begin{Bmatrix} \varepsilon_x^{cr} \\ \varepsilon_y^{cr} \\ \gamma_{xy}^{cr} \end{Bmatrix} z dz \\ \int_{-\frac{h}{2}}^{\frac{h}{2}} \begin{bmatrix} \nu & 1 & 0 \end{bmatrix} \begin{Bmatrix} \varepsilon_x^{cr} \\ \varepsilon_y^{cr} \\ \gamma_{xy}^{cr} \end{Bmatrix} z dz \\ \int_{-\frac{h}{2}}^{\frac{h}{2}} \begin{bmatrix} 0 & 0 & \frac{(1-\nu)}{2} \end{bmatrix} \begin{Bmatrix} \varepsilon_x^{cr} \\ \varepsilon_y^{cr} \\ \gamma_{xy}^{cr} \end{Bmatrix} z dz \end{Bmatrix} \tag{11}$$

The first 2 equilibrium Eqs. (1) can be satisfied if the stress function is introduced using the formulas:

$$N_x = \frac{\partial^2 \Omega}{\partial y^2}, N_y = \frac{\partial^2 \Omega}{\partial x^2}, S = -\frac{\partial^2 \Omega}{\partial x \partial y} \tag{12}$$

We represent the deformations of the middle surface from (8) taking into account (12) as

$$\begin{Bmatrix} \varepsilon_x^0 \\ \varepsilon_y^0 \\ \gamma^0 \end{Bmatrix} = \frac{1}{Eh}\left\{\begin{bmatrix} 1 & -\nu & 0 \\ -\nu & 1 & 0 \\ 0 & 0 & 2(1+\nu) \end{bmatrix}\left(\begin{Bmatrix} N_x \\ N_y \\ S \end{Bmatrix} + \begin{Bmatrix} N_x^{cr} \\ N_y^{cr} \\ S^{cr} \end{Bmatrix}\right)\right\} \tag{13}$$

or

$$\begin{Bmatrix} \varepsilon_x^0 \\ \varepsilon_y^0 \\ \gamma^0 \end{Bmatrix} = \frac{1}{Eh}\left\{\begin{bmatrix} 1 & -\nu & 0 \\ -\nu & 1 & 0 \\ 0 & 0 & 2(1+\nu) \end{bmatrix}\left(\begin{Bmatrix} \frac{\partial^2\Omega}{\partial y^2} \\ \frac{\partial^2\Omega}{\partial x^2} \\ -\frac{\partial^2\Omega}{\partial x\partial y} \end{Bmatrix} + \begin{Bmatrix} N_x^{cr} \\ N_y^{cr} \\ S^{cr} \end{Bmatrix}\right)\right\} \tag{14}$$

Substituting (14) into the deformation compatibility Eq. (6), we obtain the first resolving equation:

$$\frac{1}{Eh}\nabla^2\nabla^2\Omega - k_x\frac{\partial^2 \boldsymbol{w}}{\partial y^2} - k_y\frac{\partial^2 \boldsymbol{w}}{\partial x^2} =$$

$$= \frac{1}{Eh}\left[2(1+\nu)\frac{\partial^2 S^*}{\partial x\partial y} + \nu\left(\frac{\partial^2 N_x^*}{\partial x^2} + \frac{\partial^2 N_y^*}{\partial y^2}\right) - \frac{\partial^2 N_x^*}{\partial y^2} - \frac{\partial^2 N_y^*}{\partial x^2}\right] \tag{15}$$

Substituting Eq. (10) into (2), we obtain the second resolving equation:

$$D\nabla^2\nabla^2\boldsymbol{w} + k_x\frac{\partial^2\Omega}{\partial y^2} + k_y\frac{\partial^2\Omega}{\partial x^2} = q - \left(\frac{\partial^2 M_x^*}{\partial x^2} + 2\frac{\partial^2 H^*}{\partial x\partial y} + \frac{\partial^2 M_y^*}{\partial y^2}\right) \tag{16}$$

Thus, we obtain a system of two differential equations for two unknown functions $\boldsymbol{w}$ and Ω.The system of differential Eqs. (15) and (16) generalizes two theories of elasticity: the problem of plate bending and the plane problem [7–9].

Let us consider various cases of boundary conditions for a rectangular shallow shell, the edges of which coincide with the coordinate lines x and y, which are the lines of principal curvatures. At each edge of the shell, constraints are imposed on the functions w and φ, and there must be four such conditions - two conditions for each of the functions w and φ. Note that the boundary conditions depending on the deflection of the shell have exactly the same form as for rigid plates [9–12].

Thus, for a rigidly clamped edge of the shell, parallel to the coordinate line y (for points of this edge at $x = \pm a$– Fig. 2), the following conditions are valid:

$$w = 0|_{x=\pm a}; \quad \frac{\partial w}{\partial x} = 0|_{x=\pm a} \tag{17}$$

For a hinged edge $x = \pm a$ similarly to (17).

Let us also write down the most characteristic boundary conditions that are imposed on the function Ω. For the edge $x = \pm a$, which can move freely in the direction of the x-axis,

$$\sigma_x = 0|_{x=\pm a}; \quad \frac{\partial^2\Omega}{\partial y^2} = 0|_{x=\pm a} \tag{18}$$

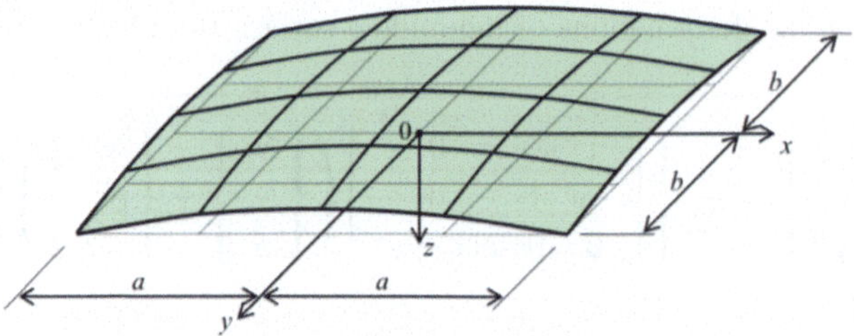

Fig. 2. Thin shallow shell.

For an edge $x = \pm a$, freely displaceable in the direction of the y-axis

$$\tau_{yx} = 0|_{x=\pm a}; \quad \frac{\partial^2 \Omega}{\partial x \partial y} = 0|_{x=\pm a} \tag{19}$$

When solving practical problems, various combinations of the listed conditions may be encountered. For example, for a rigidly clamped edge $x = \pm a$, where displacements in the direction of the x, y and z axes are excluded, one should set

$$u = v = w = \frac{\partial w}{\partial x} = 0|_{x=\pm a} \tag{20}$$

With free support of the shell edges on the end diaphragms (Fig. 3), which are absolutely rigid with respect to deformations in their plane and flexible from it, the boundary conditions take the following form:

$$v = w = \frac{\partial^2 w}{\partial x^2} = \frac{\partial^2 \Omega}{\partial y^2} = 0|_{x=\pm a} \tag{21}$$

$$u = w = \frac{\partial^2 w}{\partial y^2} = \frac{\partial^2 \Omega}{\partial x^2} = 0|_{x=\pm b}$$

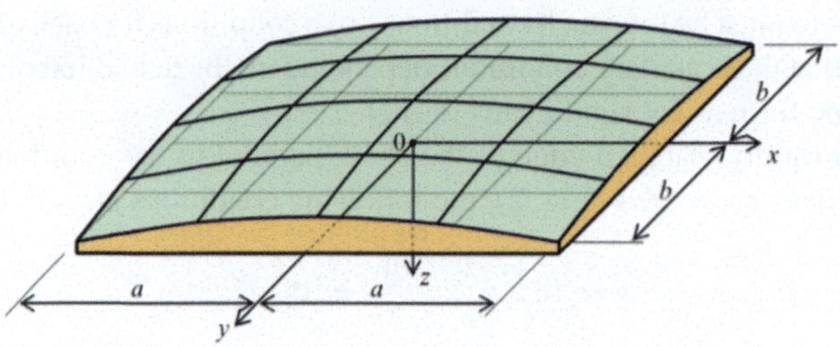

Fig. 3. Support of the shell on the diaphragms.

In the calculation we assume that on the shell contour the stress function $\Omega = 0$, then on the edges $x = 0$ and $x = a$ the derivative $\frac{\partial^2 \Omega}{\partial y^2}$, automatically vanishes, and on the edges $y = 0$, $y = b$ – the derivative $\frac{\partial^2 \Omega}{\partial x^2}$. We also require that at $x = 0$ and $x = a$ the derivative $\frac{\partial^2\Omega}{\partial x^2}$ vanishes, and on the edges $y = 0$, $y = b$ – the derivative $\frac{\partial^2\Omega}{\partial x^2}$. Then for the edges $x = 0$ and $x = a$ we will have:

$$\varepsilon_y^0 = \frac{1}{Eh}\left(N_y - \nu N_x + N_y^* - \nu N_x^*\right) = \frac{1}{Eh}\left(\frac{\partial^2\Omega}{\partial x^2} - \nu\frac{\partial^2\Omega}{\partial y^2} + N_y^* - \nu N_x^*\right) \equiv 0 \quad (22)$$

From (22) it follows that at the indicated edges the displacement v must be identically equal to a constant, which can be taken to be zero, therefore the last condition is satisfied. Similarly, for edges $y = 0$ and $y = b$:

$$\varepsilon_x^0 = \frac{1}{Eh}\left(N_x - \nu N_y + N_x^* - \nu N_y^*\right) = \frac{1}{Eh}\left(\frac{\partial^2\Omega}{\partial y^2} - \nu\frac{\partial^2\Omega}{\partial x^2} + N_x^* - \nu N_y^*\right) \equiv 0 \quad (23)$$

From (23) it follows that the displacement u at $y = 0$ and $y = b$ can be equal to zero.

2.2 Calculation Method

Let us consider an example of calculating a shallow rectangular shell, the middle surface of which is an elliptical paraboloid (Fig. 4).

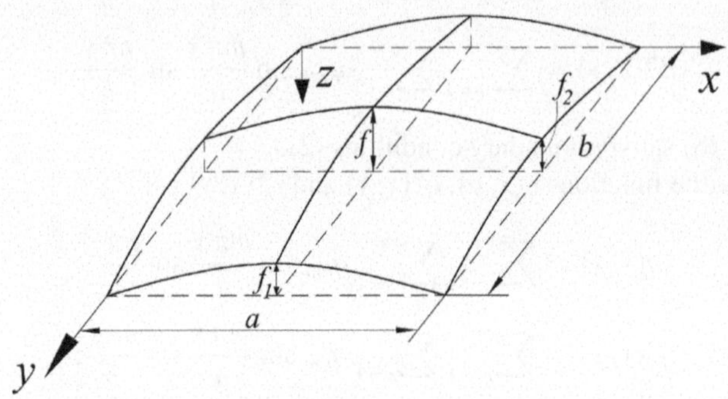

Fig. 4. Flat rectangular shell.

The surface equation is:

$$z = f\left[\frac{f_1}{f}\left(2\frac{x}{a} - 1\right)^2 + \frac{f_2}{f}\left(2\frac{y}{b} - 1\right)^2 - 1\right] \quad (24)$$

The principal curvatures of the shell under consideration are determined as follows:

$$k_x = -\frac{\partial^2 z}{\partial x^2} = -\frac{8f_1}{a^2}; \quad k_y = -\frac{\partial^2 z}{\partial y^2} = -\frac{8f_2}{b^2} \quad (25)$$

The solution of the system of resolving equations presents great difficulties, therefore it is advisable to apply variational methods of solution, for example, the Bubnov-Galerkin method, in the form developed for shells by V.Z. Vlasov or Ritz-Timoshenko. We propose a numerical-analytical solution. We will seek a solution to the equations in the form of double trigonometric series in sines. Taking into account (25), the system of resolving equations takes the form:

$$\frac{1}{Eh}\nabla^4\Omega + \frac{8f_1}{a^2}\frac{\partial^2 w}{\partial y^2} + \frac{8f_2}{b^2}\frac{\partial^2 w}{\partial x^2} = f^*$$

$$D\nabla^4 w - \frac{8f_1}{a^2}\frac{\partial^2\Omega}{\partial y^2} - \frac{8f_2}{b^2}\frac{\partial^2\Omega}{\partial x^2} = q + q^* \tag{26}$$

where

$$f^* = \frac{1}{Eh}\left[2(1+\nu)\frac{\partial^2 S^*}{\partial x\partial y} + \nu\left(\frac{\partial^2 N_x^*}{\partial x^2} + \frac{\partial^2 N_y^*}{\partial y^2}\right) - \frac{\partial^2 N_x^*}{\partial y^2} - \frac{\partial^2 N_y^*}{\partial x^2}\right] \tag{27}$$

$$q^* = -\left(\frac{\partial^2 M_x^*}{\partial x^2} + 2\frac{\partial^2 H^*}{\partial x\partial y} + \frac{\partial^2 M_y^*}{\partial y^2}\right)$$

We will look for a solution to Eqs. (26) in the form

$$w(x, y) = \sum\nolimits_{m=1}^{\infty}\sum\nolimits_{n=1}^{\infty} w_{mn}\sin\frac{m\pi x}{a}\sin\frac{n\pi y}{b} \tag{28}$$

$$\Omega(x, y) = \sum\nolimits_{m=1}^{\infty}\sum\nolimits_{n=1}^{\infty} \varphi_{mn}\sin\frac{m\pi x}{a}\sin\frac{n\pi y}{b}$$

Functions (28) satisfy boundary conditions (21).
Let us write the functions $q(x, y)$, $q^*(x, y)$ and $f^*(x, y)$ as:

$$q(x, y) = \sum\nolimits_{m=1}^{\infty}\sum\nolimits_{n=1}^{\infty} q_{mn}\sin\frac{m\pi x}{a}\sin\frac{n\pi y}{b}$$

$$q^*(x, y) = \sum\nolimits_{m=1}^{\infty}\sum\nolimits_{n=1}^{\infty} q_{mn}^*\sin\frac{m\pi x}{a}\sin\frac{n\pi y}{b}; \tag{29}$$

$$f^*(x, y) = \sum\nolimits_{m=1}^{\infty}\sum\nolimits_{n=1}^{\infty} f_{mn}^*\sin\frac{m\pi x}{a}\sin\frac{n\pi y}{b}.$$

The expansion coefficients q_{mn}, q_{mn}^* and f_{mn}^* are defined as follows:

$$q_{mn} = \frac{4}{ab}\int_0^a\int_0^b q(x, y)\sin\frac{m\pi x}{a}\sin\frac{n\pi y}{b}dxdy;$$

$$q_{mn}^* = \frac{4}{ab}\int_0^a\int_0^b q^*(x, y)\sin\frac{m\pi x}{a}\sin\frac{n\pi y}{b}dxdy; \tag{30}$$

$$f_{mn}^* = \frac{4}{ab}\int_0^a\int_0^b f^*(x, y)\sin\frac{m\pi x}{a}\sin\frac{n\pi y}{b}dxdy.$$

After substituting (29) into (26), we obtain:

$$\sum_{m=1}^{\infty}\sum_{n=1}^{\infty}\left\{\frac{\pi^4}{Eh}\left(\frac{m^2}{a^2} + \frac{n^2}{b^2}\right)^2\varphi_{mn} - \frac{8\pi^2}{a^2b^2}\left[n^2 f_1 + m^2 f_2\right]w_{mn}\right\}$$

$$\sin\frac{m\pi x}{a}\sin\frac{n\pi y}{b}=\sum_{m=1}^{\infty}\sum_{n=1}^{\infty}f_{mn}^{*}\sin\frac{m\pi x}{a}\sin\frac{n\pi y}{b}.$$

$$\sum\nolimits_{m=1}^{\infty}\sum\nolimits_{n=1}^{\infty}\left\{\pi^4D\left(\frac{m^2}{a^2}+\frac{n^2}{b^2}\right)^2 w_{mn}+\frac{8\pi^2}{a^2b^2}\left[n^2f_1+m^2f_2\right]\varphi_{mn}\right\}$$

$$\sin\frac{m\pi x}{a}\sin\frac{n\pi y}{b}=\sum\nolimits_{m=1}^{\infty}\sum\nolimits_{n=1}^{\infty}(q_{mn}+q_{mn}^{*})\sin\frac{m\pi x}{a}\sin\frac{n\pi y}{b}. \quad (31)$$

Two series are equal to each other if their coefficients are equal. As a result, for each pair of coefficients φ_{mn} and w_{mn} we have a system of two linear equations with two unknowns:

$$\begin{bmatrix} a_{11} & -a_{12} \\ a_{12} & a_{22} \end{bmatrix}\begin{Bmatrix} \varphi_{mn} \\ w_{mn} \end{Bmatrix}=\begin{Bmatrix} f_{mn}^{*} \\ q_{mn}+q_{mn}^{*} \end{Bmatrix} \quad (32)$$

where

$$a_{11}=\frac{\pi^4}{Eh}\left(\frac{m^2}{a^2}+\frac{n^2}{b^2}\right)^2, a_{12}=\frac{8\pi^2}{a^2b^2}\left[n^2f_1+m^2f_2\right]; a_{22}=\pi^4D\left(\frac{m^2}{a^2}+\frac{n^2}{b^2}\right)^2$$

From (32) the coefficients φ_{mn} and w_{mn} are determined as follows:

$$\varphi_{mn}=\frac{f_{mn}^{*}a_{22}+(q_{mn}+q_{mn}^{*})a_{12}}{a_{11}a_{22}+a_{12}^2};$$

$$w_{mn}=\frac{a_{11}(q_{mn}+q_{mn}^{*})-a_{12}f_{mn}^{*}}{a_{11}a_{22}+a_{12}^2} \quad (33)$$

When $(x, y) = const$, the first integral in (29) is written as:

$$q_{mn}=\frac{4q}{ab}\int_0^a\sin\frac{m\pi x}{a}dx\int_0^b\sin\frac{n\pi y}{b}dy=\frac{4q}{mn\pi^2}\cos\frac{m\pi x}{a}\Big|_0^a\cos\frac{n\pi y}{b}\Big|_0^b=$$

$$=\begin{cases}\frac{16q}{mn\pi^2}, ifm=2k+1andn=2p+1\\ 0, ifmorn-even\end{cases} \quad (34)$$

The last two integrals in (30) are calculated numerically. For this purpose, a uniform grid $\Delta x \times \Delta y$ is introduced. The calculation is performed using the formula:

$$\int_0^a\int_0^b f(x,y)dxdy=\sum f(x_i,y_i)\Delta x\Delta y\omega_i \quad (35)$$

where ω_i is a weighting coefficient, equal to 1 for internal grid nodes, $\omega_i = 0.5$ for nodes on the contour except for corners, and $\omega_i = 0.25$ for corner points.

When $q(x, y) = const$, the functions $q^*(x, y)$ and $f^*(x, y)$ should be expanded only for odd m and n.

The described approach was used to solve the problem presented in [1]. The number of intervals by x, y, z and time was taken to be 100.

3 Results and Discussion

The calculation was performed in the MATLAB software package. To verify the correctness of the solution, we will use the data from [1]. The calculation was performed for a polymer tank made of recycled PVC $h = 0.01m$, $a = b = 1m, f_1 = f_2 = 0.05m$, the shell material is recycled PVC, $E = 1480MPa$, $E_\infty = 5990MPa$, $m^* = 12.6MPa$, $\eta_0^* = 9.06 \cdot 10^5 MPa \cdot min$, $\nu = 0.3$. The shell is subject to a constant load of $q = 0.5kPa$ uniformly distributed over the area. The results obtained in [1] are also presented to verify the correctness of the solution. Comparison of the largest deflection values at different moments of time obtained by the finite difference method (FDM) and numerically-analytically (NAM) for $m, n = 1, 3, 5, 7$ is given in Table 1. The coincidence of the results indicates their reliability. Figure 5 shows a graph of the growth of the largest deflection value for different numbers of series members.

Table 1. Comparison of the largest displacements at different moments in time obtained by the FDM and the numerical-analytical method (NAM).

t, hour		1	2	3	4	5	6	7	8
w, mm	FDM	0.2547	0.2675	0.2759	0.2814	0.2850	0.2873	0.2889	0.2899
	NAM	0.2547	0.2675	0.2759	0.2813	0.2849	0.2872	0.2887	0.2897

The results obtained with nine and sixteen members of the series are practically the same.

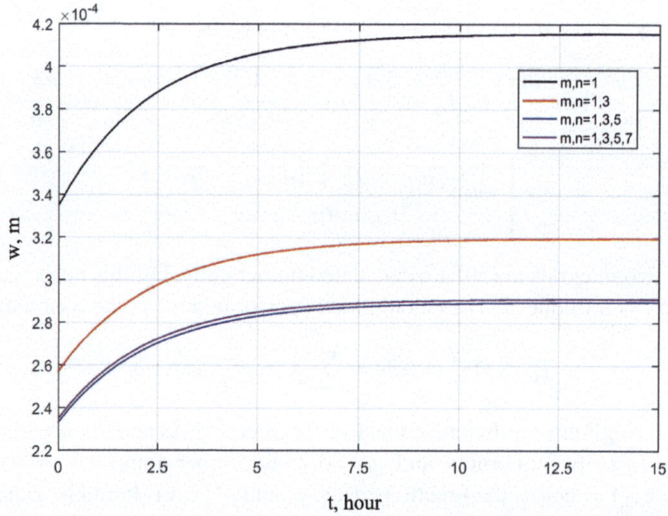

Fig. 5. Graph of the growth of the greatest value of deflection for different numbers of members of the series.

Figure 6 shows the distribution of displacements w depending on x and y: shaded surface – at $t = 0$; mesh surface at $t = 15$ h.

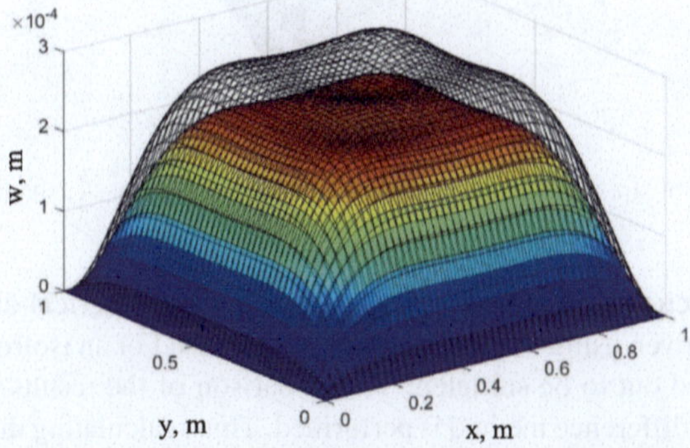

Fig. 6. Distribution of displacements w depending on x and y: shaded surface – at $t = 0$; mesh surface at $t = 15$ h.

The stresses change insignificantly during creep. The distribution of stresses σ_x and τ_{xy} as functions of x and y at $t = 15$ h is shown in Figs. 7 and 8, respectively. The grid graphs correspond to the stresses at the upper surface of the shell, the shaded ones – at the lower surface.

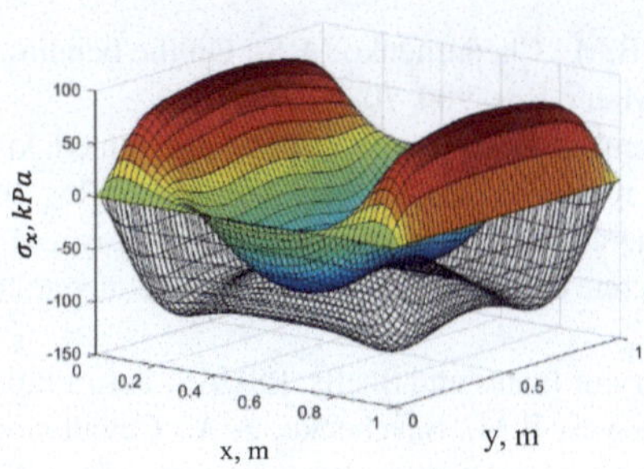

Fig. 7. Distribution of stresses σ_x at $t = 15$ h: mesh surface – $z = -h/2$, shaded surface – – $z = h/2$.

4 Conclusions

Resolving equations for calculating shallow isotropic and orthotropic shells taking into account creep are obtained. The problem is reduced to a system of two differential equations of the fourth order with respect to deflection and stress function. An example of calculating a shallow PVC shell in the form of an elliptical paraboloid is given. As a result of the calculation, it is established that the stresses and internal forces in the

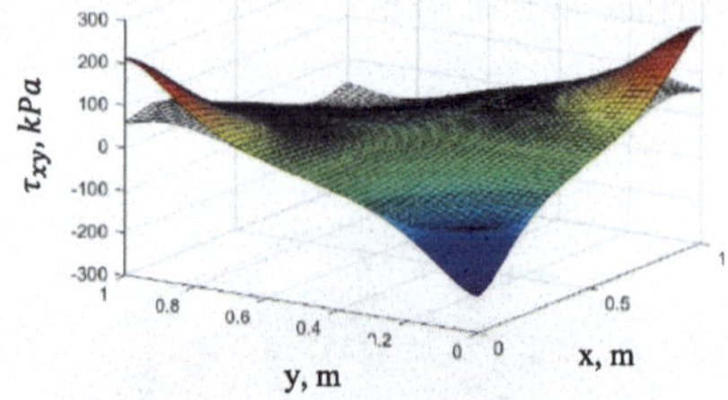

Fig. 8. Distribution of stresses τ_{xy} at $t = 15$ h: mesh surface – $z = -\frac{h}{2}$, shaded surface – $z = \frac{h}{2}$.

isotropic shell practically do not change during creep. A numerical-analytical solution of the problem is given using double trigonometric series. For an isotropic shell, 9 terms of the series turned out to be sufficient. A comparison of the results with the solution based on the finite difference method is performed. Thus, calculating the shell only in the elastic stage leads to underestimated stress values, and as a consequence, to its possible destruction during operation.

References

1. Chepurnenko, A.S., Saibel, A.V., Savchenko, A.A.: Calculation of a circular cylindrical shell using the moment theory taking into account creep. Engineering Bulletin of the Don 2 (2015)
2. Samul, V.I.: Fundamentals of the Theory of Elasticity and Plasticity. Vysshaya shkola, Moscow (1982)
3. Andreev, V.I., Yazyev, B.M., Chepurnenko, A.S.: On the bending of a thin polymer plate at nonlinear creep. Adv. Mater. Res. **900**, 707–710 (2014)
4. Andreev, V.I., Chepurnenko, A.S., Yazyev, B.M.: Energy method in the calculation stability of compressed polymer rods considering creep. Adv. Mater. Res. **1004–1005**, 257–260 (2014)
5. Yazyev, B.M., Chepurnenko, A.S., Litvinov, S.V., Kozelskaya, M.: Stress-strain state of pre-stressed reinforced concrete cylinder taking into account concrete creep. Sci. Rev. **11**, 759–763 (2014)
6. Monakhov, V.A.: Theory of Plates and Shells. PSUAS, Penza (2016)
7. Chepurnenko, A.S., Yazyev, B.M., Savchenko, A.A.: Calculation for the circular plate on creep considering geometric nonlinearity. Procedia Eng. **150**, 1680–1685 (2016)
8. Babu Gunda, J., Gandule, R.: New rational interpolation functions for finite element analysis of rotating beams. Int. J. Mech. Sci. **50**(3), 578–588 (2008)
9. Krylov, A.N.: On the Calculation of Beams Lying on an Elastic Foundation. USSR Academy of Sciences, Moscow (1931)
10. Lalin, V.V., Kolosova, G.S.: Numerical methods in construction. Publishing House of St. Petersburg State Technical University, St. Petersburg (2001)
11. Lou, T., Xiang, Y.: Numerical analysis of second-order effects of externally prestressed concrete beams. Struct. Eng. Mech. **35**(5), 631–643 (2010)
12. Milind, T.R., Date, P.P.: Analytical and finite element modeling of strain generated in equal channel angular extrusion. Int. J. Mech. Sci. **56**(1), 26–34 (2012)

Wooden Truss with Alternating Braces Using Metal Tooth Plates

Artur Zhurtov(✉) and Tolya Khezhev

Kabardino-Balkarian State University Named After Kh.M. Berbekov, Nalchik 360004, Russia
Zhurtovartur@mail.ru

Abstract. The article presents a deterministic algorithm for parametric optimization of a trapezoidal wooden truss with alternating diagonals on metal toothed plates, determining the optimal parameters of the structure with minimal wood consumption. A deterministic algorithm for parametric optimization of wooden trusses based on the minimum mass criterion was implemented in the object-oriented programming language Delphi. Implementation of the algorithm in the Borland Delphi environment provided such capabilities as automatic enumeration of variants of a given range with a given step, automatic recording of the obtained optimization results in multidimensional arrays with the possibility of their further analysis. As well as the creation of a convenient software shell. Optimal parameters were identified for a trapezoidal wooden truss with alternating diagonals with a span of 24 m at a given value of the external load, and diagrams of the dependence of the objective function on the variable parameters were compiled.

Keywords: Deterministic Algorithm · Parametric Optimization · Trapezoidal Wooden Truss with Alternating Diagonals · Metal Toothed Plates

1 Introduction

Design of construction projects is a complex organizational and technological process that requires significant expenditure of engineering labor, technical and economic resources. When designing wooden structures, it is necessary to comply with certain requirements set by standards, the main ones being ensuring the strength of structures and saving materials [1–4].

Connection of wooden structure elements using metal toothed plates (hereinafter MTP) is one of the most promising and is actively used throughout the world. This is due to the technological and economic advantages of this connection in comparison with traditional ones [5].

Thus, one of the areas of improvement and work with wooden structures is the minimization of labor and material resources in their manufacture [6–8]. The question posed comes down to creating an algorithm for optimizing the parameters of the structure, subject to minimal wood consumption [9, 10].

Below is an algorithm for parametric optimization of a trapezoidal wooden truss with alternating diagonals connected with MTP.

S. V. Klyuev et al. (Eds.): ICRAAC 2024, LNCE 627, pp. 325–332, 2025.
https://doi.org/10.1007/978-3-031-82938-3_37

2 Materials and Methods

There is a trapezoidal wooden truss with alternating diagonals with a span L, on which a useful uniformly distributed load q acts (Fig. 1). It is required to create a deterministic algorithm for optimizing the following parameters:

(a) h is the height of the truss on the support or the length of the support column,
(b) b is the thickness of the truss (the same for all elements),
(c) n is the total number of truss panels,
(d) α is the slope angle or the angle of inclination of the upper chord of the truss, from the condition of minimizing the volume of wood used in the structure.

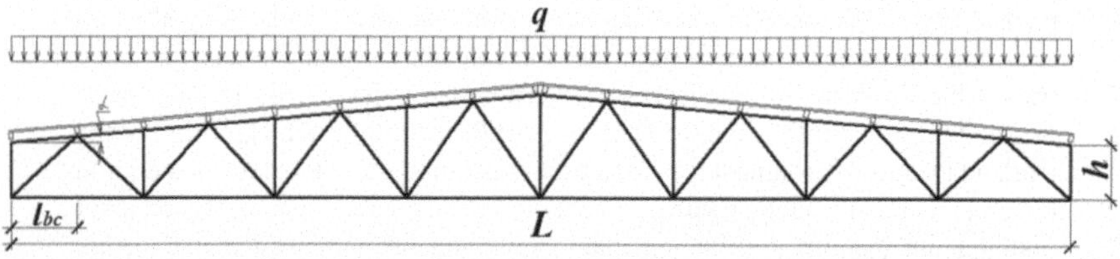

Fig. 1. Trapezoidal timber truss with alternating diagonals using MTP span L.

Since the connection of the truss rods is made on the MTP, then for a tight connection the thickness of these rods must be the same. The calculated length of the elements is determined from the conditions of fastening these elements from the plane of the truss, in the example the fastening is made at each node [5].

Based on the methodology for calculating building structures according to limit states, an algorithm for optimizing truss parameters is proposed, shown in Fig. 2.

3 Results and Discussion

The diagram for determining the geometric characteristics of the truss is shown in Fig. 3.

From the diagram shown in Fig. 3 it follows

$$L' = \frac{h}{tg\alpha}; \, l_u = \frac{L}{2 \cdot n \cdot cos\alpha}; \, l_d = \frac{L}{n}; \, l_{c,i} = h + \frac{L \cdot i}{n} \cdot tg\alpha;$$

$$\varphi_i = arctg\left(\frac{2 \cdot n \cdot h}{L} + (2 \cdot i - 1) \cdot tg\alpha\right); \, l_{p,i} = \sqrt{\left(\frac{L}{2 \cdot n}\right)^2 + \left(h + \frac{L \cdot tg\alpha}{n} \cdot (i - 0.5)\right)^2};$$

$$H_{1,i} = \left(L' + (i-1) \cdot l_d\right) \cdot \sin\varphi_i = \left(\frac{h}{tg\alpha} + (i-1) \cdot \frac{L}{n}\right) \cdot \sin\varphi_i;$$

$$H_{2,i} = \left(L' + i \cdot l_d\right) \cdot \sin\varphi_i = \left(\frac{h}{tg\alpha} + i \cdot \frac{L}{n}\right) \cdot \sin\varphi_i;$$

$$H_{3,i} = l_{c.i.} \cdot \cos\alpha = h \cdot \cos\alpha + \frac{L \cdot i}{n} \cdot \sin\alpha;$$

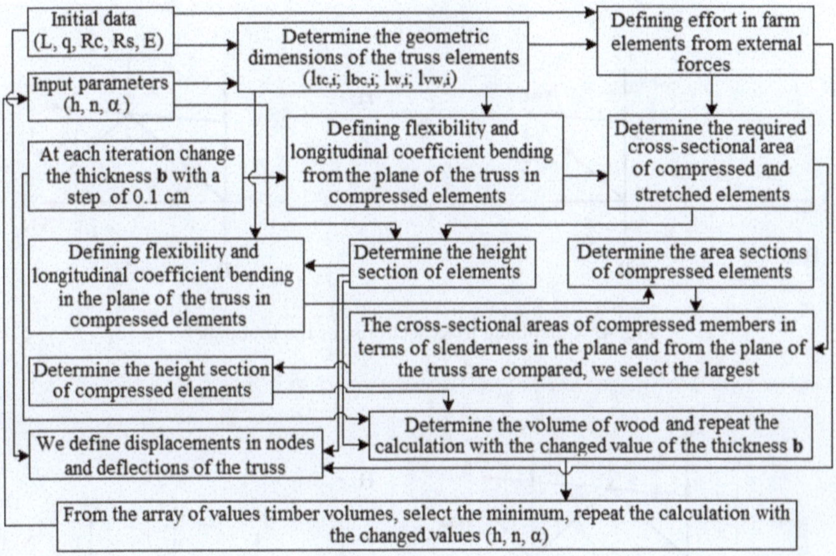

Fig. 2. Scheme based on the algorithm for finding the optimal values of h, n, α, b under the condition of minimum wood consumption.

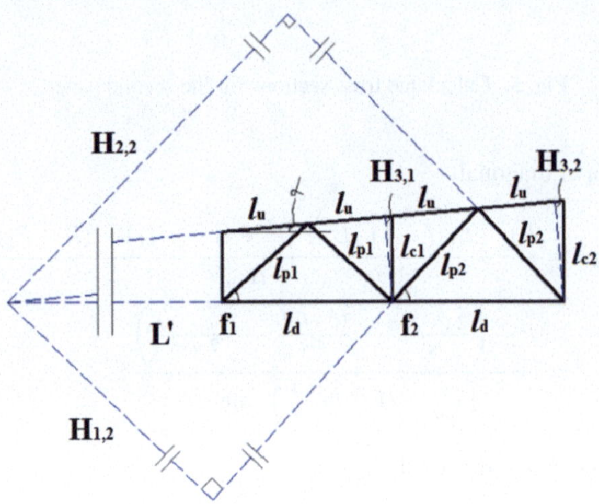

Fig. 3. Part of two panels of a trapezoidal truss with alternating diagonals.

The compressive and tensile forces that are caused in the truss by the action of external forces are determined as shown in Figs. 4 and 5.

Based on the regularities that can be determined when compiling equilibrium equations in the i-th sections of the truss, it is possible to derive equations for determining the internal forces in the truss elements in general form:

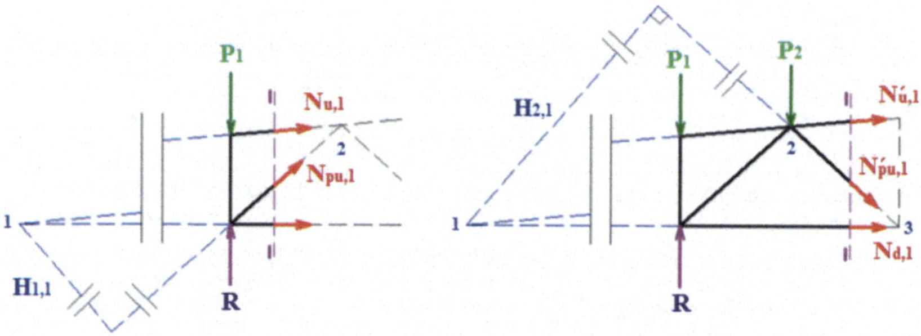

Fig. 4. Calculated truss sections for the first panel

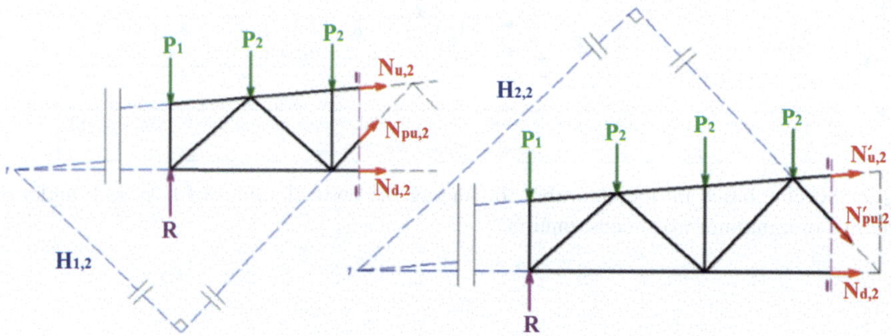

Fig. 5. Calculated truss sections for the second panel.

- for the i-th upper diagonal

$$N_{pu,i} = \frac{P_2 \cdot \left(2 \cdot (i-1) \cdot L' + \left(i^2 + i\right) \cdot \frac{1}{4} \cdot l_d\right) - (R - P_1) \cdot L'}{H_{1,i}}$$

$$= \frac{P_2 \cdot \left(\frac{2 \cdot (i-1) \cdot h}{tg\alpha} + \frac{(i^2+i) \cdot L}{4 \cdot n} - \frac{(2 \cdot n - 1) \cdot h}{2 \cdot tg\alpha}\right)}{\left(\frac{h}{tg\alpha} + (i-1) \cdot \frac{L}{n}\right) \cdot \sin\varphi_i};$$

- for the i-th descending diagonal

$$N_{pd,i} = \frac{(R - P_1) \cdot L' - P_2 \cdot \left((2 \cdot i - 1) \cdot L' + \left(i^2 + i\right) \cdot \frac{1}{2} \cdot l_d \cdot \frac{1}{2}\right)}{H_{2,i}}$$

$$= \frac{P_2 \cdot \left(\frac{(2 \cdot n - 1)h}{2 \cdot tg\alpha} - \frac{(2 \cdot i - 1) \cdot h}{tg\alpha} + \frac{(i^2+i) \cdot L}{4 \cdot n}\right)}{\left(\frac{h}{tg\alpha} + i \cdot \frac{L}{n}\right) \cdot \sin\varphi_i}.$$

– for the i-th upper chord of the truss

$$N_{u,i+1} = N_{u,i'} = \frac{P_2 \cdot (1 + (i + 1)) \cdot (i + 1) \cdot 0,5 \cdot l_d \cdot 0,5 - (R - P_1) \cdot i \cdot L\prime}{l_{c.i.} \cdot \cos\alpha}$$
$$= \frac{P_2 \cdot \left(\left((i + 1)^2 + i + 1\right) \cdot tg\alpha \cdot L - 4 \cdot \left(n^2 - 0,5 \cdot n\right) \cdot i \cdot h\right)}{4 \cdot tg\alpha \cdot (n \cdot h \cdot \cos\alpha + L \cdot i \cdot \sin\alpha)}.$$

– for the i-th lower chord of the truss

$$N_{d,i} = \frac{\left(2 \cdot (R - P_1) \cdot (2 \cdot i - 1) - P_2 \cdot \left(i^2 + i\right)\right) \cdot l_d}{2 \cdot (h + l_{c.1.})}$$
$$= \frac{\left(P_2 \cdot (2 \cdot n - 1) \cdot (2 \cdot i - 1) - P_2 \cdot \left(i^2 + i\right)\right) \cdot L}{4 \cdot n \cdot h + i \cdot tg\alpha \cdot L}.$$

– for the i-th rack

$$\begin{cases} N_{c,0} = P_1; \\ N_{c,i} = P_2,\ i \in [1; n - 1]; \\ N_{c,n} = 2 \cdot N_{u,n} \cdot \sin\alpha - P_2. \end{cases}$$
$$P_1 = \frac{q \cdot L}{2 \cdot n}; P_2 = 2 \cdot P_1 = \frac{q \cdot L}{n}; R = \frac{q \cdot L}{2}.$$

Then, a deterministic algorithm for parametric optimization of wooden trusses based on the minimum mass criterion was implemented in the Delphi object-oriented programming language.

Implementation of the algorithm in the Borland Delphi environment provides such capabilities as automatic enumeration of variants of a given range with a given step, automatic recording of the obtained optimization results in multidimensional arrays with the possibility of their further analysis. And also creation of a software shell.

The source code of the program is presented in Fig. 5.

4 Conclusions

Based on the algorithm proposed above and using the program, optimization of a wooden truss with ascending diagonals using MTP with a span of L = 24 m, with a useful uniformly distributed load of q = 3 kN/m was carried out. Based on the results of varying such truss parameters as the height on the support, the angle of inclination of the upper chord, the number of panels, diagrams of the dependence of the volume of wood on these parameters were constructed (see Fig. 7).

Thus, an algorithm was obtained and a program compiled on its basis allows for the optimization of wooden trusses on the minimum timber production line according to the criterion of minimum wood consumption, under given initial conditions.

```
unit Unit1;
interface
uses
 Windows, Messages, SysUtils, Variants,
 Classes, Graphics, Controls, Forms,
 Dialogs, StdCtrls, Math, Grids;
type
 TForm1 = class(TForm)
  Button: TButton;
  EditL: TEdit;
  EditHopora: TEdit;
  EditUgolNaklona: TEdit;
  EditKolPanel: TEdit;
  EditNagruzka: TEdit;
  Label1: TLabel;
  Label2: TLabel;
  Label3: TLabel;
  Label4: TLabel;
  Label5: TLabel;
  Label6: TLabel;
  Label7: TLabel;
  Label8: TLabel;
  Label9: TLabel;
  EditKolRaskrep: TEdit;
  Label11: TLabel;
  StringGrid1: TStringGrid;
  procedure ButtonClick(Sender: TObject);
 private
  { Private declarations }
 public
  { Public declarations }
 end;
```

①

```
var
 Form1: TForm1;
implementation
function getFi(lambda,Bferm,L:double;m:integer):double;
begin
if lambda > 70 then
 begin
  getFi:=3000 * sqr(Bferm)/sqr(L * m);
 end
else
 begin
  getFi:=1 - 0.8 * sqr((L * m)/(100 * Bferm));
 end;
end;
function getNstoika(R,P1,P2,Lshtrih,Lniz:double;i:integer):double;
begin
 getNstoika:= ((R - P1) * Lshtrih - P2 * (i * lshtrih + (sqr(i) + i)/2 * Lniz))/(Lshtrih + i * Lniz);
end;

function getNraskos(R,P1,P2,Lshtrih,H1,Lniz:double;i:integer):double;
begin
 getNraskos:= ((p1 - R) * Lshtrih + P2 * ((i - 1) * Lshtrih + (sqr(i) - i)/2 * Lniz))/H1;
end;

 function getNverh(R,P1,P2,H2,Lniz:double;i:integer):double;
begin
 getNverh:= (p1 - R) * (i - 1) * lniz/H2 + P2 * lniz * (sqr(i) - 3*i + 2)/H2/2;
end;

function getNniz(P2,h,L,alfa:double;i,n:integer):double;
begin
 getNniz:= (i * n - sqr(i)) * L * P2 /2/(h * n + L * i * tan(alfa));
end;
```

②

```
{$R *.dfm}
procedure TForm1.ButtonClick(Sender: TObject);

var
R,P1,P2,L,h,alfa1,alfa,q,Lshtrih,Lverh,Lniz:double;
k,n,m,i:integer;
Nstoikaar-ray,Nnizarray,Nverharray,Nraskosarray:array[1..1000] of double;
Lstoikaarray,Fiarray,Lraskosarray,H1array,H2array:array[1..1000] of double;
begin
L:=strtofloat(EditL.text);
h:=strtofloat(EditHopora.text);
alfa1:=strtofloat(EditUgolNaklona.text);
alfa:=alfa1/180*3.141592;
n:=strtoint(Editkolpanel.text);
q:=strtofloat(Editnagruzka.text);
m:=strtoint(EditKolRaskrep.text);
Lshtrih:=h/tan(alfa);
Lverh:=L/(2*n*cos(alfa));
Lniz:=l/n;
P2:=q*l/n;
P1:=p2/2;
R:=q*l/2;
k:=n div 2;
stringgrid1.ColCount := 8;
stringgrid1.RowCount := k+1;
 For i:=1 to k do
begin
```

③

```
 Lstoikaarray[i]:=h+l*i/n*tan(alfa);
 Fiarray[i]:=arctan(Lstoikaarray[i]/Lniz);
 Lraskosar-ray[i]:=exp(0.5*ln(sqr(Lniz)+sqr(Lstoikaarray[i])));
 H1array[i]:=(Lshtrih+(i-1)*lniz)*sin(fiarray[i]);
 H2array[i]:=Lstoikaarray[i-1]*cos(alfa);
 H2array[1]:= h*cos(alfa);
 Nraskosar-ray[i]:=getNraskos(R,P1,P2,Lshtrih,H1array[i],lniz,i);
 Nverharray[i]:=getNverh(R,P1,P2,H2array[i],Lniz,i);
 Nstoikaarray[i]:=getNstoika(R,P1,P2,Lshtrih,Lniz,i);
 Nstoikaarray[k]:=0;
 Nnizarray[i]:=getNniz(P2,h,L,alfa,i,n);
 stringgrid1.cells[1,i] := floattostr(Fiarray[i]*180/pi);
 stringgrid1.cells[2,i] := floattostr(Lraskosarray[i]);
 stringgrid1.cells[3,i] := floattostr(Lstoikaarray[i]);
 stringgrid1.cells[4,i] := floattostr(Nraskosarray[i]);
 stringgrid1.cells[5,i] := floattostr(Nstoikaarray[i]);
 stringgrid1.cells[6,i] := floattostr(Nverharray[i]);
 stringgrid1.cells[7,i] := floattostr(Nnizarray[i]);
 stringgrid1.cells[1,0] := 'Fi';
 stringgrid1.cells[2,0] := 'Ld';
 stringgrid1.cells[3,0] := 'Lu';
 stringgrid1.cells[4,0] := 'Nd';
 stringgrid1.cells[5,0] := 'Nu';
 stringgrid1.cells[6,0] := 'Npu';
 stringgrid1.cells[7,0] := 'Npd';
 stringgrid1.cells[0,i] := floattostr(i);
 stringgrid1.cells[0,0] := 'Npanel';
 end;
end;
```

④

Fig. 6. Source code of the program in Delphi language.

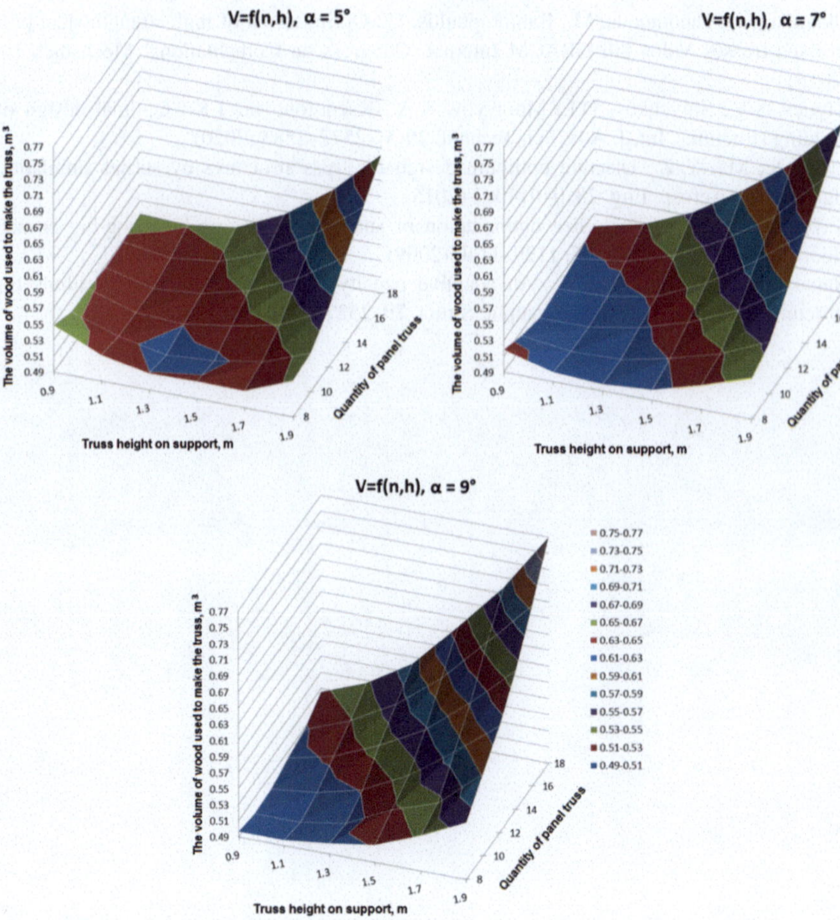

Fig. 7. Diagrams of the dependence of wood consumption on truss parameters.

References

1. Denisova, A.P., Rashchepkina, S.A.: Methods for the optimal design of building structures (Publishing House of the Association of Construction Universities (ASV) (2012)
2. Fedosov, S.V., Kotlov, V.G., Aktuganov, A.A.: Industrial wooden structures on metal toothed plates. News of higher educational institutions. Construction **11**, 39–43 (2013)
3. Klyuev, S.V., Abakarov, A.J., Lesovik, R.V., Muravyov, K.A., Tatlyev, R.Dz. Optimal engineering of rod spatial construction. J.Comput. Theor. Nanosci. **16**(1), 200–203 (2019)
4. Klyuev, S.V., Klyuev, A.V., Grishko, A.K., Trukhanov, S.V.: Management of the design parameters in optimal design problems. Mater. Sci. Forum **974**, 723–728 (2019)
5. Zhurtov, A., Hezhev, T., Kardangushev, A., Nastuyev, M., Osmanov, I.: Algorithm for parametric optimization of a trapezoidal wooden truss on metal tooth plates. E3S Web of Conferences 281, 01040 (2021)

6. Gkantou, M., Theofanous, M., Baniotopoulos, C.: Optimisation of high strength steel pre-stressed trusses Volos 8th GRACM Internat. Congress on Computational Mechanics 10, (2015)
7. Klyuev, S.V., Shlychkov, D.I., Muravyov, K.A., Ksenofontova, T.K.: Optimal design of building structures. Int. J. Adv. Sci. Technol. **29**(5), 2577–2583 (2020)
8. Li, H.-S., Ma, Y.-Z.: Discrete optimum design for truss structures by subset simulation algorithm. J. Aerosp. Eng. **28**, 04014091 (2015)
9. Kaveh, A., Talatahari, S.: Size optimization of space trusses using big bang-big crunch algorithm. Comput. Struct. **87**, 1129–1140 (2009)
10. Nanakorn, P., Meesomklin, K.: An adaptive penalty function in genetic algorithms for structural design optimization. Comput. Struct. **79**, 2527–2539 (2001)

Antiseismic Solutions at the Early Stage of an Architectural Project

Larisa Zolotareva and Anastasia Lapina(✉)

Southern Federal University, Rostov-on-Don 344006, Russia
alapi@sfedu.ru

Abstract. The complexity of modeling the structural and design scheme of an architectural object under conditions of increased seismic activity of the design site at the conceptual design stage affects the quality and timing, and can also cause inaccuracies that will require additional material and time resources of the client to correct if inefficient models are used. This paper discusses some of the most characteristic problems causing structural failure of buildings in earthquake. Structural seismic protection techniques from antiquity to the present day are analyzed. The specifics of antiseismic design principles are identified. In this study, we propose a comprehensive systematic approach to the development of primary architectural seismic protection solutions at the early stage of design of buildings and structures.

Keywords: Architecture · Design · Earthquakes · Morphometric Features · Earthquake Protection · Earthquake-Resistant Buildings · Structural Systems · Load-Bearing Systems · Enclosing Structures

1 Introduction

Structural failure occurs when a material is subjected to a load that exceeds its ultimate strength, resulting in fracture or extreme deformation. The ultimate limit state is the strength limit of a material or system. When this limit is reached, building materials can be damaged and their load-bearing capacity suddenly and permanently reduced. If the system is properly designed, localized collapse will not usually be the cause of an immediate or gradual collapse of the whole building. The ultimate strength of structural elements must be considered in the design of structures to prevent collapse.

Buildings must be designed to withstand certain loads without excessive deformation. The loads are the weight of people and objects, the dead load of the building itself (permanent loads), the weight of rain, snow and wind pressure (called temporary loads) and special loads. In buildings with a combined structural system, strength is usually combined with sufficient stiffness, and the load-bearing system is mainly the building frame and shell structures, including the building envelope.

In the Russian Federation, the most seismically vulnerable regions are Kamchatka, the Kuril Islands, Sakhalin, Transbaikalia and, to a lesser extent, the Caucasus and

S. V. Klyuev et al. (Eds.): ICRAAC 2024, LNCE 627, pp. 333–345, 2025.
https://doi.org/10.1007/978-3-031-82938-3_38

Krasnodar Krai. Seismic risks do not change over time, so they must be taken into account in the design of buildings and structures.

In order to properly design a building at the conceptual design stage, the architect, together with a structural engineer, must make a decision about the seismic protection of the design object. "In order to make design decisions at the stage of architectural design, it is necessary to obtain as much information as possible about the object and, most importantly, about the location and environment of the object in order to identify all the problems that may arise during the design process" [1]. Such information should be presented in reports on engineering geodetic, engineering geological, engineering hydrological, geophysical, meteorological and engineering environmental surveys. First of all, in areas with seismic activity, information on the actual characteristics (parameters) of earthquakes is required: acceleration of seismic activity, displacement of soil layers, dominant period of oscillations, duration of seismic activity. To calculate a real structural system, specific records of the earthquake parameters in the study area are required, not just the design value.

The task of seismic architectural design is to avoid possible irreparable building failures in the development of volumetric design solutions. It is necessary to make a decision on what structural scheme to design a building (structure): flexible or rigid. It all depends on the specific parameters of the expected earthquake.

Study Purpose. Identification of factors, parameters and principles of possible seismic protection solutions at the early stage of architectural design.

The object of the study is the modeling of a possible seismic protective structural system for the building.

The subject of the study are variants of seismic-protective load-bearing systems.

The problem can be formulated as follows: lack of results of forecasting the development of negative processes in the seismic zone, changes in the morphometric features of the soil in the zone of the proposed construction, lack of systematic studies of possible seismic protection solutions of structural systems at the initial stage of architectural design.

Relevance: development of seismic protection model of the bearing system of an architectural object at the idea stage.

Objectives of the study are:

1. To analyze the dependence of the degree of destruction of buildings or structures on their geometric parameters.
2. To identify the sequence of anti-seismic decision-making at the conceptual design stage.

2 Methods and Materials

The problem of behavior of ground structures during catastrophic earthquakes can be posed analytically. For this purpose, it is necessary to subject the investigated structure to consideration not only under the action of normally applied forces, but also under the action of inertial forces developing in it during vibrations excited by the motion of the Earth's crust during an earthquake, taking into account the configuration of the capital construction object.

Earthquake-proof building – a building that as a result of an earthquake will suffer damage that does not pose a threat to human life or the safety of material valuables contained therein.

The history of creating earthquake-resistant buildings is very short, although even in ancient and in the Renaissance period, they tried to classify earthquakes. The science of "Earthquake resistance of structures" appeared in 1900 and only since 1906 in the calculations of buildings and structures began to take into account seismic load in the form of increased wind pressure on the wall, "began to form the concept of the influence of the structural solution of the building on its operation under seismic loads. And by 1923, the basic principles of seismic load calculations were developed, mainly using empirical relationships" [2].

In the dynamics of structures, a branch of which is the theory of earthquake resistance, there is a concept of "natural period of oscillation" - this is the time for which any pendulum (building, ground surface), being deflected from the equilibrium position, makes one complete cycle of oscillations and returns to the initial position. The value of the natural period of oscillation is characterized by the flexibility of the structure.

The foundations of buildings (structures) and the ground strata below also oscillate and have their own period of oscillation layer by layer and maximum deviations, being the result of seismic waves from the epicenter of the earthquake to the ground surface. Scientists have identified some patterns that in the area close to the epicenter, oscillations with a small period predominate, and with distance from the epicenter, the period of oscillation generally increases. The effect of seismic impact on a particular building will be different, and in the case of coincidence or proximity of the prevailing parameters of the ground strata layers and the building (structure), resonance may occur and the probability of destruction of the structure increases.

Thus, an earthquake-resistant building is a building with a low probability of receiving hazardous structural changes.

2.1 Influence of Microgeological Factors

Observations over the damage elements of a number of catastrophic earthquakes have established that from the seismotectonic point of view, the nature of rocks composing the upper layers of the Earth's crust is the most important factor in the distribution of seismic force stresses on the day surface. The hazard material is the degree of conductivity of seismic waves. It has been established that rocky, solid unweathered rocks, wedging to the surface, in seismic terms appear to be the most favorable. Whereas decay products - loose, mobile, alluvial sediments should be classified as the most dangerous soils for the capital construction object erected on them.

One of the decisive moments influencing the intensity of the seismic forces manifested is the degree of mobility and wetting of rocks. The main morphological and hydrological factors on this side are the thickness of the bedrock, the degree of its density and the position of the groundwater table. On the other hand, there are geological factors, which are represented by the layering of soils when different rocks meet. The folding of the strata, tectonic fault lines and faults with possible combinations also have an impact.

In rocky soils the amplitude of oscillations usually does not exceed 2–5 mm, in earthy soils - 25 mm and only in bulk and unstable soils the amplitude reaches large values.

Comparative coefficients, which characterize the relative reliability on the seismicity side of a particular soil under the base of structures are presented in Table 1.

Table 1. Reliability coefficients on the seismicity side of the ground.

Alluvium, crushed stone, sand, peat	Clays, marls and loesses	Subsurface	Swamps and sloughs
3–6	2–10	5–12	8–16

At these ratios of coefficients reliability in seismic conditions of solid rock is taken as 1 (unit).

It should be noted that seismic properties of soils are somewhat dependent on the parameters of buildings and structures erected on them. Thus, loose mobile alluvial soils under massive buildings of large weight considerably mitigate seismic hazard for the latter: the foundation sends a counter wave towards the waves of seismic oscillation, which reduces the amplitude of the first one.

Thus, the destruction of a building (structure) from seismic impact occurs for two reasons:

1. Weak foundation;
2. Swaying of the building, when stresses in the elements of the bearing system exceed the strength limit.

2.2 Influence of Shaping Factors

At the stage of conceptual design of an architectural object it is necessary to take into account the architectural form of the building (structure). Architectural form includes both external geometric parameters and the structure of the building as a whole. "The choice of configuration in plan is based on purely geometric parameters, the required degree of seismic resistance and functional requirements for the designed building … The basis of the classification system under consideration relates to the geometric parameters of convexity and concavity, both in plan and height" [2].

The main criteria for the configuration in plan and height of the building are presented in Table 2.

Conditionally enough, we can distinguish three directions of designing earthquake-resistant buildings, that is, to program destruction and make them non-hazardous for human life, taking into account the chosen architectural form:

1. To make the building strong.
2. To tear a building off the ground by any means - passive seismic protection.
3. To allow the building to realign to the seismic load direction, active seismic protection.

Table 2. Main criteria for configuration in plan and height of the building.

In plan		By section	
Absolute dimensions	Relative dimensions	Absolute dimensions	Relative dimensions
Overall dimensions and plan area as well as diagonal area: a and b; A = a • b; c	Party ratio: a/b	Overall dimensions and coordinates of center of gravity and volume: a, b, H; V = a • b • H; xc and zc	The ratio of the height and side and free length (height) of the building to the radius of inertia of the building shape H/a and H/b; H/r

2.3 Influence of Physical and Mechanical Properties of Materials

In general, the classical materials of load-bearing systems include natural stone, brick, wood, concrete, steel and reinforced concrete. Among them we can single out the most "young" - steel, reinforced concrete, because stone, brick and concrete buildings have a history equal to the age of mankind as a whole. These materials are used in almost all load-bearing systems to a greater or lesser extent. The advantages of using these materials are 100% predictable performance of the structure in the part depending on the physical and mechanical properties of the materials throughout the life of the structure, including in case of partial loss of bearing capacity and the need for reinforcement.

Taking into account the degree of studied behavior of a building material in a load-bearing system, a conventional class of classical materials appears. For such materials there is not only a developed mathematical apparatus investigating its behavior, but also an extensive database of field studies, including the results of the behavior of these materials in extreme conditions. Since this classification is conditional, the list of materials belonging to this class also varies depending on the focus of the research direction.

In addition, the economic justification and minimization of material consumption is verified by calculation methods based on the studied behavior of classical building materials.

2.4 Historical Anti-seismic Design Solutions

Among the periods of ancient architecture in antiseismic, each has its own specificum according to the nature of the architecture, the overlap and the general state of technology of that particular period.

Antiseismic structural methods of ancient times consisted in mechanical connection of materials (pierons, pins); in deepening on the backfill or substructures of the base and foundation of the structure; in the construction of a stepped stylobate with a peculiar antiseismic belt, effectively combating seismic shifts and slides; in the easily inclined inward profile of verticals; in the reduction of intercolumnium to deliberately beam corner columns. For anti-seismic purposes, ancient Roman architecture isolates structures from different actions of the ground in order to passivate the range of vibrations, the strength of seismic shocks, organizing structures into a single monolithic organism.

So, the extensive historical-archaeological and historical-architectural material gives three main complexes of seismicity techniques for structures:

1. A system of excessive cluttering of structures with material in primitive megalithism.
2. The system of organizing mechanical structures with frame-colonnade and architrave-beam slab technique.
3. The system of organization of chemically cast structures on rigid mortars with arched-dome overlapping, united in a single unbreakable compact mass of the construction

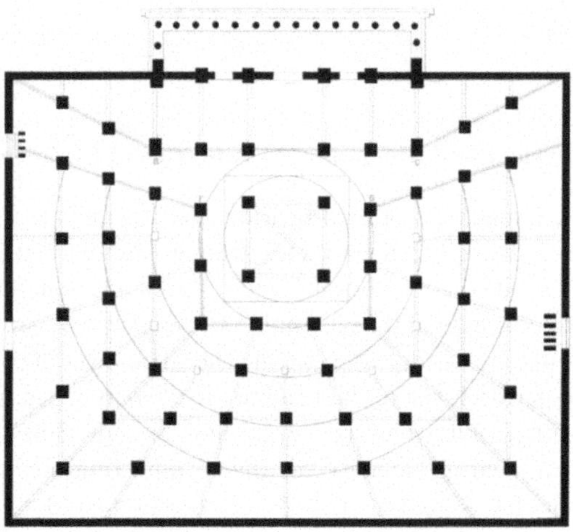

Fig. 1. Greece. Megalopolis. Tersilion. Plan (by Springer).

But we also see that in Babylonian and Assyrian architecture the asymmetry of buildings prevails in the layout when they are distributed on the same platform - the terrace (see Fig. 1). There is a variety in the change of facade lines in the places of transition from the end of buildings to verticals due to the extension of bricks (or stone) by beds or by masonry with inclined sections to resist the vertical pressure of walls and foundations. To counteract the overturning from the lateral spread of ancient architects discharged the thickness of the walls arcatures on the columns and arched niches, framed on the sides of the pilasters and columns. The capitals and entablature create elasticity or a kind of spring in the transfer forms of the floor in a system of bars organized one above the other.

The ceiling of an ancient building has its own strictly developed system based on the distribution of its gravity vertically without lateral spreading. The whole structure of the slab is based on the force of compression of individual parts and counteracting deflection, which is most characteristic as an anti-seismic protection.

The system of wall resistance is interesting, where the ancient Greek stone masonry does not use chemically bonding mortar: stone blocks, tightly fitted with their beds, are organized in the elastic body of the wall along the verticals of the masonry by a special type of pins, and horizontally - pierons; both pins and pierons were used and wooden and metal (bronze), specially organized for each material. Spikes were of the following types:

flat spike, bracket, zigzag, double-sided letter T. Pins and pierons protected masonry from shifts.

The arrangement of columns to the corners is more frequent due to the need to give the building in the corners the greatest stability; here, as if a bundle of vertical foundations is created. At seismic movement corners are antiseismic points - nodes on the perimeter of the building, which protect from excessive swaying and longitudinal and transverse parts of the building, and at that not only external, but also internal.

The system of seismic control in the ancient periods is built along the line of least resistance in order to preserve the integrity of even the smallest parts of the building. In strong earthquakes, the structure may sway and loosen, but minimally fracture and crush in parts. This is facilitated by careful material handling techniques and mechanical bonding (anchoring of parts) designed to rupture while minimizing fragmentation of the parts to be ruptured.

2.5 Modern Antiseismic Structural Solutions

Currently, technical solutions related to design in seismically hazardous areas on the territory of the Russian Federation are regulated by the set of rules "SP14.13330.2018 "Construction in seismic areas. Updated edition of SNiP II-7-81*" (approved and put into effect by the Order of the Ministry of Construction of Russia from 24.05.2018 N 309/pr)" [3] (Fig. 2).

Fig. 2. Seismic protection devices. Photo source: https://stroycomplex-5.ru/site-itselfplanradar-website.s3.amazonaws.com/production/uploads/2009/07/seism_kazakh.jpg.

Modern methods of construction of earthquake-resistant buildings:

1. "Soaring" foundation.
2. Shock absorbers.
3. Pendulum force.
4. Replaceable fuses.
5. Oscillating "core".
6. Alloys with shape memory effect.
7. Carbon fiber sheath.

3 Results and Discussion

In the depths of the Earth's crust, at a distance of about 25-60 km from the day surface, in the presence of an increasing stress state, the accumulated potential energy suddenly transforms into the kinetic energy of elastic vibrations, discharging on the surface by impacts, ruptures, slips, interrupted by friction, etc. As mentioned above, engineering structures, whether civil or artificial, in seismic areas experience during earthquakes the impact of additional seismic inertia forces in addition to the usual forces. The reality of these additional seismic forces and their specific weight in the total force field can be revealed by investigating the destructive effect of earthquakes of catastrophic nature on above-ground objects. These studies are mainly reduced to the study of materials, structures and workmanship of buildings and structures destroyed, partially damaged and preserved intact, as well as to the comparison and comparison of the data related to them.

The solution of the problem of providing seismic resistance of buildings and structures at the initial stage of their design depends to a great extent on the reliability of the information obtained during seismic effects. The main information about the movement of the base of structures during earthquakes is contained in the instrumental recording of accelerations (accelerograms).

Seismic impacts are random processes. Calculations for earthquake resistance under random vibrations are based on knowledge of the laws of distribution of seismic accelerations. For engineering solutions, these quantities are especially important because their generalization allows us to pass to the concept of random functions. Non-random quantities and characteristics can be obtained only on the basis of the study of histograms of accelerogram samples. Analyzing the conducted research, the following conclusions are obtained:

1. The most appreciable influence on the frequency of natural vibrations of underground structures is exerted by the coefficient of interaction with the ground.
2. At increase of ribs (stiffness elements) in the structures of the underground part of the building the frequency of natural vibrations increases.
3. Uniform change (increase/decrease) of the overall dimensions of the underground part does not affect the frequency of its free vibrations, but with the growth of the length of the structure the frequency of vibrations decreases.
4. The rheological properties of the material of the underground part structure lead to a shift in the phase of nonlinear oscillations.

The analysis of building failures made of different materials and structural systems allowed us to draw the following conclusions. The conditions contributing to the greatest degrees of damage in buildings from seismic effects are as follows:

- The general conclusion on chopped and frame type wooden buildings should be reduced to the fact that this type of buildings, one-storey and two-storey, with small, compared to conventional, additional costs can be made invulnerable even in the most exceptional in intensity earthquakes.
- The most sensitive place turned out to be the joints of floor beams, ties, struts. Reinforced concrete infill walls with weak cross-section cracked, brickwork infill walls

that did not have sufficient connection with the frame collapsed, infill walls made of hollow bricks completely collapsed. The behavior of reinforced concrete buildings is greatly influenced by their plan. It has been found that the destruction of a significant number of buildings, the shape of the plan, see Fig. 3, i.e., having internal corners, was more serious than in buildings with a square or rectangular plan (Fig. 4).

Fig. 3. Shape of destroyed buildings in plan.

Fig. 4. Castle of Duke Peter Frederick George of Oldenburg in Gagry. External view.

In modern conditions, at the stage of conceptual idea, architects need to analyze the data on the site of the proposed construction quite carefully. The sphere of geotechnical engineers includes comprehensive studies of soil strata, and designers - development of criteria for dimensional parameters of buildings and structures as technosphere objects, taking into account the heterogeneity of materials. “One of the modern tools to achieve this goal is the application of GIS-technologies. GIS capabilities for engineering-geological research are significant: they allow to accumulate, store and process large volumes of engineering-geological information, for example, to form an information resource about the composition and properties of the geological environment; they allow to solve the problems of typological engineering-geological zoning; architectural analysis of the design site, etc.” [4].

Information resources such as “Planet Earth”, “LAND VIEWER”, “EMSC” and “Google Map”, “GEOFON”, “EOS Land Viewer”, Central Asian Institute for Applied

Earth Research (CAIER) allow to quickly analyze the design site. The platforms provide a set of specific geospatial technologies for extracting valuable information from satellite data, such as:

- a map of the Earth's seismic activity (updates occur every 20 minutes);
- Sensor Data Storage System (SDSS), http://sdss.caiag.kg is a storage in the form of a database containing data from the sensors of the measuring stations, and serves for their use and management;
- a database of strong motion data collected by the ACROSS network. The web service provides the ability to view an interactive map with station or event locations, and to download available metadata and waveform processing results.
- The CAIIZ geodata platform (CAIIZ PD) is designed to collect, store and disseminate data and materials necessary for various works in the field of geoinformation systems and Earth remote sensing data processing.

Analyzing the data of Internet resources allows us to assess the quality of the territory to make an optimal decision on the choice of the form of the projected model of the architectural object in order to ensure the necessary safety.

4 Conclusions

Antiseismism of antique architecture was realized in general basic techniques for all its forms, and in individual, depending on the specificity of the terrain and architectural and functional appearance of the structure.

The general antiseismic techniques consisted of:

- in mechanical (pierons and pins) material connections;
- in buried sub-structures of the base and foundation;
- in the haphazard (sometimes polygonal) masonry of the base made of coarse rubble stone;
- in the construction of a two or three-stage stereobate from the hardest material with the organization of a responsible antiseismic belt in the stylobate (as a lower strapping);
- in the organization of a double antiseismic belt in the entablature (in the upper strapping at the floor level and at the roof level at the cornice);
- in differentiation (in the mass of substructure) of the base and foundation into individual vertical supports under external verticals (walls, pillars, supports) with tamping by substructures of the space between the supports;
- in the easily inclined profile of walls, pillars, columns, jambs of different openings, in the bundling of corner columns.

Pirons for horizontal bonding and pins for vertical bonding were connected to individual stone blocks in a specially cut bed (exceeding the size of pins and pirons) and filled with soft lead so that the blocks would not collapse from the metal bond during seismic movements. Ancient architects avoided rigid structures. When planning the main parts of structures, they created antiseismic complex nodes, bundles and stability points in the most important places (in corners, on short facades).

During the Early and High Middle Ages and AD, examples of seismic isolation of buildings and structures are as follows:

1. Central Asia X century - application of reed mats as seismic insulators.
2. Yalta 1727 - wooden horizontal links - belts in the array of stone masonry on clay mortar.
3. Alupka 1848 – Vorontsov's palace made of diorite on lead sheets laid in the joints of masonry.

Seismic isolation techniques consisted in the device of plastic deformations, i.e., the device of dissipative, passively absorbing earthquake energy in contrast to primitive seismic isolation systems.

Nowadays, until the 1960s, objects were built according to strict rules of classical constraints - correct shape, symmetry, calculation for small values of horizontal seismic forces.

1. After 1960, the architectural configuration of buildings, with improved construction technology and materials, changed. The structure of buildings became more open, complex in plan and height, with high flexibility. As a result - these buildings did not have a reserve of seismic resistance, because the calculations were performed according to classical low design criteria.
2. In modern conditions the use of calculation programs without analytical approach at the preliminary stage of design will not give the expected result, because at the stage of conceptual idea development the general understanding of the problem of complication of building structure and configuration is not less important than analytical calculations. And, relying only on the requirements of the normative literature of technical regulations and calculation, it is impossible to assess the situation without taking into account the dynamic impact of the environment and anticipate (foresee) the response of the designed object.
3. The development of technical solutions aimed at increasing the period of free vibrations of the building is very relevant. At different position of the center of mass of the building the frequency of horizontal oscillations changes, and, consequently, in the same movement of the base there will be different generating forces and change of overturning moment.
4. Experiments have shown that:
5. The rigid building during earthquake vibrates as well as the surrounding soil. The attenuation of oscillation is intense.
6. An elastic building oscillates according to the laws of its own oscillation. Attenuation is least when the period of natural oscillation is greatest.
7. A medium stiff building occupies a middle ground between the first two types.

It can be argued, based on observations of the behavior of buildings during earthquakes, that the shorter the period of natural vibrations, the stronger the building. The period of natural vibrations of a building decreases with increasing degree of attenuation. Asymmetrical buildings are less susceptible to collapse because the regularity of the seismic wave is broken by the irregular geometry of the structural elements of the building. It is necessary to design the structural system of a building (structure) in such a way as to provide in advance for elements that can be restored after an earthquake and that will not lead to inadequate changes in the entire load-bearing capacity of the system, will not harm human life and material values [5–11].

Thus, we can make the following conclusions and describe the first steps at the stage of development of the conceptual model of an architectural object (see Table 3):

Table 3. First steps at the stage of development of a conceptual model of an architectural object.

no	Required data	Definition	Method of data acquisition
Step 1	Seismic information about the construction site	The nature and intensity of the forces that develop during an earthquake	GIS platforms
Step 2	Engineering geodetic and geotechnical conditions	The most or least favorable conditions for the location of the structure in geologic and microgeologic terms	GIS platforms
Step 3	Type of structural system	Most or least stable structures and their parts	Combination of analytical and interactive rapid analysis
Step 4	Types of coupling	Most or least favorable conditions for interrelations of separate parts of structures	Combination of analytical and interactive rapid analysis
Step 5	Principle of static operation	Influence of mutual arrangement of connections of objects	Combination of analytical and interactive rapid analysis

Theoretical studies were related to the identification of the most rational forms and configuration of buildings (structures) taking into account the influence of elastic and plastic deformation caused by seismic impact, by processing, comparing and summarizing systematized data on the examination of elements of the main types of structures and their parts, not only of damaged buildings and structures, but also, and perhaps mainly, of surviving buildings, by comparing them under the same or allowing comparison conditions of materials and structures.

The design of facilities under seismic hazard conditions should be based on the study of relevant engineering concepts and careful analysis of design factors. Proper understanding and analysis of practical experience is necessary, which is more important than blind application of mandatory regulatory requirements and limitations. The purpose of regulations is to provide control over the extent of failure. Attempts to include in the regulations the effect of building configuration on its earthquake resistance, fail. In addition, it is obvious that mere compliance with norms without an analytical approach to solving this problem will not give the expected result either.

Based on historical assumptions that explain the nature of the constraints or boundary conditions, it can be argued that the design of facilities having the following parameters should be carried out with the scientific support of a specialized organization and feasibility study:

1. Building structure with irregular shape in plan and height;

2. Buildings with an abrupt change in bearing capacity under horizontal loads;
3. Buildings with abrupt change in horizontal stiffness;
4. Unusual or novel structural elements;
5. Difficult landscape conditions;

References

1. Zolotareva, L.A., Komarova, S.E, Shesternina, D.I., Khansivarova, N.M.: Geotechnical and architectural studies of the territory in the water protection zone of the Temernik river at the predesign analysis stage. In: IOP Conference Series: Materials Science and Engineering International Scientific Conference "Construction and Architecture: Theory and Practice of Innovative Development", vol. 1083 (2021)
2. Arnold, K., Reitherman, R.: Architectural Design of Earthquake-RESISTANT Buildings. Stroyizdat, Moscow (1987)
3. Arnold, K., Reiterman, R.: Building Configuration and Seismic Design. Wiley, New York (1982)
4. SP 14.13330.2018 "Construction in seismic areas. Updated edition of SNiP II-7-81*" (approved and put into effect by Order of the Ministry of Construction of Russia from 24.05.2018 N 309/pr)
5. Hansivarova, N.M.: Problem statement for the development of an integral indicator of the resource potential of the geological environment. Nat. Tech. Sci. **11**(149), 131–135 (2020)
6. Dominguez, D., Muñoz, P., Muñoz, L.: Choosing the most suitable antiseismic device technology for high-rise buildings, by considering the mechanical response and environmental impact. Int. J. Civ. Eng. **21**, 825–838 (2023)
7. Yu, J., Zhenzhong, S., Zhangxin, H., Haoxuan, Li.: Optimization design of anti-seismic engineering measures for intake tower based on non-dominated sorting genetic algorithm-II. Front. Struct. Civ. Eng. **17**, 1428–1441 (2023)
8. Castellano, M.G., Pigouni, A.E., Marcolin, L., Infanti, S.: Seismic Isolation of Hospitals Through Fluid Viscous Dampers Combined with Isolators. In: Sadan, B., Tuzun, C., Erdik, M. (eds.) Seismic Isolation, Energy Dissipation and Active Vibration Control of Structures. WCSI 2023, vol. 533, pp. 374–385. Lecture Notes in Civil Engineering, Springer, Cham (2023)
9. Kolev, C.: The underestimated study of the geology and historical topography of the construction site as a factor in the cracking of a new building near a small river. In: Rotaru, A. (eds.) Knowledge Transfer in the Sustainable Rehabilitation and Risk Management of the Built Environment. KNOW-RE-BUILT 2021, pp. 67–77. Springer Series in Geomechanics and Geoengineering. Springer, Cham (2024)
10. Muradyan, V., Umarov, A., Mailyan, V.: Modern methods for increasing the seismic resistance of stone buildings. In: Beskopylny, A., Shamtsyan, M., Artiukh, V. (eds) XV International Scientific Conference "INTERAGROMASH 2022". INTERAGROMASH 2022, vol 574, pp. 3101–3109. Lecture Notes in Networks and Systems, Springer, Cham (2023)
11. He, Q., Liu, D., Lei, M., et al.: Analysis of the seismic isolation characteristics of the overall friction pendulum bearing (OFPB) of a pagoda under three types of seismic actions. Herit. Sci. **12**, 326 (2024)

Experimental Analysis and Model Study of the Content of Unfrozen Water in Silty Clay Subjected to a Cycle of Artificial Freezing-Thawing of an Underwater Tunnel

Kudryavtsev Sergey[1](✉) and Zhiming Li[2]

[1] Far Eastern State Transport University, Khabarovsk 680021, Russia
olgakudr56@mail.ru

[2] School of Architecture and Civil Engineering, Heilongjiang University of Science and Technology, Harbin, China

Abstract. The presence of unfrozen water within frozen soil is the primary factor responsible for the phenomenon of segregation frost heaving. Variations in its quantity directly affect the soil's pore distribution. In this study, saturated silty clay was chosen as the subject of investigation. A nuclear magnetic resonance (NMR) testing apparatus was utilized to measure the unfrozen water content in the soil at various temperatures and dry densities under freeze-thaw cycle. By analyzing the transverse relaxation time T2 distribution curve, a qualitative and quantitative assessment of the soil's pore distribution throughout the freeze-thaw process was conducted. Considering that the most established models were empirical, a characteristic curve was developed to illustrate the changes in unfrozen water content during freeze-thaw cycle, based on capillary theory and the Gibbs-Thomson equation. The least square method and experimental results were used to ascertain the model parameters and to verify the efficacy of the freeze-thaw characteristic curve. The analysis revealed a pronounced hysteresis effect in the unfrozen water content curve during freeze-thaw cycle. At the same sub-zero temperature, as dry density increases, the unfrozen water content in frozen soil tends to decrease, and the pore distribution undergoes significant alterations with temperature fluctuations. The proposed freeze-thaw characteristic curve can accurately predict the changes in unfrozen water content with temperature under various testing conditions, which is crucial for forecasting freeze-thaw deformation in engineering projects located in cold regions.

Keywords: Nuclear Magnetic Resonance · Unfrozen Water · T_2 Distribution Curve · Freeze-Thaw Characteristic Curve · Capillary Theory

1 Introduction

The rapid expansion of urban rail transit projects stands as a pivotal marker of urbanization's swift progression. According to the authoritative "Urban Rail Transit 2023 Annual Statistics and Analysis Report," China boasts 59 cities with subway networks,

S. V. Klyuev et al. (Eds.): ICRAAC 2024, LNCE 627, pp. 346–363, 2025.
https://doi.org/10.1007/978-3-031-82938-3_39

with the cumulative operational track length having surged to an impressive 11,224.54 km. Nevertheless, the construction of subways in water-rich strata near rivers and coastlines faces risks to safety due to the highly developed groundwater systems and intricate replenishment dynamics. Against this challenging backdrop, artificial ground freezing technology has demonstrated its viability and dependability through extensive validation and widespread application across numerous engineering projects [1–3]. During the construction process that employs freezing, the migration and replenishment of water within the strata create ice lenses, which substantially influence the deformation of the ground surface and are directly correlated with the safety and stability of the ground structures.

The presence of unfrozen water is the primary cause of soil segregation frost heaving. Research has indicated [4] that even when temperatures plummet to – 80 °C, traces of unfrozen water can still be identified. Currently, numerous freeze-thaw empirical models have been proposed to forecast the variations in unfrozen water content throughout the freezing process, including the Anderson two-parameter model, the Michalowski single-parameter model, and the Dall'Amico model. However, the Anderson model suggests that the unfrozen water content approaches infinity as temperatures near 0 °C; conversely, the Michalowski and Dall'Amico models tend to underestimate the unfrozen water content in soil. In light of these discrepancies, scholars have undertaken research to refine these models [5–7], but the physical significance of the fitting parameters within these models requires further elucidation.

The determination of the moisture content of frozen soil initially located in the pores due to unfrozen water during freezing is described and implemented in the «Fem models» software package by the finite element method Kudryavtsev S.A. [19–21].

Except in the case of a pure aquatic environment, the water within the soil varies from water to ice or from ice to water over a temperature range. In other words, not all water within the soil receives a phase change at a single temperature. The percentage of the volume of soil water that remains unfrozen at a certain temperature is called the unfrozen water content.

The number of ice inclusions and humidity of frozen soil due to unfrozen water in the volume of pores of frozen soil are determined according to expression 1.

$$i = (w_{tot} - w_w) \tag{1}$$

where - i is the ice content per unit volume of soil; w_{tot} is the pre-winter soil moisture.

The characteristic of the unfrozen water content of the soil, as a function of the temperature in frozen soils in the software module "Termoground" is determined as follows.

Based on the analysis of the existing dependencies of the coefficient of unfrozen water content in clay soils, the approximating functions were selected. These approximating functions are described by a general equation of the following form

$$K_w = \frac{a + b \cdot T}{1 + c \cdot T + d \cdot T^2} \tag{2}$$

where T is the temperature of the soil; *a, b, c, d* are empirical coefficients, the values of which range from 0.25–0.95. The value of the coefficient *a, b, c, d* vary from the number of plasticity of clay soils.

At a ground temperature above the freezing point T_{bf} (usually about 0..... – 3.2 °C), all water within the cooled soil layer is in an unfrozen state. Below the freezing point of the soil, part of the water remains in an unfrozen state, gradually turning into ice with a decrease in temperature. Finally, below some negative temperature of almost frozen usually – 3.0...... – 13.0 °C) all the water within the soil layer is frozen.

In conditions of deep groundwater occurrence, when seasonally freezing soils are moistened mainly due to atmospheric precipitation, to quantify the deformations of frost heaving, the value of the calculated pre-winter humidity is determined by the formula

$$w = \frac{w_n \Omega_{oc}}{\Omega_e} \tag{3}$$

where is w_n the average humidity of freezing soils obtained during surveys in the summer and autumn period.

The values of Ω_e and Ω_{oc} are determined from long-term average data.

The duration of the period t_e, day, is determined by the ratio

$$t_e = \frac{d_{fn}}{k},\ by\, t_e \leq 90 \tag{4}$$

where: d_{fn} is the standard depth of seasonal soil freezing; k is the filtration coefficient, m/day.

The approximate values of t_e for certain types of clay soils are: for sandy loam 0.5–1 month, for loam 2 months, for clay 3 months.

The number of ice inclusions and humidity due to unfrozen water in the volume of pores of frozen soil are determined according to Eqs. (1 and 2).

When the ground environment is completely frozen (that is, the content of unfrozen water is 0%), the heat turnover of the soil is represented by the constant volumetric heat capacity of the frozen soil. Similarly, when the ground environment is completely thawed or thawed (that is, the content of unfrozen water is 100%), the heat turnover of the soil is represented by the constant volumetric heat capacity of the unfrozen soil. However, when the ground environment undergoes a phase change (that is, the function of the unfrozen water content is greater than zero, but less than 100%), the slope of the function of the unfrozen water content, the volumetric water content of the ground environment, and the latent heat of phase transitions in the range of negative temperatures absorbed or given away by the soil due to changes in the phase of groundwater should be considered in determining heat rotations of the soil [Kudryavtsev S.A.].

The implemented software module "Termoground" uses the function of the content of unfrozen water in the soil to estimate the latent heat of phase transitions in the spectrum of negative temperatures absorbed or given away by the soil due to changes in the phase of groundwater. The slope of the curve represents an indicator of the change in volumetric heat capacity in the soil.

To preserve the soil's inherent structure during the assessment of its unfrozen water content, nuclear magnetic resonance (NMR) analysis has emerged as a pivotal technology. This method allows for the indirect determination of unfrozen water content by examining shifts in the T2 distribution curve. Researchers have undertaken numerous related studies [8–14] to decipher the fundamental mechanisms underlying the role of unfrozen water in frost heaving, uncovering a substantial correlation between the migration of unfrozen water and the dimensions of soil particles, particularly in high moisture, fine-grained soils. Furthermore, the unfrozen water content is influenced by a multitude of factors, such as the soil's initial water content, permeability, temperature, mineral composition, and freezing history. Among these, the initial water content, the temperature and soil dry density exert the most significant influence. Nevertheless, there is a relative scarcity of research examining the effect of soil dry density on unfrozen water content, and the principles governing this relationship necessitate further detailed investigation.

This paper focuses on saturated silty clay as its subject of study, employing nuclear magnetic resonance (NMR) technology to assess the content of unfrozen water in soil at varying temperatures and dry densities under freeze-thaw cycle. By analyzing the transverse relaxation time T_2 distribution curve, the article offers both qualitative and quantitative insights into the soil's pore distribution throughout the freeze-thaw process. Recognizing that many classical models are empirical in nature, the study aims to construct a characteristic curve that reflects the variation in unfrozen water content during freeze-thaw cycle, grounded in capillary theory and the Gibbs-Thomson equation. The model parameters will be determined using the least squares method in conjunction with empirical data, followed by validation of the freeze-thaw characteristic curve. The findings presented in this article are intended to serve as a valuable reference for research on frost heave and the construction of infrastructure in cold regions.

2 NMR Testing of Saturated Silty Clay

2.1 Sample Selection and Preparation

The soil sample was extracted from the foundation pit of a subway station located in Qunli District, Harbin. Based on the grain size distribution curve (Fig. 1) and the Building Foundation Design Code (GB50007–2011), the soil was classified as silty clay. Subsequent to this, the fundamental physical and mechanical properties of the soil were determined in accordance with the Geotechnical Test Method Standard.

The specific parameters obtained included the liquid limit, plastic limit, unit weight, compression modulus, cohesion, and internal friction angle, all of which are detailed in Table 1.

Figure 2 illustrates the temperature and freezing time curves for saturated silty clay samples with a dry density of 1.57 g/cm3 at − 8 °C. The curve delineates four distinct stages. During the supercooling phase, as the soil temperature drops, the water within has not yet solidified into ice. Upon reaching − 3.51 °C, ice crystals start to emerge. In the abrupt temperature change stage, the latent heat is released by the ice-water phase transition, which causes the soil temperature to rise sharply. This temperature surge continues until it hits − 1.99 °C, at which point the pore water within the soil begins to freeze, and the soil temperature stabilizes at the freezing point of silty clay. The reason why the freezing temperature less than 0 °C is that no salt treatment was carried out during the sample preparation process. The presence of salt in the soil sample has decreased its freezing temperature. Once all the pore water has frozen and crystallized, the soil sample proceeds to the continuous cooling stage.

Table 1. Basic physical and mechanical parameters of silty clay.

Liquid limit	Plastic limit	Density	Compression modulus	Cohesive force	Internal friction Angle
32.8%	18%	19.3 kg/m^3	5.19 MPa	29.8 kPa	13°

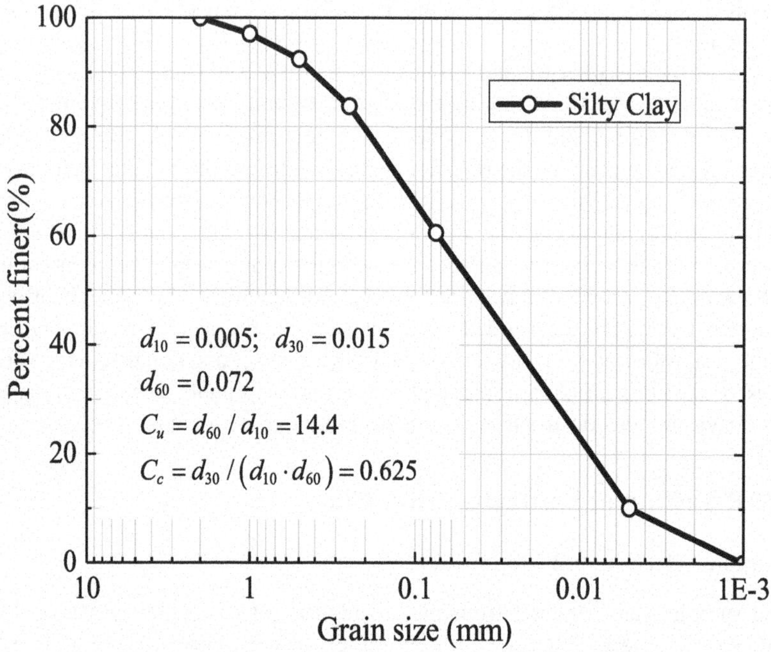

Fig. 1. Grain-size distribution of soil sample.

The preparation of soil samples for unfrozen water content determination using nuclear magnetic resonance (NMR) should adhere to the following procedure:

- Collect soil samples from the field and place them in an oven set at 105 °C for a duration of 24 h. Once dried, remove the samples and pulverize them to pass through a 2 mm standard sieve to eliminate larger particles. Subsequently, re-dry the samples to ensure complete moisture removal;
- Create saturated silty clay samples with dry densities of 1.47 g/cm^3, 1.57 g/cm^3, and 1.67 g/cm^3, which correspond to their respective saturated water contents of 37.8%, 33%, and 30.6%. The saturated water content is determined by drying samples that have undergone vacuum saturation for different dry densities;
- Compact the samples to be tested into a sample tube with an internal diameter of 9.04 mm and a height of 42.83 mm. Utilize a vernier caliper and a precise balance to regulate the sample's height and mass, thereby ensuring the desired dry density of the samples. After several trials, it is recommended to maintain the sample height at

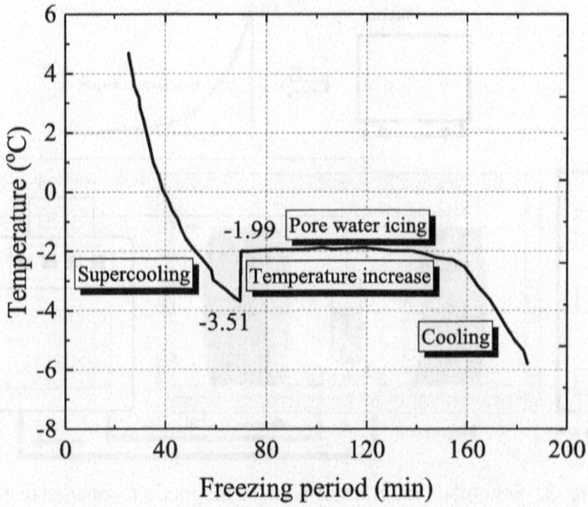

Fig. 2. Curve of the temperature and freezing time.

approximately 30 mm to guarantee adequate contact between the temperature probe and the sample.

2.2 Test Equipment

The experiment was conducted using the low-field nuclear magnetic resonance testing machine (NMRC12-010V). The test equipment's schematic diagram is depicted in Fig. 3. The testing machine is comprised of a magnet unit, radio frequency system, temperature control system (NMRC-LT35), data acquisition system, and a sample tube. The magnet unit houses a permanent magnet capable of producing a stable magnetic field across a specific temperature spectrum. Furthermore, the apparatus is equipped with a variable temperature probe, 10 mm in diameter, to ascertain the temperature of the sample under examination. The cooling liquid used is the inert and readily degradable fluorinated liquid FC-770, and the temperature control system boasts a precision of 0.01 °C.

The NMR experiments should be conducted in the following manner:

1. Gradually fill the ethylene glycol solution into the incubator to prevent bubble formation and to ensure the cooling effect is maintained;
2. Activate the radio frequency system at least 24 h prior to the commencement of the experiment to guarantee the generation of a consistent magnetic field;
3. Insert the standard sample tube into the testing apparatus to initiate the experiment, with the goal of minimizing any interference from the sample holder on the experimental outcomes;
4. Place the sample to be analyzed into the NMR magnet unit, ensuring the top of the sample is covered with cloth to avoid the dissipation of cold energy;
5. Initiate the data acquisition system and record transverse relaxation time T2 curves and free induction decay (FID) curves at the various temperatures specified.

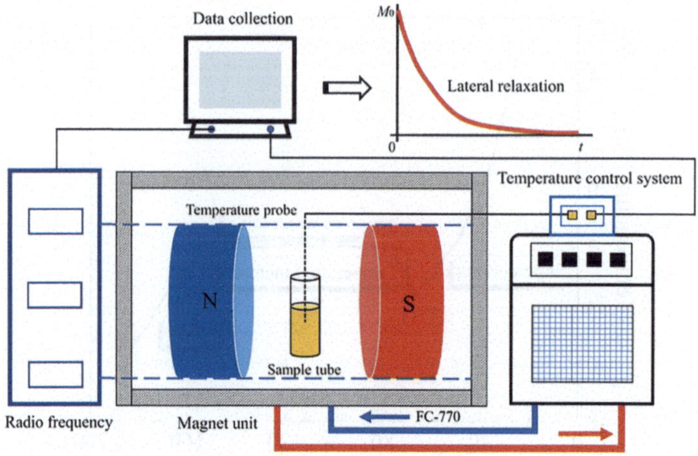

Fig. 3. Schematic diagram of nuclear magnetic resonance test.

2.3 Test Principle

Initially, align the randomly oriented hydrogen atoms in a uniform direction by applying a static magnetic field. Next, employ a radio frequency system to create an alternating magnetic field orthogonal to the static field, causing the hydrogen atoms to alter their directional alignment in response to this magnetic influence. Subsequently, when the alternating field is withdrawn, the hydrogen atoms revert to their original positions, a phenomenon referred to as relaxation. In experimental settings, the T2 transverse relaxation curve is typically selected to depict the relaxation process of hydrogen atoms, which corresponds to the expression of the free induction decay (FID) curve:

$$M_{\mathrm{XY}}(t) = M_{\mathrm{XY}}(0)e^{-t/\mathrm{T2}} \tag{5}$$

where $M_{\mathrm{XY}}(0)$ is the initial transverse magnetization vector; $M_{\mathrm{XY}}(t)$ is the transverse magnetization vector with relaxation time t.

The FID curve at each temperature can be directly derived from nuclear magnetic resonance testing. The peak intensity of the FID signal correlates with the amount of unfrozen water within the sample. However, this signal strength might encompass contributions from both liquid water and ice, potentially leading to an overestimation of the sample's unfrozen water content [15]. To address this issue, Curie's law is applied to adjust the temperature and ascertain the precise unfrozen water content. In the positive temperature range, the FID curve's peak value exhibits a linear correlation with temperature, as depicted in Fig. 4. Consequently, the calculated unfrozen water content is represented by the following expression:

$$W_u = \frac{W_0 F_m(T)}{(aT + b)} \tag{6}$$

where W_u and W0 are the unfrozen water content and initial water content, respectively; a and b are the linear fitting parameters; $F_m(T)$ is the maximum signal strength at different temperatures.

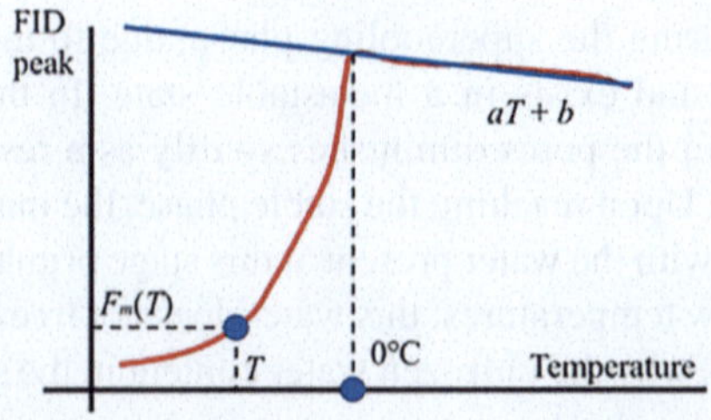

Fig. 4. Correction of unfrozen water content based on Curies law.

The T_2 distribution curve illustrates the evolution of pore water distribution during the freeze-thaw cycle, a transformation that can be derived through the Fourier transform of the FID curve. Within porous materials, the T_2 relaxation time of a hydrogen nucleus typically corresponds to the surface relaxation time, T_{2s}, and is represented by the following equation:

$$\frac{1}{T_2} = \frac{1}{T_{2s}} = \rho_2\left(\frac{S}{V}\right) = \rho_2\left(\frac{F}{R}\right) \tag{7}$$

where ρ_2 is the surface relaxation rate; R is the pore radius; F is the shape factor. When the pore is assumed to be columnar, F is equal to 2, and the above formula can be simplified as:

$$d = 48T_2 \tag{8}$$

where d is the pore diameter.

The transverse relaxation time T_2 curve in the equation above is correlated with the pore diameter, allowing for the quantification of the pore distribution in soils with varying diameters.

2.4 Test Design

The objective of this experiment is to examine how the unfrozen water content in saturated silty clay changes with varying temperatures and dry densities under the influence of freeze-thaw cycle. Throughout the testing process, the temperature was systematically reduced in the following sequence: 1, 0, – 0.5, – 1, – 1.5, – 2, – 2.5, – 3, – 4, – 5, – 6, – 8, – 11, – 14, – 17 °C. Subsequently, the temperature was increased in the reverse order, following the same sequence. Given the small dimensions of the sample, each temperature stage was sustained for a duration of only 15 min.

2.5 Test Results

Figure 5 illustrates the correlation between the unfrozen water content and temperature within saturated silty clay of varying dry densities throughout the freezing process. As depicted in the figure, the unfrozen water content variation curve during freezing can be broadly categorized into three phases: the supercooling phase, the rapid decline phase, and the stable phase. During the supercooling phase, due to the capillary effect, pore water remains unfrozen and exists in a metastable state. In the rapid decline phase, the unfrozen water within the pores diminishes swiftly as a result of the temperature-induced phase transition. Upon reaching the stable phase, the majority of the pore water has transitioned into ice, with the water present at this stage primarily being bound water. Even under extremely low temperatures, this water does not freeze due to the adsorption by soil particles, which is why the unfrozen water content in the stable phase is not zero.

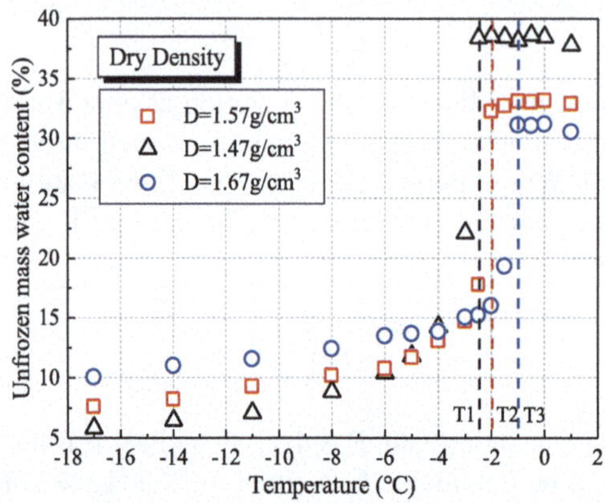

Fig. 5. Curve of unfrozen water content with temperature under different dry densities.

The initial saturated water content varies among samples with distinct dry densities. As dry density increases, soil pores become more compactly compressed, resulting in a lower initial saturated water content. Furthermore, the residual water content of samples with varying dry densities exhibits a gradual increase as dry density rises. This phenomenon occurs because smaller pores possess a relatively stronger capacity to adsorb surrounding water, causing water to transition from a free to a bound state, which is less prone to freezing. Additionally, Fig. 5 illustrates that as dry density increases, so does the freezing temperature of the soil mass. This is attributed to the fact that film water predominantly resides in larger pores, while capillary water is found in smaller ones. Given that film water has a lower freezing point than capillary water, samples with higher dry densities consequently exhibit higher freezing temperatures.

Figure 6 illustrates the variation of unfrozen water content against temperature throughout the freeze -thaw cycle. It is evident that the unfrozen water content within the sample exhibits an upward trajectory as the temperature ascends. Furthermore, the transition of unfrozen water content during the freezing and thawing phases manifests a pronounced hysteresis loop.

At identical temperatures, the unfrozen water content during the thawing phase is observed to be less than that during the freezing phase, with the hysteresis loop predominantly characterized by rapid thawing and rapid freezing stages. This phenomenon is attributed to alterations in the soil's pore structure throughout the freeze-thaw cycle, causing the thawing point during the thawing phase to exceed the freezing point during the freezing phase, a fact that can be substantiated by the Gibbs-Thomson equation.

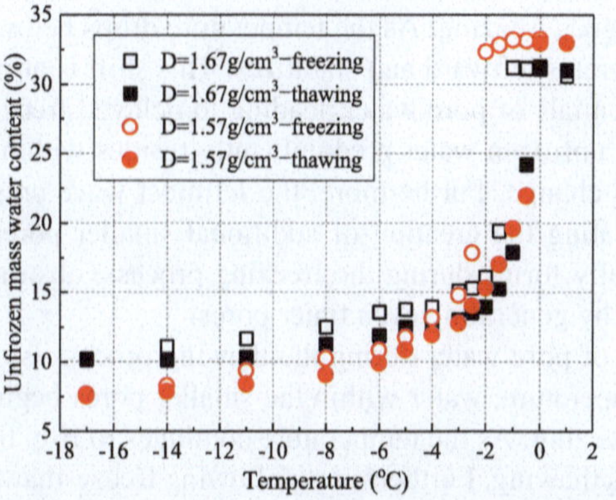

Fig. 6. Curve of unfrozen water content with temperature during freezing-thawing process.

Comparable hysteresis effects are also present in the adsorption and desorption processes of unsaturated soils. By comparing the freeze-thaw cycle with the wetting-drying cycle, hysteresis phenomena can be elucidated through concepts such as pore self-locking, capillary action, free energy barriers, contact angles, and electrolyte influences. Nonetheless, in prevalent freeze-thaw deformation calculations, most models assume a similar unfrozen water content throughout the freeze-thaw process, which may hinder wide application in the engineering.

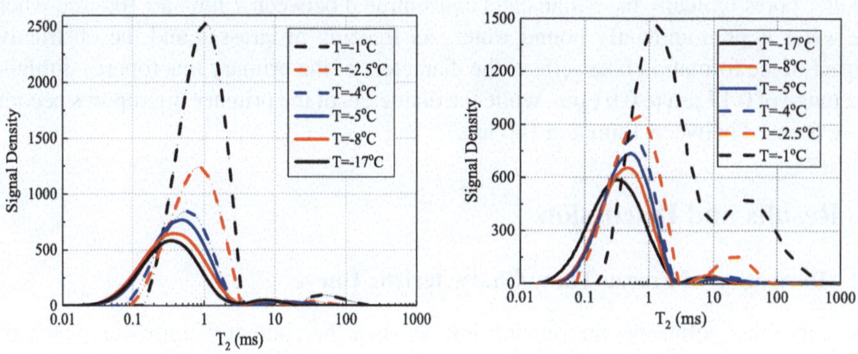

Fig. 7. Distribution of pore water in saturated silty clay during freezing and thawing.

Taking saturated silty clay with a dry density of 1.57g/cm^3 as an example, the distribution of pore water throughout the freeze-thaw process is further examined. T_2 curves at specific temperatures are depicted in Fig. 7. Notably, the T_2 value exhibits a linear correlation with pore radius. Consequently, the first and second peaks from left to right in Fig. 7 represent the alterations in small and large pores, respectively, during the freezing and thawing phases. The area delineated by the T_2 curve and the X-axis corresponds to the volume of pore water. At − 1 °C, the soil enters a supercooling phase, during which no pore water undergoes freezing. As the temperature drops below − 2.5 °C, the rightmost wave peak migrates leftward and stabilizes. This shift is attributed to the varying thermodynamic potentials of pore water, leading to delayed freezing in smaller pores. At this juncture, the unfrozen water predominantly resides within these smaller pores, exhibiting a gradual change. Furthermore, the leftmost wave crest also experiences a leftward shift, indicating the creation of additional smaller pores. This phenomenon occurs as ice gradually forms during the freezing process, obstructing portions of the pore space and thereby generating even tinier pores.

The distribution of pore water during the thawing process is depicted in Fig. 7(b). With increasing temperature, water within the smaller pores begins to thaw, a process that is contrary to freezing. As the temperature continues to rise, the water in the larger pores also undergoes thawing. Furthermore, following freeze-thaw cycle, the expansion of pores as water freezes leads to the destruction of the original pore structure, resulting in a substantial increase in the proportion of macropores and the total water content within them. The peak value of the T_2 curve associated with small pores shows a significant decrease, further confirming the impact of freeze-thaw cycle on soil structure.

The study presented above qualitatively examines soil pores throughout the freeze-thaw cycle. Utilizing the established correlation between pore diameter and transverse relaxation time, the quantitative alteration of pore size throughout the freeze-thaw process can be succinctly characterized. By integrating formula (4) with Fig. 7, the variation in pore diameter during the freezing phase is concisely elucidated. During the initial phase of freezing, larger pores predominantly exhibit diameters ranging from 1 μm to 10 μm, with the water within these pores primarily existing as free water. Conversely, smaller pores typically have diameters concentrated between 7 nm and 160 nm, where the water is predominantly bound water. As freezing progresses and the obstructive impact of ice formation takes effect, the diameters of the primary macropores diminish to a range of 0.17 μm to 0.91 μm, while the diameters of the primary micropores become more focused between 1 nm and 160 nm.

3 Results and Discussion

3.1 Derivation of Freeze-Thaw Characteristic Curve

The curve that delineates the relationship between the content of unfrozen water and temperature fluctuations throughout the freeze-thaw cycle is referred to as the freezing-thawing characteristic curve. Traditional models for this curve, including the Anderson two-parameter model [16] and the Michalowski single-parameter model [17], are characterized by their simplicity in terms of parameters and are extensively utilized in engineering applications. Nevertheless, these parameters are devoid of physical meaning and can only be determined through the least squares method. In certain instances, the correlation coefficient of the fitted results does not align closely with the experimental outcomes. Consequently, this paper proposes a new freezing-thawing characteristic curve, derived from capillary theory and taking into account the impact of soil dry density.

The water content in frozen soil that remains unfrozen is comprised of water within unfrozen pores, referred to as W_{uu}, and water within frozen pores, known as W_{uf}, expressed as:

$$W_u = W_{uu} + W_{uf} \tag{9}$$

Li et al. [15] confirmed through their research that the unfrozen water in frozen pores is negligible in comparison to that in unfrozen pores. Consequently, the subsequent derivation is exclusively for the expression of W_{uu}.

Regarding the unfrozen water within the soil's unfrozen pores, its behavior can be explained through capillary theory. This involves considering the soil's pores of varying diameters and conceptualizing them as a collection of capillary tubes, as depicted in Fig. 8. The diameter of these capillary tubes is indicative of the soil pore size, and varying capillary diameters correspond to distinct freezing points for the water within the pores. These relationships can be articulated using the Gibbs-Thomson equation, expressed as:

$$T_0 - T = \frac{2T_0\gamma_{il}}{\rho_i d L_f} \tag{10}$$

where L_f is the latent heat of ice/water; ρ_i is the ice density; γ_{il} is the free energy of ice water interface; T_0 is the freezing point of pure water.

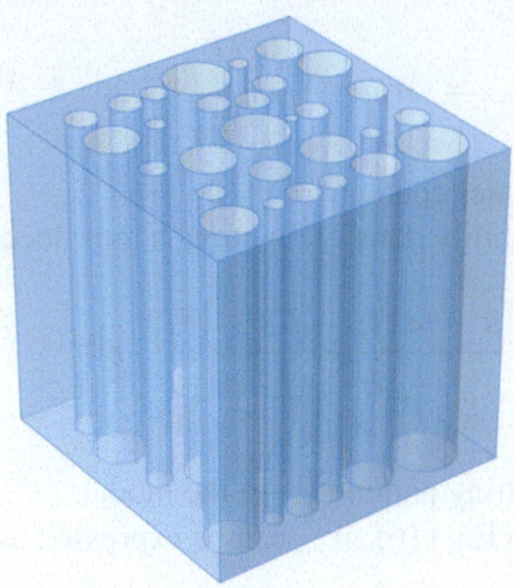

Fig. 8. Bundles of capillary tubes.

The relationship between temperature and capillary diameter can be obtained by adjusting the above formula:

$$d = g(T) = \frac{2T_0\gamma_{il}}{\rho_i L_f (T_0 - T)} \tag{11}$$

The pores within porous media are distributed in a random manner. Should the pore distribution function be represented as *f(d)*, the effective saturation S_l can thus be defined as:

$$S_l(d) = \frac{1}{V}\int_{d_{\min}}^{d} f(d)V(d)\mathrm{d}d \tag{12}$$

where *V(d)* is the pore volume corresponding to pore diameter *d*; *V* is the total pore volume of the soil.

Upon substituting formula (7) into the equation above, we obtain:

$$S_l(T) = \frac{1}{V}\int_{T_{\min}}^{T} f(g(x))V(g(x))g'(x)\mathrm{d}x \tag{13}$$

where T_{min} is the freezing temperature corresponding to pore diameter d_{min}.

Define the pore size probability density function F(x) as:

$$F(x) = \frac{1}{V} f(g(x))V(g(x))g'(x) \tag{14}$$

By substituting the above formula into Eq. (13), we get:

$$S_l(T) = \int_{T_{\min}}^{T} F(x)\mathrm{d}x \tag{15}$$

Effective saturation can also be expressed as a relationship with the unfrozen water content, and by combining Eq. (15), the expression for the unfrozen water content in the unfrozen pores can be obtained:

$$W_{uu} = (W_0 - W_r)\int_{T_{\min}}^{T} F(x)\mathrm{d}x + W_r \tag{16}$$

where W_r is the residual water content.

A proposed pore probability density function is introduced in this paper:

$$F(T) = \frac{\left[a - bc(T_0 - T)^{c-1}\right]}{\left\{1 + e^{\left[a(T_0 - T) - b(T_0 - T)^c\right]}\right\}^2} \cdot e^{\left[a(T_0 - T) - b(T_0 - T)^c\right]} \tag{17}$$

where *a, b,* and *c* are the fitting parameters in the model.

Substitute Eq. (17) into Eq. (16), W_{uu} can be expressed as:

$$W_{uu} = \frac{(W_0 - W_r)}{1 + \exp\left[a(T_0 - T) - b(T_0 - T)^c\right]} + W_r \tag{18}$$

The hysteresis effect observed in the unfrozen water content during the freeze-thaw cycle closely resembles that of water content variations in unsaturated soil undergoing wetting and drying cycles. Typically, the wetting curve for unsaturated soil is extrapolated from its drying curve. Drawing a parallel between the air-entry value of unsaturated soil and the freezing point of frozen soil, one can deduce [18]:

$$S_t(T) = 1 - \left[1 - S_f(T)\right]^p \tag{19}$$

where the feet *t* and *f* represent the freezing and thawing processes, respectively; *p* is the fitting parameter.

3.2 Verification of Freeze-Thaw Characteristic Curve. Model Parameter Determination Based on NMR Test

To validate the efficacy of the proposed freeze-thaw characteristic curve, this paper suggests beginning with the selection of freezing test outcomes depicted in Fig. 5. Subsequently, ascertain the model parameters, and proceed to compare the NMR test outcomes with the model's predictive results, as illustrated in Fig. 9. Additionally, Table 2 enumerates the model's fitting parameters.

The comparison of predictive outcomes with experimental data presented in Fig. 9 clearly demonstrates that the model introduced in this paper accurately forecasts the variations in the unfrozen water content of silty clay throughout the super-cooled, rapid decrease, and stable phases. The correlation coefficients for all scenarios exceed 0.985, thereby affirming the efficacy of the freezing characteristic curve. Turning to Fig. 10, which illustrates the comparison between experimental and predicted soil thawing characteristic curves, it is evident that the predicted and experimental curves are nearly identical. The correlation coefficients for all conditions are at least 0.959, further substantiating that the model proposed in this paper is capable of effectively predicting the changes in unfrozen water content during the freeze-thaw cycle.

Table 2. Value of the parameters in the model.

Parameter	$D = 1.67 g/cm^3$	$D = 1.57 g/cm^3$	$D = 1.47 g/cm^3$
W_0	30.6%	33.0%	37.8%
Wr	10.2%	7.7%	5.8%
a	0.3353	0.3097	0.2858
b	4848	9029	73600
c	− 23.86	− 11.02	− 10.29
R2	0.989	0.998	0.999

3.3 Relationship Between Model Parameters and Dry Density

As illustrated in Table 2 and depicted in Fig. 9, the parameter a in Eq. (14) represents the residual water content of the soil. A higher value of a signifies a greater residual water content, which ultimately tends to stabilize. The parameter b represents the freezing point, where a smaller value of b indicates a higher freezing point, but it will not exceed 0 °C. The parameter c represents both the rate of change of unfrozen water during the rapid decline phase and the freezing point. Typically less than 0, the greater the absolute value of c, the higher the freezing point and the faster the rate of change of unfrozen water.

Upon analyzing the curves depicting the unfrozen water content as a function of temperature for samples with varying dry densities, and taking into account the physical implications of the undetermined parameters in Eq. (18), it is possible to establish a

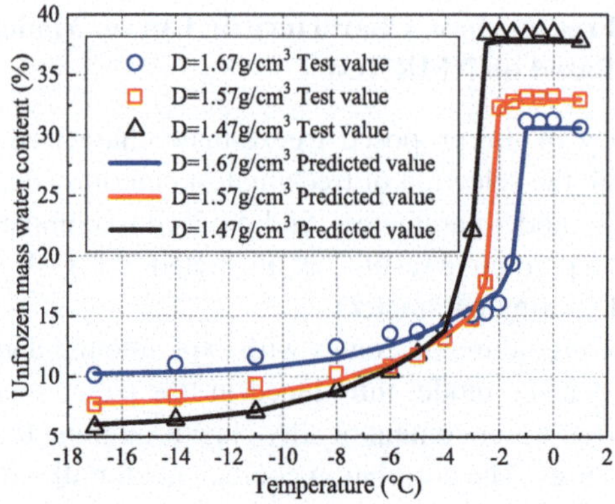

Fig. 9. Comparisons of the soil freezing characteristic curve between test and predicted results.

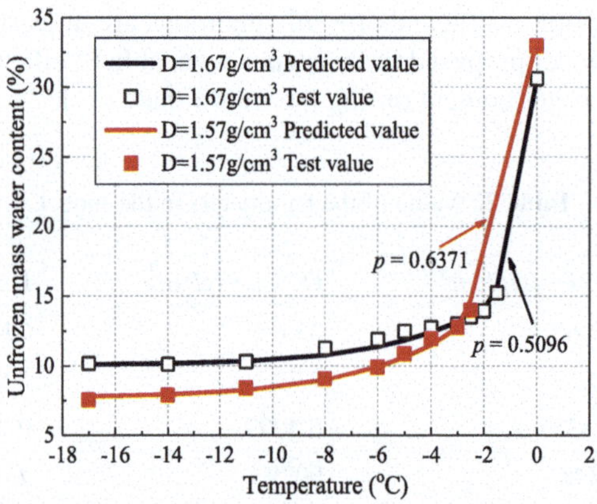

Fig. 10. Comparisons of the soil thawing characteristic curve between test and predicted results.

relationship between these parameters and the dry density. This relationship is expressed as follows:

$$a = 0.2475\rho_d - 0.0783 \tag{20}$$

$$b = 5.827 \times 10^{11}\rho_d^{-41.39} + 4496 \tag{21}$$

$$c = -5.688 \times 10^{-10}\rho_d^{46.6} - 10.25 \tag{22}$$

3.4 Effectiveness Analysis of Soil Freezing Characteristic Curve

To further validate the efficacy of the freezing characteristic curve introduced in this study, test data featuring dry densities of 1.67g/cm^3 and 1.47g/cm^3 were applied. Utilizing the Anderson two-parameter model, the Michalowski single-parameter model, and the model posited herein, the freezing characteristic curve was concurrently predicted. The outcomes of these predictions are depicted in Fig. 11. It is evident from the figure that, in comparison to other established models, the model presented in this paper more accurately predicts the unfrozen water content across various temperatures. This substantiates the effectiveness of the model proposed in this study.

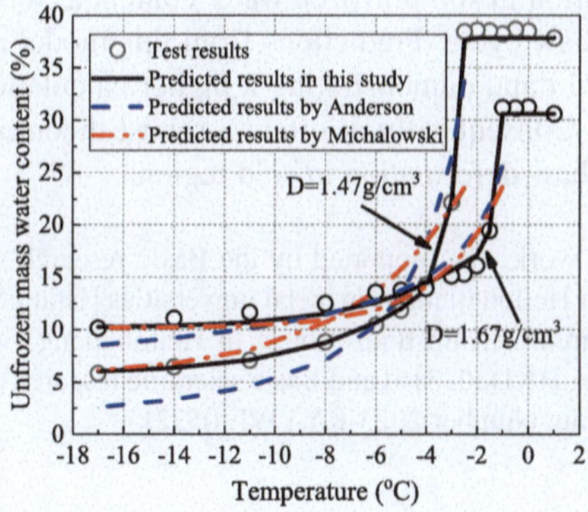

Fig. 11. Comparison of the predicted results by using different soil freezing characteristic curves.

4 Conclusions

Frost heave, caused by the migration of unfrozen water, is the primary cause of damage to buildings and infrastructure in cold regions. Consequently, investigating the variation of the unfrozen water content within frozen soil, particularly in soils prone to frost heave, holds significant theoretical importance and practical engineering value. This paper presents a freezing characteristic curve derived from capillary theory, which is substantiated by nuclear magnetic resonance testing. Additionally, a thawing characteristic curve is established by contrasting the unsaturated wetting and drying processes with freeze-thaw cycle. The physical implications of the model parameters within the freeze-thaw characteristic curve are elucidated, and the model's validity is confirmed. Drawing from these findings, the following conclusions are deduced:

The curve illustrating the variation of unfrozen water content with temperature during the freezing process exhibits three distinct phases: the sub-cooling stage, the rapid decline stage, and the stable stage. Conversely, the thawing process comprises only two phases: the rapid thawing stage and the complete thawing stage. A pronounced hysteresis loop is observed between the rapid freezing and thawing stages, highlighting the differential freeze-thaw characteristics of soil. This further underscores the necessity of accounting for the hysteresis effect in the freeze-thaw characteristic curve when calculating freeze-thaw settlement.

As the initial dry density increases, the freezing point approaches 0 °C more closely and the residual water content rises. By examining the T_2 transverse relaxation curve for a particular dry density, it becomes evident that the T_2 value correlates directly with the size of the soil pores. Moreover, the variation in pore size exhibits a pronounced temperature dependency throughout the ongoing freeze-thaw cycle.

The freeze-thaw characteristic curve, developed in accordance with capillary theory and the Gibbs-Thomson equation, incorporates the probability distribution of soil pores. The parameters specifically represent the freezing temperature, residual water content, and freezing rate, each of which holds significant physical meaning. This model effectively captures the variation in soil unfrozen water content as a function of temperature throughout the freeze-thaw cycle. Predictions from this model align closely with corresponding experimental data, demonstrating a higher calculation accuracy than classical empirical models. Consequently, it offers a robust theoretical foundation for the computation of freeze-thaw deformation in cold regions.

Acknowledgments. This work was supported by the Basic research support program for outstanding young teachers in Heilongjiang provincial universities [Grant Number YQJH2023092], the high-level talent research start-up fund project in Heilongjiang University of Science and Technology [Grant Number HKD202316] and Basic scientific research expenses of Heilongjiang provincial universities [Grant Number 2023-KYYWF-0522].

References

1. Alzoubi, M., Alzoubi, M., Xu, M., et al.: Artificial ground freezing: A review of thermal and hydraulic aspects. Tunn. Undergr. Space Technol. **104**, 103534 (2020)
2. Cui, Z.D., Zhang, L.J., Xu, C.: Numerical simulation of freezing temperature field and frost heave deformation for deep foundation pit by AGF. Cold Reg. Sci. Technol. **213**, 103908 (2023)
3. Liu, Y., Li, K., Li, D., et al.: Coupled thermal-hydraulic modeling of artificial ground freezing with uncertainties in pipe inclination and thermal conductivity. Acta Geotech. **17**, 257–274 (2021)
4. Zou, Y., Jiang, H., Wang, E., et al.: Variation and prediction of unfrozen water content in different soils at extremely low temperature conditions. J. Hydrol. **624**, 129900 (2023)
5. Wang, C., Lai, Y.M., Zhang, M.Y.: Estimating soil freezing characteristic curve based on pore-size distribution. Appl. Therm. Eng. **124**, 1049–1060 (2017)
6. Bai, R.Q., Lai, Y.M., Zhang, M.Y., et al.: Theory and application of a novel soil freezing characteristic curve. Appl. Therm. Eng. **129**, 1106–1114 (2018)
7. Kebria, M., Na, S., Yu, F.: An algorithmic framework for computational estimation of soil freezing characteristic curves. Int. J. Numer. Anal. Meth. Geomech. **46**, 1544–1565 (2022)
8. Meng, F.S., Chen, Q., Du, J., et al.: Evaluating soil freezing characteristic curve models for predicting unfrozen water content in freezing soils. Phys. Chem. Earth **135**, 103697 (2024)
9. Lu, J.G., Pei, W.S., Zhang, X.Y., et al.: Evaluation of calculation models for the unfrozen water content of freezing soils. J. Hydrol. **575**, 976–985 (2019)
10. Ming, F., Li, D., Liu, Y.: A predictive model of unfrozen water content including the influence of pressure. Permafrost Periglac. Process. **31**, 213–222 (2020)
11. Hu, G., Zhao, L., Zhu, X., et al.: Review of algorithms and parameterizations to determine unfrozen water content in frozen soil. Geoderma **368**, 114277 (2020)
12. Teng, J., Zhong, Y., Zhang, S., et al.: A mathematic model for the soil freezing characteristic curve: the roles of adsorption and capillarity. Cold Reg. Sci. Technol. **181**, 103178 (2021)

13. Chen, Y., Zhou, Z., Wang, J., et al.: Quantification and division of unfrozen water content during the freezing process and the influence of soil properties by low-field nuclear magnetic resonance. J. Hydrol. **602**, 126719 (2021)
14. Li, Z.M., Zhou, Y.D., Tang, A.P., et al.: Hydro-thermal interaction properties of unsaturated frozen soil using modified soil freezing characteristic curve. J. Cold Reg. Eng. **38**(1), 04023023 (2024)
15. Li, Z.M., Chen, J., Sugimoto, M.: Pulsed NMR measurements of unfrozen water content in partially frozen soil. J. Cold Reg. Eng. **34**(3), 04020013 (2020)
16. Anderson, D.M., Tice, A.R.: Predicting unfrozen water contents in frozen soils from surface area measurements. Highway Res. Rec. **393**, 12–18 (1972)
17. Michalowski, R.L.: A constitutive model of saturated soil for frost heave simulations. Cold Reg. Sci. Technol. **22**(1), 47–63 (1993)
18. Zhou, Y., Zhou, J., Shi, X.Y., et al.: Practical models describing hysteresis behavior of unfrozen water in frozen soil based on similarity analysis. Cold Reg. Sci. Technol. **157**, 215–223 (2019)
19. Ulitskii, V.M., Paramonov, V.N., Sakharov, I.I., Kudryavtsev, S.A.: Bed – structure system analysis for soil freezing and thawing using the Termoground program. Soil Mech. Found. Eng. **52**(5), 240–246 (2015)
20. Kudriavtsev, S.A.: Numerical investigations of the thermophysical processes in seasonally freezing soils. Kriosfera Zemli **7**(4), 76–81 (2003)
21. Sakharov, I.I., Paramonov, V.N., Kudryavtsev, S.A.: The account of frost heave and thawing processes when designing road embankments in cold regions, Transportation Soil Engineering in Cold Regions. In: Proceedings of TRANSOILCOLD 2019. Series: Lecture Notes in Civil Engineering, vol. 49, pp. 19–24. Singapore (2020)

Powder Mixtures with Highly Dispersed Technogenic Fillers with Dense Particle Packing

T. B. Novichenkova, V. B. Petropavlovskaya(✉), K. S. Petropavlovskii, B. I. Margolis, and I. Yu. Nekrasova

Tver State Technical University, Emb. Af. Nikitina, 22, 170026 Tver, Russia
victoriapetrop@gmail.com

Abstract. The study of the granulometric composition of powder mixtures with highly dispersed technogenic mineral additives - fillers introduced into the composition allows controlling the properties of composites based on them. Their introduction into the composition of the gypsum binder allows screening the impact of water, increasing the operational properties of materials. Whereas the low density of the introduced additives provides the composite with low weight and high thermal properties. Optimization of the dispersed system was carried out according to the criterion of obtaining the densest packing of particles in the composition of a multicomponent mixture based on two carbon components of bottom ash with different dispersion and expanded perlite.

Keywords: Bottom Ash · Perlite · Particle Size Distribution · Packing Density · Particle Distribution · Optimization · Computer Modeling · Structure · Properties

1 Introduction

Modern approaches to the design of building materials are based on ensuring their high quality, energy efficiency and safety. The choice of a binder, filler or aggregate is associated with the fulfillment of the conditions of the principles of sustainable development. In the process of creating innovative building materials, researchers are faced with the need to solve a number of issues, including the development of compositions, recipes, technological modes and many others [1, 2]. The formation of the internal space of dispersed systems requires a mathematically sound approach to calculating their granulometric composition [3, 4]. To do this, it is necessary to select the optimal ratio of components in the composition of dispersed systems using modeling to solve a complex mathematical problem. This approach allows for the optimization of multicomponent systems [5, 6] and the regulation of the properties of the resulting composite to improve its cost-effectiveness, durability, and energy efficiency [5, 6]. Currently, more and more mathematical and software tools are being introduced to simplify the process and reduce the time spent on solving optimization problems [7]. Correctly selected granulometry of inorganic filler when added to a composite binder provides the opportunity to improve the physical and mechanical characteristics and reduce the content of expensive and high-carbon binder [7, 8]. To impart certain properties to the composite, its internal

S. V. Klyuev et al. (Eds.): ICRAAC 2024, LNCE 627, pp. 364–371, 2025.
https://doi.org/10.1007/978-3-031-82938-3_40

structure is optimized by modeling its grain composition [9]. Modern concepts in the field of construction materials science are closely related to the development of new theoretical concepts in the field of compacted materials [3]. The selection of granulometry can determine the synthesis of compacted binder structures or the properties of the finished composite material [4, 6]. For this purpose, a cluster approach can be used to provide the necessary physical and chemical factors (Fig. 1). They can determine the formation of a structure with the maximum possible number of crystallization contacts (Fig. 2) between grains according to the selected method of their geometric arrangement (Figs. 2 and 3). For modeling such structures, it is effective to use binary systems [3], which are represented by a system of simple geometric solids – spheres (balls) of two sizes (Fig. 4). Such models describe well highly dispersed systems – industrial dusts, waste from sawing natural materials, pharmaceutical powders, etc. To study the behavior of such systems in various environments and under the influence of a large number of factors, various calculation algorithms are used, for example, the Drop and Roll algorithm (Figs. 4 and 5) [10, 11]. To optimize systems represented by grains with greatly varying sizes, other models are used – Fuller, Funk-Dinger, Belov V.V. and others [10, 12, 13]. They allow the selection of compositions of building mixtures and concretes. The aim of this work was to study the possibility of obtaining mixtures based on highly dispersed technogenic powders with the highest possible dense packing of particles along the Funk-Dinger curve using the developed computer automated complex.

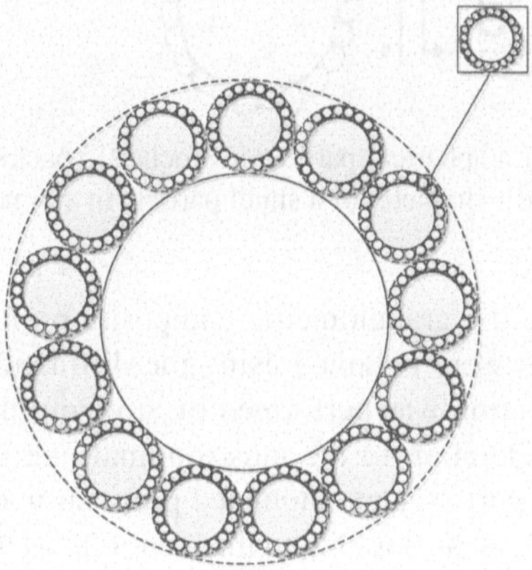

Fig. 1. Model of contact-crystallization structure based on dihydrate gypsum [3].

2 Methods and Materials

The composite included calcium sulfate hemihydrate of the α-modification. The grain size distribution was characterized by a residue on a 0.2 mm sieve according to GOST 125-2018 - 1%. A carbon component after processing the ash and slag mixture of the Kashira State District Power Plant (Fig. 5) of two degrees of dispersion and expanded perlite produced by Cemmix (Krasnoyarsk) (Fig. 6) were used as a dispersed additive.

Fig. 2. Regular dense packing of spherical particles of binder [3].

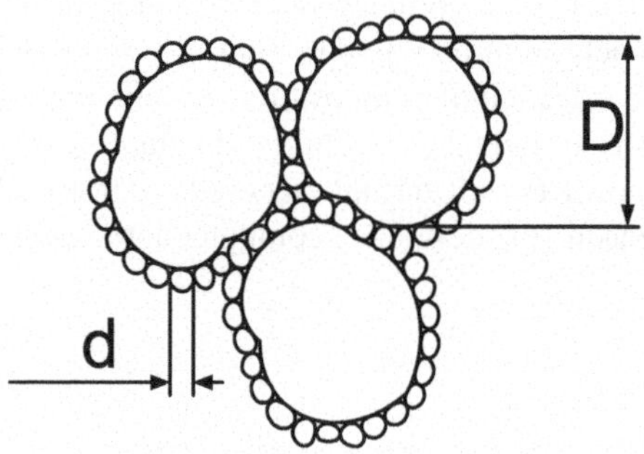

Fig. 3. Scheme of motion of a spherical particle in stochastic packing: D – diameter of a large particle in a binary mixture; d – diameter of a small particle in a binary mixture [7].

The necessary data on the granulometric composition of all the components used and the resulting mixtures were obtained using the diffraction laser particle analyzer «Analysette 22». Optimization was performed by studying the approximations of the compositions under consideration to the most optimal one according to the packing density criterion. For this purpose, an automated program was developed for selecting compositions that are as close as possible to the ideal Funk-Dinger curve.

3 Results and Discussions

Figure 8 shows a histogram of the distribution of particles of high-strength gypsum binder grade G-16 with a specific surface area of 592.2 m^2/kg, in general, it corresponds to the law of normal distribution, the integral curve is flat, without differences in size from 0.06 to 60 μm. However, the peak of the distribution is slightly shifted relative to the center towards large particles. The maximum of the distribution function is one and is within the range from 10 to 20 μm.

Figures 7, 8 and 11 show the histograms of particle distribution of two powders of technogenic carbon and expanded perlite. In all three cases, the distribution generally

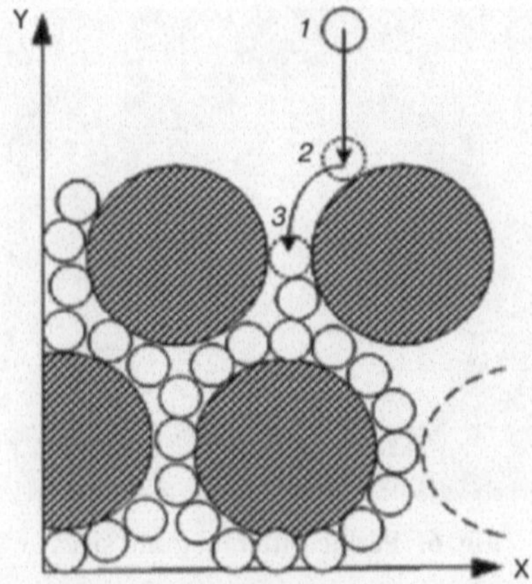

Fig. 4. Scheme of motion of a spherical particle in regular packing: 1 – initial position of the particle; 2 – position of the particle at first contact; 3 – stable position at the intersection point of three surfaces [7].

Fig. 5. Carbon concentrate.

corresponds to the normal law, the peaks of the distribution are slightly shifted relative to the center towards large particles. The maxima of the distribution functions are in the range from 20 to 40 μm. The integral curves are flat, without differences in size from 0.01 to 100 μm.

The conducted studies of optimization of granulometric composition of gypsum composites with highly dispersed fillers based on industrial waste of fuel industry are consistent with the data of studies of physical and mechanical characteristics of the obtained artificial gypsum stone. The composite is characterized by sufficient strength, low density and special internal structure (Figs. 9 and 10).

The designed compositions included gypsum binder, technogenic carbon component and expanded perlite. Optimization of the mixture compositions (Fig. 11) using an automated calculation program [14] for each of the mixture components made it possible to

Fig. 6. Perlite different grain sizes.

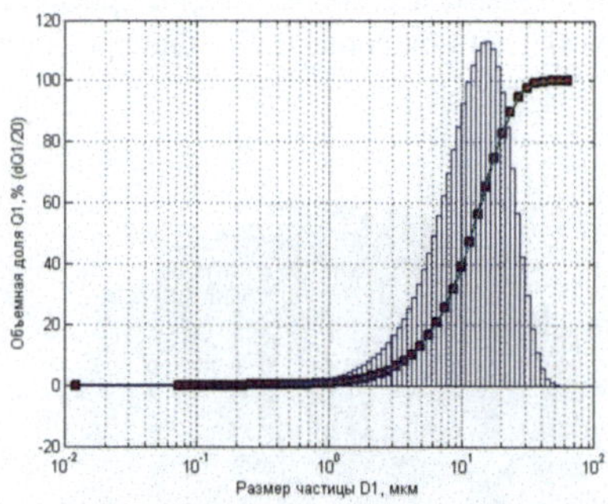

Fig. 7. Granulometric composition of gypsum binder.

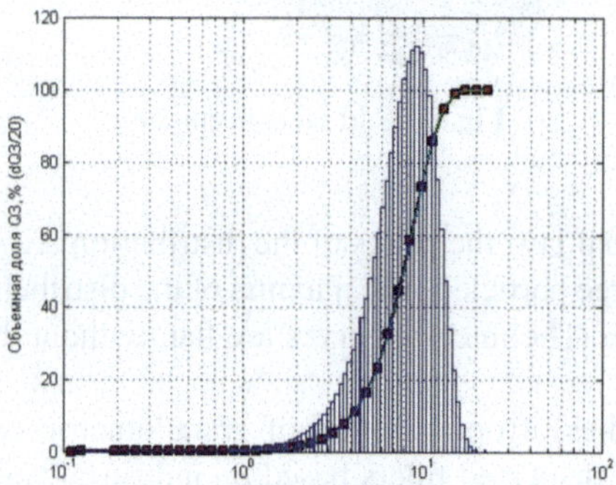

Fig. 8. Granulometric composition of unmilled carbon.

determine the best ratio between them according to the criterion of approaching the «ideal» Funk-Dinger curve. The model compositions were compared with the experimental data. The results obtained showed high convergence.

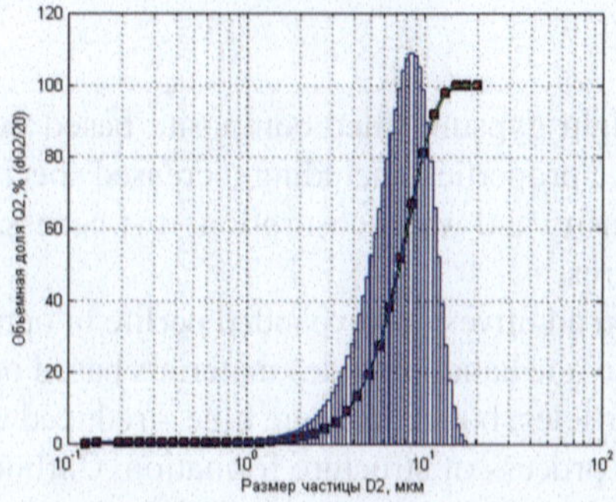

Fig. 9. Granulometric composition of ground carbon.

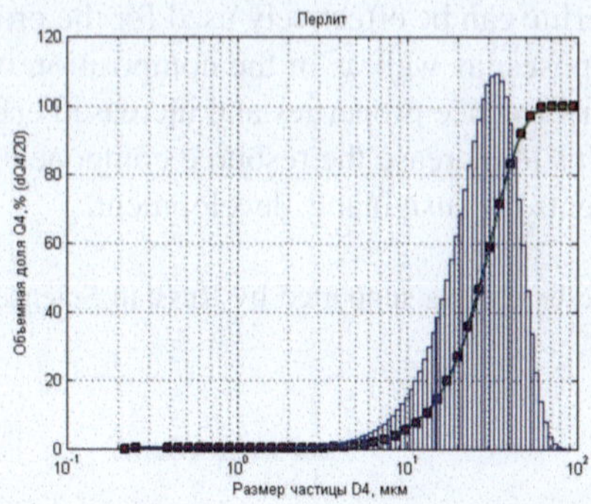

Fig. 10. Granulometric composition of perlite.

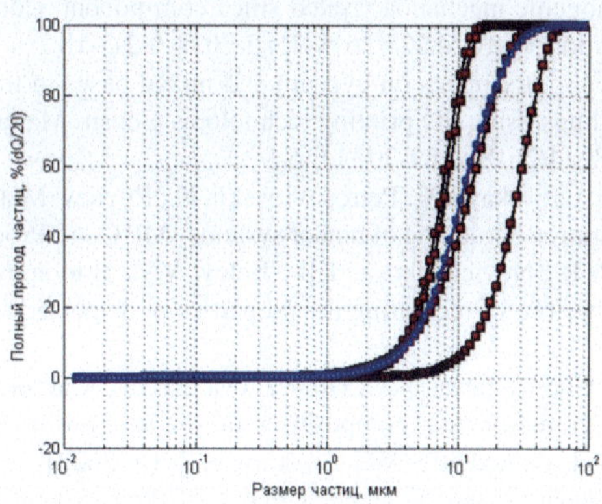

Fig. 11. Optimal granulometric composition of the mixture.

4 Conclusions

Thus, to form a lightweight gypsum filled composite based on a gypsum binder that has improved performance properties, including increased specific strength and thermal conductivity, reduced density and water absorption, it is necessary to use a complex of modifiers of various natures.

Introduction of carbon additives and expanded perlite in optimal ratios according to calculation models, on the one hand, provides materials based on the studied composite with dense packing of particles, but at the same time – reduced weight, and on the other hand – is reflected in the process of structure formation. Carbon technogenic additives contribute to formation of microporous structure of the composite with prevalence of pores less than 1 μm. This aspect is extremely for providing corrosion resistance and insulating properties of building materials. Gypsum composite with the addition of man-made carbon and perlite can be effectively used for the production of dry building mixtures for various purposes, as well as in the composition of building products for internal use with high performance properties and increased energy efficiency. The use of technogenic fillers helps to increase the resource component of the composite. This fully meets the requirements for sustainable development.

Acknowledgement. The research was supported by Russian Science Foundation (project No. 24-49-03004).

References

1. Klyuev, A.V., et al.: Experimental studies of the processes of structure formation of composite mixtures with technogenic mechanoactivated silica com-ponent. Constr. Mater. Prod. **6**(2), 5–18 (2023). https://doi.org/10.58224/2618-7183-2023-6-2-5-18
2. Kuznetsov, D.V., et al.: Dry mixes on gypsum and mixed bases in the construction of low-rise residential buildings using 3D printing technology. Constr. Mater. Prod. **6**(6), 5 (2023). https://doi.org/10.58224/2618-7183-2023-6-6-5
3. Nekrasova, I., Petropavlovskaya, V., Petropavlovskii, K.: Review: Mathematical modeling of the processes of hydration hardening of binder systems. AIP Conf. Proc. **2948**, 020049 (2023)
4. Petropavlovskaya, V.B., Novichenkova, T.B., Belov, V.V., Buryanov, A.F.: Granulometric composition as a criterion for regulating the properties of dispersed systems. Constr. Mater. **1**, 1–2 (2013)
5. Otman Azmi, S.A., Chernysheva, N.V., Drebezgova, M.Yu., Margolis, B.I., Novichenkova, T.B.: Influence of granulometric composition of the composite gypsum binder on its properties. Bull. BSTU named after V.G. Shukhov **6**, 8–16 (2023)
6. Buryanov, A., Petropavlovskaya, V., Novichenkova, T., Petropavlovskii, K.: Simulating the structure of gypsum composites using pulverized basalt waste. MATEC Web Conf. **117**, 00026 (2017)
7. Buryanov, A.F., Petropavlovskii, K.S., Petropavlovskaya, V.B., Novichenkova, T.B.: Formation of the spatial structure of a condensed system of calcium sulphate dihydrate. J. Phys: Conf. Ser. **1425**, 012194 (2020)

8. Petropavlovskaya, V.B., Zavadko, M.Yu., Korneev, A.I.: Modeling possibilities in the context of granulometry and particle packing density analysis in the development of waste modified building compositions. Innovative, information and communication technologies: proceedings of the XVIII International Scientific and Practical Conference. Moscow, 274–277 (2021)
9. Tkach, E.V., Soloviev, V., Temirkanov, R.I., Solovev, D.B.: The study of cement concrete with improved properties based on the use of astivated silica fume. Mater. Sci. Forum **992**, 228–232 (2020)
10. Klyuev, A.V., et al..: Development of alkali-activated binders based on technogenic fibrous materials. Constr. Mater. Prod. **6**(1), 60–73 (2023). https://doi.org/10.58224/2618-7183-2023-6-1-60-73
11. Lemeshkin, A.V.: Integrative methodologies in modern construction: a holistic approach to economic efficiency and technological progress. Universum: Tech. Sci. **12**(117) (2023)
12. Kastornykh, L.I., Gikalo, M.A., Kaklyugin, A.V., Serebryanaya, I.A., Kuzmenko, D.V.: Modeling of accelerated hardening of self-compacting concrete by methods of mathematical experiment planning. Constr. Mater. **3**, 25–30 (2024)
13. Petropavlovskii, K.S., Novichenkova, T.B., Petropavlovskaya, V.B.: Structure formation of self-reinforced gypsum. Mendeleev 2021. Book of abstracts XII International Conference on Chemistry for Young Scientists. Saint Petersburg (2021)
14. Margolis, B.I., Mansour, G.A.: A program for calculation of controller settings by the method of advanced frequency characteristics. Softw. Syst. **3**(31), 636–639 (2018)

Modeling of Creep of a Closed Cylindrical Shell Under Hydrostatic Pressure

Natalia Knyazeva[1], Vladimir Kuznetsov[2], Marcel Magomedov[3], Anton Popov[4], and Serdar Yazyev[5](✉)

[1] Moscow State University of Civil Engineering (National Research University), Moscow 129337, Russia

[2] Kh. Ibragimov Complex Institute of the Russian Academy of Sciences (CI RAS), Grozny City 364051, Russia

[3] Dagestan State University, Republic of Dagestan 367000, Makhachkala, Russia

[4] Kazan Federal University, Kazan 420008, Russia

[5] RUDN University, Moscow 117198, Russia

yazyev_sb@rudn.ru

Abstract. Objective. The paper presents general equations of the moment theory of a shell of zero Gaussian curvature taking into account creep deformation. The problem of the stress-strain state of a shell is considered, with the boundary conditions: rigidly fixed at the base and free edge at the top. The cylinder is subject to internal hydrostatic pressure. A resolving linear inhomogeneous differential equation of the fourth order with respect to deflection is obtained. The solution is given numerically analytically in the MATLAB software package. The nonlinear Maxwell-Gurevich equation is used as an equation of state between creep deformations and stresses. **Method.** To determine creep deformations, a linear approximation of the first derivative with respect to time was used, in other words, the Euler method. To verify the solution to the problem, a shell made of secondary PVC was calculated using the grid method [1]. The technique was tested by comparing the solution with the calculations of other well-known researchers. **Result.** A program for calculation in the MATLAB package with the ability to vary the initial data and output a graph of the dependence of displacements and stresses on time has been developed. As a result of the solution, it was found that during creep in the shell, circumferential stresses increase by 14.7%. **Conclusion.** The proposed approach can be applied to the analysis of the stress-strain state and bearing capacity of a reinforced concrete shell as well. There are no restrictions on boundary conditions and the type of loading, and the beam material can be not only polymers and composites for construction purposes, but also concrete.

Keywords: Concrete Creep · Inelastic Coefficients · Relative Deformation · Maxwell-Gurevich Constraint Equation · Bending Moments · Variable Stiffness · Krylov Functions

S. V. Klyuev et al. (Eds.): ICRAAC 2024, LNCE 627, pp. 372–384, 2025.
https://doi.org/10.1007/978-3-031-82938-3_41

1 Introduction

Polymers and composites are currently one of the promising structural materials in modern construction. This material is distinguished by a number of significant features that must be taken into account when designing building structures.

The available experimental studies have repeatedly confirmed that the effect of long-term loads leads to a change in the deformation process. The factors of force resistance of polymers and composites for construction purposes, taking into account the combined effect of nonlinearity and nonequilibrium deformation, static and dynamic loads, require additional study to solve statically indeterminate systems.

The development of new and improvement of existing methods for modeling and calculating various states and processes in polymer structures, issues of rheological calculation, especially nonlinear, for plates and shells remain relevant and are still one of the pressing problems.

2 Methods and Materials

2.1 Problem Statement

To calculate cylindrical tanks, of course, one can use the momentless theory of shells. But momentless shells only take up the load due to tension and do not assume the occurrence of bending moments and shear forces in them. However, it is far from always possible to ensure the conditions for the occurrence of a purely momentless state. Then it is necessary to apply a more complex calculation theory that takes into account, in addition to tensile deformations, the occurrence of bending deformations and creep deformations [2–5].

Let us consider a circular cylindrical tank of constant thickness (Figs. 1 and 2).

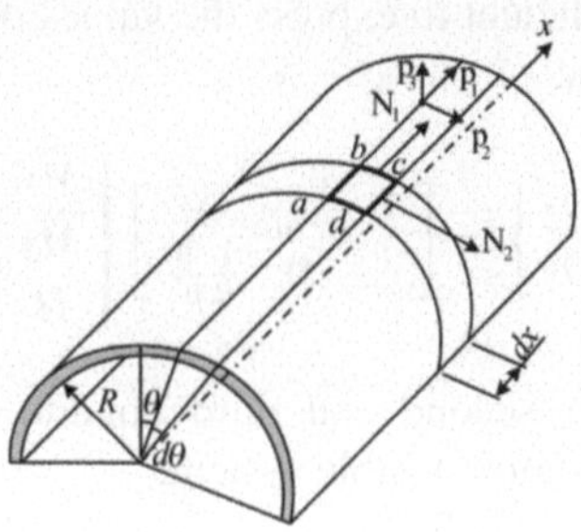

Fig. 1. Middle surface of the cylindrical shell.

This problem is axisymmetric, but first we will obtain the general equations for creep of a circular cylindrical shell under arbitrary loading.

2.2 Derivation of Resolving Equations

The Lame parameters for a circular cylindrical shell are:

$$A = \sqrt{1 + \left(\frac{dr}{dz}\right)^2}, B = r \rightarrow A = 1, B = R = const \tag{1}$$

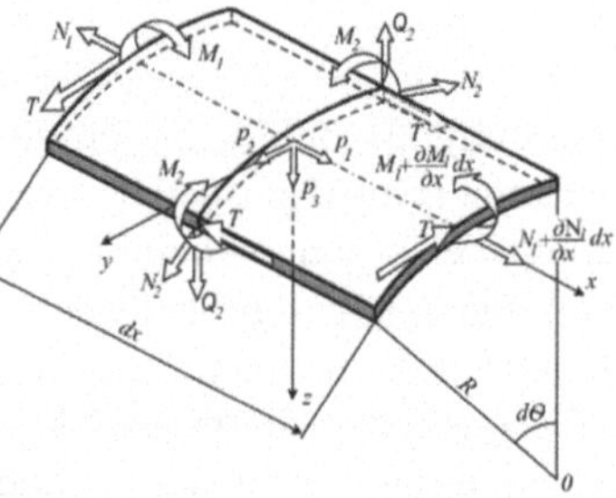

Fig. 2. Shell element in equilibrium under the action of internal forces.

Obviously, the main radii of curvature of the middle surface are the quantities $R_1 = \infty$, $R_2 = R$, where R is the radius of the shell. if we accept the values of the Lame parameters in them according to formula (1). The equilibrium equation has the form:

$$\begin{bmatrix} R\frac{\partial}{\partial x} & 0 & \frac{\partial}{\partial \theta} & 0 & 0 & 0 & 0 & 0 \\ 0 & \frac{\partial}{\partial \theta} & R\frac{\partial}{\partial x} & 0 & 1 & 0 & 0 & 0 \\ 0 & -1 & 0 & R\frac{\partial}{\partial x} & \frac{\partial}{\partial \theta} & 0 & 0 & 0 \\ 0 & 0 & 0 & 0 & -R & 0 & \frac{\partial}{\partial \theta} & R\frac{\partial}{\partial x} \\ 0 & 0 & 0 & -R & R\frac{\partial}{\partial \theta} & 0 & 0 & \frac{\partial}{\partial x} \end{bmatrix} \begin{Bmatrix} N_1 \\ N_2 \\ T \\ Q_1 \\ Q_2 \\ M_1 \\ M_2 \\ H \end{Bmatrix} + R \begin{Bmatrix} p_1 \\ p_2 \\ p_3 \\ 0 \\ 0 \end{Bmatrix} = \begin{Bmatrix} 0 \\ 0 \\ 0 \\ 0 \\ 0 \end{Bmatrix} \tag{2}$$

The system of five equations can easily be reduced to an equivalent system of three equations, for which it is sufficient to express the values of the transverse forces from the fourth and fifth conditions:

$$\begin{Bmatrix} Q_2 \\ Q_1 \end{Bmatrix} = \begin{bmatrix} 0 & \frac{1}{R}\frac{\partial}{\partial \theta} & \frac{\partial}{\partial x} \\ \frac{\partial}{\partial \theta} & 0 & \frac{1}{R}\frac{\partial}{\partial x} \end{bmatrix} \begin{Bmatrix} M_1 \\ M_2 \\ H \end{Bmatrix} \tag{3}$$

and substitute them into the second and third conditions (2). As a result of the transformations we arrive at a more visible system:

$$\begin{bmatrix} R\frac{\partial}{\partial x} & 0 & \frac{\partial}{\partial \theta} & 0 & 0 \\ 0 & \frac{\partial}{\partial \theta} & R\frac{\partial}{\partial x} & 0 & 1 \\ 0 & -1 & 0 & R\frac{\partial}{\partial x} & \frac{\partial}{\partial \theta} \end{bmatrix} \begin{Bmatrix} N_1 \\ N_2 \\ T \\ Q_1 \\ Q_2 \end{Bmatrix} + R \begin{Bmatrix} p_1 \\ p_2 \\ p_3 \end{Bmatrix} = \begin{Bmatrix} 0 \\ 0 \\ 0 \end{Bmatrix} \tag{4}$$

where N_1, N_2- longitudinal forces, $T-$ shear force, Q_1, Q_2- transverse forces, p_1, p_2, p_3- surface load components.

Geometric relationships according to [6, 7], deformations of the element of the middle surface can be established on the basis of general geometric dependencies:

$$\begin{bmatrix} \varepsilon_1 \\ \varepsilon_2 \\ \gamma \end{bmatrix} = \begin{bmatrix} \frac{\partial}{\partial x} & 0 & 0 \\ 0 & \frac{1}{R}\frac{\partial}{\partial \theta} & \frac{1}{R} \\ \frac{1}{R}\frac{\partial}{\partial \theta} & \frac{\partial}{\partial x} & 0 \end{bmatrix} \begin{Bmatrix} u \\ v \\ w \end{Bmatrix} + z \begin{bmatrix} 0 & 0 & -\frac{\partial^2}{\partial x^2} \\ 0 & \frac{1}{R^2}\frac{\partial}{\partial \theta} & -\frac{1}{R^2}\frac{\partial^2}{\partial \theta^2} \\ 0 & \frac{1}{R}\frac{\partial}{\partial \theta} & -\frac{2}{R}\frac{\partial^2}{\partial x \partial \theta} \end{bmatrix} \begin{Bmatrix} u \\ v \\ w \end{Bmatrix} \tag{5}$$

where $u, v, w-$ displacement of the middle surface in accordance with the coordinates x, θ, z.

In (5) the first term is the middle surface, the second term is the change in curvature. Let us introduce the following notations:

$$\begin{bmatrix} \varepsilon_1^0 \\ \varepsilon_2^0 \\ \gamma^0 \end{bmatrix} = \begin{bmatrix} \frac{\partial}{\partial x} & 0 & 0 \\ 0 & \frac{1}{R}\frac{\partial}{\partial \theta} & \frac{1}{R} \\ \frac{1}{R}\frac{\partial}{\partial \theta} & \frac{\partial}{\partial x} & 0 \end{bmatrix} \begin{Bmatrix} u_0 \\ v_0 \\ w \end{Bmatrix}; \begin{bmatrix} \chi_1 \\ \chi_2 \\ \chi_{12} \end{bmatrix} = \begin{bmatrix} 0 & 0 & -\frac{\partial^2}{\partial x^2} \\ 0 & \frac{1}{R^2}\frac{\partial}{\partial \theta} & -\frac{1}{R^2}\frac{\partial^2}{\partial \theta^2} \\ 0 & \frac{1}{R}\frac{\partial}{\partial \theta} & -\frac{2}{R}\frac{\partial^2}{\partial x \partial \theta} \end{bmatrix} \begin{Bmatrix} u_0 \\ v_0 \\ w \end{Bmatrix} \tag{6}$$

Physical relationships taking into account creep deformations based on the first hypothesis of the theory of calculating shells, as well as plates, the dependence of stresses on deformations is taken in the form:

$$\begin{Bmatrix} \sigma_1 \\ \sigma_2 \\ \tau_{12} \end{Bmatrix} = \frac{E}{1-\nu^2} \begin{bmatrix} 1 & \nu & 0 \\ \nu & 1 & 0 \\ 0 & 0 & \frac{(1-\nu)}{2} \end{bmatrix} \left(\begin{Bmatrix} \varepsilon_1 \\ \varepsilon_2 \\ \gamma \end{Bmatrix} - \begin{Bmatrix} \varepsilon_1^{cr} \\ \varepsilon_2^{cr} \\ \gamma^{cr} \end{Bmatrix} \right) \tag{7}$$

or

$$\begin{Bmatrix} \sigma_1 \\ \sigma_2 \\ \tau_{12} \end{Bmatrix} = \frac{E}{1-\nu^2} \begin{bmatrix} 1 & \nu & 0 \\ \nu & 1 & 0 \\ 0 & 0 & \frac{(1-\nu)}{2} \end{bmatrix} \left(\begin{Bmatrix} \varepsilon_1^0 \\ \varepsilon_2^0 \\ \gamma^0 \end{Bmatrix} + z \begin{Bmatrix} \chi_1 \\ \chi_2 \\ 2\chi_{12} \end{Bmatrix} - \begin{Bmatrix} \varepsilon_1^{cr} \\ \varepsilon_2^{cr} \\ \gamma^{cr} \end{Bmatrix} \right) \tag{8}$$

We will represent longitudinal forces as

$$\begin{Bmatrix} N_1 \\ N_2 \\ T \end{Bmatrix} = \begin{Bmatrix} \int_{-h/2}^{h/2} \sigma_1 dz \\ \int_{-h/2}^{h/2} \sigma_2 dz \\ \int_{-h/2}^{h/2} \tau_{12} dz \end{Bmatrix} = \frac{Eh}{1-\nu^2} \begin{bmatrix} 1 & \nu & 0 \\ \nu & 1 & 0 \\ 0 & 0 & \frac{(1-\nu)}{2} \end{bmatrix} \begin{Bmatrix} \varepsilon_1^0 \\ \varepsilon_2^0 \\ \gamma^0 \end{Bmatrix} - \begin{Bmatrix} N_1^{cr} \\ N_2^{cr} \\ T^{cr} \end{Bmatrix} \tag{9}$$

where

$$\begin{Bmatrix} N_1^{cr} \\ N_2^{cr} \\ T^{cr} \end{Bmatrix} = \frac{E}{1-\nu^2} \begin{Bmatrix} \int_{-\frac{h}{2}}^{\frac{h}{2}} \begin{bmatrix} 1 & \nu & 0 \end{bmatrix} \begin{Bmatrix} \varepsilon_1^{cr} \\ \varepsilon_2^{cr} \\ \gamma^{cr} \end{Bmatrix} dz \\ \int_{-\frac{h}{2}}^{\frac{h}{2}} \begin{bmatrix} \nu & 1 & 0 \end{bmatrix} \begin{Bmatrix} \varepsilon_1^{cr} \\ \varepsilon_2^{cr} \\ \gamma^{cr} \end{Bmatrix} dz \\ \int_{-\frac{h}{2}}^{\frac{h}{2}} \begin{bmatrix} 0 & 0 & \frac{(1-\nu)}{2} \end{bmatrix} \begin{Bmatrix} \varepsilon_1^{cr} \\ \varepsilon_2^{cr} \\ \gamma^{cr} \end{Bmatrix} dz \end{Bmatrix} \tag{10}$$

The relationships for bending and torsional moments are defined as:

$$\begin{Bmatrix} M_1 \\ M_2 \\ H \end{Bmatrix} = \begin{Bmatrix} \int_{-h/2}^{h/2} \sigma_1 z dz \\ \int_{-h/2}^{h/2} \sigma_2 z dz \\ \int_{-h/2}^{h/2} \tau_{12} z dz \end{Bmatrix} = -D \begin{bmatrix} 1 & \nu & 0 \\ \nu & 1 & 0 \\ 0 & 0 & (1-\nu) \end{bmatrix} \begin{Bmatrix} \chi_1 \\ \chi_2 \\ \chi_{12} \end{Bmatrix} - \begin{Bmatrix} M_1^{cr} \\ M_2^{cr} \\ H^{cr} \end{Bmatrix} \tag{11}$$

where

$$\begin{Bmatrix} M_1^{cr} \\ M_2^{cr} \\ H^{cr} \end{Bmatrix} = \frac{E}{1-\nu^2} \begin{Bmatrix} \int_{-\frac{h}{2}}^{\frac{h}{2}} [1 \; \nu \; 0] \begin{Bmatrix} \varepsilon_1^{cr} \\ \varepsilon_2^{cr} \\ \gamma^{cr} \end{Bmatrix} z dz \\ \int_{-\frac{h}{2}}^{\frac{h}{2}} [\nu \; 1 \; 0] \begin{Bmatrix} \varepsilon_1^{cr} \\ \varepsilon_2^{cr} \\ \gamma^{cr} \end{Bmatrix} z dz \\ \int_{-\frac{h}{2}}^{\frac{h}{2}} \left[0 \; 0 \; \frac{(1-\nu)}{2}\right] \begin{Bmatrix} \varepsilon_1^{cr} \\ \varepsilon_2^{cr} \\ \gamma^{cr} \end{Bmatrix} z dz \end{Bmatrix} \tag{12}$$

where $D = \frac{Eh^3}{12(1-\nu^2)}$ is the cylindrical rigidity.

Let us consider a case of practical importance – a closed cylindrical shell subject to the hydrostatic action of liquid on the walls of a vertical tank (Fig. 3).

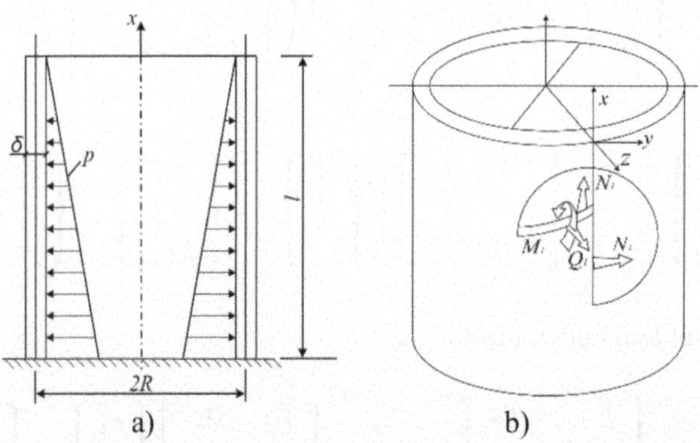

Fig. 3. Axisymmetric loading of a cylindrical shell internal forces.

In the presence of axial symmetry, the basic equations are significantly simplified. The shear force S and torque H vanish. The forces, deformations and displacements, due to symmetry, do not depend on the polar angle θ, and accordingly all derivatives with respect to θ in the indicated equations also vanish. The remaining forces shown in Fig. 3 b we assume that of the components of the surface loads, only one is non-zero: $p_3 = q = \gamma(l - x)$, where γ - specific gravity of liquid.

The equilibrium system takes the form:

$$\begin{cases} 1. \quad \frac{dN_1}{dx} = 0; \\ 2. \quad \frac{d^2 M_1}{dx^2} - \frac{N_2}{R} + q = 0 \end{cases} \tag{13}$$

From Eqs. (9) and (11) the following remain:

$$\begin{cases} 1.\ N_1 = \frac{Eh}{1-\nu^2} \cdot \left(\frac{du_0}{dx} + \nu \frac{w}{R} \right) - N_1^{cr} = 0 \\ 2.\quad N_2 = \frac{Eh}{1-\nu^2} \cdot \left(\frac{w}{R} + \nu \frac{du_0}{dx} \right) - N_2^{cr} \\ 3.\qquad M_1 = -D\frac{d^2w}{dx^2} - M_1^{cr} \\ 4.\qquad M_2 = -\nu D\frac{d^2w}{dx^2} - M_2^{cr} \end{cases} \tag{14}$$

From Eqs. (6) the following will remain for deformations:

$$\begin{bmatrix} \varepsilon_1^0 \\ \varepsilon_2^0 \end{bmatrix} = \begin{bmatrix} \frac{d}{dx} & 0 \\ 0 & \frac{1}{R} \end{bmatrix} \begin{Bmatrix} u_0 \\ w \end{Bmatrix}; \chi_1 = -\frac{d^2w}{dx^2} \tag{15}$$

From the first Eq. (14) we determine:

$$\frac{du_0}{dx} = \frac{1-\nu^2}{Eh} N_1^{cr} - \nu \frac{w}{R} \tag{16}$$

and substitute into the second Eq. (14):

$$N_2 = \frac{Ehw}{R} + \nu N_1^{cr} - N_2^{cr} \tag{17}$$

Finally, substituting (17) and the third relation from (14) into the second Eq. (13), we obtain the resolving equation for the deflection *w*:

$$D\frac{\partial^4 w}{\partial x^4} + \frac{Ehw}{R^2} = q + q^{cr} \tag{18}$$

where

$$q^{cr} = \frac{1}{R}\left(N_2^{cr} - \nu N_1^{cr}\right) - \frac{d^2 M_1^{cr}}{dx^2}$$

Border conditions:

$$if\ x = 0;\ w = 0;\ \frac{dw}{dx} = 0;$$

$$if\ x = h : M_1 = -D\frac{d^2w}{dx^2} - M_1^{cr} = 0;$$

$$Q_1 = \frac{dM_1}{dx} = -D\frac{\partial^3 w}{\partial x^3} - \frac{dM_1^{cr}}{dx} = 0 \tag{18a}$$

To calculate creep deformations, a uniform time grid is introduced. The value of $\varepsilon_{i,t+\Delta t}^*$ at time $(t + \Delta t)$ is calculated based on the Euler method [5]:

$$\frac{\partial \varepsilon_i^{cr}}{\partial t} \approx \frac{\varepsilon_{i,t+\Delta t}^{cr} - \varepsilon_{i,t}^{cr}}{\Delta t} \quad \Rightarrow \quad \varepsilon_{i,t+\Delta t}^* = \varepsilon_{i,t}^{cr} + \frac{\partial \varepsilon_i^{cr}}{\partial t}\Delta t \tag{19}$$

A similar approach to solving creep problems is used in [3, 8].

2.3 Calculation Method

The solution of the linear non-homogeneous differential equation of the fourth order (18) under the action of an axisymmetric hydrostatic load, which is distributed according to the linear law (20), can be carried out by various methods, including numerical ones [1, 10–12]:

$$p_3 = q = \gamma(l - x) \tag{20}$$

This article discusses the numerical-analytical method of integrating this equation and verifies it using a numerical method – the grid method [1].

Let us introduce the dimensionless coordinate ξ, related to the Cartesian coordinate x by the relation:

$$\xi = \psi x \tag{21}$$

Here ψ is the physical and geometric parameter of the shell.

$$\psi = \sqrt[4]{\frac{Eh}{4DR^2}} = \sqrt[4]{\frac{3(1-\nu^2)}{R^2h^2}}$$

In dimensionless coordinates, the resolving Eq. (18) takes the form:

$$\frac{\partial^4 w}{\partial \xi^4} + 4w = \frac{1}{\psi^4 D}\left\{q + \frac{1}{R}\left(N_2^{cr} - \nu N_1^{cr}\right) - \psi^2 \frac{d^2 M_1^{cr}}{d\xi^2}\right\} \tag{22}$$

When solving this problem, it is more convenient to use linear combinations, the so-called A.N. Krylov functions [10]. We write the solution to Eq. (22) as:

$$w = C_1K_1 + C_2K_2 + C_3K_3 + C_4K_4 + \overline{w}_0 + \overline{w}^{cr} \tag{23}$$

where

$\overline{w}_0$– particular elastic solution (22), $\overline{w}^{cr}$– creep-induced addition to a particular solution, K_1, K_2, K_3, K_4– A.N. Krylov functions [6].

$$K_1(\xi) = ch\xi \cdot cos\xi$$

$$K_2(\xi) = 0{,}5(ch\xi \cdot sin\xi + sh\xi \cdot cos\xi) \tag{24}$$

$$K_3(\xi) = 0{,}5ch\xi \cdot cos\xi$$

$$K_4 = 0{,}25(ch\xi \cdot sin\xi - sh\xi \cdot cos\xi)$$

The function $\overline{w}_0$ can be expressed through the A.N. Krylov functions by the integral:

$$\overline{w}_0 = \frac{1}{\psi^4 D}\int_0^{\xi} K_4(\xi - t) \cdot q(t) \tag{25}$$

Let us express the load (20) through dimensionless coordinates:

$$q(\xi) = \frac{\gamma}{\psi}(\lambda - \xi), \lambda = \psi l - \text{dimensionless shell length} \tag{26}$$

or taking into account (26), the function $\overline{w}_0$ has the form:

$$\overline{w}_0 = \frac{1}{4 \cdot \psi^5 \cdot D}(\lambda - \xi - \lambda K_1 + K_2) \tag{27}$$

For the function $\overline{w}^{cr}$ we write:

$$\overline{w}^{cr} = \frac{1}{\psi^4 \cdot D} \int_0^{\xi} \left(-\psi^2 \frac{d^2 M^{cr}}{d\xi^2} \bigg|_{\xi = t} + \frac{1}{R}\left(N_2^{cr}(t) - \nu N_1^{cr}(t)\right) \right) K_4(\xi - t) dt \tag{28}$$

The rotation angle is written as:

$$\varphi = \frac{dw}{dx} = \psi \left(-C_1 K_4 + C_2 K_1 + C_3 K_2 + C_4 K_3 + \frac{\gamma}{4 \cdot \psi^5 \cdot D}(-1 + 4\lambda K_4 + K_1) \right) + \frac{d\overline{w}^{cr}}{dx} \tag{29}$$

The bending moment is determined as follows:

$$M_1 = -D\frac{d^2 w}{dx^2} - M_1^{cr} = -\psi^2 D \left(-4C_1 K_3 + C_2 K_4 + C_3 K_1 + C_4 K_2 + \frac{\gamma}{\psi^5 \cdot D}(\lambda K_3 - K_4) \right) - -D\frac{d\overline{w}^{cr}}{dx} - M_1^{cr} \tag{30}$$

The expression for the transverse force takes the form:

$$Q_1 = -D\frac{\partial^3 w}{\partial x^3} - \frac{dM_1^{cr}}{dx} =$$

$$= -\psi^3 D \left(-4C_1 K_2 - 4C_2 K_3 - 4C_3 K_4 + C_4 K_1 + \frac{\gamma}{\psi^5 \cdot D}(\lambda K_2 - K_3) \right) - -D\frac{d^3 \overline{w}^{cr}}{dx^3} - \frac{dM_1^{cr}}{dx} \tag{31}$$

For the boundary conditions we write as follows. Substitute the boundary conditions (18a) into (23), (29)–(31), we get:

$$w(0) = 0 \rightarrow C_1 = \overline{w}^{cr} = 0 \tag{32}$$

$$\frac{dw(0)}{dx} = 0; \psi C_2 + \frac{d\overline{w}^{cr}}{dx}\bigg|_{x=0} = 0 \rightarrow C_2 = -\frac{1}{\psi}\frac{d\overline{w}^{cr}}{dx}\bigg|_{x=0} = -\frac{d\overline{w}^{cr}}{d\xi}\bigg|_{\xi=0}$$

$$M_1(\lambda) = 0 : -M_1^{cr}\Big|_{x=l} - \psi^2 \cdot D\left(\frac{4}{\psi}\frac{d\overline{w}^{cr}}{dx}\Bigg|_{x=0} K_4(\lambda) + C_3K_1(\lambda)\right.$$

$$\left. + C_4K_2(\lambda) + \frac{\gamma}{\psi^5 \cdot D}(\lambda K_3(\lambda) - K_4(\lambda))\right) - D\frac{d^2\overline{w}^{cr}}{dx^2}\Bigg|_{x=l} = 0 \quad (33)$$

$$Q_1(\lambda) = 0 : -\psi^3 \cdot D\left(\frac{4}{\psi}\frac{d\overline{w}^{cr}}{dx}\Bigg|_{x=0} K_3(\lambda) - 4C_3K_4(\lambda) + \right.$$

$$\left. C_4K_1(\lambda) + \frac{\gamma}{\psi^5 \cdot D}(\lambda K_2(\lambda) - K_3(\lambda))\right) - D\frac{d^3\overline{w}^{cr}}{dx^3}\Bigg|_{x=l} - \frac{dM_1^{cr}}{dx}\Bigg|_{x=l} = 0$$

Thus, to determine the constants C_3 and C_4 at each time step we have two equations with two unknowns.

To determine the derivatives

$$\frac{d\overline{w}^{cr}}{d\xi}, \frac{d^2\overline{w}^{cr}}{d\xi^2}, \frac{d^3\overline{w}^{cr}}{d\xi^3}$$

Let us use the method for determining the derivatives of creep deformations in [1] and present it in the same form:

$$\frac{d\overline{w}^{cr}}{d\xi} = \frac{1}{4}[e^{\xi}\cos\left(\xi - \frac{\pi}{4}\right) \cdot I_1 + e^{\xi}\sin\left(\xi - \frac{\pi}{4}\right) \cdot I_2$$

$$- e^{-\xi}\cos\left(\xi - \frac{\pi}{4}\right) \cdot I_3 - -e^{-\xi}\sin\left(\xi - \frac{\pi}{4}\right) \cdot I_4] \quad (34)$$

$$\frac{d^2\overline{w}^{cr}}{d\xi^2} = \frac{1}{4}[e^{\xi}\cos\left(\xi - \frac{\pi}{4}\right) \cdot I_1 + e^{\xi}\sin\left(\xi - \frac{\pi}{4}\right) \cdot I_2$$

$$- e^{-\xi}\cos\left(\xi - \frac{\pi}{4}\right) \cdot I_3 - -e^{-\xi}\sin\left(\xi - \frac{\pi}{4}\right) \cdot I_4] \quad (35)$$

$$\frac{d^3\overline{w}^{cr}}{d\xi^3} = \frac{1}{4}[e^{\xi}\cos\left(\xi - \frac{\pi}{4}\right) \cdot I_1 + e^{\xi}\sin\left(\xi - \frac{\pi}{4}\right) \cdot I_2$$

$$- e^{-\xi}\cos\left(\xi - \frac{\pi}{4}\right) \cdot I_3 - -e^{-\xi}\sin\left(\xi - \frac{\pi}{4}\right) \cdot I_4] \quad (36)$$

where

$$I_1 = \int_0^{\xi} f(t)e^{-t}\cos\left(t + \frac{\pi}{4}\right)dt;\ I_2 = \int_0^{\xi} f(t)e^{-t}\sin\left(t + \frac{\pi}{4}\right)dt$$

$$I_3 = \int_0^{\xi} f(t)e^{t}\cos\left(t + \frac{\pi}{4}\right)dt;\ I_4 = \int_0^{\xi} f(t)e^{t}\sin\left(t + \frac{\pi}{4}\right)dt$$

where

$$f(t) = -\alpha^2\frac{d^2M^*}{d\xi^2}\Bigg|_{\xi=t} + \frac{1}{R}\left(N_\theta^*(t) - \nu N_x^*(t)\right)$$

For the function $\overline{w}^{cr}$ we can also write:

$$\overline{w}^{cr} = \frac{\sqrt{2}}{8}\{e^{\xi}\sin\xi \cdot I_1 - e^{\xi}\cos\xi \cdot I_2 + e^{-\xi}\cos\xi \cdot I_3 + e^{-\xi}\sin\xi \cdot I_4\} \quad (37)$$

The integrals I_1, I_2, I_3, I_4 are calculated numerically using the Simpson formula.

To solve, a grid is introduced by coordinates x and z, as well as by time t. After determining the deflections from Eq. (18), the deformations are calculated using formulas (15) and (16). Then, the stresses are determined using formulas (8).

3 Results and Discussion

The calculation was performed in the MATLAB software package. To verify the correctness of the solution, we will use the data from [1].

The calculation was performed for a polymer tank made of recycled PVC:

$$h = 3\text{ cm}, l = 3\text{ m}, R = 2\text{ m}, \gamma = 10\ kN/m^3, E = 1480\text{ MPa}, \nu = 0.3.$$

The equation of state was taken as the nonlinear Maxwell-Gurevich equation [9].

Rheological parameters of PVC at a temperature of 20 °C: modulus of high elasticity $E_\infty = 5990$ MPa, modulus of velocity $m^* = 12.6$ MPa, initial relaxation viscosity $\eta_0^* = 9.06 \cdot 105 MPa \cdot min$. The shell was divided by length into $n_x = 100$ segments, by thickness into $n_z = 50$ intervals, by time into $n_t = 100$ steps. Table 1 shows the comparison performed in [1] to verify the correctness of the solution.

Table 1. Comparison of the largest displacements at different moments of time obtained by the finite difference method (FDM) and the numerical-analytical approach (NAA).

t, h		1	2	3	4	5	6	7	8
w, mm	FDM [1]	8.3732	8.7825	9.0182	9.1622	9.2532	9.3118	9.3500	9.3751
	NAA	8.3839	8.7946	9.0312	9.1758	9.2671	9.3260	9.3644	9.3896

The greatest discrepancy between the results is 0.15%, which indicates their reliability.

Figures 4 and 5 show the change in stress w_{max} over time and along the height of the cylinder at $t = 10$ hours, respectively. During creep, the maximum deflection increased by 23.5%.

Figure 6 shows the change in stress σ_1 and σ_2 at the base at $z = -h/2$, respectively. It is evident from the graphs that the stress σ_x remains practically constant, while the stress increased by 14.8%.

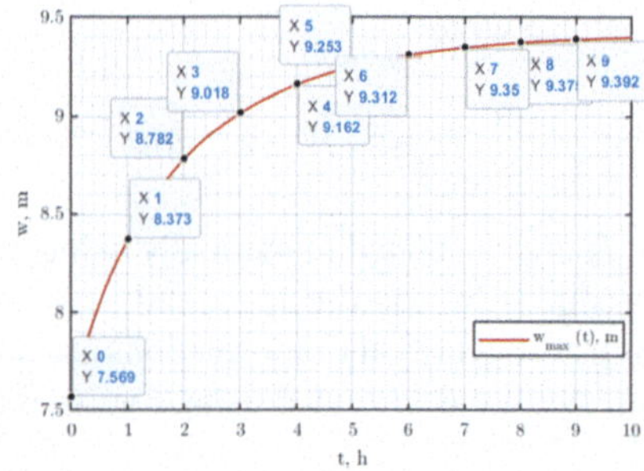

Fig. 4. Graph of the growth of the maximum deflection over time (numerical-analytical calculation).

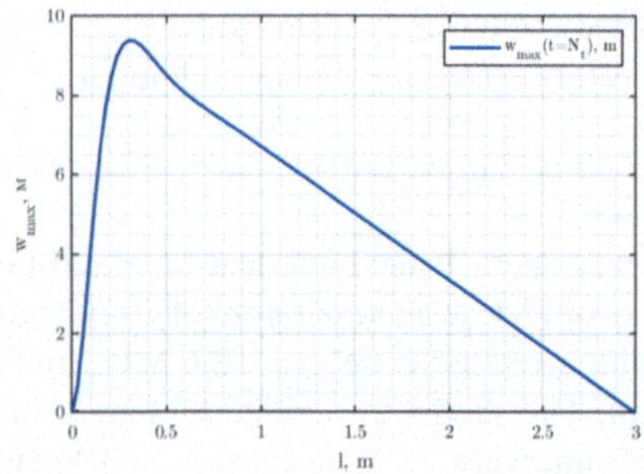

Fig. 5. Graph of change in maximum deflection w_{max} along the height of the cylinder at $t = 10$ hours.

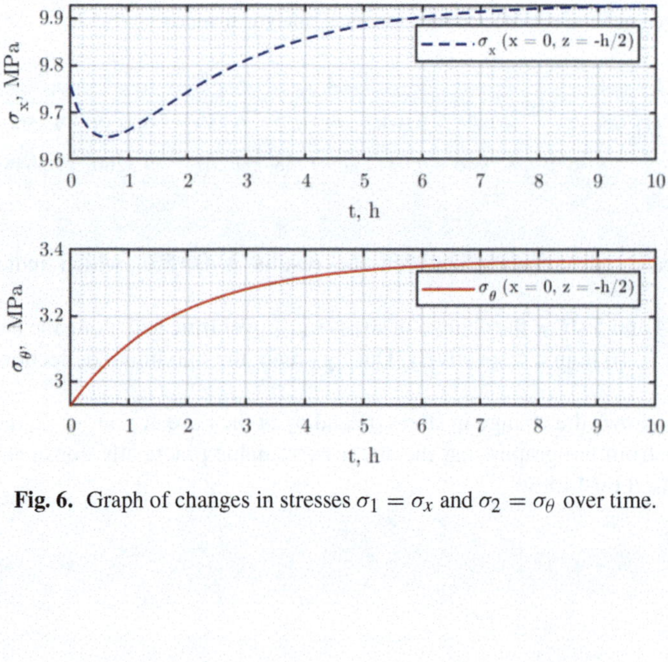

Fig. 6. Graph of changes in stresses $\sigma_1 = \sigma_x$ and $\sigma_2 = \sigma_\theta$ over time.

4 Conclusions

General equations of the moment theory of a circular cylindrical shell are given taking into account creep. The case of axisymmetric loading of a closed cylindrical shell is considered. The problem is reduced to a fourth-order differential equation with respect to deflection. The solution of the obtained equation was performed numerically using an analytical method using the example of a polymer shell made of recycled PVC.

A numerical-analytical approach to calculating axisymmetrically loaded shells for force effects taking into account creep is proposed. The solution is performed using Krylov functions, and integrals that include creep deformations are calculated numerically. Integration over time is also performed numerically using the Euler method. A comparison of the results obtained numerically and analytically and using the finite difference method is presented.

It is established that when internal hydrostatic pressure acts on the shell, the hoop stresses in it increase over time. This is due to the growth of the ring bending moments, which in turn is due to the fact that for polymers that obey the nonlinear Maxwell-Gurevich equation, the long-term Poisson's ratio is always higher than the instantaneous one. The growth of ring stresses in the base is explained by the growth in the absolute value of the ring bending moment, the graph of which over time is shown in Fig. 4. Thus, the calculation of the shell only in the elastic stage leads to underestimated stress values, and as a consequence, its possible destruction during operation.

References

1. Chepurnenko, A.S., Turina, V.S., Akopyan, V.F.: Processing of nonlinear concrete creep curves using nonlinear optimization methods. Constr. Mater. Prod. **7**(1), 2 (2024). https://doi.org/10.58224/2618-7183-2024-7-1-2
2. Chepurnenko, A.S., Turina, V.S., Akopyan, V.F.: Artificial intelligence model for predicting the load-bearing capacity of eccentrically compressed short concrete filled steel tubular columns. Constr. Mater. Prod. **7**(2), 2 (2024). https://doi.org/10.58224/2618-7183-2024-7-2-2
3. Andreev, V.I., Yazyev, B.M., Chepurnenko, A.S.: On the bending of a thin polymer plate at nonlinear creep. Adv. Mater. Res. **900**, 707–710 (2014)
4. Andreev, V.I., Chepurnenko, A.S., Yazyev, B.M.: Energy method in the calculation stability of compressed polymer rods considering creep. Adv. Mater. Res. **1004–1005**, 257–260 (2014)
5. Litvinov, S.V., Yazyev, B.M., Kuznetsov, V.V., Belyugin, N.V., Avakov, A.A.: Study of the concordance between various concrete deformation models and experimental data for uniaxial compression cases. Constr. Mater. Prod. **7**(5), 6 (2024). https://doi.org/10.58224/2618-7183-2024-7-5-6
6. Novoselov, O.G., et al.: Method for calculating the strength of massive structural elements in the general case of their stress-strain state (kinematic method). Constr. Mater. Prod. **6**(3), 5–17 (2023). https://doi.org/10.58224/2618-7183-2023-6-3-5-17
7. Chepurnenko, A.S., Yazyev, B.M., Savchenko, A.A.: Calculation for the circular plate on creep considering geometric nonlinearity. Procedia Eng. **150**, 1680–1685 (2016)
8. Babu Gunda, J., Gandule, R.: New rational interpolation functions for finite element analysis of rotating beams. Int. J. Mech. Sci. **50**(3), 578–588 (2008)
9. Krylov, A.N.: On the calculation of beams lying on an elastic foundation. USSR Academy of Sciences, Moscow (1931)

10. Lalin, V.V., Kolosova, G.S.: Numerical methods in construction. Publishing House of St. Petersburg State Technical University, St. Petersburg (2001)
11. Lou, T., Xiang, Y.: Numarical analysis of second-order effects of externally prestressed concrete beams. Struct. Eng. Mech. **35**(5), 631–643 (2010)
12. Milind, T.R., Date, P.P.: Analytical and finite element modeling of strain generated in equal channel angular extrusion. Int. J. Mech. Sci. **56**(1), 26–34 (2012)

Modeling the Strength of the Walls of I-Shaped Reinforced Concrete Beams

Tatiana Kondratieva[1(✉)], Dmitriy Vysokovskiy[1], Elizaveta Rusakova[1], Kazbek Khashkhozhev[2], and Karolina Poliakova[1]

[1] Don State Technical University, Rostov-On-Don 344000, Russia
ktn618@yandex.ru

[2] Kabardino-Balkarian State University Named After Kh.M. Berbekov, Nalchik 360004, Russia

Abstract. The article presents the research of the application of machine learning (ML) technologies for determining the strength of the walls of reinforced concrete I-beams. The current experience of using ML methods, such as neural networks, random forest and gradient boosting, is analyzed, the prospects of hybrid methods for predicting the degree of dependence and other strength characteristics of reinforced concrete structures are discussed. A hybrid model predicting the strength of the walls was developed based on the CatBoost algorithms and implemented in the Jupyter Notebook environment in Python. The model was trained on experimental data containing various parameters, designs and properties of materials, including the geometric dimensions of the beams, types of reinforcement and concrete class. Statistical analysis of the experimental sample for each instance was carried out. The regularization method was used to normalize the experimental data, optuna was used for their optimization, gridsearch, random search was used for the selection of hyperparameters. The data was checked for multicollinearity of features. Histograms of joint dependencies of variables were constructed. Practical and predicted values of the model were visualized in the form of a scatter plot. The most significant parameters affecting the strength of the walls of reinforced concrete I-beams were obtained. The constructed model passed the quality assessment according to three metrics: MAE = 0.2138; MSE = 0.4885; MAPE = 0.0033.

Keywords: Machine Learning · Catboost · Regression · Strength · I-Shaped Beams

1 Introduction

The development of machine learning models for the analysis of the strength of reinforced concrete I-beam walls allows to significantly improve the accuracy of predicting the strength characteristics of structures that withstand dynamic loads. Thus, in works [1–4] three hybrid machine learning models were developed for predicting the critical bending load of cellular I-beams. In works [5–8] when modeling the shear strength of reinforced concrete beams, artificial intelligence models are also used: classical (SVR); artificial neural network (ANN); decision trees with gradient reinforcement (GBDT).

S. V. Klyuev et al. (Eds.): ICRAAC 2024, LNCE 627, pp. 385–392, 2025.
https://doi.org/10.1007/978-3-031-82938-3_42

In work [9] a hybrid model of an artificial neural network based on the flower pollination algorithm (FPA-ANN) was built to optimize the design variables of reinforced concrete retaining walls of the cantilever type. Hybrid technologies were applied in [10] to evaluate the response of reinforced concrete walls (artificial neural network (ANN) and annealing simulation algorithm (ASA)) and in [11] to predict damage from buckling of steel columns under axial compression (adaptive neuro-fuzzy inference system optimized by genetic algorithm (ANFIS-GA) and adaptive neuro-fuzzy inference system optimized by particle swarm optimization (ANFIS-PSO).

At present, machine learning technologies are considered as one of the most promising methods for analyzing the strength of building structures, research in the field of materials science, and are used to solve problems when examining reinforced concrete structures for suitability for further operation, to predict strength under operational loads, to detect defects, etc. Taking into account the various stages of design and suitability for subsequent operation of materials in the construction field, these methods allow us to evaluate the chemical and physical properties of various structures in aggressive environments [12, 13]. However, the problem of assessing the strength of the walls of reinforced concrete I-beams has not been sufficiently studied by machine learning methods.

The objective of this study is to analyze the experimental training set and apply machine learning algorithms in predicting the wall strength of reinforced concrete I-beams.

2 Methods and Materials

Machine learning methods based on gradient boosting algorithms (CatBoost) have proven themselves well when working with numerical data of various categories [14–16]. The calculation of the strength of the walls of reinforced concrete I-beams is carried out based on the CatBoost algorithms implemented in the Jupyter Notebook environment in the Python language.

To train the model, the experimental data were prepared, including the following parameters: relative shear span a/h_0; inclination angle of transverse reinforcement α; wall thickness b; working height of the beam h_0; prism strength of concrete R_b; percentage of transverse reinforcement μ; modulus of elasticity of transverse reinforcement E; destructive transverse force Q. The total number of samples in the sample was 44 beams. For the beams, the load application point (a/h_0), section geometry (b and h_0), concrete strength R_b, reinforcement ratio μ and modulus of elasticity of transverse reinforcement E are changed. The experimental data on the beam characteristics are partially presented in Table 1.

The primary statistical analysis of the characteristics of this sample is presented in Table 2.

Figure 1 shows the correlation dependence between the features. Almost all independent variables have a very weak linear relationship, the maximum value of the correlation coefficient does not exceed 0.4. The exceptions are the variables "Wall thickness" and "Working height of the beam" $\rho_{b/h_0} = 0.94$. There is also a strong relationship between these parameters and the predictive variable of the model $\rho_{Q/b} = 0.91$ and $\rho_{Q/h_0} = 0.97$.

Table 1. Experimental data for training the model.

n/n	a/h_o	α	b, m	h_o, м	R_b, MPa	μ	E, MPa*10^6	Q, kN
1	1.60	90.00	0.03	0.26	31.90	0.59	170	88.70
2	1.60	90.00	0.03	0.26	26.90	0.59	170	68.10
3	1.60	90.00	0.03	0.26	29.80	0.59	170	81.80
4	3.20	90.00	0.03	0.26	28.00	0.59	170	52.10
5	3.20	90.00	0.03	0.26	24.80	0.59	170	48.90
6	3.20	90.00	0.03	0.26	31.90	0.59	170	61.90
7	3.20	90.00	0.03	0.26	30.80	0.59	170	59.60
8	3.20	90.00	0.03	0.26	26.90	1.64	170	62.90
9	3.20	90.00	0.03	0.26	30.80	1.64	170	78.10
10	3.20	90.00	0.03	0.26	32.40	1.64	170	82.70
11	3.20	90.00	0.03	0.26	33.00	1.64	170	84.10
12	2.00	90.00	0.06	0.35	14.25	2.78	200	120
13	2.00	90.00	0.06	0.35	22.71	3.29	200	225
…	…	…	…	…	…	…	…	…
33	3.2	45.00	0.03	0.26	25.7	1.0	65.3	46.3
34	3.2	45.00	0.03	0.26	28.7	1.3	65.3	56.2
35	2.5	90.00	0.03	0.26	28.9	1.0	170	65.1
36	2.5	90.00	0.03	0.26	28.9	1.3	170	74.5
37	3.2	90.00	0.03	0.26	29.7	1.0	170	62.5
38	2.5	90.00	0.074	0.563	34.3	2.72	200	453.1
39	2.5	90.00	0.063	0.563	29.3	3.19	200	369.1
40	2.5	90.00	0.064	0.563	33.5	1.77	200	416
41	2.00	90.00	0.04	0.33	16.00	1.75	200.00	102.50
42	2.00	90.00	0.04	0.33	16.00	1.36	200.00	95.00
43	2.00	90.00	0.05	0.33	18.00	1.63	200.00	106.90
44	3.00	90.00	0.05	0.32	20.00	1.46	200.00	114.90

Figure 2 shows the relationships between input and output parameters in the form of histograms. It shows a complete absence of multicollinearity of features, which indicates the same independence of features from the predictive variable.

When constructing the regression model in CatBoost, training data with features $X = \{x_1, x_2, x_3, x_4, x_5, x_6.x_7\}$ and the target variable Y were used, where the set of variables $x_i, i = \{1 : 7\}$ corresponds to the characteristics of the beams partially presented in Table 1, Y is the predicted parameter of the target function. The general formula for

Table 2. Statistical characteristics of the data.

index	a/h$_o$	α	b, m	h$_o$, m	R$_b$, MPa	μ	E, MPa*10^6	Q, kN
count	44.00	44.00	44.00	44.00	44.00	44.00	44.00	44.00
mean	2.46	78.75	0.04	0.32	26.28	1.28	138.41	142.22
std	0.63	19.71	0.02	0.14	5.03	0.82	59.87	211.03
min	1.60	45.00	0.03	0.26	14.25	0.40	65.30	42.70
max	3.20	90.00	0.10	0.83	34.30	3.29	200.00	1160.00

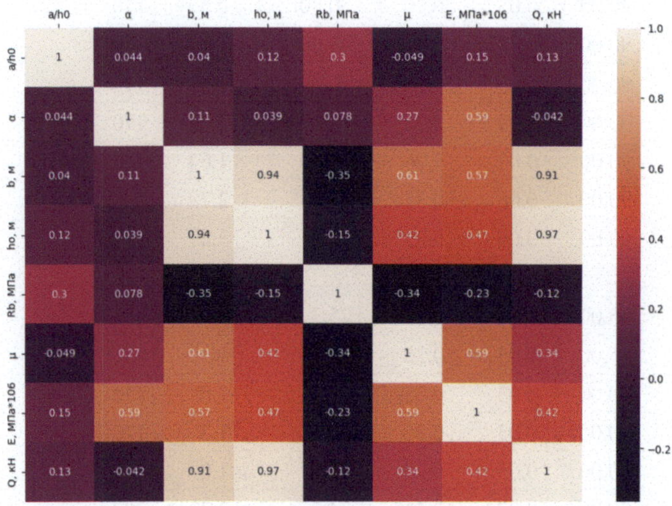

Fig. 1. Correlation matrix.

predicting the target variable is presented in formula 1.

$$\widehat{Y}_i = \sum_{i=1}^{N} M\left(Y_i^0\right) + \sum_{i=1}^{N} \beta_i W_i Y_i \tag{1}$$

$\widehat{Y}_i$ – prediction of the objective function of the i-th feature;

N—number of trees;

$M\left(Y_i^0\right)$—average value of the objective function for all features obtained as a result of the model's operation at the initial stage of testing;

β_i—learning rate coefficient that reduces the contribution of each tree given the gradient descent step;

W_i—weight of the i-th tree;

Y_i—prediction of the i-th tree for input features x_i.

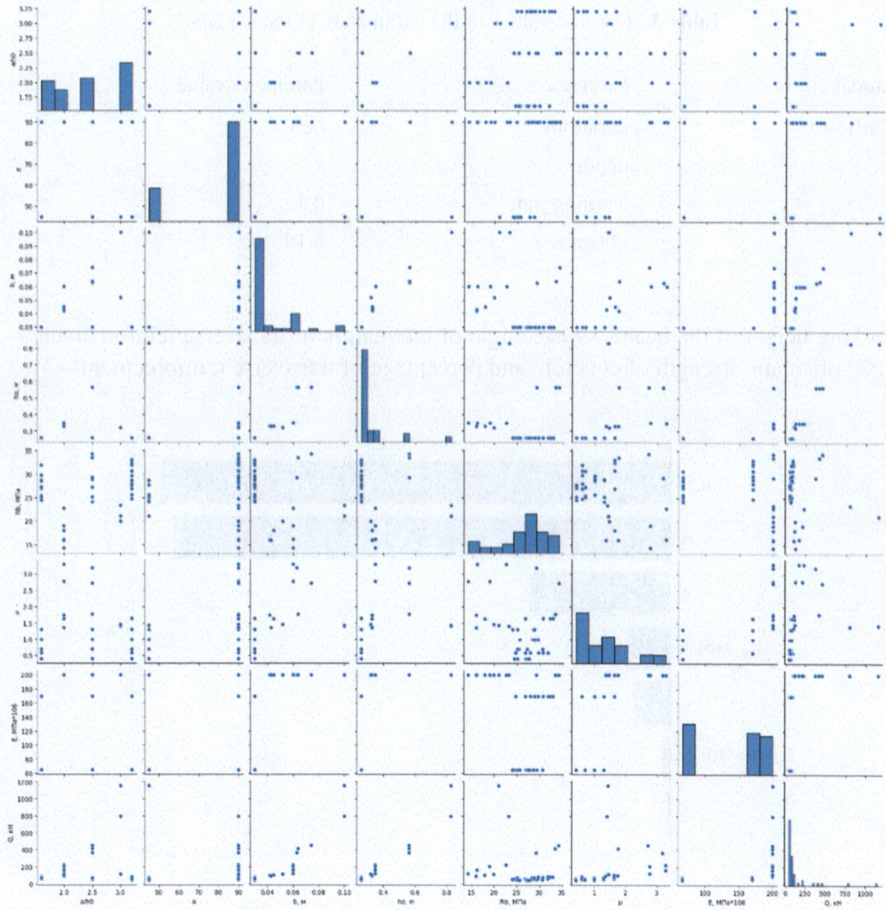

Fig. 2. Histograms of joint dependencies of variables.

3 Results and Discussion

To evaluate the efficiency of the CatBoost algorithm in predicting the strength of reinforced concrete I-beam walls, a series of computational experiments were conducted. The model was trained on data containing various parameters, designs and properties of materials, including the geometric dimensions of the beams, types of reinforcement and concrete class. The regularization method was used for normalization of the parameters, optuna for optimization, and gridsearch and random search for selection of hyperparameters. As a result of the tests, the range of parameter values took the following boundaries: iterations—500–1489; depth—4–10; learning_rate—0.1–0.9; l2 leaf reg—0.01–7.89.

The optimal parameter values obtained during the model training are presented in Table 3.

Figure 3 shows a histogram of the parameters that affect the quality of the model. The strength of the walls of reinforced concrete I-beams depends on: wall thickness by 95%;

Table 3. Optimal values of the parameters of the models.

Model	Parameter	Parameter value
CatBoost	iterations	689
	depth	8
	learning_rate	0.4
	l2 leaf reg	6.18

working height of the beam—90%; angle of inclination of transverse reinforcement—11%; prismatic strength of concrete and percentage of transverse reinforcement—3%.

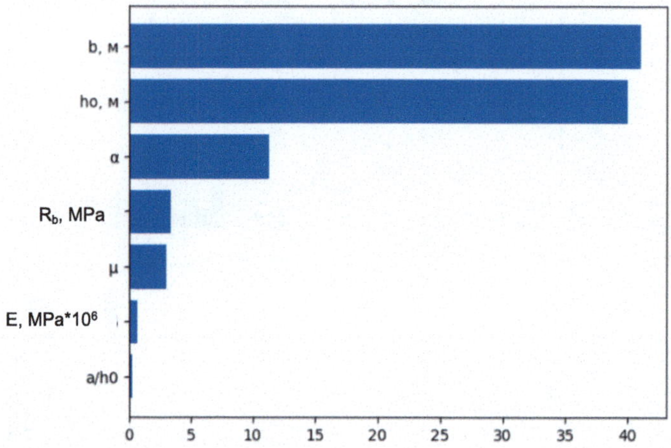

Fig. 3. Evaluation of feature importance.

During the analysis, the following indicators were used to assess the quality of the model: MAE; MSE; MAPE (%). The values of the metric readings are presented in Table 4.

Table 4. Model quality metrics.

Metrics/Model	CatBoost
MAE	0.2138
MSE	0.4885
MAPE (%)	0.0033

Figure 4 shows a scatter plot: the y-axis shows the actual values, and the x-axis shows the predicted values.

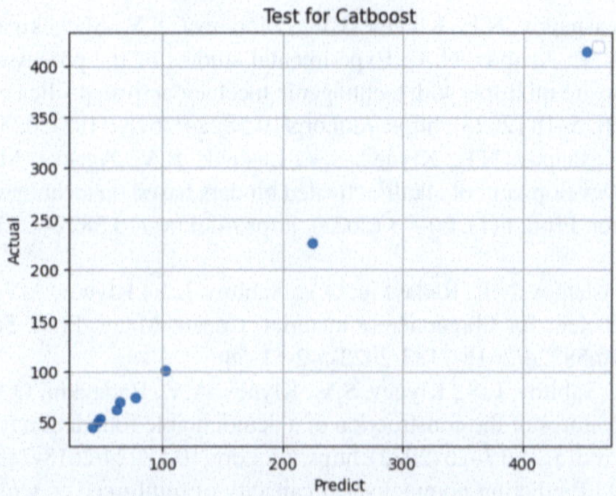

Fig. 4. Forecast Error Graph.

4 Conclusions

Machine learning technologies based on gradient boosting algorithms (CatBoost) open up new possibilities for the design and operation of reinforced concrete I-beams. Its application allows not only to improve the quality and reliability of structures, but also to reduce costs and design time. As a result of the calculations, it was possible to show that, given the initial data, the strength of the walls of reinforced concrete I-beams directly depends on the geometric parameters of the wall: wall and floor thickness; beam height and length; type and amount of reinforcement. It is possible to increase the bearing capacity of I-beams of this type by changing the cross-sections of reinforced concrete elements, thereby extending their life cycle at the construction stage. However, the following aspects must be taken into account:

1. Setting up model parameters (Table 3);
2. Determining the importance of features (Fig. 3);
3. The quality of the constructed model (Table 4);
4. Visualization of practical and predicted values (Fig. 4).

As a result, the constructed model reveals the importance of various parameters for predicting the load-bearing capacity of structures, and it is also necessary to take into account the class of concrete. This can be useful for further design decisions.

References

1. Ly, H.B., et al.: Development of hybrid machine learning models for predicting the critical buckling load of I-shaped cellular beams. Appl. Sci. **9**(24), 5458 (2019)
2. Abambres, M., et al.: Neural. Network-based formula for the buckling load prediction of I-section cellular steel beams. Computers **8**(1), 2 (2018)
3. Ferreira, F.P.V., et al.: Lateral–torsional buckling resistance prediction model for steel cellular beams generated by Artificial Neural Networks (ANN). Thin-Walled Struct. **170**, 108592 (2022)

4. Klyuev, A.V., Kashapov, N.F., Klyuev, S.V., Zolotareva, S.V., Shchekina, N.A., Shorstov,a E.S., Lesovik, R.V., Ayubov, N.A.: Experimental studies of the processes of structure formation of composite mixtures with technogenic mechanoactivated silica component. Constr. Mater. Prod. **6**(2), 5–18 (2023). https://doi.org/10.58224/2618-7183-2023-6-2-5-18
5. Klyuev, A.V., Kashapov, N.F., Klyuev, S.V., Lesovik, R.V., Ageeva, M.S., Fomina, E.V., Ayubov, N.A.: Development of alkali-activated binders based on technogenic fibrous materials. Constr. Mater. Prod. **6**(1), 60–73 (2023). https://doi.org/10.58224/2618-7183-2023-6-1-60-73
6. Klyuev, S.V., Kashapov, N.F., Radaykin, O.V., Sabitov, L.S., Klyuev, A.V., Shchekina, N.A.: Reliability coefficient for fibreconcrete material. Constr. Mater. Prod. **5**(2), 51–58 (2022). https://doi.org/10.58224/2618-7183-2022-5-2-51-58
7. Strelkov, Yu.M., Sabitov, L.S., Klyuev, S.V., Klyuev, A.V., Radaykin, O.V., Tokareva, L.A.: Technological features of the construction of a demountable foundation for tower structures. Constr. Mater. Prod. **5**(3), 17–26 (2022). https://doi.org/10.58224/2618-7183-2022-5-3-17-26
8. Chou, J.S., et al.: Predicting nominal shear capacity of reinforced concrete wall in building by metaheuristics-optimized machine learning. J. Build. Eng. **61**, 105046 (2022)
9. Yücel, M., et al.: An artificial intelligence-based prediction model for optimum design variables of reinforced concrete retaining walls. Int. J. Geomech. **21**(12), 04021244 (2021)
10. Barkhordari, M.S., Tehranizadeh, M.: Response estimation of reinforced concrete shear walls using artificial neural network and simulated annealing algorithm. Structures **34**, 1155–1168 (2021)
11. Le, L.M., et al.: Hybrid artificial intelligence approaches for predicting buckling damage of steel columns under axial compression. Materials **12**(10), 1670 (2019)
12. Beskopylny, A.N., et al.: Prediction of the properties of vibro-centrifuged variatropic concrete in aggressive environments using machine learning methods. Buildings **14**, 1198 (2024). https://doi.org/10.3390/buildings14051198
13. Beskopylny, A.N., et al.: Prediction of the compressive strength of vibrocentrifuged concrete using machine learning methods. Buildings **14**, 377 (2024). https://doi.org/10.3390/buildings14020377
14. Kondratieva, T.N., Chepurnenko, A.S.: Prediction of rheological parameters of polymers by machine learning methods. Adv. Eng. Res. (Rostov-On-Don) **24**(1), 36–47 (2024)
15. Chepurnenko, A.S., et al.: Prediction of rheological parameters of polymers using the CatBoost gradient boosting algorithm. Polym. Sci., Ser. D **17**(1), 121–128 (2024)
16. Chepurnenko, A.S., et al.: Processing of polymers stress relaxation curves using machine learning methods. Baghdad Sci. J. **20**(6), 2488 (2023)
17. Sandeep, M.S., et al.: Shear strength prediction of reinforced concrete beams using machine learning. Structures **47**, 1196–1211 (2023)
18. Mohammed, H.R.M., Ismail, S.: Proposition of new computer Artificial Intelligence models for shear strength prediction of reinforced concrete beams. Eng. Comput. **38**(4), 3739–3757 (2022)
19. Kondratieva, T.N., Chepurnenko, A.S., Poliakova, K.A., Rodionov, K.A.: CatBoost algorithms to predict the loadbearing capacity of centrally compressed short CFST columns of circular cross-section. E3S Web Conf. **583**, 06009 (2024). https://doi.org/10.1051/e3sconf/202458306009

Analysis of the Stress-Strain State of Combined Pile-Slab Foundations

Amerkhan Dzhankulaev(✉), Aslan Kaziev, Zalimkhan Likhov, and Marina Dadova

Kabardino-Balkarian State University Named After H.M. Berbekov, Nalchik 360004, Russia
dzhankulaevaj@yandex.ru

Abstract. In recent years, pile-plate foundations have been widely used in civil and industrial construction all over the world. When the grillage is turned on, the bearing capacity of the foundation increases, and the growth of sediment slows down. The article provides a study of the nature of pile-slab foundations and the development of an engineering method for calculating a slab grillage, taking into account the actual operation of the foundation and an analysis of the stress-strain state of combined pile-slab foundations from the general design scheme of interaction of the pile grillage of the soil. A finite element model of the calculation scheme was used to conduct numerical experiments, calculations were carried out to study the operation of slab grillages in pile-slab foundations, and an analysis of numerical experiments was performed. The experimental data show the role of a low grillage in the bearing capacity of pile foundations, and the results obtained serve as proof of the qualitative nature of joint work with the soil of pile foundations when resting the grillage on the base.

Keywords: Pile-Slab Foundation · Grillage · Stress-Strain State

1 Introduction

In recent years, the necessity for the construction of combined pile-slab foundations of buildings and structures has become more urgent as areas with good soil conditions for construction are created, the number of floors of buildings grows, and the loads on the foundations rise [1]. It is also known that in some circumstances, pile-plate foundations on suitably strong soils, with a layer of solid soil on the surface, prove to be more economically and technologically successful than shallow foundations. In these circumstances, when the grillage makes contact with the ground, a portion of the external weight is transmitted to the base via its sole.

In some circumstances, taking the resistance of the soil under the sole of a low grillage into consideration when estimating the bearing capacity of pile foundations may greatly lower the cost of pile foundations while also reducing the overall and uneven precipitation of the structure under severe loads. Savings are realized by lowering the number of reinforced concrete piles required in the foundation and the consumption of reinforcement in the grillage, which is caused by the unloading effect of reactive soil resistance while the grillage is operating on a transverse bend [2, 6].

S. V. Klyuev et al. (Eds.): ICRAAC 2024, LNCE 627, pp. 393–399, 2025.
https://doi.org/10.1007/978-3-031-82938-3_43

In this regard, it is still necessary to construct a design plan for the interaction of a pile-slab foundation with a base, taking into consideration the transmission of some of the weight through the sole of a low grillage [11–15].

2 Materials and Methods

To study the nature of the transfer of load by a pile-plate foundation to the ground, taking into account the work of the grillage, scientific studies were conducted, the importance of which is confirmed by publications [16–20]. In the modern world, there is an active construction, including multi-stored buildings and structures on difficult soils. This increases the requirements for the reliability and economic efficiency of foundations.

Table 1. The features and parameters of foundation.

Calculation option	Pile pitch, m	Plate thickness, cm	Load, t	Movement Z, mm	Qx, t	Qy, t	Mx, t*m	My, t*m	Mxy, t*m
1	2	3	4	5	6	7	8	9	10
1	2	30	10	0.332	11.6	11.6	2.83	2.83	1.49
			20	0.664	23.1	23.1	5.66	5.66	2.98
			30	0.997	34.7	34.7	8.49	8.49	4.47
		50	10	0.0749	11.6	11.6	2.94	2.94	1.5
			20	0.15	23.3	23.3	5.88	5.88	5.74
			30	0.225	34.9	34.9	8.82	8.82	4.5
2	3	30	10	1.35	20.1	20.1	7.79	7.79	2.84
			20	2.71	40.2	40.2	15.6	15.6	5.69
			30	4.06	60.3	60.3	23.4	23.4	8.53
		50	10	0.317	21	21	8.45	8.45	2.87
			20	0.634	42.1	42.1	16.9	16.9	5.74
			30	0.95	63.1	63.1	25.3	25.3	8.61
3	6	30	10	17.8	15.7	15.7	17.3	17.3	14.9
			20	35.6	31.3	31.3	34.7	34.7	29.9
			30	53.5	47	47	52	52	44.8
		50	10	8.03	21.7	21.7	36.1	36.1	23.7
			20	16.1	43.5	43.5	72.2	72.2	47.3
			30	24.1	65.2	65.2	108	108	71

The variety of soils in terms of strength, deformability and other characteristics makes it difficult to design foundations and requires the development of adaptive calculation methods. The use of solid foundations is often associated with overspending of materials and high cost. Methods are needed to optimize the size of the foundation plate, taking into account specific ground conditions, in order to increase the economic efficiency of projects. Modern software systems allow you to simulate the operation of soil foundations and foundations with high accuracy, which opens up new opportunities for optimizing design solutions. This paper examines the following:

- changing the VAT of the grillage depending on the distance between the piles;
- the influence of the grillage on the bearing capacity of the pile-slab foundation;
- the effect of the resistance of the grillage sole on the bearing capacity of the pile-slab foundation [3, 6, 8].

Experimental and analytical studies were carried out in this work using a finite element model based on the Lyra complex [5, 7].

Three elements were adjusted during the study: pile pitch, plate thickness, and load amplitude.

Table 1 displays the features and parameters of foundations, as well as the composition of numerical experiments modeling the functioning of pile-plate foundations.

Variants 4–6 were modeled without considering the base's rejection to estimate the percentage of the grillage in the VAT of the pile foundation [4, 9, 10].

Let's take a closer look at the results of numerical modeling.

3 Results and Discussion

The 1st version of numerical modeling. In the initial iteration of the numerical simulation, six experiments were carried out with a pile pitch of 2 m, a plate thickness of 30.50 cm, and a weight of 10,20,30 tons. Internal force diagram with deflection distribution values.

The maximum values of the deflection distribution and the internal force plot are shown in Table 1.

Calculations show that the nature of the load transfer to the soil of the pile as part of the group differs from the nature of the load transfer from a simple plate. This difference is especially pronounced between piles.

The 2nd version of numerical simulation. In the second iteration of the numerical simulation, 6 experiments were carried out with a pile pitch of 3 m, a plate thickness of 30.50 cm, and a load of 10,20,30 tons. Table 1 also shows the maximum values for deflections and internal force diagrams. The magnitude of the transverse force increases by two times, as does the bending moment. Deflections increase 4 times.

The 3rd version of numerical modeling. In the third version of the numerical simulation, 6 experiments were conducted with a pile pitch of 6 m, a plate thickness of 30.50 cm. And a load of 10, 20, 30 tons.

The maximum values of the deflection distribution and the internal force plot are also shown in Table 1. Here we see an increase in the transverse force by 1.3 times, and the bending moment by 6 times. Deflections increase by 50 times compared to the first option. The change in the proportion of the transverse force can be justified by an increase in the rebound force due to large deflections.

A comparative analysis of the results of numerical calculations is given in Figs. 1, 23, 4.

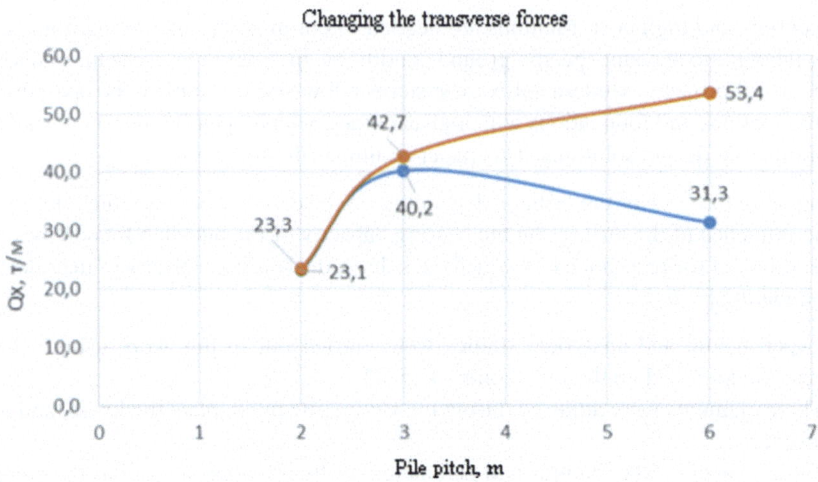

Fig. 1. Transverse forces.

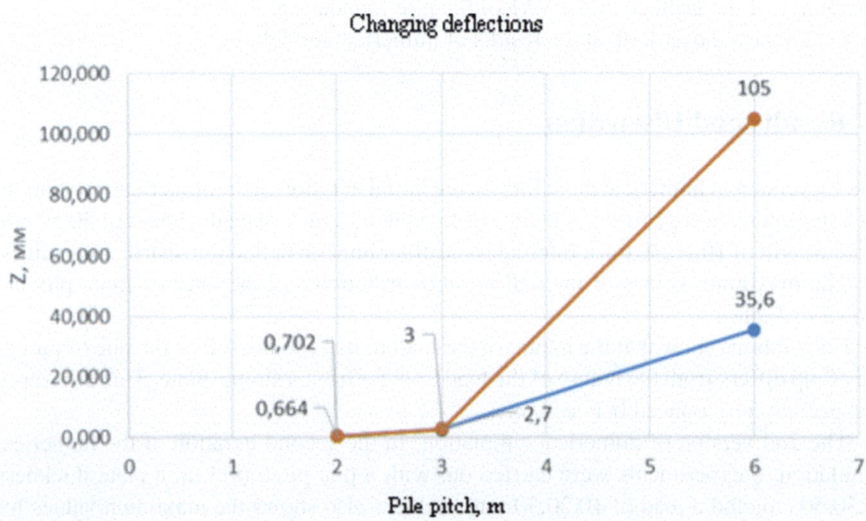

Fig. 2. Deflections.

4 Conclusions

Based on the above studies, the following conclusions can be drawn:

1. According to numerical modeling studies, the proportional involvement of the grillage in the bearing capacity can reach 5–11%, depending on the step.
2. The role of the slab element in the pile-slab foundation is significantly influenced by the size of the slab and the pitch of the piles.
3. The low grillage sole transfers a portion of the structure's exterior weight to the earth foundation;

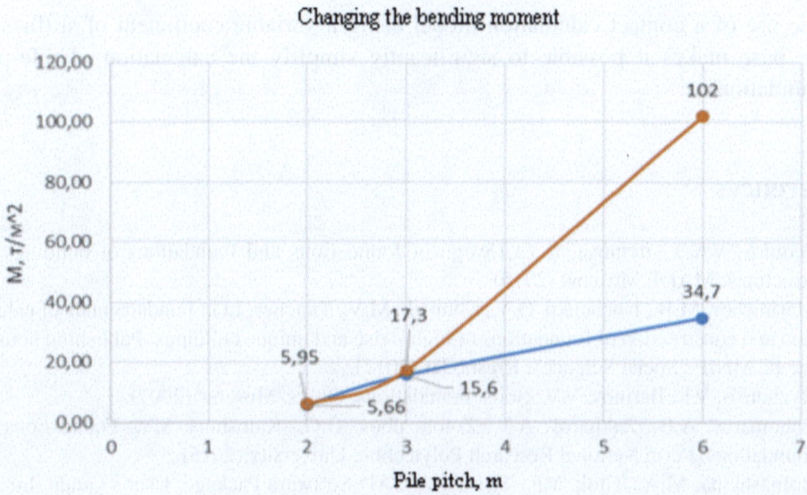

Fig. 3. Bending moments.

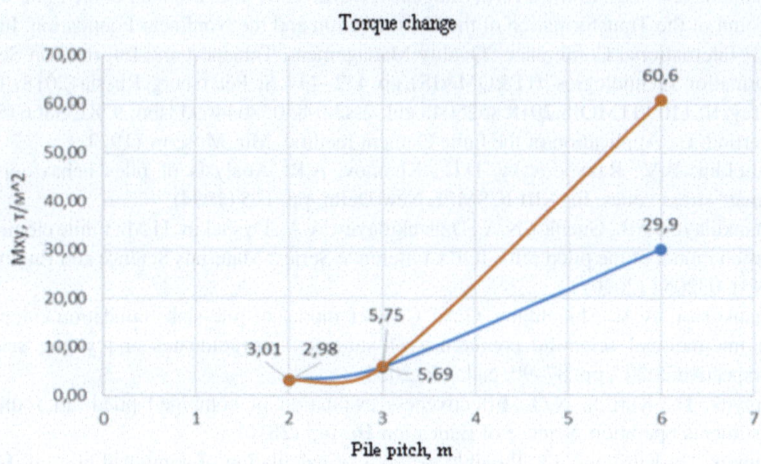

Fig. 4. Torques.

4. When the slab grillage is turned on, the combined pile-slab foundation's bearing capability improves;
5. This work contributes to the development of ideas about deformations and bearing capacity of pile-slab foundations, taking into account the work of the slab part.
6. The results of numerical experiments can serve as a basis for performing practical calculations. The resulting technique simplifies the calculation of pile-slab foundations and can be widely used in engineering practice.The results of numerical experiments can serve as a basis for performing practical calculations. The resulting technique simplifies the calculation of pile-slab foundations and can be widely used in engineering practice.

7. The use of a contact calculation model using a variable coefficient of stiffness of the base makes it possible to significantly simplify the calculation of pile-plate foundations.

References

1. Fedulov, V.K., Artemova, L.Y.: Design of foundations and foundations of buildings and structures. MADI, Moscow (2015)
2. Marinichev, M.B., Ishchenko, O.Y., Chumak, M.V., Tkachev, I.G.: Fundamentals of calculation and construction of foundations of high—rise and unique buildings. Publishing house of the IC MNIF "Social Science", Krasnodar (2017)
3. Ryzhenko, V.I., Barinov, V.V.: Soils. Foundations. Oniks, Moscow (2007)
4. Ponomarev, A.B., Zakharov, A.V., Zolotozubov, D.G., Kaloshina, S.V.: Foundations and foundations. Perm National Research Polytechnic University (2015)
5. Romashkina, M.A., Titok, V.P.: The LIRA-CAD Software Package. User's Guide. In: A.S. Gorodetsky (eds.) Academician of the Russian Academy of Sciences. Electronic edition (2018)
6. Dzhankulayev, A.Y., Likhov, Z.R., Shogenov, O.M.: The Finite Element of the Plate with the Account of the Transformation of the Cross Section and the Nonlinear Foundation. In: 2018 IEEE International Conference "Quality Management, Transport and Information Security, Information Technologies" (IT&QM&IS), pp. 432–434. St. Petersburg, Russia (2018). https://doi.org/10.1109/ITMQIS.2018.8525048 eid: 2-s2.0-85058044693 isbn: 9781538667576
7. Segerlind, L.: Application of the finite element method. Mir, Moscow (1979)
8. Bakholdin, B.V., Razvodovsky, D.E., Khamov, A.P.: Analysis of piles behaviour using ultimate stress zones. In: XIII ICSMFE New Delhi, pp. 1–5 (1994)
9. Dzhankulayev, A.J., Guenko, N.A., Dzhankulayev, A.A., Dyshekov, H.M.: Finite element calculation model of the piled rafts. IOP Conference Series: Materials Science and Engineering **913**(3), 032054 (2020)
10. Dzhankulaev, A.A., Shigalugov, G.A.: Contact model of pile-slab foundation calculation. The international scientific conference of students, postgraduates and young scientists "Perspective 2020", pp. 87–90. Nalchik (2020)
11. Chunyuk, D., Kurilin, N.O.: Effectiveness evaluation of combined piled-raft foundation constituents operation. Science of Education **16**, 1–5 (2016)
12. Chunyuk, D., Kurilin, N.O.: Possible methods of calculation of combined pile-raft foundations. MGSU **18**, 11–17 (2016)
13. Ladyzhenskiy, I.G., Sergiyenko, A.V.: Combined pile-plate foundations (KSPF) and combined plate-pile foundations – KPSF) - development prospects. High-rise buildings: high-rise technology journal **5–6**, 132–135 (2014)
14. Okyay, U.S., Dias, D., Billion, P., D. Vandeputte Impedance Functions of Slab Foundations with Rigid Piles. Geotech geological engineering **30**, 1013–1024 (2012)
15. Ter-Martirosyan, Z., Ter-Martirosyan, A., Sidorov, V., Almakaeva, A.: Pile-slab foundation's settlement evaluation in pseudo-nonlinearity formulation. E3S Web of Conferences **410**(5), 02033 (2023)
16. Ter-Martirosyan, Z.G., Sidorov, V.V., Ter-Martirosyan, K.Z.: Experience of transformation of weak water-saturated soils by their final stiffness. Bulletin of MGSU **3**(114) (2013)
17. Ter-Martirosyan, Z.G., Ter-Martirosyan, A.Z., Manukyan, A.V., Angelo, G.O.: Interaction of drainage pile with surrounding compacted clay soil and grillage work taking into account the time factor. Zhilishchnoe stroitel'stvo **11**, 1–8 (2017)

18. Ter-Martirosyan, Z.G., Ter-Martirosyan, A.Z., Angelo, G.O.: Interaction of a non-filtering crushed stone pile (column) with external consolidating soil and aggregate as part of a pile-slab foundation. Zhilishchnoe stroitel'stvo **4**, 19–23 (2019)
19. Ter-Martirosyan, Z.G., Ter-Martirosyan, A.Z., Angelo, G.O.: Interaction of a crushed stone filter pile with external water-saturated clay soil and grillage as part of a pile-slab foundation. Geotechnics **11**(1), 36–43 (2019)
20. Gotman, A.L., Gavrikov, M.D.: Study of the peculiarities of operation of vertically loaded long bored piles and their expansion. Construction and Geotechnics **12**(3), 72–83 (2021)

Fine-Grained Fiber Reinforced Concrete Using Basalt Fibers and Volcanic Ash

Alim Kazharov(✉), Tolya Khezhev, and Muhamed Kokoev

Kabardino-Balkarian State University Named After Kh.M. Berbekov, Nalchik 360004, Russia
kazharov.90@mail.ru

Abstract. Fine-grained concrete compositions have been developed using local raw materials of the Kabardino-Balkarian Republic - volcanic ash. In the studies for the development of fine-grained fiber-reinforced concrete composites the following were used: Portland cement PC500-DO produced by Belgorod Cement CJSC; sand from the Malkinskoye deposit with a maximum grain size of 5 mm; volcanic ash of fraction 0–0.14 mm from the Zayukovsky deposit; basalt fiber produced by Ivosteklo OJSC of the RNB-9–1200-4s brand. The mixture was prepared in a forced-action mixer, in which cement, volcanic ash, sand, basalt fiber were successively loaded and mixed until a homogeneous dry mixture was obtained, then water was added and mixing was continued again until a homogeneous dispersed-reinforced fine-grained concrete mixture was obtained. The mobility of the fine-grained concrete mixture was determined using a Suttard viscometer. The mixture spread was 18 cm. Samples measuring 4x4x16 cm were molded using the injection molding method and dried naturally in air-dry conditions. The compositions and properties of fine-grained fiber concretes with volcanic ash have been studied using methods of mathematical experimental planning. A second-order rotatable plan of the regular hexagon type with central points has been implemented to study the influence of fiber reinforcement parameters on the properties of fine-grained concrete. It has been revealed that reinforcement of fine-grained concrete with basalt fibers significantly increases their bending strength, the highest values of compressive strength are observed in the plan area with $\mu_v \approx 1.0$ % and ${}^{l}/_{d} = 1444$, and bending strength - $\mu_v \approx 1.25$ % and ${}^{l}/_{d} = 1444$.

Keywords: Fine-grained concrete · Pumice · Basaltic fiber · Planning experiment

1 Introduction

One of the main ways of scientific and technological progress in construction is the widespread use of new effective materials, ways to increase the operational characteristics of concrete. The construction of modern buildings and structures requires the use of concretes with high performance properties such as compressive and tensile strength, crack resistance, toughness, wear resistance, corrosion resistance, frost resistance, etc.

One of the ways to solve the problems of improving the operational characteristics of fine-grained concrete is the use of active mineral additives, its reinforcement with

S. V. Klyuev et al. (Eds.): ICRAAC 2024, LNCE 627, pp. 400–407, 2025.
https://doi.org/10.1007/978-3-031-82938-3_44

various types of metallic and non-metallic fibers of mineral or organic origin [1, 3, 4, 9–11].

We studied volcanic ash as an active mineral additive in fine-grained concrete. The expediency of their use is due to the fact that out of 60–75% of the silica that is part of volcanic rocks, 30–35% is in amorphous (active) state [2, 5–8].

The choice of basalt fiber for dispersed reinforcement can be explained by the following reasons: the presence of large reserves of basalt, diabase, gabbrodorelite for the production of basalt fibers in many regions of the country is a prerequisite for mass use in construction; environmental safety of their use; higher modulus of elasticity and low cost of basalt fibers compared to glass fibers; higher alkali resistance of basalt fibers compared to glass fibers made of aluminoborosilicate composition.

2 Methods and Materials

In the research for the development of fine-grained fiber-concrete composites, the following were used: portland cement PC500-D0 the production of JSC Belgorod Cement; sand from the Malkinskoye deposit with a maximum grain size of 5 mm; volcanic ash of the 0–0.14 mm fraction of the Zayukovsky deposit; basalt fiber produced by JSC Ivotsteklo of the RNB brand-9–1200- 4s.

The chemical composition of the volcanic ash of the Zayukovsky deposit used in the research is presented in Table 1.

Table 1. Chemical composition of volcanic ash.

The content of the main components in % by weight								
SiO_2	Al_2O_3	Fe_2O_3	CaO	MgO	TiO	$Na_2O^+ + K_2O$	SO_3	Other
73.1	13.75	1.75	1.65	1.12	0.23	3.87	0.12	2.0

The granulometric composition of the sand of the Malkinskoye deposit is shown in Table 2.

Table 2. Granulometric composition of sand.

Name of the material	Private residues on sieves, %					Passed through a sieve of 0.14
	2.5	1.25	0.63	0.315	0.14	
Sand	4.6	9.3	25.5	35.5	21	4.1

The mixture was prepared in a forced mixer, in which cement, volcanic ash, sand, basalt fiber were sequentially loaded and mixed until a homogeneous dry mixture was obtained, then water was added and mixing was continued again until a homogeneous dispersed-reinforced fine-grained concrete mixture was obtained.

The mobility of a fine-grained concrete mixture was determined by a Souttard viscometer. The spread of the mixture was 18 cm. Samples measuring 4x4x16 cm were molded by injection molding and carried out natural drying in air-dry conditions.

3 Results and Discussions

The results of studies of fine-grained concrete with the addition of volcanic ash as an active mineral additive are shown in Table 3.

It follows from Table 3 that with an increase in the amount of volcanic ash added to cement, the bending and compressive strength of fine-grained concrete decreases. Up to 20% volcanic ash with maximum grain sizes up to 0.14 mm can be injected into PC500-DO Portland cement without significantly reducing the bending and compressive strength of samples of normal hardening, which is due to the hydraulic activity of fine ash particles.

Table 3. Physical and mechanical properties of fine-grained concrete.

The ratio of cement to sand by weight	The amount of ash additive in % by weight of cement	Bending strength limit, MPa	Compressive strength, MPa	Water absorption by weight, %
1:2	–	4.3	19.5	12.4
1:2	10	4.2	19.1	12.6
1:2	20	4.1	19.0	12.9
1:2	30	3.8	18.3	13.2

Experiments were carried out to determine the effect on the characteristics of fine-grained concrete of replacing the dusty fraction ($d \leq 14$ mm) of quartz sand with volcanic ash. The test results are shown in Table 4.

It follows from Table 4 that the replacement of the pulverized fraction of quartz sand with volcanic ash leads to an increase in bending strength by 34.8–44.0%, and compression by 9.2–12.5%. The water absorption for the 1:2 composition by weight decreased by 1.59 times.

The developed fine-grained concretes have disadvantages such as brittleness, relatively low bending and compressive strength. The introduction of basalt fiber into the mixture can eliminate these disadvantages, as well as improve the physical and mechanical characteristics of fine-grained concrete.

The composition of the initial concrete matrix and its physicomechanical properties of fine-grained concrete for reinforcement with basalt fibers are shown in Table 5.

To study the effect of fiber reinforcement parameters on the properties of fine-grained concrete, a rotatable second-order plan of the type of a regular polygon with central points was implemented. The geometric interpretation of such a plan is shown in Fig. 1.

The main parameters of dispersed reinforcement were taken as the studied factors: X_1– the percentage of reinforcement by volume μ_v, %; X_2 – the ratio of the length of the

Table 4. Physicomechanical properties of fine-grained concrete with additives of volcanic ash instead of a pulverized fraction of quartz sand.

The ratio of cement to sand by weight	The amount of ash additive in % by weight of quartz sand	Bending strength, MPa	Compressive strength, MPa	Water absorption by weight, %
1:2	–	4.3	19.5	12.4
1:2	4.1	5.8	21.3	7.8
1:3	–	3.4	12.3	–
1:3	4.1	4.9	14.4	–
1:4	–	2.6	7.4	–
1:4	4.1	3.6	8.7	–

Table 5. The consumption of components in the mixture and the physicomechanical properties of fine-grained concrete.

Composition No	Consumption of components in the mixture per 1 m^3, kg			Tensile strength, MPa	
	sand	cement	volcanic ash d < 0.14 mm, 20% by weight of cement	for compression	On the bend
1	2	3	4	5	6
1	1302	521	130	19.0	4.1

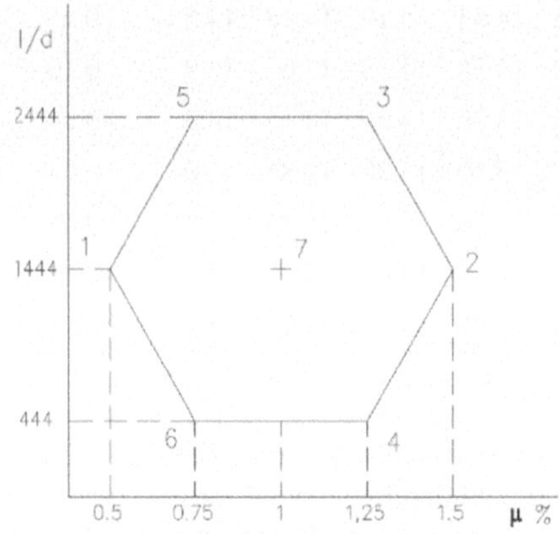

Fig. 1. The plan of the experiment.

curls to their diameter l/d. The following optimization parameters were considered: Y_1– compressive strength R t, MPa; Y_2– bending strength R b, MPa. The experiment matrix has the following form (Table 6).

Table 6. The experiment matrix.

№	Natural variables		The experiment matrix				
	x_1	x_2	X_1	X_2	X_1^2	X_2^2	X_1X_2
1	0.5	1444	−1	0	+ 1	0	0
2	1.5	1444	+ 1	0	−1	0	0
3	1.25	2444	+ 0.5	+ 0.87	+ 0.25	+ 0.75	+ 0.43
4	1.25	444	+ 0.5	−0.87	+ 0.25	+ 0.75	−0.43
5	0.75	2444	−0.5	+ 0.87	+ 0.25	+ 0.75	−0.43
6	0.75	444	−0.5	−0.87	+ 0.25	+ 0.75	+ 0.43
7	1.0	1444	0	0	0	0	0

The experimental results are shown in Tables 7 and 8.

Table 7. Results of compression testing of samples.

№	Values of parallel measurements of the response function Y1, MPa						The average value of Y1, MPa	Variance Sj2	Coef-ficient of varia-tion, %	Error Sj
	y_1	y_2	y_3	y_4	y_5	y_6				
1	14.23	14.73	14.23	14.53	15.13	14.13	14.5	0.15	0.03	0.38
2	13.80	13.80	14.20	13.56	14.35	13.60	13.9	0.10	0.02	0.32
3	13.95	14.21	14.50	14.20	14.05	14.15	14.2	0.03	0.01	0.19
4	13.10	12.95	12.80	13.10	13.20	12.85	13.0	0.03	0.01	0.16
5	14.80	15.20	14.55	14.70	14.68	14.95	14.8	0.05	0.02	0.23
6	12.90	14.40	13.10	12.95	13.60	14.20	13.5	0.43	0.05	0.65
7	15.20	15.50	15.40	14.95	15.10	14.75	15.2	0.08	0.02	0.28

After conducting the experiment and processing the obtained data, the following regression equations were obtained in encoded form:

$$Y_1 = 15,2 - 0,38X_1 + 0,71X_2 - X_1^2 - 1,43X_2^2 - 0,06X_1X_2\ ;$$

$$Y_2 = 9,8 + 0,43X_1 + 0,23X_2 - 0,5X_1^2 - 0,9X_2^2 - 0,12X_1X_2\ .$$

The response surfaces were constructed using the above equations (Fig. 2).

Table 8. Bending test results of the samples.

№	Values of parallel measurements of the click function Y2, MPa						The average value of Y2, MPa	Variance Sj2	Coefficient of variation, %	Error Sj
	y_1	y_2	y_3	y_4	y_5	y_6				
1	9.29	8.76	8.43	8.52	8.85	8.94	8.8	0.10	0.04	0.31
2	9.50	9.80	10.10	9.70	9.60	10.00	9.8	0.05	0.02	0.23
3	9.34	9.29	9.12	9.31	9.15	9.32	9.3	0.01	0.01	0.09
4	9.31	9.02	8.76	9.25	8.96	8.85	9.0	0.05	0.02	0.22
5	9.60	8.23	9.22	9.42	8.55	9.37	9.1	0.30	0.06	0.55
6	7.96	9.37	8.32	8.25	9.10	8.70	8.6	0.29	0.06	0.54
7	10.05	9.73	9.54	9.63	9.90	9.70	9.8	0.03	0.02	0.19

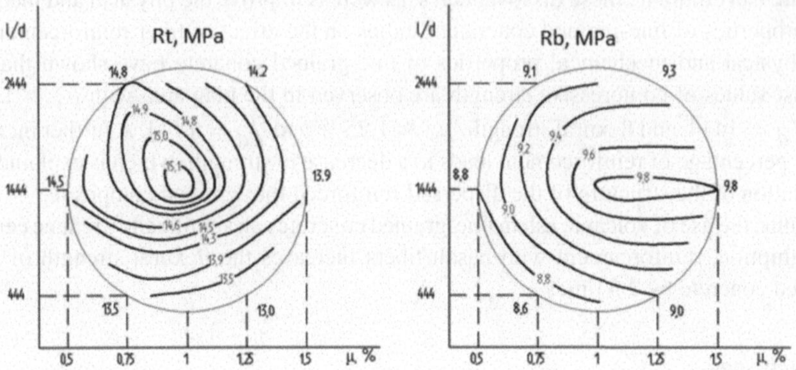

Fig. 2. Response surfaces. Here: Rt is the compressive strength, MPa; Rb is the bending strength, MPa; l/d is the ratio of the length of the fibers to their diameter; μ is the percentage of reinforcement by volume.

The analysis of the obtained equations and response surfaces showed that the highest values of compressive strength are observed in the area of plan c $\mu_\nu \approx 1.0$ % and $^l/_d = 1444$, and bending strength – $\mu_\nu \approx 1.25$ % and $^l/_d = 1444$ a further increase in the percentage of reinforcement leads to a decrease in strength, which is explained by a violation of the structure of the disc-reinforced fine-grained composite.

The compositions of the initial raw material mixtures of foamed concrete and their basic physical and mechanical properties are given in Tables 1 and 2.

4 Conclusions

Studies of fine-grained concrete with the addition of volcanic ash as an active mineral additive have shown that with an increase in the amount of volcanic ash added to cement, there is a slight decrease in the flexural and compressive strength of fine-grained concrete. Up to 20% of volcanic ash with maximum grain sizes of up to 0.14 mm can be added to Portland cement PC500-DO without a significant decrease in the flexural and compressive strength of normal hardening samples, which is due to the hydraulic activity of finely dispersed ash particles. Studies to determine the effect on the characteristics of fine-grained concrete with the replacement of the dust fraction (d $\leq$ 14 mm) of quartz sand with volcanic ash have shown that replacing the dust fraction of quartz sand with volcanic ash leads to an increase in flexural strength by 34.8–44.0%, and in compression by 9.2–12.5%. Water absorption for the 1:2 composition by weight decreased by 1.59 times.

The developed fine-grained concretes have such disadvantages as brittleness, relatively low flexural and compressive strength. The introduction of basalt fiber into the mixture can eliminate these disadvantages, as well as improve the physical and mechanical properties of fine-grained concrete. Studies on the effect of fiber reinforcement on the physical and mechanical properties of fine-grained concrete have shown that the highest values of compressive strength are observed in the plan area with $\mu_\nu \approx 1.0\ \%$ and $^l/_d = 1444$, and flexural strength -$\mu_\nu \approx 1.25\ \%$ and $^l/_d = 1444$. A further increase in the percentage of reinforcement leads to a decrease in strength, which is explained by a violation of the structure of the dispersed-reinforced fine-grained composite.

Thus, the use of volcanic ash in fine-grained concrete can significantly reduce cement consumption, reinforcement with basalt fibers increases the flexural strength of fine-grained concrete by 2.5 times.

References

1. Volkov, I.V.: Fibroblast: The state and prospects of application. Ind. Civ. Constr. **9**, 37 (2002)
2. Akhmatov, M.A.: The Effectiveness of the Use of Local Building Materials and Concrete. Elbrus, Nalchik (1986)
3. Khezhev, T., Zhurtov, A., Kazharov, A., Zrumov, T., Samgurov, A.: Fire-retardant plastering mortars based on exfoliated vermiculite and volcanic ash. Mater. Sci. Forum **1043**, 133–139 (2021)
4. Khezhev, T.A., Pukharenko, Y., Khezhev, K., Klyuev, S.V.: Fiber gypsum concrete composites with using volcanic tuff sawing waste. ARPN J. Eng. Ap-Plied Sci. **13**(8), 2935–2946 (2018)
5. Khezhev, T., Ansokov, A., Kerefov, I., Malkarov, R.: Fiber foam tuffcrete structure and properties formation on the basis of volcanic tuff dusting waste. Mater. Sci. Forum **1011**, 130–135 (2020)
6. Klyuev, S.V., Khezhev, T.A., Pukharenko, Y., Klyuev, A.V.: The fiber-reinforced con-crete constructions experimental research. Mater. Sci. Forum **931**, 598–602 (2018)
7. Khezhev, T.A., Pukharenko, Y.V., Khashukaev, M.N.: Russian Federation Patent 239600. Raw material mixture and method of manufacturing of products from foam concrete. Bulletin 33 (2008)
8. Khezhev, T, Kardanov, A, Bolotokov, E., Dottuyev, A, Mashfezh, I.: Fine-grained fibre concrete of run-of-crusher stone using volcanic ash. Mater. Sci. Forum **1043**, 61–65 (2021)

9. Klyuev, S.V., Klyuev, A.V., Khezhev, T.A., Pukharenko, Y.V.: To the question of fiber reinforcement of concrete. Mater. Sci. Forum **945**, 25–29 (2019)
10. Klyuev, S.V., Klyuev, A.V., Khezhev, T.A., Pukharenko, Y.V.: Fiber concrete for industrial and civil construction. Mater. Sci. Forum **945**, 120–124 (2019)
11. Klyuev, S.V., Klyuev, A.V., Khezhev, T.A., Pukharenko, Y.V.: Fibers and their properties for concrete reinforcement. Mater. Sci. Forum **945**, 125–130 (2019)
12. Klyuev, S.V., Klyuev, A.V., Khezhev, T.A., Pucharenko, Yu.V.: High-strength fine-grained fiber concrete with combined reinforcement by fiber. J. Eng. Appl. Sci. **13**(8 SI), 6407–6412 (2018)
13. Fediuk, R., Amran, M., Klyuev, S., Klyuev, A.: Increasing the performance of a fiber-reinforced concrete for protective facilities. Fibers **9**(11), 64 (2021)
14. Amran, M., Fediuk, R., Klyuev, S., Qader, D.N.: Sustainable development of basalt fiber-reinforced high-strength eco-friendly concrete with a modified composite binder. Case Stud. Constr. Mater. **17**, e01550 (2022)

Protection of Buildings and Structures from Sedimentary Deformations in Loess Soils Using Injection Mixtures on a Mineral Basis

A. Kharchenko, A. Panchenko, I. Kharchenko, V. Erofeev(✉), and A. Dzagov

National Research Moscow State University of Civil Engineering, 129337 Moscow, Russia
erofeevvt@bk.ru

Abstract. The article discusses methods for protecting buildings and structures from the effects of sedimentary deformations on loess soils using mineral-based injection mixtures. Loess soils are common in the south of Russia. As a result of the disruption of structural salt bonds between individual soil elements when they are moistened, significant uneven sedimentary deformations develop in loess soil. Currently, there is a list of measures to eliminate the subsidence properties of loess soils. One of the methods can be the use of injection mixtures based on mineral binders. The article analyzes the effect of alkaline activator consumption on the compressive strength of geopolymer binders. It was found that the activity of the geopolymer binder is affected by the presence of an alkaline component in its composition. As a result of the research, it was found that the greatest efficiency for increasing the activity of the geopolymer binder is shown by increasing the dispersion of the binder and the concentration of the alkaline activator. The authors conducted a comparative analysis of the compressive strength of soil concrete prepared by injection impregnation with the strength of soil concrete formed mechanically. The result showed that the strength characteristics of soil concrete differ significantly.

Keywords: Loess Subsidence Soils · Sedimentary Deformations · Injection Mixtures

1 Introduction

Loess subsidence soils are widespread in the south of Russia, including the North Caucasus, Rostov Region, Stavropol, Volgograd Region, the LPR and DPR, and the South of Western Siberia. The main distinguishing feature of loess soils is the development of significant uneven sedimentary deformations that limit or completely eliminate the operational suitability of buildings and structures.

This is a consequence of the disruption of structural salt bonds between individual soil elements when they are moistened. As the analysis of the results of industrial application of various methods and technological solutions for eliminating subsidence shows, their cost in the structure of total costs of building and construction in loess soils is 30…40% of

S. V. Klyuev et al. (Eds.): ICRAAC 2024, LNCE 627, pp. 408–417, 2025.
https://doi.org/10.1007/978-3-031-82938-3_45

the total cost of construction. At present, in order to eliminate the subsidence properties of loess soils, various measures have been developed, which can be divided into measures ensuring mechanical compaction of the soil structure by ramming, high-temperature thermal strengthening with the formation of a sintered structure, and drilling and injection technologies using liquid glass or resins. At the same time, cementation of loess soils has not been practically used to date due to the limited penetration capacity of injection mixtures based on construction cements, as well as the danger of dissolution of the structural bonds of the loess soil during the period before the injection mixture hardens. An effective solution to the problem of stabilization and strengthening of loess soils can be the use of injection mixtures based on clinker-free especially finely dispersed mineral binders, the main element of which is slag in a finely dispersed form, activated by an alkaline component. Alkaline activation of slags used to obtain high-strength materials on their basis has led to the appearance of geopolymer binders (GPB). At the same time, the alkali-containing component of GPB does not have a dissolving effect on the structural salt bonds of loess soil, the interaction of which with the slag binder has an additional activating effect. Thus, taking into account the homogeneity of the chemical and mineral composition of GPB with soil, geotechnical and underground construction should be considered as the most effective area of application of GPB in construction.

2 Methods and Materials

The study of the physical, mechanical and structural characteristics of soil concrete was performed using injection mixtures based on GPV with different dispersion: GPV-B (basic, with dispersion up to 160 μm); GPV-T (finely ground, with dispersion up to 40 μm); GPV-U (ultrafine, with dispersion up to 5 μm). The properties of GPV were studied using a normal consistency test. An aqueous solution of NaOH was used as an alkaline activator. The consumption of the alkaline activator varied in the amount of 1%; 5.0% and 9% of the binder weight. The studies of the properties of soil concrete were performed on models of loess soils formed in cylinders with a diameter of 100 mm and a height of h = 300 mm by layer-by-layer laying and stamping of natural loess soil. Comparative analysis of the properties of the prepared model and natural loess soil showed that the main physical and mechanical characteristics of soil samples with natural and artificial structure are almost the same and had a convergence level of more than 90%. The formation of soil concrete samples was carried out by impregnating them with injection mixtures prepared on the basis of GPV with different slag dispersion and alkali activator consumption. After keeping the formed soil concrete samples for 3 days, they were removed from the cylinders and placed in a normal curing chamber. The main characteristics of injection-stabilized loess soils were the compressive strength R_{comp} and the deformation modulus E_CS at the age of 3, 7 and 28 days of normal curing.

3 Results and Discussion

Analysis of the results of the studies showed that the main factors determining the intensity of hardening and the final strength of the GPV are the dispersion of the binder and the content of the alkaline activator in its composition. With an increase in the consumption of the alkaline activator from 1 to 5%, the compressive strength of the GPV increases from 15 to 64 MPa, which is especially important to consider when designing injection mixtures, which are used with a high water-binding ratio of up to 3.5 in the cuff technology. The results of the study of the effect of the consumption of the alkaline activator on the compressive strength of the GPV are presented in Fig. 1.

As follows from the data in Fig. 1, with an increase in the alkali activator content from 1 to 9% in the GPV composition, the compressive strength increases by 70%. Moreover, the efficiency of introducing the activator into the GPV composition increases with an increase in slag dispersion.

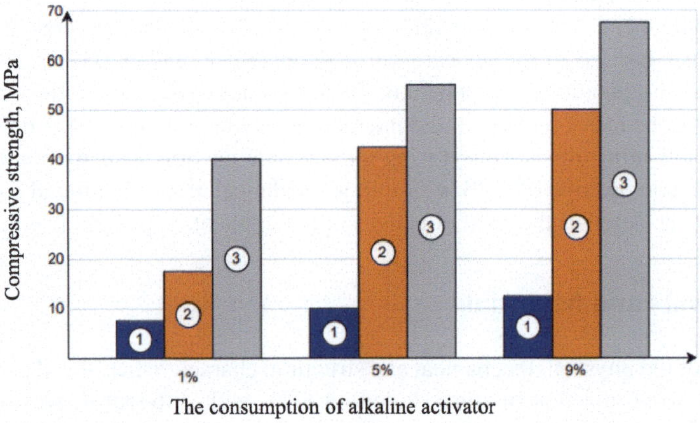

Fig. 1 Compressive strength of GPV depending on the fineness of grinding and consumption of alkaline activator: 1- $d_{95} < 160$ μm; 2-$d_{95} < 40$ μm; $d_{95} < 5$ μm; dough of normal density, 28 days

The GPV strength after 28 days of hardening is within the range of 25 … 65 MPa, depending on the ratio of components, with a flexural strength of 4.2 … 7.5 MPa, an elastic modulus of 2.2 … 3.1 GPa, and water absorption of 2.2 … 3.7%. It was found that GPV is characterized by a significant decrease in compressive strength with an increase in the W / W values, which is typical for all types of hydraulic binders. (Fig. 2).

Thus, the factor that largely determines the activity of the GPV is the content of the alkaline component in the composition of the composite binder, and the dispersion of the binder. At the same time, simultaneously with the acceleration of hardening and an increase in the final strength, the setting time of the GPV is significantly reduced, which must be taken into account when performing injection work in the process of forming soil concrete massifs. Based on the analysis of the results of studies of soil concrete samples obtained as a result of impregnation of loess soils with injection mixtures based on GPV, it was found that the compressive strength of soil concrete samples varies in a wide range from 1.5 … 2.0 MPa to 20 … 25 MPa, depending on the composition of the GPV and W / W injection mixtures based on them (Fig. 3).

The development of the process of structuring of GPV is preceded by the dissolution of aluminosilicate components in an alkaline medium. In this case, the dissolution of glassy aluminosilicate is accompanied by the exchange of H+ ions for Ca2+ and Na+,

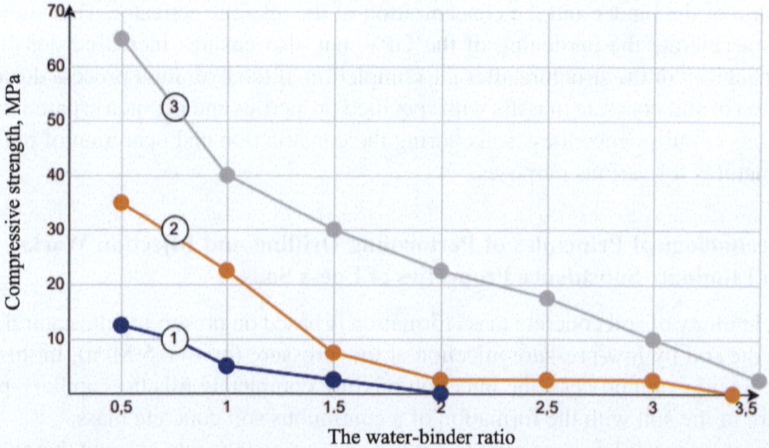

Fig. 2 Compressive strength of GPV depending on the water-binder ratio: 1-GPV-B; 2-GPV-T; 3-GPV-U; alkaline activator consumption - 5%; 28 days of normal hardening

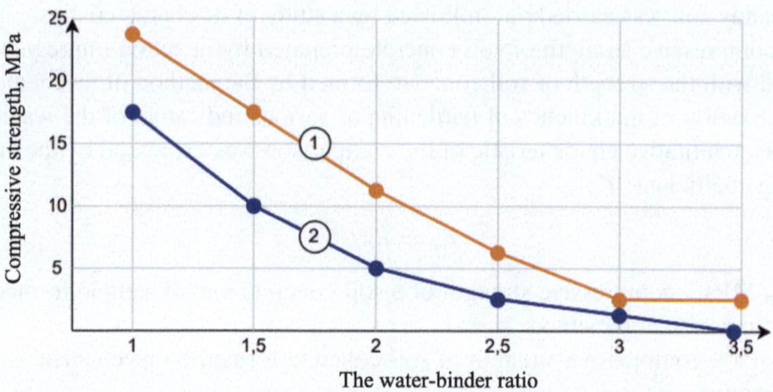

Fig. 3 Compressive strength of soil concrete depending on the W/C ratio of the injection mixture: 1-GPV-U; 2-GPV-T; alkaline activator consumption – 5%; 28 days of normal hardening

with subsequent hydrolysis of aluminosilicate compounds, destruction of the depolymerized glassy structure and splitting of Si and Al compounds into monomeric structures. The geopolymerization process begins with the dissolution of finely dispersed, heat-treated aluminosilicate raw materials in an alkaline medium. In this case, the degree of dissolution of aluminosilicate raw materials is affected by its dispersion and the reactivity of mineral components. Dissolution of aluminosilicate raw materials by alkaline hydrolysis occurs with a sufficient amount of water and is accompanied by the destruction of aluminum and silicon compounds, which pass into solution and accumulate in the form of individual particles on the surface of mineral components in monomeric form. As solid particles accumulate in the solution, the process of geopolymerization and strengthening of the GPV develops. From the presented research results it is evident that the most effective way to increase the activity of the GPV is to increase the dispersion of the binder and the concentration of the alkaline activator. This allows not only to accelerate the hardening of the GPV, but also ensures increased density and impermeability of the structure after the completion of the hydration process during the formation of soil concrete massifs with specified properties and geometric parameters in conditions of subsidence loess soils during the construction and operation of buildings and structures for various purposes.

3.1 Technological Principles of Performing Drilling and Injection Works to Eliminate Subsidence Properties of Loess Soils

The technology of soil concrete mass formation is based on preserving the natural structure of the soil by low-pressure injection at low pressure (up to 0.5 MPa). In this case, during the injection process, the injection mixture completely fills the capillary-porous structure of the soil with the formation of a continuous soil concrete mass.

Compositions of injection mixtures it is necessary to take into account that the main laws of concrete science are not acceptable for soil concretes formed in the mode of impregnation of the soil structure.

For comparative analysis, mechanical mixing of the injection mixture with the studied sandy soil was carried out, followed by a study of its characteristics.

The compressive strength of soil concrete prepared by injection impregnation was compared with the strength of soil concrete formed by the method of mechanical mixing with fixation of the kinetics of hardening at various indicators of the water-binder ratio. The quantitative characteristic of the comparison was expressed by the empirical hardening coefficient "K_u":

$$K_u = R_p/R_s$$

where R_p MPa – compressive strength of a soil concrete massif sample formed by the injection impregnation method;

R_s, MPa – compressive strength of soil concrete formed by mechanical mixing of soil and mixture.

According to the classical theory of concrete science, the characteristics of hardened concrete are determined mainly by the ratio of the components of the concrete mixture in the system "binder - coarse and fine aggregate - water - additives". When analyzing the properties of soil concrete, it was found that with an equal amount of binder in the composition of soil concrete prepared using injection technology and mechanical mixing, the strength characteristics of soil concrete differ significantly and do not correlate with each other.

The high strength of soil concrete obtained in the mode of impregnation of the structure of loess soil is explained by the fact that the microstructure of the soil massif is a three-component system consisting of solid particles of naturally formed packaging, strong salt bonds between individual particles of soil, which are covered with films of X-ray amorphous minerals and have inclusions of dusty and dispersed particles of various compositions with varying degrees of reactivity. The intergranular volume of soil usually consists of open pores and capillaries. In a natural dry state, loess soil has a fairly high natural strength due to the formed structural salt bonds. When filling the pore space with an injection mixture, the natural structure of the soil receives additional strengthening.

In this case, salt bonds are preserved, and the dusty parts on the surface of the soil grains are involved in the process of structuring of the GPV, forming a high-quality contact zone between the injection mixture and the soil elements stages, which should be reflected in the regulations for injection work (Fig. 4).

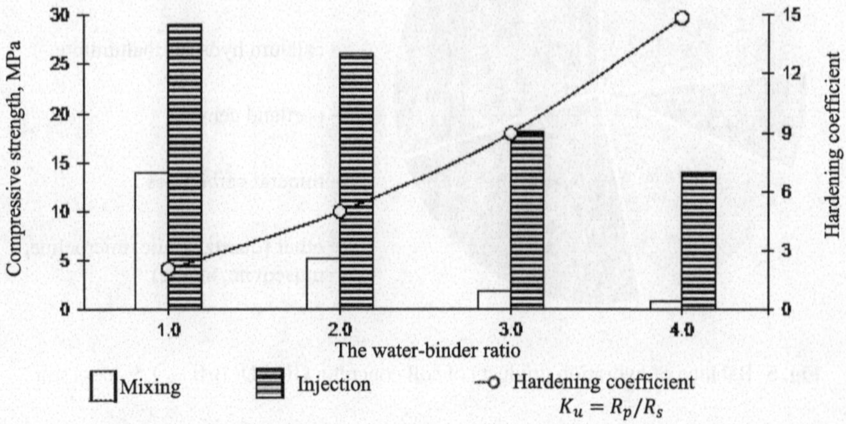

Fig. 4 Compressive strength of soil concrete prepared by injection impregnation and mechanical mixing: GPV-; hardening 28 days under normal conditions

The conducted studies of the hydration products of the fixed injected soil (soil concrete) in the capillary-porous structure of the injected soil showed the participation of amorphous sand grains in the process of structure formation along with hydrosilicate, calcium hydrosulfoaluminate and portlandite (Fig. 5).

After completion of the intensive hydration process, a fixed massif with specified physical, mechanical and geometric parameters is formed in the soil concrete.

The conducted analysis of the structure of the soil concrete confirmed that the hydrated injection mineral composition filling the system of capillaries and pores of the intergranular space of the soil is a sufficiently dense hardened binder composition that tightly connects adjacent grains, forming a strong and dense structure similar to the microstructure of fine-grained concrete, where the role of individual soil particles is played by fine aggregate (Fig. 6).

The study of the contact zone "cement stone - soil particles" showed a fairly homogeneous dense defect-free microstructure of the impregnated soil massif, providing a joint mechanism for perceiving loads as a dense conglomerate. In the soil concrete massif obtained by mixing, an increased pore volume and specific volume of capillary porosity are noted (2.5–3 times compared to impregnated samples) - on average 85% of open pores and capillaries and 15% closed. The high degree of homogeneity of the dense structure of the injection-impregnated consolidated soil massif is formed due to the features of the impregnation technology, when the mineral mixture based on high-penetration GPV penetrates into the intergranular space of soil particles without causing grain separation, connecting the incoherent soil into a single framework.

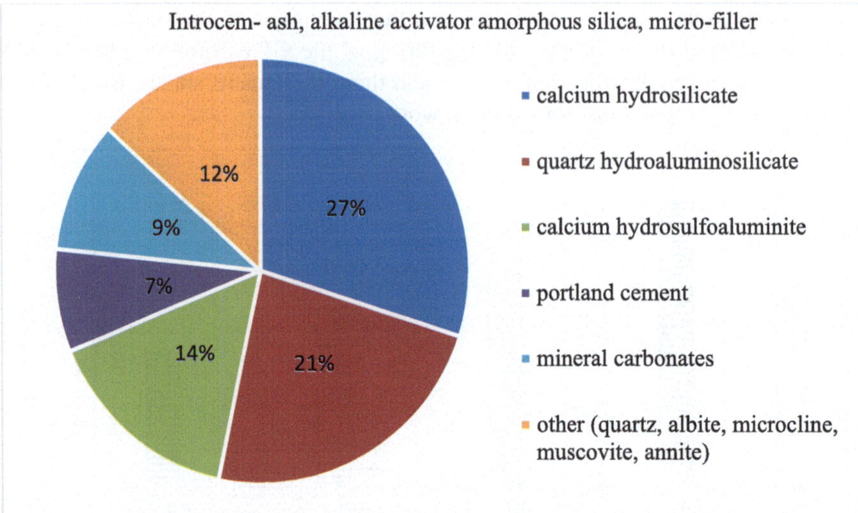

Fig. 5 Balance of hydration products of soil concrete: GPV-U: B/B = 1.5; loess soil

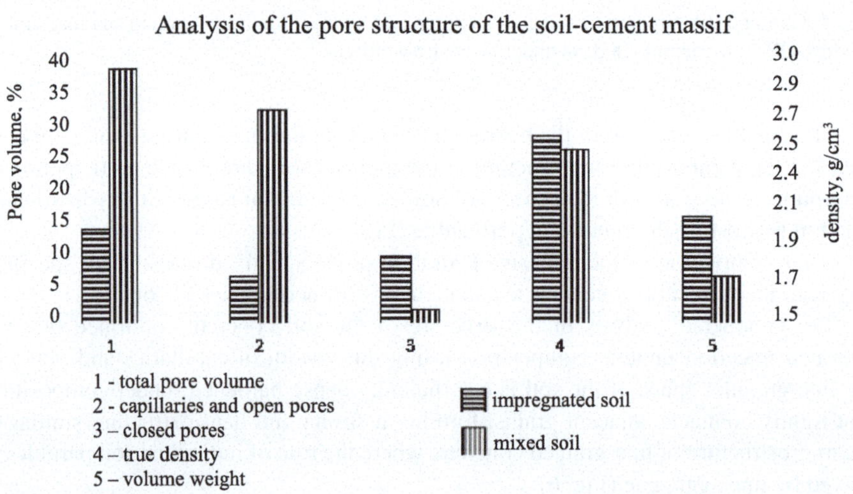

Fig. 6 Characteristics of the porosity of the structure of soil concrete samples: GPV-U; B/binding = 1.5, loess soil model

3.2 Technological Principles of Performing Drilling and Injection Works to Eliminate the Sedimentary Deformations that Have Appeared

Operation of buildings and structures for various purposes in loess soils is often accompanied by the development of uneven, excessive sedimentary deformations, limiting or completely eliminating their operational suitability. An effective solution to the problem of eliminating the sedimentary deformations that have appeared is the use of compensatory injection (CI) technology using injection mixtures based on GPV. The essence of CI technology is to compensate for the soil deficiency formed in the foundation of the building due to the development of sedimentary processes. The implementation of bored injection works according to the KN technology is preceded by a set of preparatory measures, providing for a thorough examination of the state of the main load-bearing structures of buildings and structures that need protection from the development of possible deformations or uplift, engineering-geological and geophysical studies of the foundation bases with the determination of the actual physical and mechanical characteristics of the soil, the development of a mathematical model that adequately describes the formation and development of the SSS of the soil mass at all stages of bored injection works during KN, the calculation justification of the main process parameters (amount of injection mixture, pressure and injection intensity, distribution of the injection mixture by the area and height of the soil base), the forecast of the development of movements of the main building structures during bored injection works. As a rule, compensatory injection is implemented in 5 stages.

Stage 1 - preparatory work. At this stage, preparatory work is performed, including the organization of the construction site, preparation of drilling and mixing equipment, assembly of collar pipes and injection materials, preparation of equipment for laboratory control and adjustment of the quality of the applied injection mixtures and shell solution, arrangement of the facility with automated monitoring tools, formation of the Lifting Control Center (LCC). At the stage of preparatory work, marking and fixing of the well grid on the daylight surface is performed. In areas of work with the location of underground utilities, it is necessary to mark the wells in the presence of representatives of the operating organizations.

Stage 2 - execution of drilling operations with the arrangement of wells. Drilling is performed using drilling equipment that ensures the arrangement of wells with a diameter of up to 150 mm with a design length of 15 … 30 m with a controlled and monitored drilling route. The well body is filled with a special casing solution with high sedimentation resistance and adhesive strength with the surface of the sleeve pipes, with adjustable viscosity and hardening kinetics of the Solidur type or according to TU 5734-201-43260-800-2016.

In order to fix the sleeve injectors along the well axis, it is recommended to use centralizers. The quality of the well arrangement largely determines the efficiency of injection work during compensation injection. After installing the sleeve pipes in the well body, before starting the injection work, a technological break must be provided, necessary for the hardening of the collar solution to a value of 1.0 MPa, which is from 2 to 3 days. In case of violation of reliable contact of the collar solution with the surface of the injector, the injection mixture is distributed in the base of buildings and structures in a disorganized manner, up to manifestations at the wellhead, which excludes control of the compensation injection process.

Stage 3, preparation of the enclosing volume of soil. It is carried out in order to bind the unstable structure of the soil by impregnating its capillary-porous structure with a highly permeable solution "KN-1". In this case, the values of the deformation modulus and Poisson's ratio of the massif increase compared to its initial values due to a decrease in the porosity coefficient. This reduces the risk of uncontrolled hydraulic fracturing during injection, reduces the consumption of the injection mixture for the formation of a compressible soil layer at the base of the structures, which ultimately leads to an increase in the injection efficiency coefficient.

Stage 4, injection of the "jack" injection mixture. It is performed as sedimentary deformations appear due to construction work near the structure or stress relaxation in the volume of the "matrix", which ensures that the design position of the structure is preserved. The injection sequence is determined by the direction of transitions from the lower horizons towards the upper ones. The technological parameters of injection (quantity, pressure and intensity of injection, injection location) are determined in accordance with the calculation justification.

Stage 5 - liquidation or conservation of injection wells. It is performed after compensation for possible sedimentary deformations and complete planned - high-altitude stabilization of the structure. In this case, it is advisable to perform not liquidation, but conservation of injection wells, which will allow performing KN works at all stages of the life cycle of the structure based on the analysis of the results of geotechnical monitoring.

Due to the fact that the development of sedimentary deformations in the conditions of loess soils develops as a result of their soaking and is distributed over the area of the building site extremely unevenly, a mandatory condition for the beginning of work to eliminate sedimentary deformations is the elimination of the subsidence properties of the soils under the entire building.

4 Conclusions

As a result of the conducted research, the main patterns of the influence of the ratio of components on the properties of GPV and injection mixtures based on them were established.

The use of injection mixtures based on GPV ensures not only the elimination of subsidence properties of loess soils, but also the formation of soil-concrete massifs with a compressive strength of soil-concrete massifs of up to 20 … 25 MPa, which allows them to be considered as a structural element of a building. The developed compositions of injection mixtures are an important element of the technology of compensatory injection in eliminating the manifested subsidence deformations of buildings and structures.

References

1. Glukhovsky, V.D., Pakhomov, V.A.: Slag-Alkali Cements and Concrete. Budivelnik, Kyiv (1978)
2. Krivenko, P., Kavalerova E.: Performance of alkali-activated cements - prospective ways for carbon dioxide emissions reduction. Proceed. 3rd International Symposium "Non-traditional cement & concrete", Brno, pp. 389–398 (2008)
3. Rakhimova, N.R., Rakhimov, R.Z., Stoyanov, O.V.: Slag-alkali composite materials for protection against radioactive radiation and immobilization of radioactive waste. Bull. Kazan Tech. Univ. **7**, 129–131 (2013)
4. Davidovits, J.: Chemistry of Geopolymeric Systems Terminology. Geopolymer, pp. 9–40 (1999)

5. Davidovits, J.: Geopolymer chemistry and sustainable development. The Poly(sialate) terminology: a very useful and simple model for the promotion and understanding of green-chemistry. Geopolymer, pp. 9–15 (2005)
6. Davidovits, J.: Geopolymer Chemistry and Applications. Geopolymer Institute (2008)
7. Palomo, A., Glasser, G.: Chemical-Bonded cementitious materials based on metakaolin. Br. Ceram. Trans. J. **91**(4), 107–112 (1992)
8. Palomo, A., Alonso, S., Fernández-Jiménez, A., Sobrados, I., Sanz, J.: Alkaline activation of fly ashes: NMR study of the reaction products. J. Am. Ceram. Soc. **87**(6), 1141–1145 (2004)
9. Eroshkina, N.A., Korovkin, M.O., Tymchuk, E.I.: Structural formation of geopolymers. Young Sci. **7**, 123–126 (2015)
10. Barbosa, V.F.F., Mackenzie, K.J.D., Thaumaturgo, C.: Synthesis and characterization of materials based on inorganic polymers of aluminum and silica: sodium polysialate polymers. Int. J. Inorg. Mater. **2**(4), 309–317 (2000)

Defining Eigenvalues and Functions in Problems About Vibrations of a Vertical Stand

Lyalusya Baragunova, Maryana Shogenova(✉), Valentina Vodakhova, and Astemir Shinakhov

Kabardino-Balkarian State University Named After Kh.M. Berbekov, Nalchik 360004, Russia
shogenova_mar@mail.ru, v.a.vod@yandex.ru, Lilala07@mail.ru

Abstract. Vibrations of buildings and structures, transport and construction equipment, various mechanical engineering and instrument making objects are of undoubted interest. Analytical methods are in most cases ineffective. The finite difference method and other numerical methods for calculating elastic elements with simple geometry are very relevant and attractive. Numerical methods are distinguished by the simplicity and transparency of algorithms and computer programs [1–3]. The formulation of the initial-boundary value problem on the dynamic loading of an elastic rod is presented [4–6]. With its hinged ends, the spectrum of eigenvalues (frequencies) and functions (shapes) of vibrations is determined, which are used in problems of vibrations under the action of an external harmonic load. The vertical stand is compressed by an axial force under its own weight or the weight of the equipment. The d'Alembert principle is used. In problems about eigenvalues and functions, boundary conditions are added to the equation. The upper end of the rack is free, and the lower end is rigidly fixed. Eigenvalues cannot be found in closed form.

Keywords: Rack · Vibration Functions · D'alembert's Principle · Analytical Methods · Gravitational Acceleration · Initial-Boundary Value Problem · Spectrum of Eigenvalues · Young's Modulus of the Material · Geometric Characteristics of the Section

1 Introduction

Vibrations of buildings and structures, transport and construction equipment, various mechanical engineering and instrument making objects are of undoubted interest. In particular, vertical rods are widely used as masts, transmission line supports, antenna devices, etc.

With an integrated approach to solving complex problems in the theory of oscillations and its applications in technology, analytical methods in most cases do not give the expected effect. For the calculation of elastic elements with simple geometry, the finite difference method is very attractive, the meaning of which is to replace partial derivatives in differential operators with their approximate values, represented through discrete values of functions at the nodes of the computational grid.

S. V. Klyuev et al. (Eds.): ICRAAC 2024, LNCE 627, pp. 418–425, 2025.
https://doi.org/10.1007/978-3-031-82938-3_46

The finite difference method is distinguished by the simplicity and transparency of algorithms and computer programs [1–3], although skill is required to create a suitable mesh. The most frequently solved problems are in the field of mechanics. In this case, using a method widely used in modern numerical analysis, a complex problem in problems of the theory of oscillations was solved, which is difficult to solve analytically.

In [4–6] the formulation of the initial-boundary value problem on the dynamic loading of an elastic rod is given. With its hinged ends, the spectrum of eigenvalues (frequencies) and functions (shapes) of vibrations is determined, which are used in problems of vibrations under the action of an external harmonic load.

2 Methods and Materials

A vertically standing rod (stand) under the influence of its own weight or the weight of equipment is compressed in Fig. 1 by an axial force $N(x)$ variable in length. For this purpose, consider the elementary section shown in Fig. 2. Here it is indicated: $u(x,\ t)$– rod deflection, $N(x)$ – longitudinal compressive force, $Q(x)$– shear force, $M(x)$– bending moment in the section, m– linear (linear) mass of the rod material, $m\,\ddot{u}(x,\ t)$– distributed inertia force, $q = mg$– uniformly distributed own weight of the rod, $m = \rho A$, A – the cross-sectional area of the rod, ρ– the density of the material.

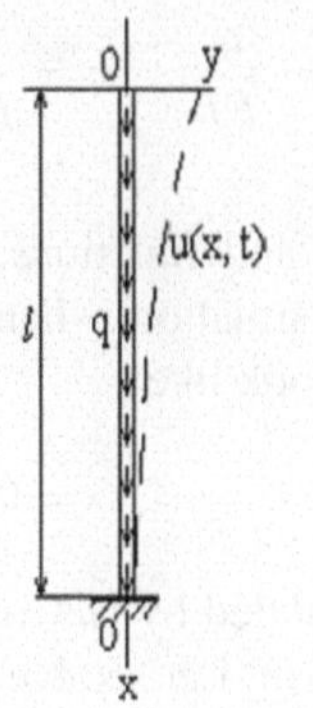

Fig. 1. Calculation scheme.

Using d'Alembert's principle, write out the equation in projection onto the y-axis, taking into account the smallness of linear and angular displacements

$$Q - Q - dQ - Nu' + (N + dN)\left(u' + du'\right) - m\ddot{u}\,dx = 0 \tag{1}$$

Let's take into account that $N = qx, \quad Q = EJ\,u'''$, Let's make obvious reductions, eliminate quantities of higher orders of smallness and arrive at a partial differential equation

$$EJ\,u^{IV} + q\left(u' + x\,u''\right) + m\ddot{u} = 0 \tag{2}$$

where E is the Young's modulus of the material, J- moment of inertia of the rack section, A- cross-sectional area, g- acceleration of gravity. Primes and Roman numerals

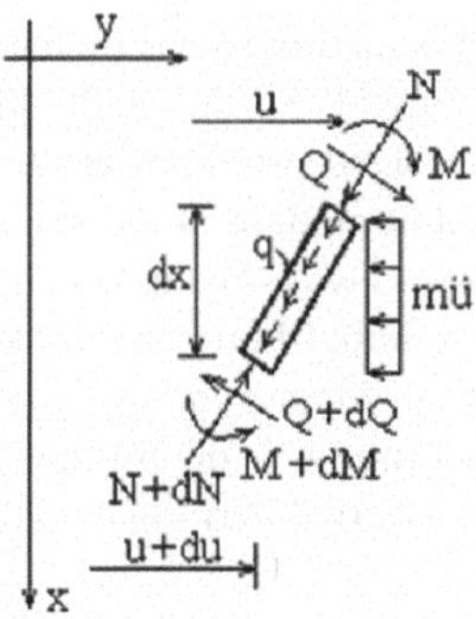

Fig. 2. Compressed bar element.

correspond to differentiation with respect to x, dots above the functions mean differentiation with respect to time t. Equation (2) is divided by EJ and transform it into a more convenient form

$$u^{IV} + aq\left(u' + x\,u''\right) + b\ddot{u} = 0, \quad x \in (0,\ l), \quad t > -\infty, \tag{3}$$

where

$$\mathrm{a} = \frac{q}{EJ}, \quad b = \frac{m}{EJ}$$

.

In problems about eigenvalues and functions, boundary conditions are added to Eq. (3). In a frequently encountered situation, when the upper end of the rack is free and the lower end is rigidly fixed, they look like

$$u''(0, t) = 0, \ \ u'''(0, t) = 0, \ \ u(\mathrm{l}, t) = 0, \ \ u'(\mathrm{l}, t) = 0, \ \ t > -\infty \tag{4}$$

Problem (1), (2) was first formulated by L. Collatz [7–9], indicating that its eigenvalues cannot be found in closed form. Let's solve the problem using a combination of analytical, numerical and graphical methods.

Using the method of separation of variables, we represent the function $u(x,\ t)$ as a work

$$u(x, t) = X\,(x)\,e^{\mathrm{j}\omega t} \tag{5}$$

where

ω - characteristic indicator, frequency of free oscillations, j - imaginary unit.

Substituting (5) into (3), (4) gives

$$X^{IV}(x) + \mathrm{a}[X' + \mathrm{x}X''(x)] - \omega^2 \mathrm{b}X\,(x) = 0 \quad x \in (0,\ \mathrm{l}) \tag{6}$$

$$X''(0) = 0, \ \ X'''(0) = 0\,, \ \ X(l) = 0\,, \ \ X'(l) = 0\,. \tag{7}$$

In problem (6), (7) ω is an eigenvalue at which the eigenfunction will be non-trivial, i.e. $X\,(x) \not\equiv 0$. To determine them, we move from the differential Eq. (4) to the difference equation. For this purpose, we replace the region of continuous change in the argument $0 \le x \le 1$ discrete set of points (grid) with a step $h = l/(n-1)$ and many nodes $\overline{\Omega} = \{x_i = (i-1)\,h,\;\; i = 1,\;\; 2,\; \ldots\;,\; n\}$. Instead of the function of continuous argument X(x), we introduce the function.

$y(x_i) = y_i$ discrete argument, i – grid function node number. Thus, further X(x) will be represented by a vector $Y = \{y_1,\;\; y_2,\; \ldots,\;\; y_n\}$. In Eq. (6), the derivatives at the point x_i approximated using difference operators of the 2nd order of accuracy

$$X''(x_i) \approx \frac{1}{\mathrm{h}^2}(y_{i-1} - 2y_i + y_{i+1}),$$

$$X^{IV}(x_i) \approx \frac{1}{h^4}(y_{i-2} - 4y_{i-1} + 6y_i - 4y_{i+1} + y_{i+2}),\quad i = 3,\;\; 4,\;\; \ldots,\;\; n-2. \tag{8}$$

Replacement introduced here $y_i \approx X(x_i)$. After substituting derivatives and simple transformations, we write instead of (6) the finite difference equation

$$y_{i-2} +_{\mathrm{i}}\, y_{i-1} + d_i y_i + c_i\; y_{i+1} + y_{i+2} = 0,\quad i = 3,\;\; 4,\;\; \ldots,\;\; n-2, \tag{9}$$

where indicated

$$c_i = -4 - 0,5\mathrm{ah}^3 + ah^2x_i\,,\quad d_i = 6 - 2ah^2x_i - \beta h^4\omega^2,\quad e_i = -4 + 0,5\mathrm{ah}^3 + ah^2x_i$$

Let us make similar replacements in boundary conditions (7) and obtain

$$\begin{gathered} 2y_1 - 5y_2 + 4y_3 - y_4 = 0,\quad -5y_1 + 18y_2 - 24y_3 + 14y_4 - 3y_5 = 0, \\ y_n = 0,\quad y_{n-2} - 4y_{n-1} + 3y_n = 0. \end{gathered} \tag{10}$$

Combining Eqs. (9), (10) leads to a homogeneous algebraic system of equations, which can be rewritten in matrix-vector form

$$\mathrm{G}(\omega)\,\mathrm{Y} = 0 \tag{11}$$

where G(ω) is a square matrix of order n. It can be easily shown that the coefficient matrix is band and six-diagonal

$$\mathbf{G}(\omega) = \left(\begin{array}{cccccccccccc} 2 & -5 & 4 & -1 & & \vdots & & & & & \\ -5 & 18 & -24 & 14 & -3 & \vdots & & & & & \\ 1 & {}_3 & \mathrm{d}_3 & e_3 & 1 & \vdots & & & & & \\ & 1 & {}_4 & d_4 & e_4 & \vdots & & & & & \\ \cdots & \cdots & \cdots & \cdots & \cdots & \cdots & \cdots & \cdots & \cdots & \cdots & \cdots \\ & & & & & \vdots & 1 & {}_{n-2} & d_{n-2} & e_{n-2} & 1 \\ & & & & & \vdots & & & 1 & -4 & 3 \\ & & & & & \vdots & & & & & 1 \end{array} \right)$$

Here the zero elements are not written out. Elements c_i, d_i, e_i the matrices are functions of the characteristic exponent ω, i.e., the eigenvalue of the problem (6), (7). The condition for the existence of a nontrivial solution to the system of Eqs. (11) gives the characteristic equation

$$\det\mathbf{G}() = 0 \tag{12}$$

from which the eigenvalues are determined, i.e. the spectrum $\{\omega_1, \ \omega_2, \ldots\}$.

Graph of the function on the left side (12) in the form $\omega \ - \ \det G$ easily visualized on a monitor screen (for example, in the MatLab system). Points of intersection with the axis ω next, the required values are determined, which can be counted from the screen with a high degree of accuracy.

3 Results and Discussions

Example 1. We will test the proposed computational schemes on a rack with conditional parameters

$$1 = 1, \mathrm{n} = 101, \mathrm{EJ} = 1, \mathrm{A} = 1, \rho = 1, \mathrm{q} = 0.$$

The exact values of the natural frequencies of such a rod are known [10–12]. Values ω_k, obtained by accurate and numerical methods are presented in Table 1.

Table 1. Values of natural frequencies of the vertical column.

Exact values	3.516	22.034	61.701
Approximate values	3.517	22.058	61.794

The three eigenvalues of the table in the second row are directly taken from the graph displayed on the computer monitor in Fig. 3.

It is obvious that there is a high degree of accuracy in determining the eigenvalues. The finite difference method, in combination with the capabilities of visualizing the calculation results of the MatLab system, as tests have shown, allows one to determine the eigenvalues of partial differential equations, which have numerous applications in engineering and construction [11–13].

$$\begin{pmatrix} -5 & 4 & -1 & & & \vdots & & & & \\ 18 & -24 & 14 & -3 & & \vdots & & & & \\ c_3 & d_3 & c_3 & 1 & & \vdots & & & & \\ 1 & c_4 & d_4 & c_4 & 1 & \vdots & & & & \\ \cdots & \cdots & \cdots & \cdots & \cdots & \vdots & \cdots & \cdots & \cdots & \cdots \\ & & & & & \vdots & 1 & c_{n-2} & d_{n-2} & c_{n-2} & 1 \\ & & & & & \vdots & & & 1 & -4 & 3 \end{pmatrix} = \begin{pmatrix} z_1 \\ z_2 \\ z_3 \\ z_4 \\ \cdots \\ z_{n-2} \\ z_{n-1} \end{pmatrix} = \begin{pmatrix} -2 \\ 5 \\ 1 \\ 0 \\ 0 \\ 0 \end{pmatrix} \tag{13}$$

The spectrum of eigenvectors corresponding to the eigenfunctions can be established using the system of Eqs. (11). Due to the fact that the determinant of the matrix G is equal to zero, the eigenforms can be found only up to a factor. Then we can accept, for example, that $y_1 = 1$, and find the remaining unknowns from the system of equations formed from (9) by eliminating the last or any other line. Note that the right side of the system of equations is formed from the elements of the first column of the matrix G. Between vector elements Y and Z there is a one-to-one correspondence

$$\{y_1, y_2, y_3, \ldots, y_{n-1}, y_n\} = \{1, z_1, z_2, \ldots, z_{n-1}, 0\}.$$

Alternately substituting eigenvalues into formulas for calculating matrix elements G and then solving the system of Eqs. (11) will lead to obtaining eigenvectors [14–16]. Note that the right side of the system of equations is formed from the elements of the first column of the matrix G.

Example 2. Based on the data in Example 1, calculate the first three eigenvectors Yk(x).

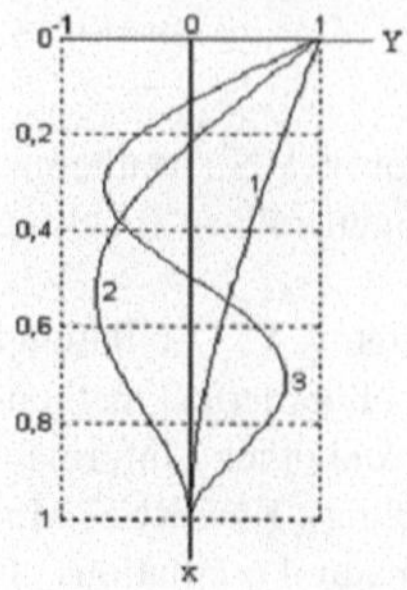

Fig. 3. First eigenvectors.

The calculation results in the form of graphs are shown in Fig. 3. The numbers of the curves coincide with the numbers of the eigenvalues; normalization of eigenvectors was carried out, satisfying the condition

$$\max_i |y_i| = 1$$

The graphs shown here coincide with those given in the technical literature [17, 18].

Example 3. Consider a steel rack with real input data

$$l = 2\,\text{m},\ \ d = 1\,\text{cm},\ \ \rho = 7810\,\text{kg/m}^3,\ \ E = 210\,\text{GPa},\ J = \pi d^4/64,\ \ n = 101.$$

The results obtained by the computer program are shown in Fig. 3.

4 Conclusions

Solving engineering problems comes down to considering systems of equations that have a unique solution only if the value of some parameter included in them is known. This parameter is its own value. Eigenvalue problems are often solved in the field of mechanics. In the dynamic analysis of mechanical systems, the eigenvalues correspond to the natural frequencies of oscillations, and the eigenvectors characterize the modes of these oscillations.

The choice of the most effective method for determining eigenvalues and eigenvectors for engineering problems depends on the type of equations. The spectrum of eigenvectors corresponding to the eigenfunctions can be established using a system of equations.

The finite difference method, combined with the capabilities of visualizing calculation results of the MatLab system, allows you to determine the eigenvalues of partial differential equations, which have numerous applications in engineering and construction.

Testing of the proposed computing circuits was carried out on a rack with conditional parameters. As tests have shown, for calculating elastic elements, using the finite difference method is the simplest and most convenient way to determine natural frequencies and natural vectors.

References

1. Zolotov, A.B., Akimov, P.A., Sidorov, V.N.: Mathematical methods in structural mechanics (with the basics of the theory of generalized functions). ASV Publishing House, Moscow (2008)
2. Ilyin, M.M., Kolesnikov, K.S., Saratov, Y.: Oscillation theory. MSTU, Moscow (2003)
3. Verzhbitsky, V.M.: Fundamentals of numerical methods. High school, Moscow (2002)
4. Akimov, P.A., Belostotsky, T.B.: And others: Informatics in construction (with the basics of mathematical and computer modeling). KNORUS, Moscow (2017)
5. Akulenko, L.D.: High-frequency natural oscillations of mechanical systems. Prikl. mat. i fur. Moscow **64**(5), 817–832 (2000)
6. Zolotov, A.B., Akimov, P.A., Sidorov, V.N., Mozgaleva, M.L.: Numerical and analytical methods for calculating building structures. ASV Publishing House, Moscow (2009)
7. Kashevarova, G.G., Permyakova, T.B., Laishcheva, M.E.: Numerical methods for solving construction problems. Part 1. Perm National Research Publishing House, Perm (2015)
8. Kashevarova, G.G., Permyakova, T.B., Laishcheva, M.E.: Numerical methods for solving construction problems. Part 2. Perm National Research Publishing House, Perm (2015)
9. Ilyin, V.P., Karpov, V.V., Maslennikov, A.M.: Numerical methods of solving problems of structural mechanics. Publishing house ASV, Moscow, Saint-Petersburg State University of Civil Engineering and Architecture, Saint-Petersburg (2005)
10. Samarsky, A.A., Mikhailov A.P.: Mathematical modeling. Ideas. Methods. Examples: 2nd ed., rev. FIZMATLIT, Moscow (2005)
11. Formalev, V.F., Reviznikov, D.L.: Numerical methods, 2nd edn. Fizmatlit, Moscow (2006)
12. Baragunova, L.A., Shogenova, M.M.: Stability of multi-span rods. Bull. Dagestan State Tech. Univ. Tech. Sci. **49**(3), 116–122 (2022)
13. Kulterbaev, K., Lafisheva, M., Baragunova, L.: Solving the Euler problem for a flexible support rod base on the finite difference method. In: Tchernykh, A., Alikhanov, A., Babenko, M., Samoylenko, I. (eds.) Mathematics and its applications in new computer systems MANCS 2021, LNNS, vol. 424, pp. 143–151. Springer, Cham. (2022)
14. Kulterbaev, K., Baragunova, L.A., Shogenova, MM: Random Oscillations of the vertical continual and discrete rod under seismic influences. Mater. Sci. Forum **931**, 66–71 (2018)

15. Kulterbaev, Kh.P., Shogenova, M.M., Baragunova, L.A.: On the influence of the characteristic frequency and broadband of seismic effects on the vertical rod oscillations. IOP Conf. Ser. Mater. Sci. Eng. **753**, 042040 (2020)
16. Kulterbaev, Kh.P., Baragunova, L.A., Shogenova, M.M., Senov, Kh.M.: About a high-precision graphoanalytical method of determination of critical forces of an oblate rod. Proceedings 2018 IEEE International Conference "Quality Management, Transport and Information Security, Information Technologies" (IT&QM&IS), pp. 794–796. St. Petersburg. Russia (2018)
17. Kulterbayev, Kh.P.: Determination of eigenvalues in problems on the buckling of compressed rods (part I). ISTC-EARTHSCI. IOP Publishing. IOP Conf. Ser. Earth Environ. Sci. **988**, 052072 (2022)
18. Kulterbayev, Kh.P.: Determination of eigenvalues in problems of loss of stability of compressed rods (part II). ISTC-EARTHSCI. IOP Publishing. IOP Conf. Ser. Earth Environ. Sci. **988**, 052072 (2022)

Simulation of Creep of a Rectangular Wooden Beam Under Prolonged Static Load

Batyr Yazyev[1(✉)], Marcel Magomedov[2], Mikhail Petrov[1], Kirill Chernenko[1], and Anastasia Lapina[3]

[1] Don State Technical University, Rostov-On-Don 344000, Russia
ps62@yandex.ru
[2] Dagestan State University, Republic of Dagestan, Makhachkala 367000, Russia
[3] Southern Federal University, Rostov-On-Don 344006, Russia

Abstract. Depending on the stress level, the operation of wooden beams under prolonged load is characterized by both linear and nonlinear creep. This has been shown by numerous experimental studies of wood. However, it should be noted that the theoretical description of these processes has been little studied or is extremely rarely presented. It follows that one of the priority directions in the calculations of the elements of wooden structures is the derivation of resolving equations of linear or nonlinear creep for various types of stress-strain state. The Maxwell-Thomson equations are used as relations establishing the relationship between stresses and deformations. The technique was tested by comparing the solution with the calculation of well-known researchers. An example of calculation is given for various boundary conditions for fixing a beam of rectangular cross-section loaded with a uniformly distributed load. The deflection value is determined by the grid method. A program has been developed for calculating in the MATLAB package with the possibility of varying the initial data and displaying a graph of the dependence of displacement, bending moment on time. The comparison of the maximum deflection value with the analytical solution is given. It is noted that the stresses practically do not change during creep. The proposed approach can be applied to the analysis of the stress-strain state and bearing capacity of wooden beams of arbitrary cross-section. There are no restrictions on the boundary conditions and the type of loading, and the material of the beam can be not only wood, but also fiberglass.

Keywords: Wood Creep · Relative Deformation · Maxwell-Thomson Equation · Bending Moments · Variable Stiffness

1 Introduction

Researching of wood performance over time can be divided into two dimensions: long-term strength of wood and creep of wood.

Moreover, both of these areas of research are developing quite separately from each other, and most of the works are devoted to long-term strength. Long-term nonlinearity of wood deformation is little-investigated. Eurocode 5 [1, 2] establishes a mandatory

S. V. Klyuev et al. (Eds.): ICRAAC 2024, LNCE 627, pp. 426–435, 2025.
https://doi.org/10.1007/978-3-031-82938-3_47

principle in the calculation of limit states of wooden structures - taking into account the creep of wood. The domestic standards do not take into account the creep of wood in the calculation models of structures. Modern models of creep of wood are based on linear theories. Instantaneous properties of wood in the specified creep studies are established on the basis of Hooke's law, which does not correspond to the nonlinear diagram of the elastic-plastic type inherent in compressed wood. This circumstance was noted by K.P. Pyatikrestovsky [3, 4]. Taking into account instantaneous elastic-plastic deformations in models of creep of wood is a complex problem [5–9].

It is impossible to solve it within the framework of those models of wood creep that have been developed in recent years abroad and are based on the sequential connection of various linear models.

2 Methods and Materials

2.1 Problem Statement

A method that takes into account the complexity of the process of deformation of the material over time by dividing the deformation processes into corresponding stages, which allows replacing the complex nonlinear relationship between stresses and deformations with piecewise linear dependencies often use in the equations of connection (state) for wooden beams [10]. Such stages with their characteristic features can be clearly demonstrated in Fig. 1.

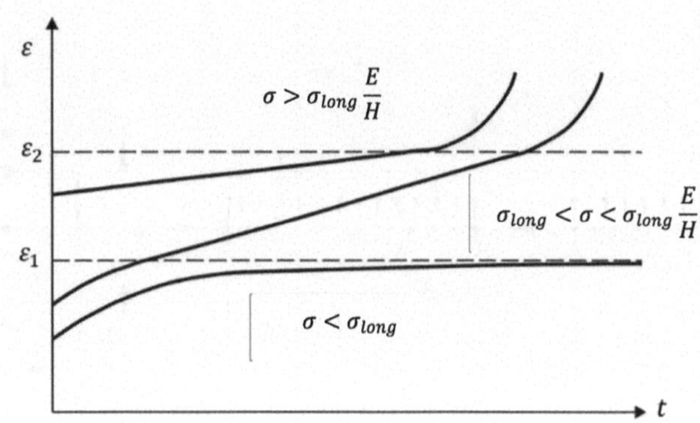

Fig. 1. Deformation curves under prolonged action of static load.

It can be noted from the from the Fig. 1 [11, 12]:

The first stage of deformation is creep, it is reversible, obeys the basic principles of the linear theory of creep and is described by an equation called the simplified law of linear creep (the Maxwell-Thomson equation)

$$n \cdot E_0 \cdot \frac{\partial \varepsilon}{\partial t} + H\varepsilon = \sigma + n \cdot \frac{\partial \sigma}{\partial t} \tag{1}$$

where E_0 is the instantaneous initial deformation modulus equal to the average value of the elastic modulus (appendix SP 64.13330.2017);

$H = (0.6 + 0.75) \cdot E_0$ is a constant that has the physical meaning of the long-term deformation modulus;

σ, ε are stresses and relative deformations (current);

n is the relaxation time.

The second stage of deformation is the stage of steady-state creep

$$n \cdot E_0 \cdot \frac{\partial \varepsilon}{\partial t} = \sigma + n \cdot \frac{\partial \sigma}{\partial t} - \sigma_{long} \tag{2}$$

The third stage of deformation is the stage with increasing deformation rate.

$$n \cdot E_0 \cdot \frac{\partial \varepsilon}{\partial t} - B \cdot \varepsilon = \sigma + n \cdot \frac{\partial \sigma}{\partial t} - \sigma_{pp} \tag{3}$$

where B is a constant that has the meaning of the deformation modulus at critical deformation growth;

σ_{pp} is the short-term ultimate strength equal to the standard resistance of wood (appendix B of SP 64.13330.2017);

$\sigma_{long} = 0.5 * \sigma_{pp}$ is the long-term ultimate strength.

This article considers the first stage of deformation.

2.2 Derivation of Resolving Equations

Let us consider the calculation schemes shown in Fig. 2.

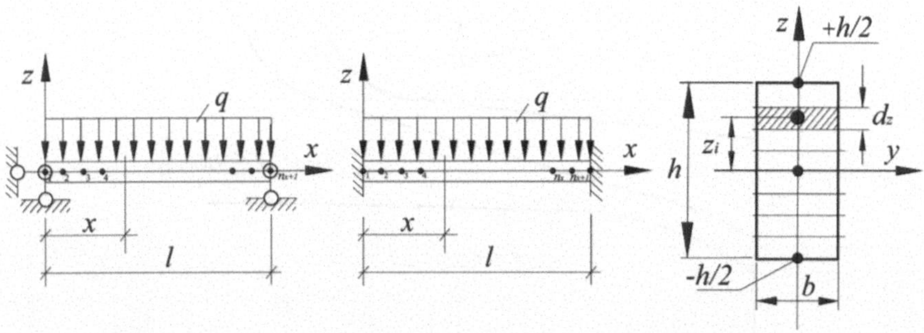

Fig. 2. Calculation schemes of the problem for a rectangular beam.

We will represent the total deformations over time as the sum of the components of elastic ε_{el} and creep deformation ε_{cr}.

$$\varepsilon = \varepsilon_{el} + \varepsilon_{cr} = \frac{\sigma}{E} + \varepsilon_{cr} \tag{4}$$

Considering the geometric aspect of the problem, the total deformation can be represented as:

$$\varepsilon = -z \frac{d^2 x}{dx^2} \tag{5}$$

Stresses can be represented in terms of total deformations using relations (4) and (5):

$$\sigma = -E \cdot \left(z\frac{d^2w}{dx^2} + \varepsilon_{cr} \right) \tag{6}$$

The integral relationship between the bending moment and the normal stress has the form:

$$M = -\int_A \sigma \cdot z dA = -\int_A -E \cdot \left(z\frac{d^2x}{dx^2} + \varepsilon_{cr} \right) z dA \tag{7}$$

Opening the brackets in the integrand (7), we'll get:

$$M = E\left(\int_A z^2 dA \cdot \frac{d^2w}{dx^2} + \int_A \varepsilon_{cr} z dA \right) \tag{8}$$

The first integral is nothing more than the fundamental moment of inertia:

$$I_y = \int_A z^2 dA = b \int_{-h/2}^{+h/2} z^2 dz = \frac{b \cdot h^3}{12}$$

We can write the second integral as:

$$M_{cr} = \int_A \varepsilon_{cr} z dA = Eb \int_{-h/2}^{+h/2} z \cdot \varepsilon_{cr} dz \tag{9}$$

As a result, the resolving equation takes the form:

$$M = EI \cdot \frac{d^2w}{dx^2} + M_{cr} \tag{10}$$

Substituting into the equation of state (1), the total deformations from relation (4), we obtain:

$$n \cdot E \cdot \left(\frac{1}{E}\frac{\partial\sigma}{\partial t} + \frac{\partial\varepsilon_{cr}}{\partial t} \right) + H \cdot \left(\frac{\sigma}{E} + \varepsilon_{cr} \right) = \sigma + n \cdot \frac{\partial\sigma}{\partial t}$$

Or, finally, the creep strain rate:

$$\frac{\partial\varepsilon_{cr}}{\partial t} = \frac{1}{n \cdot E}\left[\sigma\left(1 - \frac{H}{E} \right) - H \cdot \varepsilon_{cr} \right]$$

A uniform time grid is introduced to calculate creep deformations. The value of ε_{cr} at time $(t + \Delta t)$ is calculated based on the Euler method [6]:

$$\frac{\partial\varepsilon_{cr}}{\partial t} \approx \frac{\varepsilon_{cr,t+\Delta t} - \varepsilon_{cr,t}}{\Delta t} \Rightarrow \varepsilon_{cr,t+\Delta t} = \varepsilon_{cr,t} + \frac{\partial\varepsilon_{cr}}{\partial t}\Delta t \tag{11}$$

A similar approach to solving creep problems is used in [7–10].

2.3 Calculation Method

The finite difference approximation of Eq. (10) based on the figure (Fig. 2.a) will take the form:

$$\frac{w_{i+1} - 2w_i + w_{i-1}}{\Delta x^2} = \frac{1}{EI}(M_i + M_{cr,i}) \quad (12)$$

The boundary conditions are the equality of the deflection at the supports to zero:

$$w_1 = 0; \; w_{n+1} = 0$$

By dividing the beam into n segments, we reduce the problem to a system of linear algebraic equations of the form:

$$[A] \cdot \{w\} = \{B\} \quad (13)$$

where [A] is the matrix of coefficients for w_i with dimension $(n + 1; n + 1)$.

$$[A] = \frac{1}{\Delta x^2}\begin{bmatrix} 1 & 0 & 0 & \ldots & 0 & 0 & 0 \\ a_2 & b_2 & c_2 & \ldots & 0 & 0 & 0 \\ 0 & a_3 & b_3 & \ldots & 0 & 0 & 0 \\ \ldots & \ldots & \ldots & \ldots & \ldots & \ldots & \ldots \\ 0 & 0 & 0 & \ldots & a_n & b_n & c_n \\ 0 & 0 & 0 & \ldots & 0 & 0 & 1 \end{bmatrix}$$

where

$$a_i = 1; \; b_i = -2; \; c_i = 1$$

$\{w\}$– vector of unknown nodal displacements;

$$\{w\} = \{w_1 \quad w_2 \quad \ldots \quad w_n\}^{\mathrm{T}}$$

$\{B\}$– vector of the right hand side (free terms);

$$\{B\} = \{0 \;\; f_1 \;\; f_2 \quad \ldots \quad w_n \quad 0\}^T$$

$$f_i = \frac{M_i}{EI} - \frac{M_{cr}}{EI}; \; M_i = \frac{qx}{2}(l - x), \; (0 < x < l)$$

As for M_{cr}, having determined from (11) the value of $\varepsilon_{cr,i}$ for time $t + \Delta t$, we can calculate the integral (9) numerically:

$$\int_A \varepsilon_{cr} z dA = \Delta z\left(\frac{\varepsilon_{cr,0}}{2} + \sum_{i=1}^{n-1} \varepsilon_{cr,i} + \frac{\varepsilon_{cr,n}}{2}\right)$$

where

Δz– integration step;

$\varepsilon_{cr,0}$ and $\varepsilon_{cr,n}$– initial and final values of the function;

ε_{cr}; $\varepsilon_{cr,i}$– intermediate values of a function.

As for the diagram in Fig. 2b, the boundary conditions for such a problem are as follows:

$$w_1 = 0; \varphi_1 = \left.\frac{dw}{dx}\right|_1 = 0; w_{n+1} = 0; \varphi_{n+1} = \left.\frac{dw}{dx}\right|_{n+1} = 0$$

Four boundary conditions do not allow the use of the second-order differential Eq. (10). To satisfy all boundary conditions, Eq. (10) must be differentiated twice with respect to x:

$$\frac{d^4 w}{dx^4} = \frac{1}{EI}\left(-q - \frac{d^2 M_{cr}}{dx^2}\right) \tag{14}$$

Note. In Eq. (14), the load q is considered positive if it is directed upwards. In the case under consideration, the load is directed downwards, so the first term in brackets (14) is preceded by the "-" sign. The difference approximation of Eq. (14) is written as:

$$\frac{w_{i+2} - 4w_{i+1} + 6w_i + 4w_{i-1} + w_{i-2}}{\Delta x^4} = \frac{-q_i}{EI} - \frac{1}{EI}\left(\frac{M_{cr,i+1} - 2M_{cr,i} + M_{cr,i-1}}{\Delta x^2}\right);$$

$$i = 2, 3, 4, \ldots n \tag{15}$$

When compiling Eq. (14) for nodes 2 and n, it becomes necessary to determine the deflections at the so-called contour points, which have indices *0* and $n + 2$ according to [10]. This also applies to the boundary conditions.

As for a beam with a hinged sling, the problem is reduced to a system of linear algebraic equations of the form (13):

where [A] is the matrix of coefficients for w_i with dimensions $(n + 1; n + 1)$.

$$[A]^* = \begin{bmatrix} 1 & 0 & 0 & 0 & 0 & 0 & 0 & 0 & 0 & 0 & 0 & 0 & 0 \\ 0 & 7 & -4 & 1 & 0 & 0 & 0 & 0 & 0 & 0 & 0 & 0 & 0 \\ 0 & -4 & 6 & -4 & 1 & 0 & 0 & 0 & 0 & 0 & 0 & 0 & 0 \\ 0 & 1 & -4 & 6 & -4 & 1 & \ldots & 0 & 0 & 0 & 0 & 0 & 0 \\ 0 & \ldots & \ldots & \ldots & \ldots & \ldots & \ldots & \ldots & \ldots & \ldots & \ldots & 0 & \ldots \\ 0 & 0 & 0 & 0 & 0 & 0 & \ldots & 1 & -4 & 6 & -4 & 1 & 0 \\ 0 & 0 & 0 & 0 & 0 & 0 & \ldots & 0 & 1 & -4 & 6 & -4 & 0 \\ 0 & 0 & 0 & 0 & 0 & 0 & \ldots & 0 & 0 & 1 & -4 & 7 & 0 \\ 0 & 0 & 0 & 0 & 0 & 0 & 0 & 0 & 0 & 0 & 0 & 0 & 1 \end{bmatrix}$$

where

$$[A] = \frac{1}{\Delta x^4} \cdot [A]^*$$

$\{w\}$– vector of unknown nodal displacements;

$$\{w\} = \{w_1 \quad w_2 \quad \ldots \quad w_n\}^{\mathrm{T}}$$

$\{B\}$– vector of the right hand side (free terms);

$$\{B\} = \{0 \;\; f_1 \;\; f_2 \;\; \ldots \;\; w_n \;\; 0\}^T$$

$$f_i = -\frac{q_i}{EI} - \frac{1}{EI}\left(\frac{M_{cr,i+1} - 2M_{cr,i} + M_{cr,i-1}}{\Delta x^2}\right); \; (0 < x < l)$$

3 Results and Discussion

The calculation was performed in the MATLAB software package. From the given dependence (Fig. 3) it is evident that the deformations increase sharply in the period from 0 to 80 days, and then reach the final value.

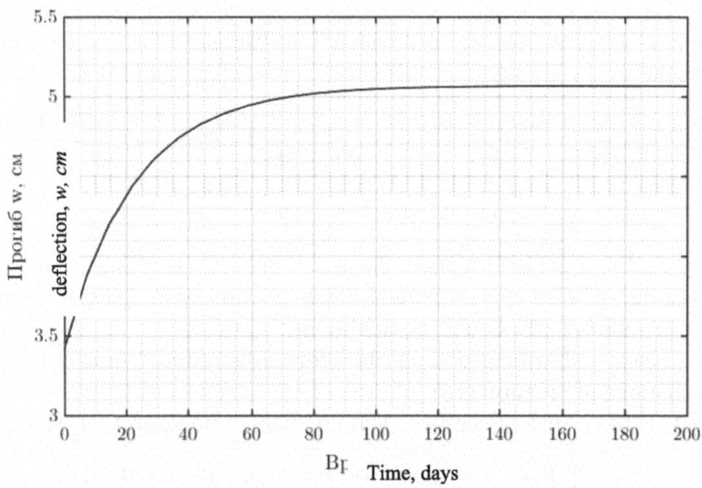

Fig. 3. Graph of maximum deflection growth over time. "Hinge-hinge" scheme.

To determine the maximum deflection in an elastic setting for a hinged beam under the action of a uniformly distributed load from the course of resistance of materials:

$$w_{max} = \frac{5}{384} \cdot \frac{q \cdot l^4}{E \cdot I} \quad (16)$$

The authors of [1] believe that to determine the deflection at the end of the creep process it is sufficient to replace the instantaneous elastic modulus E with the long-term modulus H. One can agree with this statement, but it does not work for all dependencies of the equation of the relationship between deformations and stresses.

When substituting the given dimensions of the beam, we obtain:

$$w_{max} = \frac{5}{384} \cdot \frac{2 \cdot 6^4}{10^7 \cdot \left(\frac{0.1 \cdot 0.2^3}{12}\right)} = 0.0506m \quad (17)$$

The graphical value of w_{max} and the value of w_{max} calculated according to (16) have an error of $\Delta = 1 \cdot 10^{-7}\%$.

The obtained graphs of the dependence of the maximum deflection on time and the distribution of the bending moment over time are shown in.

Figures 4 and 5, both for the "Hinge-Hinge" condition and for "Seal-Seal".

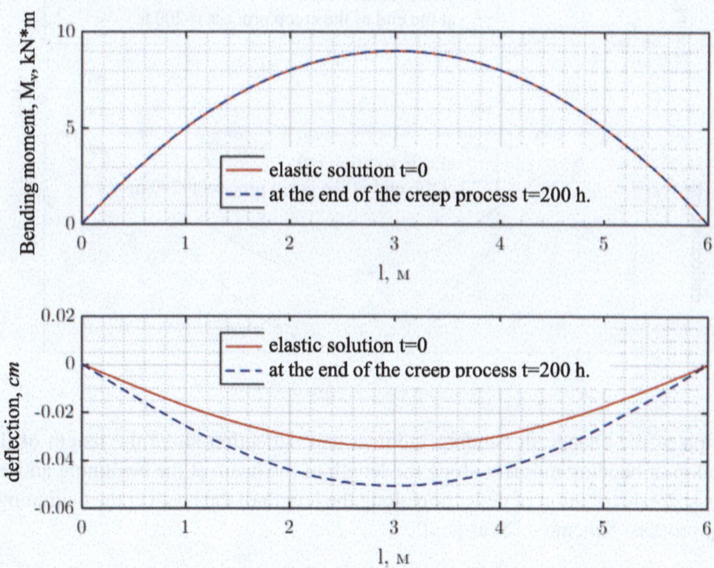

Fig. 4. Graph of the change in bending moment and deflection along the length of the beam: 1 – distribution of bending moment along the beam length at the beginning and end of the creep process; 2 – distribution of deflection along the beam length at the beginning and end of the creep process. Scheme – "Hinge-Hinge".

The graph of the change in deflection over time is shown in Fig. 6. Here we note that the deflection reaches a constant plateau only at 120 h.

We also note that with the chosen creep law, the stresses do not change over time (Fig. 7).

4 Conclusions

If we draw a conclusion based on linear creep, then in this region at $\sigma = const$ we can accept the dependence proposed by Yu. M. Ivanov [6]:

$$\varepsilon(t) = \varepsilon(t_0) \cdot \left(1 + bt^{0.21}\right), whereb = \frac{10^{-2}}{0.735 - 0.02086 \cdot W} \quad (18)$$

W – wood moisture content, %

This dependence can be adopted instead of solving Eq. (1).

If $\sigma < \sigma_{long}$ then the deformations will not exceed the value $\varepsilon_1 = \sigma_{long}/H$ for any long duration of the load, then the second stage of deformation will not occur at all.

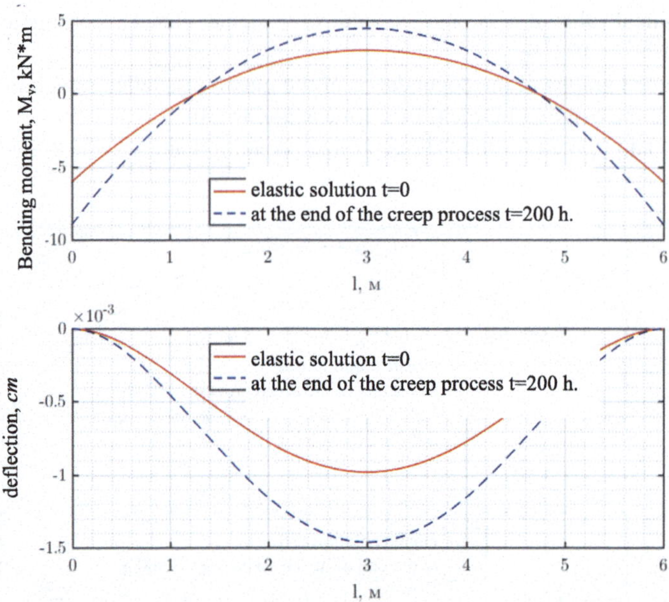

Fig. 5. Graph of the change in bending moment and deflection along the length of the beam: 1 – distribution of bending moment along the length of the beam at the beginning and end of the creep process; 2 – distribution of deflection along the length of the beam at the beginning and end of the creep process. Scheme – "Seal-Seal".

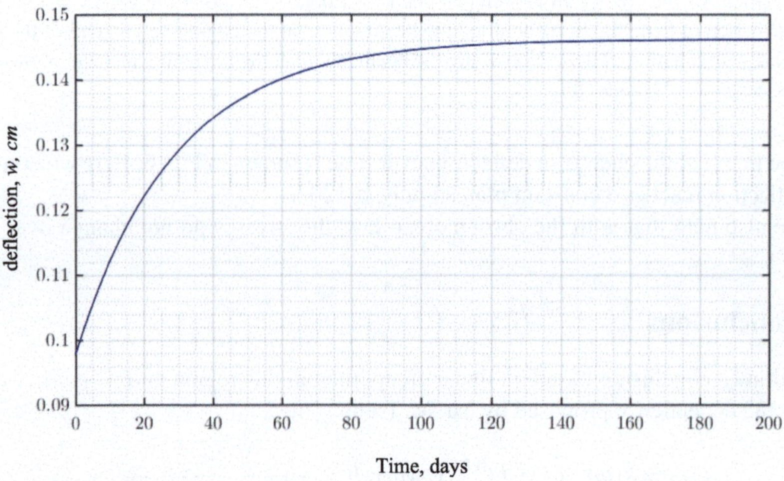

Fig. 6. Graph of maximum deflection growth over time. "Hinge-hinge" scheme.

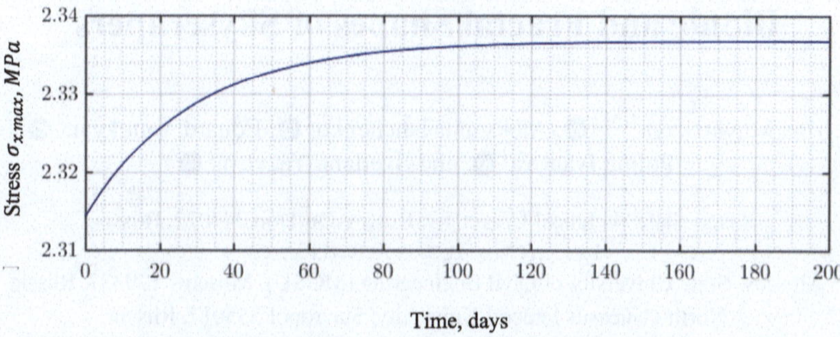

Fig. 7. Dependence of the normal stress on time.

References

1. Yagnuk, B.N.: Theoretical foundations of the design of wooden structures according to the standards of the European Union — Eurocode 5. Direct Media, Moscow-Berlin (2015)
2. BS EN 1995–1–1:2004+A2:2014. Year 5. Design of wooden structures
3. Pyatikrestovsky, K.P., Sokolov, B.S.: Numerical studies of the stress-strain state of the tent roof model under prolonged stepwise increasing loads. Construction and reconstruction. Orel GTU, Izvestia (2009)
4. Pyatikrestovsky, K.P.: Nonlinear methods of mechanics in designing modern wooden structures (Ministry of Education and Science of the Russian Federation, MGSU, 2004)
5. Varenik, K.A.: Calculation of centrally compressed wooden elements taking into account creep (Veliky Novgorod, 2015)
6. Shorstov, R.A., Yaziev, S.B., Chepurnenko, A.S.: Improving the calculation for the stability of compressed wooden rods of variable rectangular cross-section in length. System technologies **1**, 140–150 (2023)
7. Lapina, A.P., Chepurnenko, A.S., Zotov, I.M., Yazev, B.M.: Implementation of the algorithmic method in the calculation of beams on the stability of the aquatic environment. Bulletin of the Volgograd State University of Architecture and Civil Engineering. Series: Construction and Architecture 4(77), 5–16 (2019)
8. Lukash, P.A.: Fundamentals of nonlinear structural mechanics. Stroyizdat, Moscow (1978)
9. Ivanov, Y.: Long–term bearing capacity of wooden structures. Izv. universities. Construction and architecture **11**, 6–12 (1972)
10. Pogoreltsev, A.A., Pyatikrestovsky, K.P.: Issues of further development of structures made of wood and plastics and improvement of standards for designing structures made of wood. PGS **2**, 14–18 (2019)
11. Rzhanitsyn, A.R.: Theory of creep. Stroyizdat, Moscow (1968)
12. Gatiev, M.Sh., Yazyev, B.M., Ivanova, Yu.P., Klyuev, S.V.: Creep of a closed cylindrical tank at hydrostatic pressure. Bulletin of the Dagestan State Technical University. Technical sciences 50(4), 184–190 (2023)

Bionic and Fractal Shapes of Skyscrapers

Irina Mayatskaya[1(✉)], Aleksandr Ishchenko[2], Eduard Yanukyan[3], Polina Ivanova[1], and Svetlana Yazyeva[1]

[1] Don State Technical University, Rostov-On-Don 344022, Russia
irina.mayatskaya@mail.ru
[2] Moscow State University of Civil Engineering (MGSU), Moscow 129337, Russia
[3] North-Caucasus Federal University, Stavropol 355017, Russia

Abstract. Bionic architecture is a part of architecture that uses the forms and organization of biological objects based on mathematical modeling methods and using digital technologies. It is high technology and computer systems that allow modern architects to design skyscrapers, each of which has its own unique style. Such unique structures harmoniously fit into the urban and environmental environment. Bionic and fractal methods allow us to find architectural elements that are interesting in shape and structure. Fractality allows you to create unique forms of high-rise structures that are durable and stable under the influence of external load. The properties of fractality include shear, torsion, sliding, and motion trace. The article discusses the stages of the creation of skyscrapers, the history of their construction in the world. It is shown how bionic shapes help to find the optimal design solution, creating a unique project of a high-rise structure. The achievements of fractal geometry make it possible to create mathematical models of the structure of the artificial environment that a person creates through architecture. It was the development of bionic and fractal architecture that made it possible to create a unique style in architecture. It is necessary to widely introduce mathematical modeling methods into practice to create conditions for more efficient design of modern urban space. Modeling the design processes and the search for optimal solutions using the principles of fractal geometry allows you to create a comfortable urban environment for harmonious human development.

Keywords: Architecture · Mathematical Modeling · Bionics · Fractal · Space · Skyscraper

1 Introduction

The architects turned their attention to the perfect natural forms and the functionality of living organisms. Bionics is the use of principles and methods of organization of living organisms and natural structures in the design of a variety of structures and structures from the point of view of harmony and rationality [1–3]. There are two directions in bionics: architectural bionics and building bionics.

If the first direction is a science that creates architectural forms and structures based on the study of the structure and shape of natural objects, then the second direction is a

S. V. Klyuev et al. (Eds.): ICRAAC 2024, LNCE 627, pp. 436–448, 2025.
https://doi.org/10.1007/978-3-031-82938-3_48

science that studies the structure of natural objects and allows you to find rational forms and properties for use in the design of structures.

The structure of natural objects in the process of evolution acquired an optimal structure, which was able to be realized on the principle of minimizing mass and on the choice of a rational form and design [1, 4]. Living organisms have increased strength, rigidity, stability and high load-bearing capacity. But when designing, engineers do not copy the object, but create structures that only resemble the original image, and make their projects unique and amazing. Thus, the application of mesh shells with a complex lattice structure or with an optimal arrangement of block elements in space allows you to create a strong and stable structure. Both the methods of mathematical modeling and the capabilities of computer systems allow you to choose the optimal option using a numerical experiment.

So, bionics is characterized by the direct study of natural objects and the transferring of more rational solutions to the projected structure; the creation of structures with modeling stages of development of living objects, improving the shape, structure and processes; the creation of robotic systems based on bionic principles and other areas of research.

2 Methods and Materials

The principle of mathematical modeling based on bionic and fractal regularities lies in the design of modern buildings with a higher number of floors. These principles make it possible to create durable and resistant to external influences building structures. At the same time, the shape of such skyscrapers is unique and meets the modern requirements of architects.

The history of the creation of high-rise buildings and skyscrapers is very interesting. At the turn of the XIX and XX centuries, all the prerequisites for the growth of city buildings were ripe: the invention of elevators and steel frames, the use of reinforced concrete. The almost complete absence of places for new construction within the city limits has led to an increase in the cost of land and, consequently, to the need to increase the profitability of the usable area of construction facilities. At the same time, the main argument for the appearance of high-rise buildings was the growth of numerous banks and other large companies that sought to demonstrate their image by creating the most prominent buildings. And the building becomes noticeable precisely due to its height.

It so happened that in the USA of the second half of the XIX century, the city of Chicago became a financial and industrial center. Considerable resources have been concentrated in it. The great fire that occurred in Chicago in 1871 literally destroyed the city's buildings. Thus, there was a platform for an absolutely experimental approach to the design and construction of a new type of buildings – skyscrapers, which led to the creation of the famous "Chicago school" of designers led by Louis Sullivan. In practice, the American approach to high-rise construction was formed, where skyscrapers, competing in height and pretentiousness of facades, formed not only the face of the city, but also demonstrated the strength of the state system as a whole. Therefore, this principle of building very soon moved to New York, where the rocky rocks of the base of Manhattan Island made it possible to significantly increase the load on the soil, which made it possible to build floors again and again.

3 Results and Discussion

The blossoming of high-rise construction in America occurred in the first third of the twentieth century [5]. The use of reinforced concrete, the improvement of the structural system, the invention of new technologies and materials allowed architects to show more imagination when working on facades and the shape of a skyscraper as a whole, developing a silhouette and using compositional techniques to divide facades into large components. It was this image of a skyscraper that developed as a traditional one until the 1950s.

In Russia at that time, high-rise buildings were being erected, for example, the building of the Moscow State University, the Ministry of Foreign Affairs and other structures (Figs. 1, 2).

Fig. 1. High-rise building, Krasnopresnenskaya metro station, Moscow.

Fig. 2. High-rise building of the Leningradskaya Hotel, Moscow.

In the second half of the twentieth century, the rapid development of engineering innovations led to the appearance of constructive solutions in the form of shells with load-bearing external walls.

Already in the 1980s, a huge number of "ordinary" high-rises appeared, the main purpose of which was functional justification and the desire for individuality disappears and the interest around high-rise construction in America has faded.

By this time, the problem with ecology comes to the fore. They began to pay special attention to the environmental friendliness of the constructed structures and their engineering support with energy-saving technologies. This again created an impetus for new scientific developments in building materials and technologies.

With the advent of the new millennium, individual high-rise buildings began to appear in European countries. But the most ambitious construction of high-rise buildings has moved to Asia [5–7]. The world's first building above 500 m Taipei (Fig. 3) was built in Taiwan already in 2004. And in 2010, a record height of 828 m was set by Burj Khalifa (Fig. 4) in Dubai.

Fig. 3. Taipei in Taiwan.

The design, construction and process of operation of ultra-high buildings, as well as the tightening of environmental requirements, forced architects-designers to turn to wildlife, which throughout the development of architectural and construction activities of man helped to solve a variety of problems. Only nature, thanks to many millions of years of evolution, was able to create plants and animals.

In 1960, at a symposium in Daytona, researcher Jack Ellwood Steele said that "living prototypes of artificial systems are the key to a new technique" and suggested naming a new field of knowledge – "Bionics", the purpose and task of which is to study the characteristics of living organisms and plants [6, 7].

Fig. 4. Burj Khalifa in Dubai.

The bionic shapes of modern skyscrapers are characteristic of modern skyscrapers. An ultra-high skyscraper is an autonomous organization of existence based on energy-efficient independent life support systems with comfortable conditions for people. The scientific minds of the construction industry find and skillfully apply in practice those methods that make it possible to solve complex, seemingly unrealizable tasks that arise when creating ultra-high skyscrapers, while they skillfully use the knowledge of bionics researchers in relation to the development of shaping and autonomy of the life support of skyscrapers, and using eco-friendly building materials and structures [1–4, 8].

When designing load-bearing structural systems of skyscrapers, the greatest attention is paid to ensuring its stability. In many ways, stability depends on the shape of the building itself. And here bionics comes to the rescue as a system of knowledge of the structure of living organisms, for example, the structure of the stems of cereal crops or the spiral shape of a snail shell [4].

In the skyscraper, all floors are built around the core, which is built of reinforced concrete. To reduce the vibrations of the structure of skyscraper, reinforced concrete struts are organized every fifty to seventy meters, connecting them with external columns, according to the principle of fiber structure in the trunks of tall plant stems.

The question of the shape of a high-rise building always remains actual. According to architect Erich Stolz, it should be designed "from the inside out". Architect Simon Alford argues that "the question of form is a question of creative freedom and it is important that the design idea is based on economy and efficiency." From the point of view of resisting the form of influences and ensuring stability, designers turn to the study of the structure of both the stem itself and the root system of cereal plants.

To reduce wind pressure, which, with a significant increase in the height of the building, can exceed the total weight of the building, form zones of increased turbulence and vortex flows, it is necessary to resort to simple streamlined natural forms.

Interesting, from the point of view of ensuring resistance to external influences of skyscrapers, is the shape of the spiral. Unlike the classical forms of parallelepipeds and elongated pyramids of skyscrapers, the spiral development of the body of a skyscraper advantageously distinguishes it against the background of urban environment, the dynamics of the structure animates the material environment, merging it with the nature of air, wind, sky and clouds.

For the first time when designing a skyscraper, architect Santiago Calatrava applied this dynamic form in 2005, drawing an analogy with a twisting torso on the theme of an antique discobolus (Fig. 5). "Nature is my guide. It inspires the creation of buildings that reflect natural shapes and rhythms," says Calatrava, giving an interview to journalists. He always drew architectural ideas from the foundations of the universe.

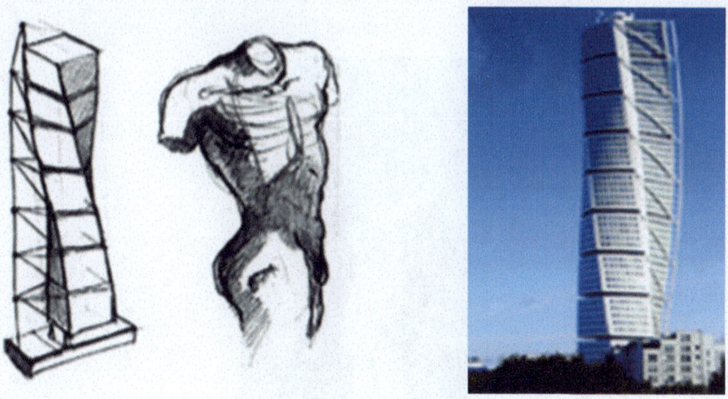

Fig. 5. «Twisting Torso».

The 36-storey spiral tower, erected in 2008 in the extremely active seismic zone of Japan in the city of Nagoya, is equipped with one of the most advanced seismic systems, which, combined with the shape of the spiral of a skyscraper, is able to withstand strong earthquakes. An important role is also assigned to environmental sustainability - during construction, air flow systems with double window glass and natural ventilation were used (Fig. 6).

In Canada, in 2012, towers were built, which, with their surprisingly plastic shape, evoked a human feeling with obvious natural sexuality among local residents (Fig. 7). High-rises began to be called "Marilyn Monroe".

The Evolution Tower was built in 2015 and is part of the Moscow City complex (Fig. 8). Each floor of the tower is rotated by 3 degrees relative to the previous one, as a result of which the building "twists" by more than 150 degrees. The shape of the skyscraper stands out spectacularly among the other skyscrapers of the complex (Fig. 9).

Table 1 shows the parameters of the skyscrapers of the complex «Moscow City»: the height and number of floors of buildings.

Fig. 6. Mode Gakuen Spiral Tower.

Fig. 7. Absolute World Towers.

Table 2 shows the parameters of the skyscrapers of New York: the height and number of floors of buildings (year of construction from 2016 to 2019 and height above 250 m). Figure 10 shows a graph of the number of skyscrapers built in New York with a height of more than 180 m from 2001 to 2019. Since 2020, there have been changes in the economy due to the pandemic and the pace of construction has slowed down. No data were considered for the period from 2020 to 2023. To identify patterns of commissioning of buildings with high floors, it is necessary to consider a longer period and take into account the economic conditions of the development of the construction industry.

In the capital of Taiwan, Taipei, in 2018, the Agora Garden Tower skyscraper was erected, the design of the skeleton of which repeats the spiral system of the structure of the DNA molecule. Each of the floors of the 246-m high-rise is rotated three degrees

Fig. 8. Moscow City.

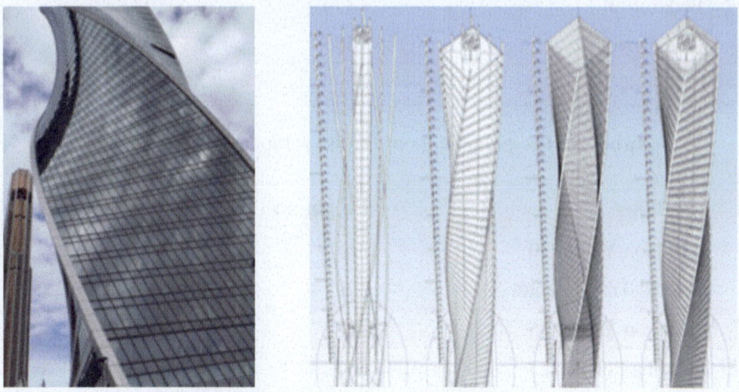

Fig. 9. Evolution Tower. Moscow City.

relative to the previous one. The spectacular bend of the double helix maximizes the stability of the skyscraper. The geometry of the cornices, made according to the terraced principle, increases the useful area of the building.

The skyscraper has implemented systems that convert solar and wind energy, all ultramodern intelligent technologies for monitoring and controlling the processes of generation, preservation, consumption and economy of electricity have been applied, effective heating, air conditioning and ventilation systems for all rooms have been created.

The skyscraper is equipped with its own system for collecting, cleaning and supplying rainwater. And since it is also a vertical garden, a large amount of water is required for watering plants and these requirements are fully met precisely with the help of a rainwater collection system. An autonomous in-house system for the disposal of organic waste allows their reuse for the care of plants. In connection with these qualities, the "Agora Garden Tower" is called the eco-tower "Green DNA" (Fig. 11).

On this small number of considered examples of skyscrapers in the design, which clearly read the bionic motives of shaping, it is possible to state the influence of shape on all the qualitative characteristics of skyscrapers [3, 4].

Table 1. The parameters of the skyscrapers of the complex «Moscow City».

N	Building	Height, m	Number of floors
1	Northern	108	27
2	Evolution	225	53
3	Empire	239	60
4	Tower on the embankment	268	60
5	City of Capitals	275	73
6	Eurasia	309	72
7	Mercury	339	75
8	Eye	354	85
9	Federation	374	97
10	One Tower	405	101

Table 2. The parameters of the skyscrapers of New York.

N	Building	Height, m	Number of floors
1	Hudson Yards, 30	315	73
2	World Trade Center 3	329	80
3	Hudson Yards, 35	308	72
4	Manhattan West,1	303	67
5	Central Park South, 220	290	69
6	Park Place, 30	286	82
7	Hudson Yards, 15	278	70
8	Hudson Yards, 10	268	52
9	One Manhattan-Garden	256	72
10	Leonard Street, 56	250	57

The interrelation of bionic and fractal forms is characteristic of modern structures. The use of the laws of nature in the shaping of architectural forms made it possible to create historical structures with a fractal structure in structures in the style of architectural bionics. Intuitive fractality is present in many historical architectural monuments by architects Antoni Gaudi, Zaha Hadid, Santiago Calatrava, Frank Gehry, Frank Lloyd Wright, Renzo Piano. At the same time, bionic structures of structures have a fractal principle of shaping, creating simply unique buildings [9].

To set a fractal structure means to set the principle of development, the law of shape change, the organization of the object itself [10–12]. Fractal forms in architecture create a holistic appearance that ensures the unification of structural elements using such properties as self-similarity, similarity, trace of movement, shift or slip [13–15]. As can

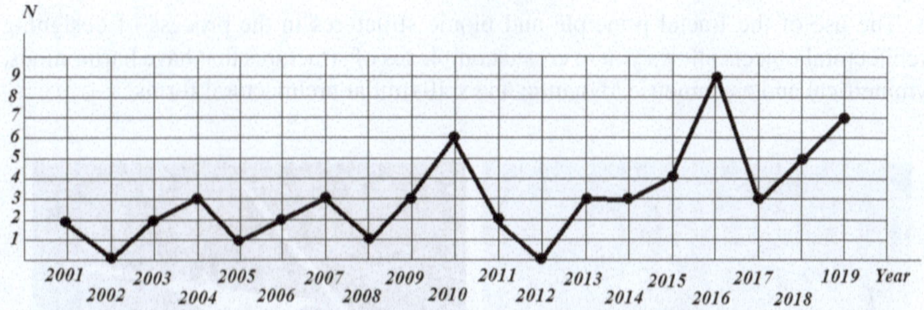

Fig. 10. Graph of changes in the number of skyscrapers built in New York

Fig. 11. Agora Garden Tower.

be seen, these properties may occur in the design of structures based on bionic principles. Figures 9 and 10 show the elements of the structure with shifting and sliding, as well as with a trace of movement.

Shear, bending, fracture of surfaces, trace of movement – these are the elements of shaping in the design of unique structures [14–16]. The use of complex structures and compositions in shaping leads to a new stage in architecture. But these techniques do not interfere with ensuring the unity of the external and internal space of this building and the creation of a modern architectural appearance of the city.

The application of bionic and fractal principles in the creation of a building structure is only a stage of optimal design and the search for an interesting solution. The Art Museum in Milwaukee, USA was built by architect Eero Saarinen, David Kaler and Santiago Calatrava (Fig. 12). Unfortunately, the name of the second architect is rarely mentioned in the literature. It is this structure that shows the dynamism of the object, the trace of movement in reality (Fig. 13).

Fractality is used both in the construction of the structure itself and in the creation of fractal facades. This is clearly visible in the buildings of the Moscow City complex (Fig. 14).

The use of the fractal principle and bionic structures in the process of designing architectural objects allows you to create complexes of structures that have harmonious, symmetrical and asymmetric, dynamic and self-similar architectural forms.

Fig. 12. Milwaukee Art Museum, architect Santiago Calatrava.

Fig. 13. Milwaukee Art Museum in dynamics.

Fig. 14. Fractal structures in the structures of the Moscow City complex.

4 Conclusions

The main purpose of bionic research is to study the laws of formation of living organisms, their structure, design features, principles of operation in order to use this knowledge in architecture [17, 18]. To ensure reliability, to find the optimal shape, energy-saving methods, to design unique structures, for example, skyscrapers, it is necessary to use such bionic principles as the principle of combining architectural elements based on the laws of the structure of natural objects; the principle of optimality of design solutions; the principle of searching for new structural structures and new materials; the principle of saving materials taking into account the strengthening of strength; the principle of saving energy resources; the principle of maximum environmental friendliness; the principle of harmonious development of the urban environment and its surrounding nature; the principle of innovative development of the entire surrounding space.

Of course, the aesthetics of the bionic form creates visual comfort when looking at a skyscraper. But the main thing is that the bionic forms of skyscrapers significantly reduce the wind load on the building, optimally distributing it along the facade, and increase the stability index of the skyscraper. All other equally important indicators, such as the reduction of the facade area, which entails savings in cladding and a reduction in the number of steel structures and, as a result, significantly reduces the weight of the building as a whole, are also derived from the use of bionic principles of shaping skyscrapers [19, 20].

Methods of bionic research and fractal geometry, as well as modern software systems allow you to find the most amazing architectural elements in shape and original in structure. It was the development of fractal architecture that made it possible to create a unique style in architecture. The combination of bionic principles and fractal geometry methods creates the basis for the development of unique architectural structures and amazing structures. The future remains for a creative approach in design, for the harmony of man and nature, for the creation of unique high-rise structures.

References

1. Mayatskaya, I.A., Yazyeva, S.B., Yazyev, B.M.: Optimality of structures from the point of view of architectural bionics. Constr. Technog. Saf. **9**(61), 7–11 (2017)
2. Vityuk, E.Y., Humorina, J.E.: Natural technologies as a new principle of shaping in architecture. Bull. Tomsk. State Univ. Arch. Civ. Eng. **20**(4), 55–64 (2018)
3. Sugar, V., Leczovics, P., Horkai, A.: Bionics in architecture. J. Built Environ. **5**(1) (2017). https://doi.org/10.1515/jbe-2017-0003
4. Mayatskaya, I., Eremin, V.: Bionics and the choice of rational structural form. E3S Web Conf. **110**, 01042 (2019). https://doi.org/10.1051/e3sconf/201911001042
5. Gnedich, P.P.: World Architecture. Publishing House Eksmo-Press (2012)
6. Villesenor, D.: Architecture and Nature. Publishing Rizzoli (2015)
7. Mayatskaya, I.A., Yazyev, B.M., Yazyeva, S.B., Kulinich, P.B.: Building construction: architecture and nature. MATEC Web Conf. **106**, 01031 (2017). https://doi.org/10.1051/matecconf/201710601031
8. Mayatskaya, I.A., Yazyeva, S.B., Lapina, A.P., Davydova, V.V.: Architectural bionics and the search for optimal solutions in the unique structures' design. IOP Conf. Ser.: Mater. Sci. Eng. **913**, 022070 (2020). https://doi.org/10.1088/1757-899X/913/2/022070

9. Batty, M., Longley, P.: Fractal Cities: A Geometry of Form and Function. Publishing Academic Press, San Diego (1994)
10. Mayatskaya, I., Yazyev, B., Kuznetsov, V., Tetenkov, N., Klyuev, S., Nabiullina, K.: Features of the development of architectural bionics in the modern world. Lect. Notes Civ. Eng. **436**, 285–293 (2024)
11. Mayatskaya, I., Yazyev, B., Kuznetsov, V., Rusakova, E., Klyuev, S., Sabitov, L.: Use of modern composite materials in construction and repair of structures. Lect. Notes Civ. Eng. **436**, 346–355 (2024)
12. Yazyev, B.M., Mayatskaya, I.A., Yazyeva, S.B., Yazyev, S.B.: Fractality in architectural forms and in organization of space in buildings. Mater. Sci. Eng. **698**(2), 022087 (2019). https://doi.org/10.1088/1757-899X/698/2/022087
13. Mayatskaya, I., Yazyeva, S., Gatiev, M., Kuznetsov, V., Klyuev, S., Sabitov, L.: Application of fractal methods in the design of modern structures. Lect. Notes Civ. Eng. **436**, 414–422 (2024)
14. Mayatskaya, I.A., Yazyev, B.M., Eremin, V.D.: Urban environment and the theory of deterministic chaos. Constr. Arch. **9-1**(30), 76–80 (2021). https://doi.org/10.29039/2308-0191-2021-9-1-76-80
15. Mayatskaya, I.A., Yazyev, B.M., Kashina, I., Gerlein, N.: Fractal geometry and design of modern structures. E3S Web Conf. **281**, 02018 (2021). https://doi.org/10.1051/e3sconf/202128102018
16. Mayatskaya, I.A., Yazyev, B.M.: Methods of mathematical modeling and urban space organization. E3S Web Conf. **281**, 02029 (2021). https://doi.org/10.1051/e3sconf/202128102029
17. Vishnevskaya, E.V., Tolynbekova, G.A., Nurkusheva, L.T.: Global and regional aspects of sustainable design in the frame of innovative technologies. IOP Conf. Ser.: Mater. Sci. Eng. **698**(3), 033009 (2019). https://doi.org/10.1088/1757-899X/698/3/033009
18. Dymchenko, M.E., Dakoro, M.F., Dadiyan, D.G.: The problem of form in modern architecture. E3S Web Conf. **281**, 02026 (2021). https://doi.org/10.1051/e3sconf/202128102026
19. Dmitriev, D.S., Belostosky, A.M., Nagibovich, A.I.: Scientific and technical support of the high-rise buildings design. IOP Conf. Ser.: Mater. Sci. Eng. **698**(2), 022064 (2019). https://doi.org/10.1088/1757-899X/698/2/022064
20. Mayatskaya, I.A., Yazyev, B.M., Demchenko, D.B., Yazyeva, S.B.: Creation of a comfortable environment in urban and rural settlements based on bionic principles. IOP Conf. Ser.: Earth Environ. Sci. **937**, 042026 (2021). https://doi.org/10.1088/1755-1315/937/4/042026

Possibilities of Applying Artificial Intelligence in Supplying a Construction Site in Arctic Conditions

Alan Khubaev, Dmitry Topchy(✉), Egor Pomytko, and Nikita Marchenko

National Research Moscow State University of Civil Engineering, 129337 Moscow, Russia
TopchiyDV@mgsu.ru

Abstract. In recent years, the Arctic has garnered increasing significance as a region for economic development, driven by the discovery of new resources and opportunities for infrastructure expansion. However, construction in the Arctic region presents unique challenges, notably logistical and supply-related issues. Therefore, the development of a new supply concept based on the use of modern technologies and consideration of regional specifics is of utmost importance. The research employed a systematic approach, ensuring a comprehensive examination of the supply process as a holistic system, considering the interconnections between its elements and the impact of external factors. Key findings revealed that extreme climatic conditions, limited transportation infrastructure, and remoteness from primary production centers significantly complicate the organization of supply for construction projects in the Arctic. A comprehensive supply condition indicator was developed to assess the degree of complexity and difficulty in providing a construction project with all necessary resources. The implementation of AI technologies is associated with certain risks, such as high implementation costs, lack of source data for system training, and the need for highly qualified personnel. These factors require careful analysis when deciding to implement AI in each specific case. The success of AI implementation will largely depend on the ability to consider the region's specific conditions and the readiness of construction organizations for technological transformation.

Keywords: Artificial Intelligence · Arctic Construction · Supply Chain Management · Logistics · Process Automation · Inventory Management · Incoming Quality Control

1 Introduction

In the context of the active development of the Arctic Zone of Russia, driven by the region's rich resource potential and strategic importance, the issues of effective construction organization in extreme climatic conditions are becoming increasingly relevant. One of the key aspects of successful implementation of construction projects in the Arctic is the organization of a reliable supply system, taking into account the region's specific features. [1].

S. V. Klyuev et al. (Eds.): ICRAAC 2024, LNCE 627, pp. 449–463, 2025.
https://doi.org/10.1007/978-3-031-82938-3_49

Existing research in the field of construction organization in the Far North and Arctic primarily focuses on the technological aspects of construction production, while the issues of supply organization of construction sites remain understudied. Traditional approaches to supply organization, effective in regions with developed infrastructure and favorable climate, prove to be ineffective in the Arctic, where construction organizations face a complex of specific problems: extremely low temperatures, difficult transportation accessibility, short construction season, and the need to comply with strict environmental requirements. [2].

In recent years, significant progress in the field of artificial intelligence has opened up new opportunities for optimizing supply processes. However, the application of AI technologies in the Arctic has its own specifics and requires careful study. Existing research on the implementation of AI in logistics processes is primarily oriented towards regions with developed infrastructure and does not take into account the peculiarities of Arctic conditions.

The purpose of this study is to develop a conceptual model for the application of artificial intelligence to optimize the supply system of construction sites in the Arctic. Particular attention is paid to the analysis of specific factors affecting the organization of supply in this region and the identification of potential opportunities and limitations of using AI technologies to solve the identified problems.

The scientific novelty of the research lies in the comprehensive approach to analyzing the possibilities of applying artificial intelligence in the supply system of construction sites, taking into account the specific conditions of the Arctic region, which allows for innovative solutions to be proposed for improving the efficiency of construction production in extreme conditions.

2 Methods and Materials

The object of the research is the supply process of construction sites in Arctic conditions. Special attention is paid to specific factors affecting supply efficiency in this region, including extreme climatic conditions, limited transport infrastructure, seasonal nature of work, and remoteness from major production centers.

The research employs a systematic approach, ensuring a comprehensive examination of the supply process as an integrated system, taking into account the interrelations between its elements and the influence of external factors. The research was conducted in several stages, each contributing to the achievement of the set objectives. The research stages and tasks for each stage are described in Table 1.

The Arctic represents a unique region with extreme climatic conditions that significantly complicate the supply process for construction sites. Low temperatures, extended polar nights, lack of developed infrastructure, and complex logistics routes - all these factors require a special approach to supply organization.

Given the global interest in Arctic resources and the increasing number of infrastructure projects, research into the specifics of construction site supply in this region becomes particularly relevant. [3–5].

3 Results and Discussion

The main identified factors affecting the organization of construction site supply in Arctic conditions are presented below.

Table 1. Research Stages.

Research Stage	Research Stage Objectives
1. Collection and analysis of regional characteristics data	Search and review of scientific articles, reports, statistical data, and regulatory documents related to construction and logistics in the Arctic
	Identification of current challenges and practices in construction site supply under extreme conditions
2. Identification of influencing factors	Identification and Evaluation of the Significance of Climatic, Geographic, Economic, and Logistical Factors Specific to the Arctic
3. Analysis and synthesis of obtained data	Systematization and analysis of collected information
4. Development of conceptual model	Development of a conceptual model for AI implementation in construction site supply management in the Arctic
	Description of AI systems' functions and capabilities in various aspects of supply management
5. Evaluation and analysis of results	Assessment of AI systems implementation feasibility
	Analysis of advantages and disadvantages of proposed solutions

The Arctic region is characterized by extreme climatic conditions that significantly complicate the organization of construction site supply. [6, 7] The main climatic features include:

- *Low temperatures:* average annual temperatures in the Arctic can drop below −20 °C, with winter minimums reaching −50 °C and lower. This leads to increased requirements for equipment and machinery used for cargo transportation and storage.
- *Polar night and polar day:* extended periods without sunlight or with continuous daylight affect personnel biorhythms and visibility conditions for transport operations.
- *Strong winds and blizzards:* frequent storms and blizzards reduce visibility and can completely halt transport operations, leading to supply delays.

- *Ice conditions:* ice-covered sea routes and rivers limit navigation and require the use of special ice-class vessels.

These climatic characteristics of the region give rise to the following factors affecting the organization of construction site supply in the Arctic:

1. Level of development of transportation infrastructure [8, 9].

- Level of road network development:

In most Arctic regions, roads are absent or in poor condition, making it difficult to transport goods by land.

- Proportion of logistical routes that pass through seasonal winter roads:

Temporary winter roads, laid out on frozen rivers and lakes, are only available during the cold season and require special precautions.

- Possibility of using sea communication:

The use of sea transport is limited by the short navigation period in the summer months, when the ice recedes. This requires advance planning and large-scale deliveries within limited timeframes.

- Possibility of using air communication:

Aerial transport is the most reliable but also the most expensive way to deliver goods, mainly used for urgent and valuable deliveries.

2. Materials, equipment, machines, and mechanisms used

- Need for special building materials:

Materials must be used that are resistant to low temperatures and extreme weather conditions. Additionally, the materials used must be environmentally friendly to minimize the impact on the Arctic ecosystem, protected by both Russian and international laws. [2, 10].

- Need for special equipment and maintenance:

Only specialized construction equipment can be used, capable of working in low-temperature conditions. In addition, a broader range of maintenance activities must be carried out compared to standard construction equipment. [11].

3. Conditions for storing materials and equipment

- Need for thermal insulation:

Warehouses must be equipped with effective thermal insulation to protect materials from low temperatures. This includes the use of insulated containers, the construction of insulated warehouses, and the application of heating systems.

- Need for anti-corrosion protection:

Metal structures and equipment require special anti-corrosion coatings to prevent damage from moisture and low temperatures. The use of stainless materials and protective paints helps to extend the service life of equipment.

- Need to reduce air humidity in warehouses:

In conditions of high humidity and frequent temperature fluctuations, it is necessary to control humidity in warehouses to prevent material spoilage.

4. Conditions for accommodating workers

- Need for expeditionary and shift camps:

Since a large part of the Arctic is located at a significant distance from major cities where construction organizations are based, construction is often carried out using expeditionary or shift methods, which requires providing workers with temporary accommodation. Due to the extreme climatic conditions, increased requirements are imposed on the components of expeditionary and shift camps for heat protection, wind and snow load resistance.

To assess the impact of the listed factors on the supply system of a construction site, the concept of the complexity of supply of a construction site is introduced.

The conditions of supply of a construction site are a comprehensive characteristic that reflects the degree of complexity and difficulty of ensuring the construction site with all the necessary materials, equipment, machines, mechanisms, and labor resources.

Absolutely unimpeded (ideal) supply conditions of a construction site are considered to be:

- The presence of a developed network of all-season roads, ensuring unhindered delivery of goods at any time of the year.
- The possibility of using standard, widely available materials and equipment, not requiring special adaptation to local conditions.
- The absence of the need for special conditions for storing materials and equipment, and the possibility of using standard warehouse facilities.
- The presence of a sufficient number of local qualified labor resources, not requiring special conditions for accommodation and transportation.
- Favorable climatic conditions, not having a significant impact on the supply and construction process.
- The location of the object near developed populated areas and industrial centers, ensuring the availability of necessary resources and infrastructure.

The comprehensive indicator of the conditions of supply of a construction site, which allows for the assessment of the conditions of supply organization, is proposed to be determined by the formula:

$$y = \sum_{i=1}^{4} X_i P_i$$

y – comprehensive indicator of supply conditions
X_i– influence of a factor
P_i– factor impact

The relative weight of each factor was determined through an expert survey. The results are summarized in Fig. 1.

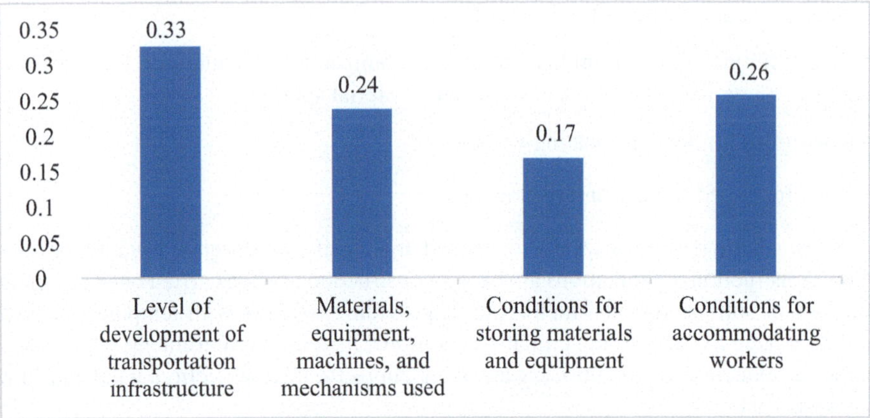

Fig. 1. Relative Weight of Influence Factors.

To evaluate the impact of each factor on the comprehensive indicator of supply conditions for a construction site in Arctic conditions, the following scale of values is proposed, presented in Table 2:

Based on the value of the comprehensive indicator of supply conditions, a qualitative assessment of the supply conditions of the construction site can be made, according to Fig. 2:

Accordingly, a system for the qualitative assessment of supply conditions for a construction site is proposed, as shown in Fig. 3.

Considering the identified features of the region, it is proposed to implement artificial intelligence into the supply organization system across various aspects of the supply system. Table 3 lists the tasks that are expected to be solved through the application of artificial intelligence within specific aspects of the supply system. [12–16].

1. Optimization of Logistics Processes

- Utilization of Route Optimization Algorithms:

AI can analyze a multitude of variables, such as weather conditions, road conditions, the presence of ice obstacles, and time constraints, to determine optimal delivery routes. Algorithms can dynamically adjust routes in real-time, considering changes in conditions.

- Weather Forecasting:

AI systems can utilize data from meteorological satellites and sensors for accurate weather forecasting. This allows for planning deliveries during favorable periods and avoiding hazardous situations like blizzards and storms.

2. Automation of Inventory Management Processes

Table 2. Scale for evaluating the impact of identified factors on the supply system.

Factor	Value	Description
Level of development of transportation infrastructure	4	Developed network of all-season roads, possibility of using sea and air communication all year round
	3	Presence of all-season roads, seasonal restrictions on the use of sea and air communication
	2	Presence of seasonal winter roads, significant restrictions on the use of sea and air communication
	1	Absence of land routes, possibility of using only sea or air communication
	0	Absence of land routes, extremely limited possibilities of using sea and air communication
Materials, equipment, machines, and mechanisms used	4	Use of standard materials and equipment, not requiring special conditions
	3	Need for partial use of special materials and equipment
	2	Most materials and equipment require consideration of Arctic conditions
	1	Use of exclusively special Arctic materials and equipment
	0	Need for development of unique materials and equipment for a specific project
Conditions for storing materials and equipment	4	Storage possible in standard conditions, without additional measures
	3	Standard insulation and anti-corrosion protection required
	2	Enhanced insulation, anti-corrosion protection, and humidity control required
	1	Special warehouses with heating and air-drying systems required
	0	Unique storage facilities with complex monitoring and maintenance systems required
Conditions for accommodating workers	4	Local labor resources can be attracted, without the need for temporary settlements

(*continued*)

Table 2. (*continued*)

Factor	Value	Description
	3	Part of the personnel needs to be accommodated in temporary structures
	2	Most personnel are accommodated in shift camps
	1	All personnel live in temporary settlements, requiring enhanced protection from climatic conditions
	0	Construction of permanent settlements with developed infrastructure required for long-term stay of personnel

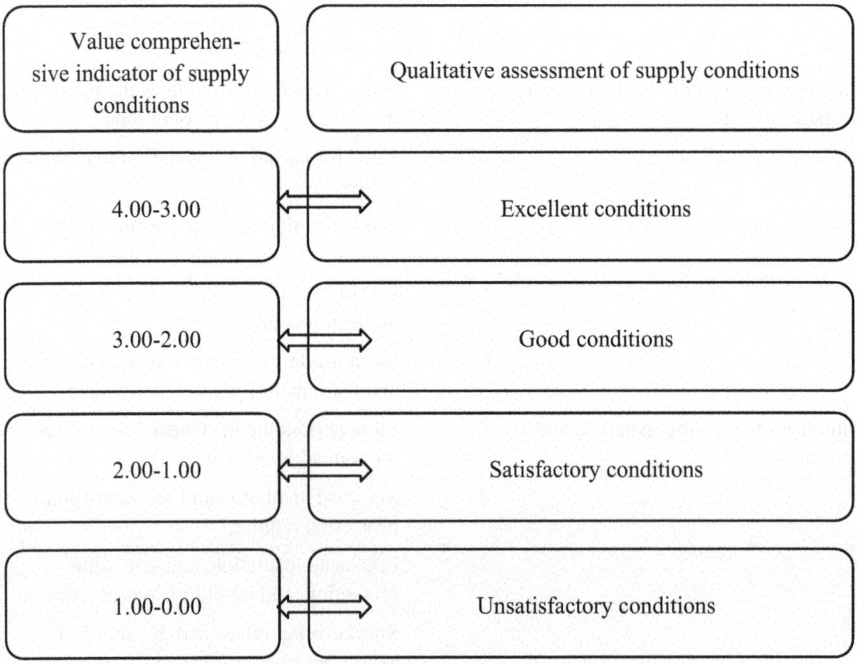

Fig. 2. Qualitative assessment of supply conditions.

- Inventory and Stock Monitoring:

AI systems can automatically track the inventory levels of materials and equipment using sensors and RFID tags. This enables timely ordering of new batches and avoids shortages or surpluses of resources.

- Demand Forecasting:

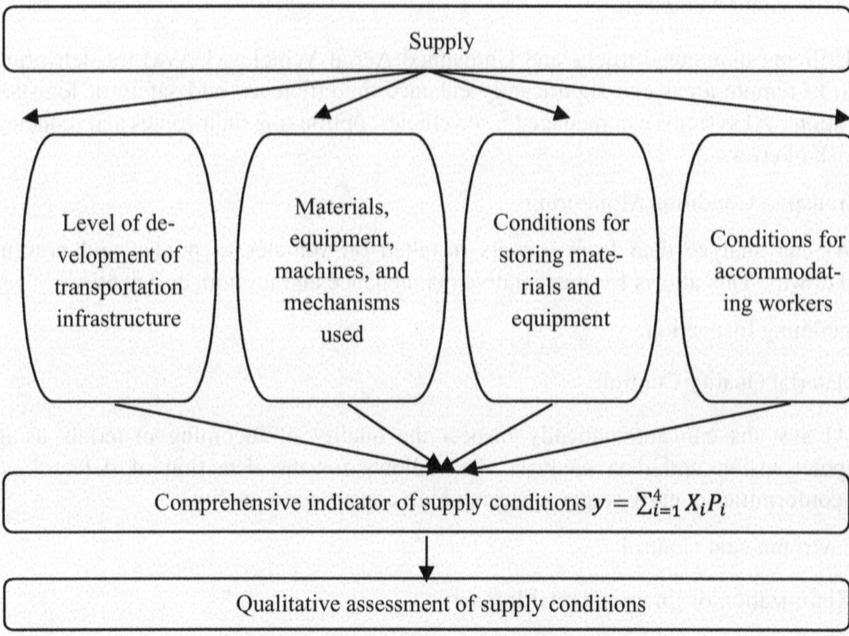

Fig. 3. Supply conditions assessment system.

Table 3. Expected Effect of AI Implementation in the Supply System.

Aspect of the Supply System	Expected Effect of AI Implementation
Logistics	Optimization of delivery routes
	Reduction of the risk of accidents and other emergencies on delivery routes
Inventory Management	Automation of inventory and stock monitoring processes
	Demand forecasting
Fleet Management	Use of unmanned vehicles
	Monitoring of the technical condition of transport
Incoming Inspection	Automation of the incoming material quality control process
Environmental Control	Minimization of environmental impact
	Waste management

AI can analyze historical data and current trends to forecast future needs for materials and equipment. This allows for more accurate planning of deliveries and reduces downtime due to resource shortages.

3. Fleet Management

- Autonomous Vehicles:

Utilizing unmanned trucks and Unmanned Aerial Vehicles (UAVs) for delivering cargo to remote areas can significantly enhance the efficiency and safety of logistics operations. AI systems can manage these vehicles, optimizing their routes and reducing the risk of errors.

- Transport Condition Monitoring:

AI can analyze data from sensors installed on vehicles to predict and prevent breakdowns. This allows for preventative maintenance and minimizes downtime.

4. Incoming Inspection

- Material Quality Control:

AI systems can automatically inspect the quality of incoming materials using computer vision and data analysis. This allows for the detection of defects and non-conformities at early stages, ensuring high construction quality.

1. Environmental Control

- Minimization of Environmental Impact:

AI systems can analyze data on emissions and waste, suggesting measures for their reduction. Utilizing environmentally friendly technologies and materials, as well as optimizing logistics processes, will help minimize the negative impact on the Arctic ecosystem. [17].

- Waste Management:

AI can optimize the processes of waste collection, sorting, and recycling, ensuring their environmentally safe disposal. This technology will help minimize environmental pollution.

Assessment of the feasibility of using AI in supply:

The feasibility of implementing AI in the Arctic supply system depends on the specific project conditions. With a thorough analysis of all factors (potential benefits and costs), and considering the unique conditions of the Arctic region, it can be determined whether the implementation of AI will be economically viable. A detailed assessment is necessary, including the calculation of expected costs and potential savings, as well as taking into account the risks and uncertainties inherent in the Arctic region. If there is a sufficient database and AI systems prove to be sufficiently adaptive, then implementation may be justified. Otherwise, the project may be ineffective due to significant costs and uncertainty.

The expected positive impact of artificial intelligence has been discussed above. In order to be able to assess the feasibility of implementing AI systems, it is also necessary to consider the risks that may arise from the use of artificial intelligence systems. The main such risks, relevant today, are presented in Table 4. [18–20].

The evaluation of the feasibility index for AI application (F) is proposed to be conducted based on the value of the comprehensive supply conditions index. The relationship

Table 4. Matrix of ranks for factors affecting labor productivity during major repairs.

Potential Problem	Description	Potential Consequences
High Implementation Cost	The need to utilize expensive equipment and attract highly qualified personnel	The need to attract additional funding to the detriment of investment and construction projects implemented in parallel with construction in the Arctic zone
Lack of Data	Insufficient information about a process or object for training the AI system	Inability to apply AI in certain aspects of the supply system
Need for Highly Qualified Personnel	Additional training of employees is required for the correct interpretation of AI results	Inability to apply AI due to a shortage of qualified personnel capable of working with it
		High probability of misinterpreting the results of AI systems if personnel with insufficient qualifications are involved, which can lead to emergencies or a decrease in the quality of the final construction product

between the feasibility index of artificial intelligence application and the value of the comprehensive supply conditions index will be as follows:

$$F = \frac{1}{1 + e^{-k(y-b)}}$$

F – feasibility index for AI application in supply.

k – the coefficient determining the steepness of the curve in the graph of dependence (determined based on statistical data from project implementation).

b – The threshold value of the comprehensive supply indicator at which the feasibility of AI application begins to significantly increase.

The value of parameter b is set to 0.25 (based on expert assessment) and remains constant.

Therefore, the general form of the relationship between AI feasibility and the comprehensive supply index is as follows (Fig. 4):

The dependence of the appropriateness of AI application in the supply of a construction object on the value of the comprehensive supply condition indicator is nonlinear and characterized by a monotonic increase. The logistic function allows for a quantitative assessment of the degree of necessity and the expected effectiveness of AI usage depending on specific supply conditions, which facilitates a more substantiated approach to decision-making regarding the implementation of AI in the supply system of Arctic construction projects.

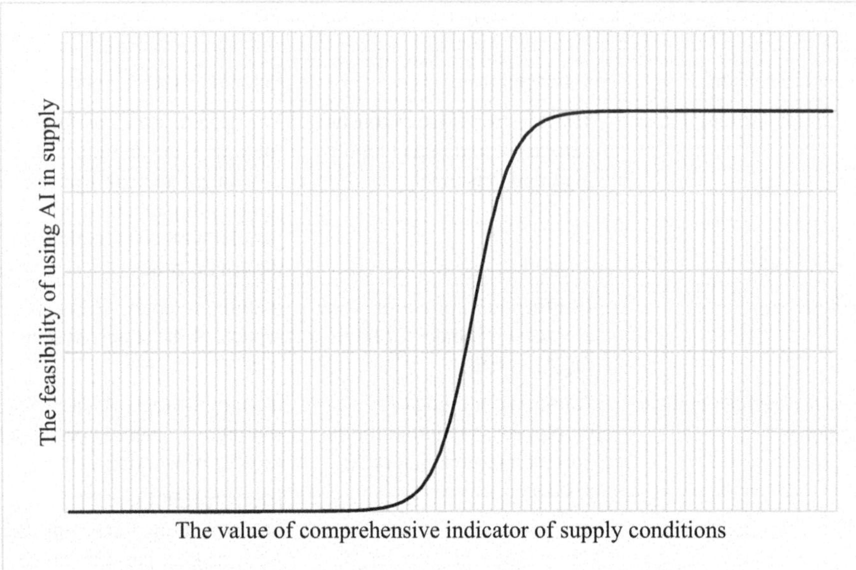

Fig. 4. Graph illustrating the relationship between the feasibility of AI application in construction project supply and the comprehensive supply conditions indicator.

4 Conclusions

The conducted research allowed for a comprehensive examination of the specifics of supply organization for construction projects in the Arctic and an assessment of the potential application of artificial intelligence to optimize this process.

The study identified key factors that determine the characteristics of supply in the Arctic region: extreme climatic conditions, a low level of transport infrastructure development, increased requirements for the characteristics of materials and equipment, specific conditions for resource storage, and the specifics of worker accommodation. Analysis of these factors allowed us to determine the main areas of application of artificial intelligence in the supply system.

The developed conceptual model for AI implementation covers five key aspects: optimization of logistics processes, automation of inventory management, fleet management, incoming inspection, and environmental monitoring. For each aspect, specific tasks and expected effects from the implementation of AI technologies have been defined.

The practical significance of the research lies in the formation of an integrated approach to modernizing the supply system for construction projects in the Arctic using modern artificial intelligence technologies. The proposed solutions are aimed at increasing the efficiency of logistics operations, reducing risks, and minimizing the negative impact on the environment.

However, it is important to note that the implementation of AI technologies is associated with certain risks, such as high implementation costs, a lack of initial data for training systems, and the need to attract highly qualified personnel. These factors require careful analysis when making decisions about implementing AI in each specific case.

Further research in this area can be directed towards:

- Developing specialized AI algorithms adapted to Arctic conditions;
- Creating methods for assessing the economic efficiency of implementing AI systems in Arctic projects;
- Investigating the possibilities of integrating various AI technologies into a unified supply management system;
- Developing training programs for specialists to work with AI systems in Arctic conditions.

In the future, we can predict the increasing role of artificial intelligence in the organization of construction in the Arctic zone as technologies develop and experience in their application accumulates. At the same time, the success of AI implementation will largely depend on the ability to take into account the specific conditions of the region and the readiness of construction organizations for technological transformation.

References

1. Istomin, A.V.: Main directions and priorities for sustainable development of the Arctic zone of Russia. North and Market: Formation of Economic Order **1**(22), 3–7 (2009)
2. Rudenko, A.A.: Resource supply of construction as an element of its organizational-technological reliability. Bulletin of V.N. Tatishchev Volga University **2-1**(51), 130–139 (2023). https://doi.org/10.51965/2076-7919_2023_2_1_1_130
3. Shipilova, N.A., Mikheev, G.V., Yesina, E.S., Volkov, A.A.: Features of construction management in the Far North and permafrost regions. Econ. Entrepreneurship **10**(147), 696–699 (2022). https://doi.org/10.34925/EIP.2022.147.10.132
4. Savinova, V., Brodach, M.M.: Features of design and construction in the Arctic region. High Technol. Buildings **4**, 50–57 (2018)
5. Belyaev, I.S.: Problems of Arctic construction: trends and prospects. Bull. Civil Eng. **2**(85), 248–255 (2021). https://doi.org/10.23968/1999-5571-2021-18-2-248-255
6. Porfiryev, B.N., Khlebnikova, E.I.: Construction in the context of climate change in the Arctic: risks and opportunities for their mitigation. Econ. Constr. **6**(36), 4–15 (2015)
7. Borisov, V.N., Pochukaeva, O.V.: Relationships between development factors of the Arctic Zone of the Russian Federation. Stud. Russ. Econ. Dev. **27**(2), 159–165 (2016). https://doi.org/10.1134/S1075700716020040
8. Toropushina, E.E.: Trends in the development of social infrastructure in the Arctic regions of Russia. North and Market: Formation of Economic Order **4**(41), 78a–784 (2014)
9. Dolgov, A.P., Rybnov, E.I.: Logistics of supplies and stocks in construction: strategies, methods, models. St. Petersburg, ASV Publishing House, Moscow (2003)
10. Khubaev, A.O.: Features of concrete work production at negative ambient temperatures. In Current Problems of the Construction Industry and Education - 2023: Proceedings of the IV National Scientific Conference, Moscow, December 15, 2023, pp. 218–224. Moscow State University of Civil Engineering (National Research University, Moscow (2024)
11. Khubaev, A.O.: Improvement of the production process of winter concreting on the basis of the potential of organizational and technical solutions (2022)
12. Kryukov, K.M., Metlev, A.M.: Possibilities of using artificial intelligence in construction. Eng. Bull. Don **10**(94), 51–60 (2022)

13. Abramyan, S.G., Klyuev, S.V., Polyakov, V.G., Sabitova, T.A., Akopyan, G.O., Guseynov, K.M.: Specifics of information model development for functional conversion of offshore oil platforms. Constr. Mater. Prod. **6**(4), 42–57 (2023). https://doi.org/10.58224/2618-7183-2023-6-4-42-57
14. Gorodnova, N.V.: Application of artificial intelligence and nanotechnology in the investment and construction sector of Russia. Bulletin of NSUEU 3, 81–95 (2021). https://doi.org/10.34020/2073-6495-2021-3-081-095
15. Shalina, D.S., Stepanova, N.R.: Theory and practice of using artificial intelligence in real estate. Bull. Altai Acad. Econ. Law **5-1**, 193–200 (2020). https://doi.org/10.17513/vaael.1128
16. Ginzburg, A.V., Ryzhkova, A.I.: Capabilities of artificial intelligence to enhance the organizational and technological reliability of construction production. Bull. MGSU **13**(1), 7–13 (2018). https://doi.org/10.22227/1997-0935.2018.1.7-13
17. Lapidus, A.A., Kuzmina, T.K., Bolshakova, P.V.: Developing administrative and process-related solutions for the benefit of residential construction projects. Constr. Mater. Prod. **7**(6), 2 (2024). https://doi.org/10.58224/2618-7183-2024-7-6-2
18. Kolchin, V.N.: Specifics of applying "artificial intelligence" technology in construction. Innov. Investment **3**, 250–253 (2022)
19. Lapidus, A., Bidov, T., Khubaev, A.: The study of the calibration dependences used when testing the concrete strength by nondestructive methods. MATEC Web of Conf. **117**, 00094 (2017). https://doi.org/10.1051/matecconf/20171170009
20. Lapidus, A., Khubaev, A., Bidov, T.: Organizational and technological solutions justifying use of non-destructive methods of control when building monolithic constructions of civil buildings and structures. MATEC Web of Conf. **251**, 05014 (2018). https://doi.org/10.1051/matecconf/201825105014
21. Fedosov, S.V., Lapidus, A.A., Narmaniya, B.E., Ayzatullin, M.M.: Cauchy problem for modeling of unsteady mass transfer processes in an unbounded plate by the integral Laplace transform method. Constr. Mater. Prod. **7**(5), 4 (2024). https://doi.org/10.58224/2618-7183-2024-7-5-4
22. Khubaev, A., Bidov, T., Bzhienikov, A., Nesterova, V.: Modern Russian high-tech construction materials and their application in domestic construction industry (on example of metal-ceramic panels Hardwall). IOP Conf. Ser. Mater. Sci. Eng. **365**, 032005 (2018). https://doi.org/10.1088/1757-899X/365/3/032005
23. Lapidus, A., Khubaev, A., Bidov, T.: Development of a three-tier system of parameters in the formation of the organizational and technological potential of using non-destructive testing methods. E3S Web of Conf. **97**, 06037 (2019). https://doi.org/10.1051/e3sconf/20199706037
24. Lapidus, A.A., Kuzmina, T.K., Bolshakova, P.V.: Improving the procedure for interaction between participants in the construction control system in order to increase labor productivity. Constr. Mater. Prod. **7**(4), 1 (2024). https://doi.org/10.58224/2618-7183-2024-7-4-1
25. Lapidus, A., et al.: Development software for the non-destructive control of monolithic structures in housing construction. E3S Web of Conf. **258**, 09003 (2021). https://doi.org/10.1051/e3sconf/202125809003
26. Khubaev, A.O.: Determination of the efficiency of the production process of winter concreting based on field studies. E3S Web of Conf. **389**, 06012 (2023). https://doi.org/10.1051/e3sconf/202338906012
27. Zholtikov, A., Zinovkin, A., Khubaev, A.: Improving the efficiency of applying the method of slab and floor lifting. E3S Web of Conf. **389**, 01067 (2023). https://doi.org/10.1051/e3sconf/202338901067
28. Bidov, T.H., Khubaev, A.O., Shabanova, A.A.: Organizational and technical modeling of the complex system of concrete works production in winter period during the construction of

residential buildings. Constr. Geotech. **12**(2), 15–25 (2021). https://doi.org/10.15593/2224-9826/2021.2.02

29. Bidov, T.Kh., Khubaev, A.O., Pomytko, E.A., Kotelnikova, A.D.: Methodology of evaluating of factors affecting the efficiency of construction project implementation by general contracting organization in Fit-out direction. Vestnik MGSU [Monthly J. Constr. Archit.] **19**(9), 1562–1569 (2024). https://doi.org/10.22227/1997-0935.2024.9.1562-156

Author Index

S. V. Klyuev et al. (Eds.): ICRAAC 2024, LNCE 627, pp. 465–467, 2025.
https://doi.org/10.1007/978-3-031-82938-3

The manufacturer's authorised representative in the EU is Springer Nature Customer Service Centre GmbH, Europaplatz 3, 69115 Heidelberg, Germany. If you have any concerns regarding our products, please contact ProductSafety@springernature.com

Printed and bound by CPI Group (UK) Ltd, Croydon, CR0 4YY
30/09/2026
02225600-0018